TROPICAL TREES AS LIVING SYSTEMS

Tropical trees as living systems

The proceedings of the Fourth Cabot Symposium held at Harvard Forest, Petersham Massachusetts on April 26–30, 1976

Edited by

P. B. TOMLINSON and MARTIN H. ZIMMERMANN

Harvard University, Harvard Forest
Petersham, Massachusetts

with a foreword by
LAWRENCE BOGORAD
Director, Maria Moors Cabot Foundation for
Botanical Research of Harvard University

CAMBRIDGE UNIVERSITY PRESS

Cambridge

London - New York - Melbourne

CAMBRIDGE UNIVERSITY PRESS
Cambridge, New York, Melbourne, Madrid, Cape Town, Singapore,
São Paulo, Delhi, Dubai, Tokyo

Cambridge University Press
The Edinburgh Building, Cambridge CB2 8RU, UK

Published in the United States of America by Cambridge University Press, New York

www.cambridge.org
Information on this title: www.cambridge.org/9780521142472

First published 1978
This digitally printed version 2010

A catalogue record for this publication is available from the British Library

Library of Congress Cataloguing in Publication data
Main entry under title:
Tropical trees as living systems.
Sponsored by the Maria Moors Cabot Foundation for Botanical Research.
Includes index.
1. Trees – Tropics – Congresses.   2. Trees – Physiology – Congresses.
3. Forest ecology – Tropics – Congresses.   4. Rain forests – Congresses.
I. Tomlinson, Philip Barry, 1932-   II. Zimmermann, Martin Huldrych, 1926-
III. Harvard University. Maria Moors Cabot Foundation for Botanical Research.
QK493.5.T76   582'.1609'093   77-8579

ISBN 978-0-521-21686-9 Hardback
ISBN 978-0-521-14247-2 Paperback

Contents

Contents

Contributors

Fredrick T. Addicott
Department of Botany
University of California
Davis, CA 95616 U.S.A.

Paulo de T. Alvim
Centro de Pesquisas do Cacau
Caixa Postal 7
Ilhéus/Itabuna
Itabuna Bahia, Brazil

P. S. Ashton
Botany Department and Institute
of Southeast Asian Biology
University of Aberdeen
St. Machar Drive
Aberdeen AB9 24D
Scotland

Herbert G. Baker
Botany Department
University of California
Berkeley, CA 94720 U.S.A.

Rolf Borchert
Department of Botany, Physiol-
ogy, and Cell Biology
University of Kansas
Lawrence, KA 66045 U.S.A.

P. Champagnat*
Centre National de la Recherche
Scientifique
91 Gif-sur-Yvette, Essone
France

James A. Doyle
Department of Ecology and Evolu-
tionary Biology
Museum of Paleontology
University of Michigan
Ann Arbor, MI 48109 U.S.A.

John Dransfield
The Herbarium
Royal Botanic Gardens
Kew, Richmond, Surrey
England

Jack B. Fisher
Fairchild Tropical Garden
10901 Old Cutler Road
Miami, FL 33156 U.S.A.

Thomas J. Givnish
Department of Biology
Harvard University
16 Divinity Avenue
Cambridge, MA 02138 U.S.A.

Francis Hallé
Université du Languedoc
Institut de Botanique
5 rue de A. Broussonet
34000 Montpellier
France

Gary S. Hartshorn
Organization for Tropical Studies
Universidad de Costa Rica
San José, Costa Rica

Daniel H. Janzen
Department of Biology
University of Pennsylvania
Philadelphia, PA 19174 U.S.A.

Jan Jeník
Institute of Botany
Czechoslovak Academy of
Sciences
25243 Pruhonice
Czechoslovakia

Tatuo Kira
Department of Biology
Faculty of Science
Osaka City University
Sugimoto-Cho, Sumiyoshi-Ku
Osaka 558, Japan

K. A. Longman
Institute of Terrestrial Biology
Unit of Tree Biology
Bush Estate, Penicuik
Midlothian, Scotland EH26 OQB

F. S. P. Ng
Forest Research Institute
Kepong, Selangor
Malaysia

R. Nozeran
Université de Paris-Sud
Laboratoire de Botanique II
Bâtiment 360
91405 Orsay
France

Roelof A. A. Oldeman
Agr. University
"Hinkeloord"
Box 342
Wageningen, Netherlands

W. R. Philipson
Botany Department

*Not attending

University of Canterbury
Christchurch 1, New Zealand

Marie-Françoise Prévost
O.R.S.T.O.M.
B.P. 165
97301 Cayenne Cedex
French Guiana

José Sarukhán
Departamento de Botánica
Instituto de Biología
Universidad Nacional Autónoma
de México
México 20 D. F.
Mexico

P. B. Tomlinson
Harvard University
Harvard Forest
Petersham, MA 01366 U.S.A.

Jean-Marie Veillon
O.R.S.T.O.M.
B.P. A5
Nouméa, New Caledonia

Trevor Whiffin
Department of Botany
La Trobe University
Bundoora, Victoria
Australia 3083

T. C. Whitmore
c/o British Museum (Natural
History)
Cromwell Road
London SW 7
England

Martin H. Zimmermann
Harvard University
Harvard Forest
Petersham, MA 01366 U.S.A.

Chairmen of sessions and discussants

Kamaljit Bawa, *Department of Biology, University of Massachusetts – Boston, Boston, Massachusetts 02125 U.S.A.*

Brian Bowes, *Harvard University, Harvard Forest, Petersham, Massachusetts 01366 U.S.A. Now at University of Glasgow, Glasgow G12 8QQ, Scotland*

Gordon Browning, *John Innes Institute, Norwich NOR 7OF, England*

Robert C. Cook, *Biological Laboratories, Harvard University, Cambridge, Massachusetts 02138 U.S.A.*

Peter Del Tredici, *Harvard University, Harvard Forest, Petersham, Massachusetts 01366, U.S.A.*

Ernest M. Gould, *Harvard University, Harvard Forest, Petersham, Massachusetts 01366 U.S.A.*

D. Roger Lee, *Harvard University, Harvard Forest, Petersham, Massachusetts 01366 U.S.A. Now at Memorial University of Newfoundland, St. John's, Newfoundland, Canada A1C 5S7*

Walter H. Lyford, *Harvard University, Harvard Forest, Petersham, Massachusetts, 01366, U.S.A.*

Thomas A. McMahon, *Division of Engineering and Applied Physics, Harvard University, Cambridge, Massachusetts 02138 U.S.A.*

David Policansky, *Department of Biology, University of Massachusetts – Boston, Boston, Massachusetts 02125 U.S.A.*

Hugh M. Raup, *Harvard University, Harvard Forest, Petersham, Massachusetts 01366 U.S.A.*

Otto Stein, *Department of Botany, University of Massachusetts, Amherst, Massachusetts 01002 U.S.A.*

Peter Stevens, *Arnold Arboretum, Harvard University, Cambridge, Massachusetts 02138 U.S.A*

John G. Torrey, *Harvard University, Harvard Forest, Petersham, Massachusetts 01366 U.S.A.*

Foreword

Most of our knowledge of the growth of trees is based upon a study of their temperature representatives, particularly those in Europe and North America, where foresters and botanists are concentrated. In contrast, tropical trees, especially those of the wet lowlands with nonseasonal climates, are relatively little studied, and the task of cataloguing their floristic richness is far from complete. The diversity of tree species in the lowland tropics is insufficiently appreciated by biologists, but may be indicated by a recent assessment of an area 23 ha in extent in the Jengka Forest Reserve, Malay Peninsula, where only trees with a basal circumference greater than 91 cm were measured (M. D. Poore, *J. Ecol.*, *56*, 143–96, 1968): 375 species in 139 genera representing 52 families were recorded. To achieve about the same figures in North America, one would have to survey a large part of the continental United States (including tropical Florida).

It seems appropriate that the woody plants of the tropics be given attention at a time when the world has been made acutely aware that fossil fuel resources are dwindling rapidly, and, in contrast, the appreciation of trees as a renewable source of energy is greatly heightened. It was Godfrey Lowell Cabot's early appreciation of this situation that led to his endowment of the foundation under whose auspices we meet. One of his concerns was the genetic improvement of forest trees, and appropriately we give emphasis here to tropical trees because they represent the largest genetic resource of woody plant material available to foresters.

Clearly, we cannot deal in detail with all topics deserving the attention of biologists; indeed, the gaps in our coverage are as numerous and obvious as the gaps in the forest canopy itself. However, in trying to emphasize to the organization and growth of the individual tree, as the unit of which the forest is made, we hope to establish principles that will guide the future advancement of research on tropical woody plants.

Seen in this light, our responsibility is a considerable one, and it is one we must accept with a feeling of urgency. The destruction and exploitation of tropical forests proceed at an alarming rate. If we are to appreciate how tropical trees function within forest communities, our research must proceed apace while there is still a tropical forest to analyze.

We hope the exchange of ideas and information that this symposium permits not only will achieve one other of Mr. Cabot's wishes – the dissemination of knowledge about trees – but will rapidly lead to the efficient management of tropical forests in a way useful to mankind, not only presently but for the benefit of later generations.

Lawrence Bogorad
Professor of Biology, Harvard University
Director, Maria Moors Cabot Foundation
for Botanical Research

Editorial preface

This volume is the outcome of a symposium held at Harvard Forest, Petersham, Massachusetts, in April 1976, the fourth in the series made possible by the Cabot Foundation of Harvard University, but in this instance with additional support from the Atkins Garden Fund of Harvard University. The theme of the symposium was a consideration of the individual tree as an integrated living system and its interaction with similar individuals in tropical vegetation. Participants were selected because of their expert knowledge of tree biology or their extensive field knowledge of tropical woody plants; the goal was to effect interchange across the unnatural and unnecessary barriers that so often exist between the scientists of temperate and tropical countries. Chapters by each of the invited participants are included, together with a summary of the discussion of the majority of the papers. One paper submitted by a participant who was unable to attend is also included.

In choosing so general a theme as tropical trees and in attempting to cover the subject from many different aspects, we were aware of the dangers of omission. One obvious deficiency is in the coverage of periodicity of cambial activity, which relates directly to wood production and the absence of growth rings from the trunks of most tropical trees – a matter of utter frustration to tropical ecologists. However, this aspect had been covered in an earlier Cabot Symposium (*The Formation of Wood in Forest Trees*, ed. M. H. Zimmermann, Academic Press 1964). Despite this omission, we feel attention has been given to most of the important topics.

Contributors were asked not merely to present detailed research data, but to offer a summary of the state of knowledge of the area of their special interest so that a general overview would be presented. It is evident that our speakers responded magnificently to this demanding task and have placed our present knowledge of tropical trees within the wider framework of plant biology so that this volume becomes more of a source book than is usual for specialized symposia. The need for such a synthesis is clear to anyone who has attempted to do research on woody plants in the tropics and despaired of assessing information that is currently widely scattered, both geographically and linguistically, in botanic and forestry journals. No one author could fairly cover the field either from a knowledge of the literature or from personal experience. Standard reference books on trees refer overwhelmingly to temperate examples, and this information may not be wholly appropriate to the circumstances of the lowland tropical rain forest. That temperate trees offer only a very limited sampling of the range of morphologic diversity of woody plants becomes evident when a cosmopolitan view is taken – to the extent, for example, that the currently accepted terminology for branch organization can be shown to be inadequate. The present volume, adopting a cosmopolitan viewpoint, provides a corrective influence. It also demonstrates the remarkable impetus to the study of woody plants given by recent work in the tropics that has advanced knowledge of their construction, reproductive biology, productivity, and community interaction. When similar attention is given to adaptive morphology, morphogenesis, translocation physiology, and photosynthetic capability of tropical trees, the discipline will further burgeon. Clearly, we are dealing with an area of rapid future research expansion.

This volume begins by placing tropical trees in an evolutionary context: Their likely origins are now being traced with increasing reliability as angiosperm phylogeny is traced via the study of leaf and pollen fossils. Modern methods of chemical analysis are then shown to be useful in measuring geographic variation, which may represent incipient material for evolutionary advance. Four important aspects of reproductive biology are considered, because the sexual process is the medium for evolutionary change. The chemical significance of floral rewards, mainly nectar, is examined from the point of view both of pollinating agents and of the physiology of the tree itself. The resulting seed crop is a major attraction to many predators, and the way in which the periodicity of seed production may provide a mechanism for minimizing losses is considered in detail. But seeds do survive this onslaught to germinate and renew tree populations. The examination of seedlings in a large sample of tropical tree species shows a wider range of morphologic categories than previously recognized, with some indication of their adaptive significance and relation to seed size. As a population of new recruits is admitted to the forest undergrowth, mortality continues, and a demographic study of selected examples can be presented, despite the difficulties of estimating tree age indirectly.

A subsequent series of chapters emphasizes morphologic aspects that are appropriate in view of the diversity of tree form, evident to even the most casual observer in the tropics. Recent research has begun to catalog accurately this diversity in terms of recognizable genetic growth plans (architectural models) that often include modular construction and marked differentiation between kind of axis. The classic example of *Terminalia* is analyzed in quantitative detail for the first time. Two distinctive tropical plant families, the Palmae (monocotyledons) and the Araliaceae (dicotyledons) together with the gymnospermous genus *Araucaria* are selected for detailed case studies. It is useful to contrast these accounts with intraspecific variations that result from "mutation," although this is not necessarily genetic. Two opposite parts of the tree are next drawn to our attention. First, the greatly neglected root system, which is seen to be almost as diverse as the shoot system – often conspicuously so, as aerial roots are common in tropical trees. Second, compound leaves are considered from the point of view of their adaptive significance, with the suggestion that they provide a means for rapid adjustment to water stress because they are relatively large, "throwaway" units of construction. This chapter can be linked to the next, which surveys processes of abscission, emphasizing that loss of parts is as important a biologic process as their accretion.

Earlier chapters having emphasized constructional diversity, it is now appropriate to examine the physiologic mechanisms that determine form. A tree develops a trunk because it limits basal branching by a process variously described as apical control or acrotony, and what we know of temperate examples suggests that the controlling mechanism is complex and not explicable solely by "apical dominance." This leads naturally to a discussion of the evidence for multiple correlations that determine branch expression in woody plants. A feature of many tropical climates is the absence of marked seasonality, so that the obvious constraints of cold winters or extended dry periods on shoot extension might appear to be lacking. However, tree growth in the tropics is frequently periodic, and the factors that govern rhythmic behavior are shown to be subtle and dependent on elaborate hormonal interactions. This subject is discussed from several points of view, with experimental analysis strongly emphasized. Evidence for feedback control mechanisms in rhythmic growth is supported by computer simulation models. Because water supply to shoots appears to be a limiting factor, it becomes appropriate to discuss hydraulic conductivity in tropical trees, which in turn depends on a detailed knowledge of their anatomy.

So far, the tropical tree has been analyzed largely at the level of the individual and its parts, but the final chapters address the problem of how these individuals interact within natural communities. Overviews that indicate how the forest is constructed in qualitative and quantitative terms are presented, and the concept of the forest as a mosaic of successional stages is established by the last three contributors. The smallest segment of the mosaic is obviously the individual tree. The conclusion is that not only the tropical tree, but also the tropical forest itself, is a living system.

As the discussion of tropical trees progresses from parts, to whole, and finally to community, it is evident that a linear sequence is inappropriate and artificial: The topics dwelt upon can be interrelated in a reticulate fashion.

We expect the reader will move backward and forward through this volume along paths of his own choosing, but we hope the ways we have charted are clear and well sign-posted. The tropical forest is an awesome place into which biologists need to venture in increasing numbers. We will be satisfied if our efforts provide helpful guidance in the future.

The success of our meetings at the Harvard Forest was the result of the willing collaboration of its staff, to whom we express our collective thanks. All played a significant part, but perhaps we dare single out Mrs. Sandra K. Weidlich and Mrs. Dorothy R. Smith for individual mention.

P. B. Tomlinson
Martin H. Zimmermann

Harvard Forest
Petersham, Massachusetts
December 1977

Part I: Origins and variation

1

Fossil evidence on the evolutionary origin of tropical trees and forests

JAMES A. DOYLE

Museum of Paleontology and Department of Ecology and
Evolutionary Biology, University of Michigan, Ann Arbor, Michigan

Since early in this century, recognition of "tropical" features in the angiosperm-dominated Late Cretaceous and Early Tertiary fossil floras of the present north temperate zone has contributed to the realization that angiosperm evolution must be viewed from a tropical rather than a temperate perspective (Bews, 1927). However, it is primarily in the past two decades that advances in interpretation of the first Early Cretaceous records of angiosperms have begun to provide a solid paleobotanic (as opposed to strictly comparative morphologic) framework in which to examine such problems as the origin and specific evolutionary role of tropical trees in the initial adaptive radiation of the angiosperms. In the same period, studies of the morphology and stratigraphic distribution of early land plants of the Silurian and Devonian have shed new light on the origin and early evolution of the pteridophyte and gymnosperm groups that dominated forests prior to the rise of angiosperms, and on the origin of the basic organs and morphogenetic relations among them, which form the elements of the wide variety of architectural models seen in both living and fossil plants (Hallé & Oldeman, 1970).

Despite these encouraging advances, paleobotany is still far from providing a detailed account of the origin and evolutionary history of the vast number of lineages that make up modern tropical forests. Although studies of fossil pollen and spores (palynology) are beginning to reveal the general features of Cretaceous and Tertiary tropical floras, the most intensively studied palynofloras and essentially all the megafossil floras (which might be expected to yield the most evidence on evolution of vegetative features) are still from extratropical regions of the Northern Hemisphere. Furthermore, because the sorts of data and the problems most amenable to solution are so different in the study of living and fossil plants, the answers the paleobotanist

3

feels prepared to give (e.g., evolution of pollen and leaf venational characters) often do not correspond to the questions of most topical interest to the neobotanist (e.g., floral biology, growth patterns), and the paleobotanist may want answers to questions on the adaptive significance of pollen and leaf features that have been relatively little considered by neobotanists. Hence, instead of attempting a comprehensive review of the still fragmentary data, I prefer to concentrate on some selected facts and speculations about the architecture of fossil plants (in part updating the pioneering discussion of Hallé & Oldeman, 1970) and about the early ecological evolution of angiosperms that have emerged from recent paleobotanic studies, and to point out some areas where greater communication and collaboration between students of living and fossil plants might provide answers of interest to both.

Evolution prior to the rise of angiosperms
Architecture and evolution of the earliest land plants

Recent studies of spores and plant megafossils from Silurian and Devonian rocks, summarized by Chaloner (1967, 1970), Banks (1968, 1970, 1975), Beck (1970, 1976), and Scheckler & Banks (1971), have helped correct errors in dating and morphologic interpretation that contributed to the older belief that land plants must have had a long pre-Silurian history and permit some general conclusions about the origin of organs and taxa of lower vascular plants (Fig. 1.1). Most vascular plants of the Late Silurian and Early Devonian, including both the oldest definite vascular plant, *Cooksonia* from the Late Silurian, and such Early Devonian forms as *Rhynia*, were small, herbaceous, perhaps sometimes still semiaquatic, and characterized by an organography radically different from that of modern plants, but departing only in minor respects from the prototype postulated by the telome theory of Zimmermann (1953): rootless, leafless, dichotomously branched, with terminal sporangia. The most notable exceptions, both younger than the oldest *Cooksonia*, are the order Zosterophyllales (appearing in the earliest Devonian), with lateral, reniform sporangia, and the first Lycopsida (middle Early Devonian), with both lateral sporangia and microphyllous leaves.

Examination of the sequence of Devonian floras reveals complex and varied modifications of the telome plan toward the normal differentiation of organs seen in modern plants, loosely correlated with a gradual increase in size and eventual attainment of the tree habit by several independent lines. Some of these modifications correspond to the elementary processes of the telome theory; others do not. For example, most groups appear to have escaped from the limitations on size inherent in a leafless, dichotomous system (mechanical support, light-gathering efficiency) by the telome process of "overtopping" – a shift to unequal dichotomous (pseudomonopodial) and eventually monopodial branching, resulting in differentiation of a main trunk and lateral branches, sometimes later modified into leaves. This trend is already evident in *Psilophyton* from the middle and late Early Devonian, with pseudomonopodial branching of the main axes and dichotomous laterals (Banks, 1968, 1970). Monopodial or pseudomonopodial branching of all but the ultimate orders of branching (which often remain dichotomous) is almost universally established in Middle Devonian and younger groups of vascular plants except for Lycopsida – i.e., in all the groups Banks (1968, 1970) has postulated were derivatives of *Psilophyton*-like ancestors on the basis of branching, sporangial morphology, and anatomy.

Fig. 1.1. Stratigraphic distribution, changes in relative abundance (highly generalized), and suggested phylogenetic relations of major groups of land plants.

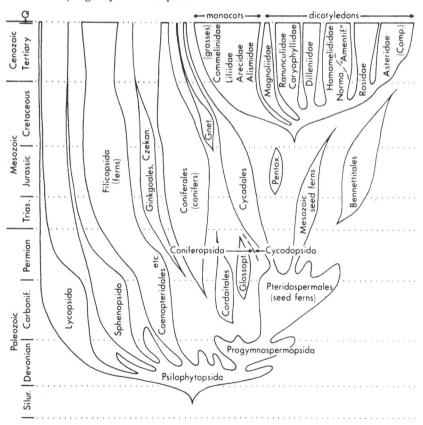

The most important Early Devonian example of a nontelome process appears to be the origin of the leaf (microphyll) of the Lycopsida. Their exarch stelar anatomy and sporangial morphology suggest that the lycopsids were derived from the Zosterophyllales, many of which have spines or enations that differ from microphylls mostly in the absence of vascular tissue (Banks, 1968; Chaloner, 1970). Interestingly, later arborescent members of the Lycopsida (the predominantly Carboniferous order Lepidodendrales) are noteworthy in retaining much of the dichotomous branching of their ancestors while enlarging their microphylls to as much as 1 m in length, as if the enation origin of the leaf allowed them to compensate for the limitations of a dichotomous system without requiring the modifications in branching patterns seen in other groups.

Progymnosperms and the origin of gymnosperms

Among the arborescent groups appearing in the Devonian, the most important for our purposes are the Middle and Late Devonian Progymnospermopsida – plants still pteridophytic in reproduction, but with typically gymnospermous anatomic innovations (secondary xylem, secondary phloem, periderm) and/or tendencies (stelar anatomy, branching patterns) that support the concept that they were ancestral to both cycadopsid and coniferopsid gymnosperms (Beck, 1960, 1970, 1971, 1976; Scheckler & Banks, 1971).

The older of the two principal progymnosperm orders, the Middle and Late Devonian Aneurophytales, was characterized by several orders of spirally or decussately arranged, orthotropic, pseudomonopodial branches, of which the last order bore dichotomous ultimate branchlets or "leaves" (Scheckler, 1976). Thus, except for the fact that the branches had no relation to subtending leaves, the Aneurophytales correspond to Attims's model; the assignment by Hallé & Oldeman (1970) of one genus, *Eospermatopteris*, to Corner's model is based on older misinterpretations of the three-dimensional lateral branch systems as fronds (Beck, 1970, 1976; Scheckler & Banks, 1971). It may be noted that if the progymnosperms are indeed ancestral to the gymnosperms, the almost complete suppression of dichotomous organization (except in the ultimate appendages) even in the Aneurophytales strongly suggests that later occurrences of dichotomous branching in seed plants (e.g., the palm *Hyphaene*: Hallé & Oldeman, 1970; *Flagellaria*: Chap. 7) represent secondary specializations rather than primitive holdovers, occasional claims of overenthusiastic telome theorists notwithstanding.

The abundant Late Devonian genus *Archaeopteris* (*Callixylon*), type of the order Archaeopteridales, is of even greater interest as one of the first forest-forming trees, as the first clear example of plagiotropy, and for its bearing on the origin of the leaf architectural and habital differences between cycadopsid and coniferopsid gymnosperms. Detailed anatomic and morphologic studies by Carluccio, Hueber, & Banks (1966) and Beck (1971) have shown that the apparently bipinnate "fronds" of *Archaeopteris* (which led earlier authors to consider it a prefern) were actually plagiotropic branch systems: a primary axis bearing spirally arranged, dichotomously veined, and variously dissected wedge-shaped "leaves," and, replacing leaves at positions in the same ontogenetic spiral, distichously arranged secondary axes, themselves bearing spirally and/or decussately arranged leaves (Fig. 1.2). Plagiotropy was manifested in the distichous branching pattern, a certain dorsiventrality in leaf arrangement resulting from deviations from normal spiral phyllotaxy (Beck, 1971), and possibly twisting of the leaf bases (Banks, 1970). Hence, Hallé & Oldeman (1970) are correct in assigning *Archaeopteris* to Roux's model, even though the older restoration they cite still interpreted the phyllomorphic lateral branch systems as fronds.

As was first clearly proposed by Meeuse (1966) and elaborated and modified by Beck (1970, 1971), further flattening and differentiation from the trunk of a plagiotropic branch system like that of *Archaeopteris* might lead to a compound frond of the sort seen in the earliest cycadopsid gymnosperms, the seed ferns or pteridosperms of the latest Devonian and Carboniferous (Carluccio et al., 1966), whereas

Fig. 1.2. Plagiotropic lateral branch system ("frond") of Late Devonian progymnosperm *Archaeopteris macilenta*. (Reproduced with permission from Beck, 1971)

reduction of the leaves without modification of the whole system to appendicular status might result in a plagiotropic branch system with needlelike leaves of the sort seen in the earliest conifers of the Late Carboniferous and Permian and in living species of *Araucaria* (Massart's model: see Florin, 1951; Hallé & Oldeman, 1970; Beck, 1970, 1971; Chap. 10). This hypothesis has many complex corollaries, not all of which have received the attention they require if Devonian fossil evidence is to be related to the morphology of modern plants. For instance, it is interesting to note that the first set of transformations postulated – essentially from Roux's model to Cook's model to Corner's model (Hallé & Oldeman, 1970) – would be impossible in modern angiosperms without violating the apparently obligatory association between a branch and a subtending leaf (i.e., no matter how phyllomorphic the branch systems of *Phyllanthus* become, they retain their axillary position), but this would be less of a problem in Devonian plants, where branching was rarely, if ever, axillary. Ironically, *Archaeopteris* itself may constitute an exception to this last generalization, as there are cases where the branch system appears to be associated with a juvenile leaf on the main trunk (see Fig. 1.2). This only underlines the arguments of Beck (1971) that *Archaeopteris* itself has distinctively "coniferopsid" wood anatomic and morphologic features (e.g., decurrent leaf bases) that indicate it is already too specialized to be the common ancestor of both cycadopsid and coniferopsid gymnosperms. On the other hand, any theory of the origin of coniferopsid gymnosperms based on similarities in plan between the compound strobili and the branch systems of *Cordaites* and early conifers and the branch systems of *Archaeopteris* (e.g., Meeuse, 1966) must take into account the fact that branching was strictly axillary in the former (Florin, 1951), but predominantly or wholly nonaxillary in the latter. These problems are less central to evolution of early cycadopsid gymnosperms, only a few of which are known to have had axillary rather than adventitious branching (e.g., the Pennsylvanian seed fern *Callistophyton poroxyloides*: Delevoryas & Morgan, 1954), but axillary branching was to become an important feature of later, Mesozoic cycadopsids (see below).

Whatever the exact resolution of these problems, the morphology of known progymnosperms and the trends inferred from *Archaeopteris* (if at all representative of trends in other progymnosperms) go far toward explaining how the contrasting sparsely branched, compound-leaved cycadopsid and the profusely branched, simple-leaved coniferopsid patterns might both have been derived from a single common ancestry. Their bearing on angiosperms is more obscure because of the uncertainties concerning angiosperm ancestors and the long time separating the Devonian and the Cretaceous. It may be worth noting, however, that although the plagiotropy of *Archaeopteris* may be related to plagiotropy in modern conifers, it is unlikely to have anything to do with plagiotropy in angiosperms. If the anatomic evidence that suggests the angiosperms are cycadopsid rather than coniferopsid derivatives (Bailey, 1949; Takhtajan, 1969) is correct, the branch systems of *Archaeopteris* should be homologous with the leaves of angiosperms, not their branches.

Carboniferous tropical forests

Although many areas covered by the *Archaeopteris*-dominated forests of the Late Devonian may have had tropical climates, the best known and first distinctively tropical Paleozoic forests were the coal swamp floras of the Late Carboniferous (Pennsylvanian) of Europe and North America. Geologic and anatomic evidence (e.g., paleomagnetic and other geologic reconstructions of positions of poles and continents, lack of growth rings) indicates that these areas were located in the wet equatorial tropics (Frederiksen, 1972; Chaloner & Lacey, 1973; Chaloner & Meyen, 1973). It is, therefore, significant that they supported a remarkable variety of architectural models known from the modern flora, many of them restricted to, or most characteristic of, wet tropical environments today, but, of course, belonging to totally unrelated, nonangiospermous groups (Hallé & Oldeman, 1970). Thus, the pachycaul tree fern or palmlike growth pattern (Corner's model), often associated with stilt roots, was exuberantly represented by *Psaronius* in the Marattiales, various coenopterid ferns, and gymnospermous seed ferns such as *Medullosa*; the models of Rauh, Attims, and Tomlinson, combined with horizontal rhizomes, were represented by arborescent sphenopsids (Calamitales); and the dichotomously branched model of Schoute, seen today in the palm *Hyphaene* and a few other monocotyledons, was represented by the arborescent lycopsid order Lepidodendrales (see Hallé & Oldeman, 1970, and above). At least some (though not all!) members of the genus *Cordaites*, the oldest definite representative of the coniferopsid gymnosperms, had stilt roots with a combination of anatomic features found only in mangroves today (Cridland, 1964). *Cordaites* corresponded to Attims's model (Hallé & Oldeman, 1970); its long, strap-shaped, parallel-veined leaves contrast with the leaves of both Paleozoic and living conifers, but they are less different from the presumed juvenile leaves (*Eddya*) of *Archaeopteris* (Beck, 1971).

Most of the bizarre groups making up European–North American Carboniferous coal swamp forests became extinct with the gradual disappearance of coal swamp conditions and evidence of increasing aridity in the following Permian period (White, 1936; Frederiksen, 1972). Their disappearance coincides with the expansion of apparently more drought-resistant members of both great gymnosperm groups – coniferopsids represented by primitive conifers (e.g., *Walchia, Lebachia, Ernestiodendron:* Florin, 1951) and Ginkgoales, cycadopsids represented by new seed fern groups and forms with simple but pinnately veined leaves (*Taeniopteris,* some of which had megasporophylls suggesting relation with modern Cycadales: Mamay, 1969) – as well as ferns. As noted above, many early conifers conform to Massart's model (Hallé & Oldeman, 1970), with pseudo-whorled plagiotropic lateral branch systems as in modern *Araucaria* species, from which, however, they differed in their cone structure and the forked leaves on the main trunk. Interestingly, the influx of conifers is not seen in the then-temperate Angara province of Siberia (Chaloner & Meyen, 1973), suggesting that conifers were primitively adapted to hot and dry climates, rather than the cool, wet climates where they are most diverse today.

Mesozoic gymnosperms

Coniferopsids, cycadopsids, and ferns (represented, of course, by a variety of successively evolving and radiating subgroups) continued to dominate land floras until, and in some areas even after, the mid-Cretaceous rise of angiosperms. Understanding of their growth patterns and ecology is important both in searching for possible ancestors of the angiosperms and in attempting to explain the pattern of angiosperm occupation of Mesozoic communities. Although Mesozoic coniferopsids did include large, monopodial trees with needlelike, simple leaves, and Mesozoic cycadopsids included squat, unbranched, manoxylic forms with crowns of pinnately compound leaves (e.g., *Cycadeoidea*), it would be a serious mistake to visualize them solely in terms of modern conifers and cycads. As Hughes (1976) emphasizes in his stimulating, though controversial, review of Jurassic and Early Cretaceous seed plants, modern gymnosperms are clearly relict and restricted to their present niches by competition with angiosperms; before the rise of angiosperms, they undoubtedly covered a much greater spectrum of ecologic and architectural types.

Among the coniferopsids, a prime example of Hughes's concept is provided by the great group of Jurassic and Early Cretaceous conifers represented in the pollen record by *Classopollis* and in the megafossil record by several genera, including *Brachyphyllum* (branch systems with tightly appressed, rhomboidal, xeromorphic scale leaves) and *Frenelopsis* (with the exserted portions of the opposite-whorled leaves so reduced that the stems have a jointed appearance: Hluštík & Konzalová, 1975). These "Mesozoic brachyphylls" are often but misleadingly compared with Araucariaceae or Cupressaceae, from which they are quite distinct in cone and pollen morphology (Hughes, 1976). In the Jurassic, *Classopollis* dominates palynofloras from most areas of the world except Siberia and adjacent Arctic regions (Vakhrameev, 1970). In the Early Cretaceous, it becomes somewhat less abundant at middle paleolatitudes, but it is still dominant, along with newly appearing "ephedroid" pollen grains, in an African–South American belt straddling the paleoequator (Kuyl, Muller, & Waterbolk, 1955; Herngreen, 1974; Jardiné, Kieser, & Reyre, 1974; Brenner, 1976). These plants were clearly the most successful group of tropical forest-forming trees in the Jurassic and Early Cretaceous, but they were so completely replaced during the phenomenal mid-Cretaceous radiation of angiosperms that their closest tropical analogs on which to base speculations about growth habit are in totally unrelated groups (e.g., *Casuarina?*). At the same time, the Ginkgoales, which dominated northern latitudes, included genera with leaf architecture (long, ribbonlike leaves) quite removed from that of modern *Ginkgo biloba;* however, even some of the ribbon-leaved forms are known to have had a pronounced long-shoot–short-shoot dimorphism (Florin, 1936) that, together with the abundance of detached leaves, suggests a winter-deciduous habit.

The Mesozoic cycadopsids included an even greater variety of extinct groups – the orders Bennettitales and Pentoxylales; the seed fern families Corystospermaceae, Peltaspermaceae, and Caytoniaceae; and extinct Cycadales (Nilssoniales) – which exhibited a far greater range of morphology than the relict living Cycadales. In particular, as emphasized by Harris (1961), Delevoryas & Hope (1971, 1976), and Hughes (1976), many had smaller leaves, more slender stems, and more profuse branching than modern cycads. Examples are Cycadales such as the Late Triassic *Leptocycas* (Delevoryas & Hope, 1971) and the Cretaceous *Nilssoniocladus,* with a remarkable *Ginkgo*-like shoot dimorphism (Kimura & Sekido, 1975); Bennettitales such as the Late Triassic *Wielandiella* (Nathorst, 1909), the Jurassic *Williamsoniella* (Thomas, 1915), and the Late Triassic *Ischnophyton* (Delevoryas & Hope, 1976); and, judging from the small size of its palmately compound leaves and a single slender, branched twig with scale leaves, the Jurassic seed fern *Caytonia* (Harris, 1964). However, there are intriguing hints that one of the most distinctive architectural features of modern Cycadales (and many angiosperms!) – sympodial branching, as opposed to the predominantly (though not exclusively) monopodial branching of both Mesozoic and modern conifers – was widely distributed in other, less closely related Mesozoic cycadopsid groups. In the bennettitalian genera *Wielandiella* and *Williamsoniella,* upward growth of each stem segment was terminated by flowering, but upward growth of the plant

12   James A. Doyle

was continued by pairs of immediately subjacent lateral relay axes, resulting in a classic example of Leeuwenberg's model (Hallé & Oldeman, 1970).

Early ecologic evolution of angiosperms

The recent reawakening of interest in the Cretaceous fossil record as a source of critical evidence on the course of early angiosperm evolution has been the result of conceptual and methodologic advances similar to those that stimulated Devonian paleobotany. Indeed, the inferred adaptive radiations of early land plants and early angiosperms show striking analogies with each other and with paleontologically documented radiations of animal groups (Simpson, 1953; Doyle, 1977). Previously, poor stratigraphic resolution and the attempts of paleobotanists to assign early angiosperm fossils (mostly leaves) to modern families and genera, without either sufficient background information on the systematic usefulness and limitations of leaf characters or recognition of the effects of extinction and evolution, led to an exaggerated impression of the systematic diversity and modernity of Cretaceous angiosperm floras (see discussions in Wolfe, 1973; Dilcher, 1974; Wolfe, Doyle, & Page, 1975; Doyle & Hickey, 1976; Hughes, 1976). This, together with reports of supposed pre-Cretaceous angiosperm fossils, led logically enough to theories that the angiosperms must have originated and diversified to the familial or generic level in some region isolated from areas of the known fossil record (e.g., tropical uplands: Axelrod, 1952, 1960, 1970; Southeast Asia and Australiasia: Takhtajan, 1969; Smith, 1973) prior to their general appearance in the late Early Cretaceous. Although some workers voiced doubts concerning the leaf-identification methods on which this interpretation was largely based, the first substantive challenges to the prior diversification theory came as a result of studies of the fossil pollen record, which might be expected to provide a more complete picture of regional floras, including even previously hidden upland elements, than the record of less readily transported leaves. As was recognized by Scott, Barghoorn, & Leopold (1960), Hughes (1961), Pierce (1961), Pacltová (1961), and Brenner (1963), palynologic studies reveal no records of distinctively angiospermous pollen types before the middle or late Early Cretaceous, and a far lower morphologic diversity of angiospermous types in late Early and early Late Cretaceous pollen floras than would have been predicted from previous identifications of modern taxa based on leaves from the same or correlative beds. At the same time, more critical examination shows that supposedly pre-Cretaceous angiosperm fossils are stratigraphically misplaced, belong to gymnospermous or other groups poorly known at the time they were described, or lack sufficient diagnostic characters to determine whether they are gymnosperms or angiosperms (Scott et al., 1960; Hughes, 1961, 1976; Doyle, 1973; Wolfe et al., 1975; Doyle, Van Campo, & Lugardon, 1975).

More detailed studies of Cretaceous pollen sequences, concentrating on morphologic interrelations and stratigraphic distribution of angiosperm pollen types, have strengthened the concept that the angiosperms were undergoing their primary adaptive radiation in the Early Cretaceous and have led several authors to propose schemes for the evolution of angiosperm pollen types (Doyle, 1969, 1973; Muller, 1970; Jarzen & Norris, 1975; Singh, 1975; Hughes, 1976). Although these schemes themselves are independent of supposed systematic relations with modern taxa, they have been used to evaluate and modify phylogenetic hypotheses derived from comparative studies of extant plants (Doyle, 1969, 1973; Muller, 1970; Wolfe et al., 1975; Walker & Doyle, 1975). Recently, reexamination of the Cretaceous angiosperm leaf record using a similar approach has confirmed the suspicion that older attempts to identify modern low-rank taxa in the mid-Cretaceous record were grossly premature and has shown that the leaf record too is consistent with the concept of a rapid adaptive radiation of the angiosperms in the Cretaceous (Wolfe et al., 1975; Doyle & Hickey, 1976). Both pollen and leaf records show the first appearances of rare and generalized forms in the middle Early Cretaceous (Barremian-Aptian), acceleration to maximum evolutionary rates in the late Early Cretaceous (Albian), and attainment of widespread ecologic dominance and intermediate levels of morphologic advancement by the most specialized forms in the early Late Cretaceous (Cenomanian-Turonian) – i.e., over a period of some 20–30 million years (Doyle, 1969, 1977; Muller, 1970; Wolfe et al., 1975; Doyle & Hickey, 1976).

In the present discussion, I wish to concentrate on those aspects of the Cretaceous fossil record that help clarify the role of tropical forest trees in the general pattern of ecologic and habital radiation of the angiosperms. Unfortunately, to an even greater degree than with lower plant groups, where leaves often remain attached longer to the stems, the fossil record of angiosperms consists largely of isolated leaves, pollen grains, and, less commonly, wood, fruits, and seeds, rather than large portions of whole plants or branch systems that might give direct evidence on vegetative architecture. However, I will attempt to draw some indirect inferences about habit and ecology from (1) functional-morphologic analysis based on analogies between fossil and modern pollen and leaves (which need not be related taxonomically) and consideration of physiologic-ecologic principles, and (2) the distribution of fossils in various sediment types (facies) and geographic areas, arguing from geologic evidence about environments to the ecologic preferences of the plants, rather than vice versa. This procedure is deliberately intended to avoid arguments based on the ecology and habit of supposedly related modern taxa and the attendant risk of errors due to incorrect identifications and mosaic evolution (i.e., where one set of features associated with a modern group evolved before another set, such as secondary growth and the tree habit before seeds in the progymnosperm-gymnosperm transition; see above). Some comments on possible relations to modern high-rank taxa (classes, subclasses) will be made in other contexts, however. Of course, because of the present rudimentary understanding of the functional significance of pollen and leaf features and the factors controlling the distribution and preservation of leaves in sedimentary environments, many of these inferences are admittedly rather speculative; this is an area where comparative studies of modern plants might permit more solid conclusions. Although I will summarize a model for the early ecologic-adaptive radiation of the angiosperms previously proposed by L. J. Hickey and myself (Doyle & Hickey, 1976), I do not claim this is the single best model that can be derived from the existing data. Rather, my intent is to illustrate the point that the fossil record is definitely more consistent with some existing models of the ecologic evolution of angiosperms than it is with others (e.g., Bews, 1927; Stebbins, 1974).

Another limitation of the present discussion is the fact that although Early Cretaceous pollen sequences from tropical areas are now becoming fairly well known, Early Cretaceous leaf floras from the tropics remain virtually unknown. In discussing the leaf record I will necessarily be limited to sequences from North America, Europe, and the USSR (at the time parts of the single continent of Laurasia), and particularly the sequence that has been restudied in greatest detail from the perspective outlined above, the Potomac Group of the Atlantic Coastal Plain of the United States (Fig. 1.3; Wolfe et al., 1975; Doyle & Hickey, 1976). Luckily, although latitudinal climatic and floristic zonation existed in the Early Cretaceous, it was far less pronounced than it is today (as indeed it was throughout most of the geologic past). Oxygen isotope data (Bowen, 1961) and distributions of marine faunas (Sohl, 1971) indicate that the late Early Cretaceous (Albian) was a major climatic optimum. This agrees with the very warm temperate to subtropical aspect of the Potomac flora (diverse cycadophytes and ferns of the families Schizaeaceae, Gleicheniaceae, and Cyatheaceae: Brenner, 1963, 1976)

Fig. 1.3. Principal angiosperm pollen and leaf types in Potomac Group (Cretaceous) of Atlantic Coastal Plain, United States. At left are proposed correlations between standard European stages of the Cretaceous (4 of 12) and rock units recognized in Maryland; at right are palynologic zones and subzones of Brenner (1963), Doyle & Hickey (1976), and Doyle & Robbins (1977). Dashed lines indicate extensions of ranges of leaf types inferred from other sedimentary sequences. Pollen types indicated: *a*, generalized tectatecolumellar monosulcates (*Clavatipollenites, Retimonocolpites, Stellatopollis*); *b*, monocotyledonoid monosulcates (*Liliacidites*); *c*, reticulate to tectate tricolpates; *d*, tricolporoidates; *e*, small, smooth, prolate tricolporoidates; *f*, small, smooth, oblate-triangular tricolporoidates. Leaf types indicated: *g*, narrowly obovate, monocotyledonoid, with apical vein fusion (*Acaciaephyllum*); *h*, pinnately veined, reniform; *i*, pinnately veined, serrate; *j*, pinnately veined, narrowly obovate-lanceolate; *k*, pinnately veined, large, elliptic (*Ficophyllum*); *l*, palmately veined, reniform, lobate; *m*, pinnately veined, obovate; *n*, palmately veined, peltate (*Nelumbites*); *o*, palmately veined, ovate-cordate; *p*, pinnately veined, serrate; *q*, pinnately lobed (*Sapindopsis*); *r*, palmately lobed, irregular fine venation; *s*, pinnately compound, with tendency for regular tertiary venation; *t*, palmately lobed, with very regular tertiary venation (typical platanoid). (After Doyle & Hickey, 1976)

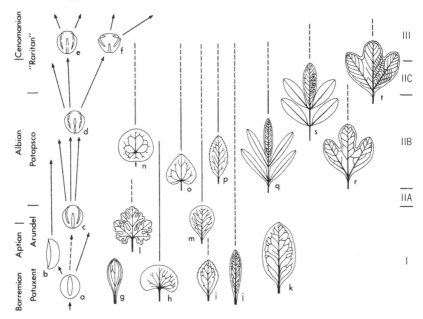

and the existence of many common elements (hence interchange) between the palynofloras of the Potomac Group and Early Cretaceous tropical areas (Africa, South America: see below).

Early angiosperm pollen record

In the Laurasian area, the oldest definite pollen records of angiosperms so far reported are several species of rare monosulcate grains (*Clavatipollenites*, *Retimonocolpites*, *Liliacidites*, *Stellatopollis*) from the basal Potomac Group (the lower part of Brenner's pollen zone I: Fig. 1.3a; Brenner, 1963; Doyle, 1969, 1973; Wolfe et al., 1975; Doyle et al., 1975; Doyle & Hickey, 1976; Doyle & Robbins, 1977) and from the better-dated upper Wealden of England (Barremian: Couper, 1958; Kemp, 1968; Hughes & Laing, unpubl. data presented at XII International Botanical Congress, Leningrad, 1975). The principal criteria for distinguishing these forms from more common gymnospermous monosulcates are their reticulate, columellar exine structure and/or the absence of the laminated inner exine layer (endexine) characteristic of gymnosperms (Wolfe et al., 1975; Doyle et al., 1976). Interestingly, a few of these monosulcates (*Liliacidites* spp.: Fig. 1.3b) have exine sculpture features found today in monocotyledons but not in the magnoliid dicotyledons, which also retain monosulcate pollen (Doyle, 1973; Walker & Doyle, 1975; Wolfe et al., 1975), but they are far too generalized to be assigned to any specific modern monocot subgroup (to say nothing of habital type).

In the early Albian of England (Kemp, 1968; Laing, 1975, 1976) and near the top of zone I of the Potomac sequence (Wolfe et al., 1975; Doyle & Hickey, 1976; Doyle & Robbins, 1977), angiospermous monosulcates are joined by the first tricolpate pollen (Fig. 1.3c), a type found today only in the six subclasses of dicots exclusive of Magnoliidae (*sensu* Takhtajan, 1969; Walker & Doyle, 1975). Later in the Albian (Potomac subzones IIB and IIC), tricolpate pollen becomes locally dominant for the first time (Doyle, 1969; Doyle & Hickey, 1976), and many lines show the first tendencies toward the tricolporate condition (Fig. 1.3d–f), the most common condition in modern dicots (Doyle, 1969; Wolfe & Pakiser, 1971; Pacltová, 1971; Wolfe et al., 1975). Some of the resulting early Cenomanian tricolporates – specifically, smooth, triangular forms (Fig. 1.3f; Doyle, 1969; Wolfe et al., 1975) – have been proposed as the ancestors of the triangular triporate Normapolles that appear in middle Cenomanian rocks and undergo a major radiation in the later Cretaceous of Europe and eastern North America (Góczán, Groot, Krutzsch, & Pacltová, 1967; Doyle, 1969; Wolfe & Pakiser, 1971; Pacltová, 1971). The Normapolles type is extinct as such, but later Cretaceous Normapolles have been considered ancestors of some if not all modern triporate Amentiferae (Góczán et al., 1967; Doyle, 1969; Wolfe, 1973; Wolfe et al., 1975).

Although the pollen record can hardly be expected to supply evidence about vegetative morphology of early angiosperms, geographic differences in early angiosperm pollen floras and functional interpretations of pollen morphology might give hints about their climatic preferences and floral biology, respectively. As has been documented in greatest detail by Brenner (1976; see Doyle, 1977), significant deviations in diversity, abundance, and timing of appearance of angiosperm pollen types (though not great enough to contradict the evolutionary scheme outlined above) occur both to the north and the south of the most intensively studied southern Laurasian belt. In the Canadian Arctic (northern Laurasia), and to a lesser extent in western Canada and in Australia (southern Gondwana: Dettmann, 1973), angiosperm pollen is not reported or is very rare before the late Albian or even Cenomanian (Brenner, 1976), i.e., after it has already become locally abundant in southern Laurasia. On the other hand, angiosperms appear to be more abundant and diverse in pollen floras from the Early Cretaceous equatorial belt (northern Gondwana, just beginning to separate into the two continents of Africa and South America) than at correlative horizons in Laurasia. Although recent studies on Gabon and the Congo (Doyle, Jardiné, & Doerenkamp, 1976) have shown that the earliest phases of the angiosperm record consist of monosulcates comparable to, and possibly contemporaneous with, those from the basal Potomac and the Barremian of England, tricolpates appear well below marine beds whose fauna date them as late Aptian in Brazil (Müller, 1966; Brenner, 1976), Africa (Jardiné et al., 1974; Doyle et al., 1976), and Israel (Brenner, 1976), predating their first early Albian occurrences in Laurasia. Furthermore, although the Albian-Cenomanian interval in Africa–South America shows the same order of appearance of major types as in Laurasia (tricolpate, tricolporate, triporate), it also contains endemic types known only later (polyporates) or not at all (tricolpodiorates) from other areas (Herngreen, 1974; Jardiné et al., 1974; Brenner, 1976).

These data clearly support the concept that the angiosperms are a basically tropical group, initially poorly adapted to colder climates, and that some major types (especially tricolpates) originated in Africa–South America and spread poleward into Laurasia and Australia (Axelrod, 1959, 1970; Brenner, 1976). Perhaps more surprising, however, is the fact that independent paleoclimatic indicators – predominance of *Classopollis* (noted above to be produced by xeromorphic ''brachyphyll'' conifers) and *Ephedripites* (whose two most probable modern derivatives are the xerophytic genera *Ephedra* and *Welwitschia*), low diversity of pteridophyte spores, and presence of extensive late Aptian salt deposits – suggest that conditions in the African–South American area were at least semiarid (Brenner, 1968, 1976; Jardiné et al., 1974). This initially appears to lend strong support to Stebbins's (1974) concept that the angiosperms originated and diversified most actively under semiarid conditions, as opposed to the wet tropical conditions postulated by most authors (e.g., Bews, 1927; Axelrod, 1952, 1960; Takhtajan, 1969). But the fact that the angiosperm element at the earliest stages of the African–South American record is comparable to that in Laurasia indicates that the situation was more complex. It seems reasonable to infer that semiarid climates were generally more favorable for early angiosperm diversification than might have been predicted from Bews's (1927) model, but it would be premature to conclude that the angiosperms originated in Africa–South America or that aridity was necessarily the key selective factor in their origin.

If generalizations based on correlations between pollen morphology and pollination syndromes in modern plants are valid (Faegri & van der Pijl, 1966; Heslop-Harrison, 1971), the reticulate sculpture of the first recognizably angiospermous monosulcates of the Barremian and the first tricolpates of the Aptian can be taken as evidence that insect pollination was well established at the early stages of angiosperm radiation (Doyle & Hickey, 1976). Likewise, as the presence of more than one aperture has been interpreted as an adaptation to ensure more efficient germination on the exposed surface of a stigma (Hughes, 1961, 1976; Doyle & Hickey, 1976), the appearance of tricolpates in the Aptian suggests that the basic angiosperm feature of carpel closure had also evolved by this early stage. In contrast, combinations of characters believed adaptive for wind pollination – smooth, thin exine and reduced, porelike apertures (Faegri & van der Pijl, 1966; Whitehead, 1969) – do not appear until relatively late in the radiation; e.g., in the first mid-Cenomanian Normapolles, or, more definitely, the rounder, thinner-walled triporates of the middle Late Cretaceous (Doyle, 1969; Wolfe, 1973; Doyle & Hickey, 1976).

Early angiosperm leaf record

The angiosperm leaf record in the Potomac Group, recently redescribed and compared with Early Cretaceous floras elsewhere in Laurasia by Wolfe et al. (1975) and in greater detail by Doyle & Hickey (1976), shows remarkable analogies with the pattern of angiosperm pollen diversification in the same sequence (Fig.1.3). Leaf localities are now known from enough different sedimentary facies (hence environments) at several horizons to allow confidence that sampling of the flora is reasonably complete and that the vertical changes seen are not simply the effects of shifting environments.

In pollen zone I (Barremian–early Albian?: Fig. 1.3*g–m*), angiosperm leaves show a restricted facies distribution and never dominate the flora. They occur in coarser-grained sediments believed deposited on stream levees and point bars, but are absent from fine-grained floodplain and swamp facies dominated by ferns, conifers, and cycadopsids (Doyle & Hickey, 1976). Morphologically, they are characterized by pinnate or related secondary venation patterns; range in shape from lanceolate to obovate to reniform, with or without blunt marginal serrations or lobes; and tend with the exception of one group of large, broad forms (*Ficophyllum:* Fig. 1.3*k*) to be small. Their most striking features are the generally poor differentiation of blade and petiole and the unusual irregularity in course and differentiation of all orders of venation. Although it is impossible to be sure whether these fossils are whole simple leaves or leaflets of compound leaves without finding them attached to stem or rachis, it may be significant that there is no positive evidence of compound leaves in zone I, in contrast to the situation higher in the section. In agreement with the pollen record, there is at least one zone I leaf type, known attached to slender, apparently flexible stems, which has apically fusing "parallel" venation suggestive of monocotyledonous affinities (*Acaciaephyllum:* Fig. 1.3*g*; Doyle, 1973; Doyle & Hickey, 1976).

In the next group of Potomac leaf localities (lower-middle subzone IIB: middle Albian?), angiosperm leaves occupy a wider variety of sedimentary facies, are occasionally locally abundant, and include several important new morphologic types (Fig. 1.3*n–r*). One new complex consists of palmately veined cordate to peltate leaves (Fig. 1.3*n,o*), of which especially the latter have features suggestive of a rooted aquatic habit (veins dichotomizing and looping well within the margin; long, apparently flexible petioles; and occurrence in apparently in situ clumps in fine-grained pond facies: Doyle & Hickey, 1976). Another consists of highly variable, seemingly compound, but actually pinnately lobed leaves (*Sapindopsis:* Fig. 1.3*q*), which, judging from their local abundance, must have formed dense stands, and the first palmately lobed platanoids (Fig. 1.3*r*). Higher in subzone IIB and in subzone IIC, these types are largely replaced by the first truly pinnately compound leaves, related to *Sapindopsis* on details of cuticle and venation structure but tending toward more regular alignment of the tertiary veins (Fig. 1.3*s*), and by palmate platanoids that show an even more pronounced tendency for rigid orientation of the finer venation (Fig. 1.3*t*). The latter in particular are widely reported from rocks near the Lower–Upper Cretaceous boundary elsewhere in

Laurasia (e.g., the Dakota Group of Kansas) and are typically associated with sandy stream levee deposits (Doyle & Hickey, 1976).

On the basis of the pollen evidence cited above that some Potomac angiosperms represented groups distributed contemporaneously over both wet subtropical and dry tropical belts, and others were recent immigrants from hotter and drier regions of Africa–South America, these leaf data may be related to models for the early ecologic-habital radiation of the angiosperms. As is discussed in detail elsewhere (Doyle & Hickey, 1976), several aspects of the observed diversification pattern seem easier to reconcile with Stebbins's (1974) hypothesis that the earliest angiosperms were fast-growing, weedy shrubs adapted to colonizing disturbed habitats in semiarid environments, than with the more conventional postulate that they were magnolialian rain forest trees (e.g., Bews, 1927; Axelrod, 1952, 1960, but not 1970; Takhtajan, 1969). First, the restriction of zone I angiosperm leaves to coarser fluvial facies and their absence from fine-grained floodplain deposits dominated by conifers and ferns (the presumed climax vegetation) would be consistent with disturbed stream margin habitats, one of the principal environments to which Stebbins's semixeric shrubs would be preadapted in a wet lowland area. Second, assuming such a starting point makes it easier to explain in selectionist terms the early origin of small-leaved monocotyledonoid forms (not palms, which are not reported until well into the Late Cretaceous: Muller, 1970; Read & Hickey, 1972; Doyle, 1973) and the appearance not long after of dicots with aquatic specializations. Third, the functional morphology and subordinate role of the largest zone I angiosperm leaves (Fig. 1.3k) are consistent with the expectation that one of the forest niches most accessible for stream margin shrubs would be the shaded understory (Doyle & Hickey, 1976). Their broad, elliptic shape and large size suggest a monolayer light-gathering strategy advantageous for maximum interception of low-intensity light (Horn, 1971); their poorly differentiated petioles and loose, disorganized venation would seem mechanically disadvantageous except in conditions of low wind and rain stress, such as the understory habitats of modern Winteraceae, which retain the most similar leaf architecture (Carlquist, 1975; Wolfe et al., 1975; Doyle & Hickey, 1976).

Aspects of the later Potomac leaf record are consistent with the concept that subsequent evolution of woody angiospermous types tended to run up ecologic succession (cf. Margalef, 1968), first producing early successional trees (gap phase and riparian species?), and only later large trees capable of competing with the conifers that dominated the climax forest canopy (Doyle & Hickey, 1976). In contrast to the situation in zone I, the local abundance and morphology of subzone IIB and IIC pinnately and palmately lobed and pinnately compound leaves (Fig. 1.3*q–t*) are suggestive of an open, early successional tree niche. Whether the lobed or compound leaf is interpreted as part of a multilayer light-gathering strategy for most effective utilization of high-intensity light (Horn, 1971) or as a means of producing a cheap, throwaway branch advantageous for rapid upward growth (Chap. 15), it would seem particularly advantageous for early successional trees. Hence, it is interesting that the trend toward more rigid tertiary venation, presumably mechanically advantageous in exposed situations, is most marked in the pinnately compound and palmately lobed groups. Finally, the association of subzone IIC platanoids with stream margin facies and the continued predominance of conifer pollen in deposits of Cenomanian age suggest that the postulated restriction of dicot trees to early successional stages continued until well into the Late Cretaceous (Pierce, 1961; Doyle & Hickey, 1976; Doyle, 1977).

General implications
Although I have emphasized that this interpretation of the early angiosperm record does not depend on supposed systematic affinities between living and fossil forms, it does have some general implications for the evolutionary status of certain modern high-rank taxa (Wolfe et al., 1975; Doyle & Hickey, 1976). First and most notably, there are no modern angiosperms that can be considered direct, unmodified survivors of the inferred original small-leafed, semixerophytic, weedy adaptive type; presumably the original occupants of this niche have long since been replaced by a succession of competitively superior, younger groups (Stebbins, 1974). Second, although larger zone I pinnately veined leaves (e.g., *Ficophyllum*) could be interpreted as early representatives of lines leading to modern woody Magnoliidae (Hickey & Wolfe, 1975; Wolfe et al., 1975; Doyle & Hickey, 1976), it is interesting that they have not been shown to be older than leaves with probable monocot affinities (*Acaciaephyllum*) and apparent early representatives of the trend leading to aquatic herbs of possible protonymphaealian affinities (the reniforms: Fig. 1.3*h*; Doyle, 1973; Doyle & Hickey, 1976). Finally, the younger pinnately lobed to compound and palmately lobed groups interpreted as early successional and riparian trees have been considered transitional to the "higher" dicot subclasses Rosidae and Hamamelididae (Takhtajan, 1969), respectively, though their first members exhibit a grade of leaf architectural evolution more primitive than any living rosids or hamamelidids (Hickey & Wolfe, 1975; Wolfe et al., 1975; Doyle & Hickey, 1976). Differentiation and specialization of all these groups into recognizable modern orders and

families and attainment of the full spectrum of modern adaptive types were presumably phenomena of the later Cretaceous and Tertiary (Muller, 1970).

The model for early ecologic evolution of the angiosperms summarized above – beginning with pioneer shrubs of disturbed and/or semiarid environments, radiating into aquatic and other herbaceous niches in one direction, and running up ecologic succession through early successional trees to climax canopy trees in the other (Doyle & Hickey, 1976) – has several intriguing implications that may help explain the origin and diversity of tropical angiospermous trees and their success in displacing earlier tropical plant groups. One such corollary is that much of the diversity of architectural models seen in modern tropical forests may reflect multiple, opportunistic origins of the tree habit from shrubby and in some cases even herbaceous ancestries (e.g., palms?), rather than adaptive radiation of models at the tree level. Another, more ecologic corollary is that the initial spread and success of the angiosperms were the result, not of any superior ability to compete with the conifers as climax forest dominants, but rather of an ability to occupy disturbed, early successional, aquatic, and understory habitats to which conifers, cycadopsids, and ferns were relatively poorly adapted. Here, especially advantageous features might have included (1) the ability to produce broad, reticulate-veined leaves by intercalary growth (Doyle & Hickey, 1976); (2) more flexible growth habits (Stebbins, 1974), perhaps reflecting a new, superficially coniferopsid-like capacity for more profuse branching (related to reduction in leaf size during the postulated semixeric phase, or simply a continuation of the trend begun among Mesozoic cycadopsids?) superimposed on a capacity for sympodial growth inherited from a cycadopsid ancestry (see above); and particularly (3) more rapid and efficient reproduction, involving protection and telescoping of vulnerable stages of the life cycle and throwaway parts, some aspects of which (insect pollination, closed carpels) are indirectly inferred from the fossil record (Stebbins, 1974; Doyle & Hickey, 1976; and above). As Stebbins (1974) argues, many of these features can be reasonably (though not all necessarily!) explained as original adaptations to seasonally arid conditions; they form part of the basis for his critique of the hypothesis of Bews (1927), Takhtajan (1969), and others that the tropical rain forest was the cradle of the angiosperms. However, the fact that the later stages of the angiosperm radiation attested by the fossil record can be readily interpreted as occurring in mesic, lowland environments belies to some extent Stebbins's alternative hypothesis that the rain forest is primarily a "museum" in which there has been relatively little major adaptive diversification.

Upon attainment of the tree habit, many of the innovations that, it is suggested, contributed to the initial success of the angiosperms as early successional species may have proved advantageous in replacing the conifers. In the reproductive sphere, rapidity of reproduction itself might not be particularly advantageous for a tropical tree (consider the long developmental periods cited in Chap. 4), but I suggest that insect pollination, as one factor permitting hyperdispersed distributions and resulting escape from the predation, parasitism, and disease that are so severe in nonseasonal tropical climates (Janzen, 1970), may have given the angiosperms a critical edge over wind-pollinated tropical conifers, constrained to form denser stands (Whitehead, 1969). This advantage would have been far less significant in temperate areas; hence, it is interesting that the replacement of conifers, particularly *Classopollis*-producing brachyphylls, was earliest (Turonian) and most complete in Africa–South America (Jardiné et al., 1974), whereas it is primarily in certain temperate areas that conifers have remained dominant the longest and the angiosperm pollen record shows the first evidence of reversion to wind pollination (Doyle & Hickey, 1976; and above).

Finally, it must be noted for perspective that the Early Cretaceous and Cenomanian events discussed in detail here were followed by another 90 million years of later Cretaceous and Tertiary angiosperm evolution. Whatever the causal factors and processes involved in the initial rise of the angiosperms, the present composition of tropical forests is the result of a complex pattern of secondary radiations, extinctions, and shifting geographic distributions of younger angiosperm subgroups, some hints of which can be gleaned from discussions of Late Cretaceous and Tertiary tropical palynology (e.g., Germeraad, Hopping, & Muller, 1968; Muller, 1970; Wolfe, 1975). Likewise, whether angiosperms originated and underwent their most rapid initial radiation in the tropical rain forest or in drier tropical environments, Bews (1927) was clearly correct in believing that the fossil record provided evidence for his view (and that of other tropical botanists: e.g., Corner, 1949; Tomlinson & Gill, 1973) that present temperate zone biomes are highly anomalous and peripheral to the main theater of angiosperm evolution. Some temperate biomes, particularly broad-leaved deciduous forests and temperate grasslands, seem to have originated only with a marked climatic deterioration and increase in annual temperature extremes in the Oligocene (Wolfe, 1971, 1972; Leopold & MacGinitie, 1972). On the other hand, Late Cretaceous and Early Tertiary temperate forests, apparently characterized by less annual variation in temperature (Wolfe, 1971), may have been more "tropical" in many aspects of their biology, as Tomlinson & Gill (1973) suggest is the case for the analogous temperate forests of New Zealand and Australia.

If nothing else, I hope this brief and speculative review has shown that the fossil record does yield much information potentially relevant to such problems of plant evolution as the origin of tropical trees and forests, even though far more work is needed before it can be unambiguously interpreted. On the paleobotanic side, studies of Early Cretaceous leaf floras from areas outside Laurasia, particularly Africa and South America, should receive high priority, and paleobotanists should be alert for evidence, however fragmentary, of branching patterns, growth habit, anatomy, and ecologic associations of early angiosperms and their contemporaries. On the neobotanic side, students of tree architecture and reproductive biology should devote more attention to possible correlations among leaf architectural features (which are visible in the fossil record) and growth patterns, role in community structure, and environment, and between pollen morphology and pollination biology. More direct evidence is also needed on factors controlling the distribution of leaves and pollen in modern sedimentary environments. With such evidence, it may be possible to evaluate more realistically what inferences can be extracted from the fossil record and to provide solid answers to evolutionary questions many have considered in the realm of pure speculation.

Acknowledgments

I wish to thank C. B. Beck, T. J. Givnish, L. J. Hickey, and D. H. Janzen for valuable discussions that helped me formulate many of the ideas expressed in this chapter, and P. Kimmel and G. Newton for assistance in preparing the manuscript.

General discussion

Ng: Have you looked at fossil seeds? Among present-day tropical rain forest seeds, those of pioneer or nomadic species in ephemeral habitats are less than 1 cm long; often they are not longer than 0.3 cm. In contrast, the vast majority of seeds of climax forest trees are longer than 1 cm. If the primitive angiosperms arose in a tropical humid environment, a comparative study of fossil seed sizes might give us a clue to whether or not they were components of ephemeral communities.

Doyle: It would be very nice to know. We have a few infructescences that have not been analyzed yet. It will be rather difficult from a technical point of view to draw as general conclusions from them as can be drawn from pollen and leaves, where you have a larger sample and a better concept of what you are looking at. One of the big problems is in distinguishing angiosperm fructifications from those of the many gymnosperm groups that existed then, especially without structural preservation.

Ashton: First, your pictures of the first putative angiosperm leaves unfortunately lack any base to the midrib, so it seems very difficult to say dogmatically that they were not part of compound leaves. Can you be certain that these leaves were not pinnate? Second, there are only limited circumstances under which leaves, fruits, and, indeed, pollen are going to be preserved. One of them is under the conditions you mentioned, in river valleys, which gives you information about a river valley flora. But it is not then valid to deduce that this is where the early angiosperms are likely to be found, in the wet tropics. One would like, for instance, to examine volcanic deposits from hillsides or ridgetops and look at those from parts of the world that have been suggested as the sites of the early evolution of the angiosperms (e.g., Southeast Asia). One would like to see somebody looking at late Jurassic deposits or volcanic deposits in that section of Eurasia that was abutting onto this still major ocean in the "Pacific." Presumably, the world has always been spinning in the same direction, with the trade winds always going in the same direction. Alternative sources of speculation about the evolution of angiosperms still have a very important part to play, and we should not underestimate them, especially as the geologic evidence can never be sufficiently complete.

Doyle: First, it is hard to rule out with certainty the existence of compound leaves in the lower Potomac. However, there is good evidence of compound, and slightly earlier than that, pinnately lobed leaves, much higher in the Potomac sequence. These were among the first really abundant remains and were readily preserved as whole leaves. There is no evidence of this type in the older deposits. In the lower Potomac, it is often hard to say what the morphology of the leaf base is, particularly where the lamina narrows down to the petiole. This again contrasts with the situation in many of the younger deposits, where there are leaves with well-defined petioles, quite distinct from the bases of the laminar portion. It would be a contradiction to the modern distribution of systematic characters if you found compound leaves in groups with monosulcate pollen. As to whether what we see is a representative record, I would argue that the pollen record does reflect more than what is right there at the site of deposition where sediment is accumulating. It is the pollen evidence that makes us confident that the angiosperms were in fact diversifying in the mid-Cretaceous rather than coming in fully developed from some other area and that we are, therefore, in the right part of the geologic column to look for evidence of angiosperm evolution. I would prefer first to look at the evidence that exists from the habitats we do have in order to see if it makes any sort of sense, bearing in mind that we are seeing only one part of the flora, rather than assuming from the start that it is misleading. As more evidence accumulates about various environments – and I would emphasize that we have leaves from several different environments at each stratigraphic level (e.g., pond facies, stream margin facies) – and continues to show a reasonably consistent pattern, it becomes less and less defensible to say that there is some entirely different pattern in some area for which we have no fossil record.

Givnish: You showed that the peltate, cordate, aquatic leaves did not appear in your area until fairly well into the sequence. Is that because the facies they are in do not occur until that time or are there similar facies available at earlier times that lack this particular sort of leaf; if so, what is in those facies?

Doyle: I do not think we are lacking in pond facies from the lower Potomac, nor are we lacking in them from other parts of the world. It is quite striking that in several geographic areas these peltate and pseudopeltate leaves appear at about the same time, a little way up in the angiosperm record, i.e., middle Albian. As to what was in pond facies before, careful sedimentologic analysis would be needed to establish that one was dealing with exactly the same sedimentary environment that the peltate leaves occur in later. I have the impression that little else was growing in place in pond facies even in the Albian. Mixed in with the *Nelumbites* leaves, as they are called, are bits of twigs of conifers, small pieces of fern foliage, and so on – things that presumably floated out into the pond. I find it interesting that no early Cretaceous pteridophytic or gymospermous groups seem to have had the habit of aquatics with horizontal leaves stuck at the ends of petioles; I suspect this was an adaptive zone that the angiosperms were the first to occupy.

Givnish: You have shown that there is an increase in the regularity of the venation of leaves as you move up in the chronologic sequence. Is there any correlation of the regularity of the venation with leaf thickness in existing plants? Do you not tend to find rather disorganized venation among leathery leaves, as in some primitive Ranales, in the Proteaceae or Epacridaceae, for example?

Doyle: Hickey (1971) found in looking at leaf venation in modern angiosperms that there are several situations where there is a very disorganized type of venation comparable to what is seen in the early angiosperm record. One of them, which may be a truly primitive condition, is in such Magnoliid families as Winteraceae and Canellaceae. But there is also a situation in "higher" groups that may be interpreted in a standard comparative morphologic way as the result of secondary reversion. This apparent breakdown of the trend is associated with thicker leaves (e.g., Ericaceae, Arctic-alpine and xerophytic plants). The regularity of venation seen in the related mesic groups is lost. Comparative morphology suggests that it is not as necessary to have a rigid venational network in a leathery leaf with a thick cuticle. Whether early Cretaceous leaves had disorganized venation because they were leathery or because they had not "learned" how to become organized remains a question.

References

Axelrod, D. I. (1952). A theory of angiosperm evolution. *Evolution, 6,* 29–60.
– (1959). Poleward migration of early angiosperm flora. *Science, 130,* 203–207.
– (1960). The evolution of flowering plants. In *The Evolution of Life,* ed. S. Tax, pp. 227–305. Chicago: University of Chicago Press.
– (1970). Mesozoic paleogeography and early angiosperm history. *Bot. Rev., 36,* 277–319.
Bailey, I. W. (1949). Origin of the angiosperms: Need for a broadened outlook. *J. Arnold Arbor., 30,* 64–70.
Banks, H. P. (1968). The early history of land plants. In *Evolution and Environment,* ed. E. T. Drake, pp. 73–107. New Haven: Yale University Press.
– (1970). *Evolution and Plants of the Past.* Belmont, Calif.: Wadsworth. 170 pp.
– (1975). The oldest vascular land plants: A note of caution. *Rev. Palaeobot. Palynol., 20,* 13–25.
Beck, C. B. (1960). The identity of *Archaeopteris* and *Callixylon. Brittonia, 12,* 351–68.
– (1970). The appearance of gymnospermous structure. *Biol. Rev., 45,* 379–400.
– (1971). On the anatomy and morphology of lateral branch systems of *Archaeopteris. Amer. J. Bot., 58,* 758–84.

- (1976). Current status of the Progymnospermopsida. *Rev. Palaeobot. Palynol., 21,* 5–23.

Bews, J. W. (1927). Studies in the ecological evolution of angiosperms. *New Phytol., 26,* 1–21, 65–84, 129–48, 209–48, 273–94.

Bowen, R. (1961). Paleotemperature analyses of Mesozoic Belemnoidea from Germany and Poland. *J. Geol., 69,* 75–83.

Brenner, G. J. (1963). The spores and pollen of the Potomac Group of Maryland. *Maryland Dept. Geol. Mines Water Resources Bull., 27,* 1–215.

- (1976). Middle Cretaceous floral provinces and early migrations of angiosperms. In *Origin and Early Evolution of Angiosperms,* ed. C. B. Beck, pp. 23–47. New York: Columbia University Press.

Carlquist, S. (1975). *Ecological Strategies of Xylem Evolution.* Berkeley: University of California Press. 259 pp.

Carluccio, L. C., Hueber, F. M., & Banks, H. P. (1966). *Archaeopteris macilenta,* anatomy and morphology of its frond. *Amer. J. Bot., 53,* 719–30.

Chaloner, W. G. (1967). Spores and land-plant evolution. *Rev. Palaeobot. Palynol., 1,* 83–93.

- (1970). The rise of the first land plants. *Biol. Rev., 45,* 353–77.

Chaloner, W. G., & Lacey, W. S. (1973). The distribution of Late Palaeozoic floras. In *Organisms and Continents through Time,* ed. N. F. Hughes. *Palaeontological Association of London, Special Papers in Palaeontology, 12,* 271–89.

Chaloner, W. G., & Meyen, S. V. (1973). Carboniferous and Permian floras of the northern continents. In *Atlas of Palaeobiogeography,* ed. A. Hallam, pp. 169–86. Amsterdam: Elsevier.

Corner, E. J. H. (1949). The Durian theory or the origin of the modern tree. *Ann. Bot. N.S., 13,* 367–414.

Couper, R. A. (1958). British Mesozoic microspores and pollen grains. *Palaeontographica, 103B,* 75–179.

Cridland, A. A. (1964). *Amyelon* in American coal-balls. *Palaeontology, 7,* 186–209.

Delevoryas, T., & Hope, R. C. (1971). A new Triassic cycad and its phyletic implications. *Postilla, 150,* 1–31.

- (1976). More evidence for a slender growth habit in Mesozoic cycadophytes. *Rev. Palaeobot. Palynol., 21,* 93–100.

Delevoryas, T., & Morgan, J. (1954). A new pteridosperm from Upper Pennsylvanian deposits of North America. *Palaeontographica, 96B,* 12–23.

Dettmann, M. E. (1973). Angiospermous pollen from Albian to Turonian sediments of eastern Australia. *Geol. Soc. Australia, Special Publication, 4,* 3–34.

Dilcher, D. L. (1974). Approaches to the identification of angiosperm leaf remains. *Bot. Rev., 4,* 1–157.

Doyle, J. A. (1969). Cretaceous angiosperm pollen of the Atlantic Coastal Plain and its evolutionary significance. *J. Arnold Arbor., 50,* 1–35.

- (1973). Fossil evidence on early evolution of the monocotyledons. *Quart. Rev. Biol, 48,* 399–413.

- (1977). Patterns of evolution in early angiosperms. In *Patterns of Evolution,* ed. A. Hallam. Amsterdam: Elsevier.

Doyle, J. A., & Hickey, L. J. (1976). Pollen and leaves from the mid-Cretaceous Potomac Group and their bearing on early angiosperm evolution. In *Origin and Early Evolution of Angiosperms,* ed. C. B. Beck, pp. 139–206. New York: Columbia University Press.

Doyle, J. A., Jardiné, S., & Doerenkamp, A. (1976). Evolution of angiosperm pollen in the Lower Cretaceous of equatorial Africa. *Botanical Society of America, Abstracts of Papers,* p. 25.

Doyle, J. A., & Robbins, E. I. (1977). Angiosperm pollen zonation of the continental Cretaceous of the Atlantic Coastal Plain and its application to deep wells in the Salisbury Embayment. *Proc. Amer. Assoc. Strat. Palynol., 1,* 501–46.

Doyle, J. A., Van Campo, M., & Lugardon, B. (1975). Observations on exine structure of *Eucommiidites* and Lower Cretaceous angiosperm pollen. *Pollen Spores, 17,* 429–86.

Faegri, K., & van der Pijl, L. (1966). *The Principles of Pollination Ecology.* Oxford: Pergamon Press. 248 pp.

Florin, R. (1936). Die fossilen Ginkgophyten von Franz-Joseph-Land nebst Erörterungen über vermeintliche Cordaitales mesozoischen Alters. *Palaeontographica, 81B,* 71–173; *82B,* 1–72.

– (1951). Evolution in cordaites and conifers. *Acta Horti Berg., 15,* 285–388.

Frederiksen, N. O. (1972). The rise of the Mesophytic flora. *Geosci. Man, 4,* 17–28.

Germeraad, J. H., Hopping, C. A., & Muller, J. (1968). Palynology of Tertiary sediments from tropical areas. *Rev. Palaeobot. Palynol., 6,* 189–348.

Góczán, F., Groot, J. J., Krutzsch, W., & Pacltová, B. (1967). Die Gattungen des "Stemma Normapolles Pflug 1953b" (Angiospermae). *Paläontol. Abhandl., 2B,* 429–539.

Hallé, F., & Oldeman, R. A. A. (1970). *Essai sur l'architecture et la dynamique de croissance des arbres tropicaux.* Paris: Masson. 178 pp.

Harris, T. M. (1961). The fossil cycads. *Palaeontology, 4,* 313–23.

– (1964). *The Yorkshire Jurassic Flora, II, Caytoniales, Cycadales and Pteridosperms.* London: British Museum (Natural History). 191 pp.

Herngreen, G. F. W. (1974). Middle Cretaceous palynomorphs from northeastern Brazil. *Sci. Géol. Bull. Strasbourg, 27,* 101–16.

Heslop-Harrison, J. (1971). Sporopollenin in the biological context. In *Sporopollenin,* ed. J. Brooks, P. R. Grant, M. Muir, P. van Gijzel, & G. Shaw, pp. 1–30. New York: Academic Press.

Hickey, L. J. (1971). Evolutionary significance of leaf architecture features in the woody dicots. *Amer. J. Bot. (abstr.), 58,* 569.

Hickey, L. J., & Wolfe, J. A. (1975). The bases of angiosperm phylogeny: Vegetative morphology. *Ann. Missouri Bot. Gard., 62,* 538–89.

Hluštík, A., & Konzalová, M. (1975). *Frenelopsis alata* (K. Feistm.) Knobloch from the Cenomanian of Bohemia: A new plant producing *Classopollis* pollen. *Abstracts, International Conference on Evolutionary Biology, Liblice, Czechoslovakia* (unnumbered pages).

Horn, H. S. (1971). *The Adaptive Geometry of Trees.* Princeton, N.J.: Princeton University Press. 144 pp.

Hughes, N. F. (1961). Fossil evidence and angiosperm ancestry. *Sci. Prog. 49,* 84–102.

– (1976). *Palaeobiology of Angiosperm Origins.* Cambridge: Cambridge University Press. 242 pp.

Janzen, D. H. (1970). Herbivores and the number of tree species in tropical forests. *Amer. Nat., 104,* 501–28.

Jardiné, S., Kieser, G., & Reyre, Y. (1974). L'individualisation progressive du continent africain vue à travers les données palynologiques de l'ère secondaire. *Sci. Géol. Bull. Strasbourg, 27,* 69–85.

Jarzen, D. M., & Norris, G. (1975). Evolutionary significance and botanical relationships of Cretaceous angiosperm pollen in the western Canadian interior. *Geosci. Man, 11,* 47–60.

Kemp, E. M. (1968). Probable angiosperm pollen from British Barremian to Albian strata. *Palaeontology, 11,* 421–34.

Kimura, T., & Sekido, S. (1975). *Nilssoniocladus* n. gen. (Nilssoniaceae n. fam.), newly found from the early Lower Cretaceous of Japan. *Palaeontographica, 153B,* 111–18.

Kuyl, O. S., Muller, J., & Waterbolk, H. T. (1955). The application of palynology to oil geology, with special reference to western Venezuela. *Geol. Mijnbouw, N.S., 17,* 49–76.

Laing, J. F. (1975). Mid-Cretaceous angiosperm pollen from southern England and northern France. *Palaeontology, 18,* 775–808.

– (1976). The stratigraphic setting of early angiosperm pollen. In *The Evolutionary Significance of the Exine,* ed. I. K. Ferguson & J. Muller. *Linnean Soc. Symp. Ser., 1,* 15–26.

Leopold, E. B., & MacGinitie, H. D. (1972). Development and affinities of Tertiary floras in the Rocky Mountains. In *Floristics and Paleofloristics of Asia and Eastern North America,* ed. A. Graham, pp. 147–200. Amsterdam: Elsevier.

Mamay, S. H. (1969). Cycads: Fossil evidence of late Paleozoic origin. *Science, 164,* 295–6.

Margalef, R. (1968). *Perspectives in Ecological Theory.* Chicago: University of Chicago Press. 111 pp.

Meeuse, A. D. J. (1966). *Fundamentals of Phytomorphology.* New York: Ronald Press. 231 pp.

Müller, H. (1966). Palynological investigations of Cretaceous sediments in northeastern Brazil. In *Proceedings of the Second West African Micropaleontological Colloquium (Ibadan, 1965),* ed. J. E. van Hinte, pp. 123–36. Leiden: Brill.

Muller, J. (1970). Palynological evidence on early differentiation of angiosperms. *Biol. Rev., 45,* 417–50.

Nathorst, A. G. (1909). Ueber *Williamsonia, Wielandia, Cycadocephalus* und *Weltrichia.* Paläobotanische Mitteilungen, 8. *Kgl. Svenska Vetenskap. Handl., 45,* 1–38.

Pacltová, B. (1961). Zur Frage der Gattung *Eucalyptus* in der böhmischen Kreideformation. *Preslia, 33,* 113–29.

— (1971). Palynological study of Angiospermae from the Peruc Formation (?Albian-Lower Cenomanian) of Bohemia. *Ústřední ústav geologický, Sborník geologických věd, paleontologie, řada P, 13,* 105–41.

Pierce, R. L. (1961). Lower Upper Cretaceous plant microfossils from Minnesota. *Bull. Minnesota Geol. Surv., 42,* 1–86.

Read, R. W., & Hickey, L. J. (1972). A revised classification of fossil palm and palm-like leaves. *Taxon, 21,* 129–37.

Scheckler, S. E. (1976). Ontogeny of progymnosperms. I. Shoots of Upper Devonian Aneurophytales. *Can. J. Bot., 54,* 202–19.

Scheckler, S. E., & Banks, H. P. (1971). Anatomy and relationships of some Devonian progymnosperms from New York. *Amer. J. Bot., 58,* 737–51.

Scott, R. A., Barghoorn, E. S., & Leopold, E. B. (1960). How old are the angiosperms? *Amer. J. Sci., 258-A (Bradley Volume),* 284–99.

Simpson, G. G. (1953). *The Major Features of Evolution.* New York: Columbia University Press. 434 pp.

Singh, C. (1975). Stratigraphic significance of early angiosperm pollen in the mid-Cretaceous strata of Alberta. *Geol. Assoc. Canada, Special Paper, 13,* 365–89.

Smith, A. C. (1973). Angiosperm evolution and the relationship of the floras of Africa and America. In *Tropical Forest Ecosystems in Africa and South America,* ed. B. J. Meggers, E. S Ayensu, & W. D. Duckworth, pp. 49–61. Washington, D.C.: Smithsonian Institution Press.

Sohl, N. F. (1971). North American Cretaceous biotic provinces delineated by gastropods. *Proceedings of the North American Paleontological Convention (Chicago, 1969), 2,* 1610–38.

Stebbins, G. L. (1974). *Flowering Plants: Evolution above the Species Level.* Cambridge, Mass.: Harvard University Press. 399 pp.

Takhtajan, A. L. (1969). *Flowering Plants: Origin and Dispersal.* Washington, D.C.: Smithsonian Institution Press. 310 pp.

Thomas, H. H. (1915). On *Williamsoniella,* a new type of bennettitalian flower. *Phil. Trans. Roy. Soc., Lond. Ser. B, 207,* 113–48.

Tomlinson, P. B., & Gill, A. M. (1973). Growth habits of tropical trees: Some guiding principles. In *Tropical Forest Ecosystems in Africa and South America,* ed. B. J. Meggers, E. S. Ayensu & W. D. Duckworth, pp. 129–43. Washington, D.C.: Smithsonian Institution Press.

Vakhrameev, V. A. (1970). Yurskie i rannemelovye flory. In *Paleozoyskie i Mezozoyskie Flory Yevrazii i Fitogeografia Etogo Vremeni,* ed. V. A. Vakhrameev, I. A. Dobruskina, Ye. D. Zaklinskaya, & S. V. Meyen, pp. 213–81. Moscow: Nauka.

Walker, J. W., & Doyle, J. A. (1975). The bases of angiosperm phylogeny: Palynology. *Ann. Missouri Bot. Gard., 62,* 664–723.

White, D. (1936). Some features of the early Permian flora of America. *Report XVI International Geologic Congress, Washington, D.C., 1933,* pp. 679–89.

Whitehead, D. R. (1969). Wind pollination in the angiosperms: Evolutionary and environmental considerations. *Evolution, 23,* 28–35.

Wolfe, J. A. (1971). Tertiary climatic fluctuations and methods of analysis of Tertiary floras. *Palaeogeog. Palaeoclimatol. Palaeoecol., 9,* 27–57.

– (1972). An interpretation of Alaskan Tertiary floras. In *Floristics and Paleofloristics of Asia and Eastern North America,* ed. A. Graham, pp. 201–33. Amsterdam: Elsevier.

– (1973). Fossil forms of Amentiferae. *Brittonia, 25,* 334–55.

– (1975). Some aspects of plant geography of the Northern Hemisphere during the Late Cretaceous and Tertiary. *Ann. Missouri Bot. Gard., 62,* 264–79.

Wolfe, J. A., Doyle, J. A., & Page, V. M. (1975). The bases of angiosperm phylogeny: Paleobotany. *Ann. Missouri Bot. Gard., 62,* 801–24.

Wolfe, J. A., & Pakiser, H. M. (1971). Stratigraphic interpretations of some Cretaceous microfossil floras of the Middle Atlantic States. *U.S. Geol. Surv. Professional Papers, 750-B,* B35–47.

Zimmermann, W. (1953). Main results of the "telome theory." *Palaeobotanist, 1,* 456–70.

2

Geographic variation in tropical tree species

TREVOR WHIFFIN

Department of Botany, La Trobe University, Bundoora, Victoria, Australia

"Geographic variation" is the pattern of variation present within a species over its entire range. The study of geographic variation is an integral part of any detailed systematic or evolutionary study. Such a study provides valuable insights into the pattern of variation present, into possible modes of speciation, and into the historical biogeography and lines of migration of the species.

The study of geographic variation has played, and continues to play, an important role in investigations into the nature of species and of speciation processes (Gould & Johnston, 1972). In this respect, geographic variation studies have provided valuable information about temperate species. However, because geographic variation studies (and, indeed, systematic and evolutionary studies in general) have usually been conducted on temperate species, little is known about the detailed patterns of variation in tropical species.

The tropical rain forest is, in general, the richest of all communities in terms of number of species and differs from most other communities in that many closely related species are found in the same general area, although each species is represented by only a small number of individuals. Discussions of speciation in the tropics variously emphasise genetic drift in small populations (Federov, 1966) or natural selection (Ashton, 1969). The mode of speciation predominant in a species, whether genetic drift or selection, will be affected by, and reflected in, the extent and type of variation present within and among populations. Thus, any study, as well as providing information about the geographic variation within a species, will also provide valuable evolutionary and biogeographic data.

Geographic variation studies should, as far as possible, examine variation over a large number of different characters and provide numerical integration of changes over all characters taken together.

In this way, spurious results depending on only one or a few characters can be avoided. For this reason, modern studies of geographic variation in plant species have often emphasised the collection of relatively large sets of data and the subsequent use of numerical techniques to aid in the simplification and visualisation of the overall pattern of variation.

In this chapter, some of the methods of geographic variation study will be reviewed, and an account will be given of the study of geographic variation in the Australian rain forest species *Flindersia bourjotiana* as an example.

Review of methods

For an area of study as large as geographic variation, the variety of methods available is correspondingly large. It is not possible here to review all such methods. Instead, emphasis is placed on those techniques that have proved useful in the study of geographic variation in temperate tree species and that should prove equally applicable to tropical tree species. Two major techniques are discussed: the use of chemical characters, particularly volatile oils, and the use of numerical methods of analysis. Even within these two areas, it is impossible to review all available techniques. Instead, in each case, those techniques most likely to be useful in studying tropical tree species are introduced and discussed, with particular emphasis on their utility and validity in any such study.

Volatile oils

The use of chemical characters in the study of infraspecific variation is now well established. Turner (1970) and Mabry (1973) provide a number of examples of the utility of the chemical approach in such studies. In particular, the use of leaf volatile oils is now well established as a convenient and accurate method for obtaining significant amounts of data characterising individuals or populations. Modern analytic methods, such as gas-liquid chromatography with an associated automatic digital integrator, are capable of producing large sets of accurate data relatively quickly, with no preselection or bias on the part of the investigator (Adams & Turner, 1970; Turner, 1970). The provision of such large sets of data has necessitated the use of computer programs for the various numerical procedures employed in data analysis. In this way, the numerical analysis of volatile oil data on a population basis has proved a strong tool in the study of geographic variation in plant species.

Recently, von Rudloff (1975) has discussed the utility and validity of volatile oils as taxonomic characters. With stated and standardised extraction and analysis techniques, volatile oils are capable of fulfilling all the requirements necessary for a group of compounds to be of taxonomic value. Work is continuing on the biosynthesis of volatile oils (Loomis & Croteau, 1973) and on the mechanism of their genetic control (Irving & Adams, 1973).

The mechanism of genetic control of volatile oils has not yet been fully established. In a number of cases, the inheritance of a given compound appears to be under the control of one or a few genes; for other compounds the mode of inheritance appears to be more complex (Hanover, 1966a; Squillace & Fisher, 1966; Zavarin, Critchfield, & Snajberk, 1969). The estimates by Irving & Adams (1973) of the minimal number of genes involved in the inheritance of monoterpenes in the *Hedeoma drummondii* complex indicate that most monoterpenes are controlled by a relatively small number of genes, although there are some indications of more complex genetic control. Basically, however, their results are in agreement with previous work.

Whatever the actual mechanism of genetic control, recently published work establishes that volatile oils are under strict genetic control (Hanover, 1966a; von Rudloff, 1972). Provenance trials show that environmental factors have little or no effect on the volatile oil composition (Hanover, 1966b). Although the actual amount of oil produced may be influenced by various environmental factors, the percentage composition is found to be more stable and thus more useful taxonomically (Powell & Adams, 1973; von Rudloff, 1975). The results obtained using volatile oils are repeatable; in a study of clinical variation in *Juniperus virginiana*, samples taken in two successive years gave essentially the same infraspecific variation pattern (Flake, von Rudloff, & Turner, 1973). A similar result was obtained by Adams (1975a) in a study of variation in *Juniperus ashei*.

It is of interest to compare chemical characters, especially volatile oils, with morphologic characters, as these are the two main sources of data in geographic variation studies, and both may equally be subjected to the numerical techniques discussed below. The major advantage of volatile oils is that they provide a large number of different characters for each individual studied, a feat often difficult to achieve with morphologic characters. This is especially true where floral and reproductive morphologic characters are being used, as it is often difficult to collect the plants in the appropriate condition; leaf volatile oils, on the other hand, do not suffer in this respect. If morphologic characters from the vegetative parts alone are used, there is generally a major restriction on the number of characters available. However, morphologic characters have the advantage of being easier to measure, and it is thus generally easier to collect data from a large number of different trees than is the case with volatile oils. Volatile oils have the advantage that they are under fairly simple genetic control and are not markedly affected by the environment. Morphologic characters, on the other hand, are generally under more complex genetic control and often show marked phenotypic variation. The major disadvantages of volatile oils are that they require a complex procedure for collection and analysis and that samples must be collected specially for the study at hand. In the case of morphologic characters, normal collection procedures may be used, and it is sometimes possible to use herbarium material or material collected for another purpose. In the current context, volatile oils prove useful because they allow study of a large number of characters that appear to be under strong genetic control.

Numerical methods

The techniques discussed here are those most likely to prove useful in the study of geographic variation in tropical tree species. These techniques may be applied to any set of quantitative data, derived from any selected group of characters. The characters most commonly used are those from morphology and chemotaxonomy. Sampling is on a population basis, and numerical techniques are employed to test the significance of inter- and intrapopulation variation. Then, significant interpopulation variation is projected onto a geographic map. Further techniques attempt to find correlations between the observed geographic variation and known environmental variables such as climate and locality details. When combined with other biologic and biogeographic data, this may allow the formulation of hypotheses about processes such as species origin and migration.

Geographic variation may be studied character by character, in a univariate manner, or by considering the variation in all characters simultaneously, in a multivariate manner. The multivariate approach is, in general, more useful and more valid, as geographic variation is generally the result of the complex action of a series of environmental and genetic factors on the whole genotype. Nevertheless, some univariate procedures prove useful prerequisites for later multivariate methods, as they provide information on the nature of the variation of the characters. In addition, the univariate methods may be extended by considering mathematically created composite variables instead of individual characters.

A number of univariate significance tests will indicate which characters are useful in distinguishing among populations. In general, geographic variation is concerned with that component of the total variation which is significantly different among populations, after the extent of variation within populations has been accounted for.

Thus an F test (as variance among populations/variance within populations) on each of the measured characters will indicate which of the characters show significant interpopulation variation. In a similar manner, it is also possible to use a multiple-range test, such as the Student-Newman-Keuls (SNK) test or Duncan's New Multiple Range test, to determine which population means are significantly different (Steel & Torrie, 1960; Sokal & Rohlf, 1969).

Another useful univariate procedure is contour mapping, which produces a geographic presentation of the regional trends within the species based on the population means. Numerical methods can be used to produce smoothed and simplified contours (or isophenes), as shown by Adams (1970). This may then allow correlation of the observed geographic trends with known environmental factors.

The utility of both the significance tests and the contour maps may be extended by the use of composite variables. These composite variables are mathematically derived in such a way that each new variable accounts for more of the total variation than does any individual original character. The most commonly used composite variables are the projections of the individuals or the populations upon the new axes produced by an ordination technique.

A multivariate extension of contour mapping is the mapping of the composite differential produced by Adams (1970). Here, through the technique of differential systematics, the rates of change of all characters are summed, so that areas of rapid change are indicated.

Multivariate procedures are generally more useful, although, as indicated above, valuable results can also be obtained by employing composite variables in univariate procedures. In attempting to determine the phenetic relations among the populations sampled using all available characters, two main multivariate approaches are used: classification and ordination (Gould & Johnston, 1972). "Classification" is the process of ordering samples (whether individuals, populations, or species) into a hierarchic and usually nonoverlapping system of clusters. "Ordination" is the process of placing samples in relation to one or more axes in such a way that the placement (or projection) of the samples on these axes conveys the maximum information concerning the relations between these samples.

In a geographic variation study, the samples are divided a priori into groups (the populations). It is, therefore, permissible to weight the characters by some measure of their ability to distinguish among the groups: i.e., by their F value or some derivative of their F value. In this way, those characters showing the most significant interpopulation variation carry most weight in the further analyses. In a study comparing different weighting parameters, Adams (1975b) found empirically that the $F - 1$ value was the best weight.

In a numerical study of geographic variation, the population means are generally used. The F values for each character are determined from an analysis of variance and used, in one form or another, as character weights. These data may then be subjected to various classification or ordination techniques. These techniques may be used on the individual samples, but the analysis provided is often difficult to interpret. More readily understood is the classification or ordination of the populations based on the weighted population means for each character.

For classification, a suitable measure of similarity is computed among the various populations, either directly or via a distance measure. This is then used to group the populations on the basis of their similarities to one another, usually in the form of hierarchic, non-overlapping clusters. This is often expressed graphically as a dendrogram, but it is also possible to project the clustering levels onto a geographic map of the populations. Jardine & Sibson (1971) recommend for various theoretic reasons that clustering be according to the single-linkage criterion (Sneath & Sokal, 1973). However, this recommendation has not been put to any empiric test in studies of geographic variation, and some workers would disagree (Clifford & Stephenson, 1975).

The pattern of geographic variation within a species is often not well represented in the form of hierarchic, nonoverlapping clusters (Gould & Johnston, 1972; Jardine & Edmonds, 1974). Recently, Jardine & Edmonds (1974) have suggested the use of a nonhierarchic clustering method with the allowance of a limited amount of overlap among clusters. Compared with the standard hierarchic, nonoverlapping clustering schemes as used in numerical taxonomy, this form of data analysis has been little studied; a full study of the utility of nonhierarchic and overlapping clustering schemes in the representation of geographic variation would prove most interesting.

Ordination generally has the aim of projecting the populations upon new axes, where these new axes are composites of the original characters. As the new axes are formed in order of descending information content, the first few axes often account for a fairly large percentage of the original variation. Thus, to account for the majority of the variation, a smaller number of new axes is required than original characters, and this reduction often proves a valuable simplification of the data. The projection of the populations upon the first few axes often provides valuable insights into the relations among the populations. There is a series of ordination techniques often known under the general term of factor analysis, although this term is increasingly being used in a more restricted sense. The procedures most commonly used in systematics are principal components analysis (Seal,

1964) and principal coordinates analysis (Gower, 1966). Sneath & Sokal (1973) provide a convenient summary of these two methods.

It is sometimes found that the axes produced by the ordination technique, although leading to the recognition of clusters of populations, do not effect the best or most meaningful separation of these clusters. Rotation of the axes may be undertaken, and the new axes may be produced under different criteria from the first ordination axes. Projection of the populations upon these rotated axes may provide additional information about the relations among the populations and about the pattern of geographic variation.

A final ordination technique, multiple discriminant or canonical analysis, is much used in the study of geographic variation (Gould & Johnston, 1972). Unlike the classification and ordination procedures discussed above, it uses the individual samples and their unweighted character values. In a similar way, however, the samples are divided a priori into groups, the populations. Multiple discriminant analysis produces axes (actually a series of discriminant functions) that best separate the groups; these new axes are extracted so as to maximise the between-group variance relative to within-group variance. As is true of other ordination techniques, the first few axes contain most of the information, and the axes may be rotated if desired.

To detect possible causes of the pattern of geographic variation present, a number of different techniques may be used (Gould & Johnston, 1972). Often the simple projection of the results of a classification or ordination procedure onto a geographic map will suggest possible explanations for the observed pattern of variation. This approach may provide possible associations among environmental variables and the pattern of geographic variation, especially where additional information such as topographic or climatic details are overlayed on the map.

A more direct approach is to attempt to find correlations among the various characters and environmental variables. The various characters may be plotted against such environmental variables as altitude, latitude, longitude, and various measures of the local climate. In most cases, a plot involving one of the individual characters is not particularly informative, whereas a plot involving a composite variable may prove so.

A final technique in the study of causes is to determine, through multiple regression analysis, the relative importance of each of the measures of the environment in explaining the observed variation among the populations.

These numerical techniques lead into the final stage of geographic variation study, in which additional biologic and biogeographic knowledge is used to formulate hypotheses about the origin of the pattern of variation.

Geographic variation in Flindersia bourjotiana

The genus *Flindersia* (Rutaceae) contains 17 species and is found in the Moluccas, New Guinea, New Caledonia, and eastern Australia, with an evident concentration in the last region (Hartley, 1969; Hartley & Hyland, 1975). Of the 17 species and 1 variety, only 2 species and 1 variety are absent from Australia; with the possible exception of the New Caledonian endemic *F. fournieri* these taxa have closely related Australian counterparts. The majority of the species are rain forest trees, and they form an ecologically and commercially important group in the Australian tropical and subtropical rain forests. The genus has recently been revised (Hartley, 1969), and this revision forms a convenient base for further systematic and evolutionary study of the genus.

Fig. 2.1. Map of northeast Queensland, showing approximate distribution of *Flindersia bourjotiana* and populations sampled.

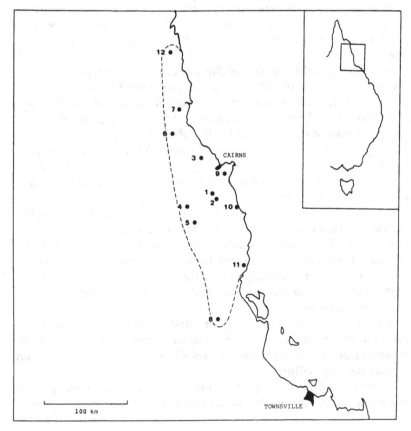

Flindersia contains both wide-ranging and narrow endemic species and thus constitutes a convenient system for the comparison of the variation present within each. It is to be expected that a study of variation and evolution in selected *Flindersia* species will provide information on the nature of species and the pattern of speciation within Australian rain forest trees. The present study on geographic variation in *F. bourjotiana* is the first stage in such a study.

F. bourjotiana is endemic to the area of northeast Queensland, where it is one of the relatively more common rain forest tree species. It is found in rain forests generally between Cooktown to the north and Ingham to the south, both in the coastal lowlands and in the eastern ranges and tablelands of the Dividing Range. The approximate known distribution of *F. bourjotiana* is shown in Fig. 2.1; within the distribution indicated the range is not continuous, owing especially to recent clearing of rain forests.

Materials and methods

Twelve populations of *F. bourjotiana* were sampled throughout its known natural range. The location of these populations is shown in Fig. 2.1, and the locality details are provided in Table 2.1. Populations 9, 10, and 11 are coastal and at low altitude; populations 3 and 7 are slightly further inland and at intermediate altitudes. The remaining populations are basically inland and at varying but higher altitudes.

Five trees per population were sampled, except for population 1, from which only three trees were sampled. Fresh foliage samples from each tree were sealed in polyethylene bags, kept as cool as possible, and air-freighted to Melbourne, where they were stored at 2° C

Table 2.1. *Locality details of populations of* Flindersia bourjotiana

Population	Latitude	Longitude	Altitude (m)
1	17°12'	145°42'	800
2	17°15'	145°45'	680
3	16°50'	145°35'	350
4	17°20'	145°25'	1120
5	17°30'	145°30'	1140
6	16°35'	145°15'	925
7	16°20'	145°20'	220
8	18°30'	145°45'	685
9	17°00'	145°50'	80
10	17°20'	146°00'	80
11	17°55'	146°05'	30
12	15°45'	145°15'	660

until processed. Voucher plant material for each tree was collected, and the specimens are deposited in the herbarium of the Queensland Regional Station, CSIRO Division of Forest Research, Atherton, Queensland (QRS).

Mature adult leaves only were taken (Whiffin & Hyland, unpubl.) and were steam-distilled in an all-glass apparatus to remove the volatile oils. The oil was extracted into twice-distilled Freon 11 (trichlorofluoromethane, b.p. 23° C) following the suggestion of Hardy (1969), and this was dried with anhydrous sodium sulfate. After filtration, the Freon was partially removed by distillation, and then the oil sample was concentrated with a jet of high-purity nitrogen. The oil samples were stored under nitrogen in capped vials at −20° C until required for analysis.

The oils were analysed on a Shimadzu GC4B gas-liquid chromatograph with dual columns and flame-ionisation detectors. The glass-lined metal columns (3.5 m by 1.8 mm i.d.) were packed with acid-washed, DMCS-treated 80/100 Chromosorb G, coated with 5% Carbowax 20 M. The chromatograph was temperature-programmed from 80° C to 240° C at 4° C/minute and held at 240° C for a further 30 minutes. The peaks were quantified with an Infotronics CRS-204 automatic digital integrator.

Each individual component of the volatile oils was assigned a unique number by superimposition of the chromatograms and by cross-comparison of the retention times. Although errors of miscomparison may occur at this stage, they would not generally be significant (Adams, 1972). In this way, 91 consistently separable components were recognised as characters; neighbouring peaks that could not be consistently separated were treated as single characters. For each plant sampled, the characters were expressed in terms of their percentage of the total oil derived from that sample.

The volatile oil characters were subjected to a number of different statistical and numerical procedures, the utility and validity of which were discussed in general terms above. The 91 characters were each subjected to an analysis of variance, including calculation of the F value (as variance among populations/variance within populations). The characters were also subjected to the SNK multiple-range test to determine which population means were significantly different (at the 0.01 level). These two statistical tests determine the extent to which the various volatile oil characters exhibit significant interpopulation variation (as compared with the intrapopulation variation).

Those characters that were significant (at the 0.01 level) in both the F test and the SNK test were then subjected to contour mapping of the population means. Then all these significant characters were taken together in the differential systematics procedure; each character was F-weighted in the production of the composite differential.

The above analyses were performed on the percentage composition data. For the remaining analyses, the original 91 characters were standardised so that for each character the mean was 0 and the standard deviation 1. This form of standardisation is known as centring followed by standardisation by the standard deviation (Noy-Meir, 1973; Clifford & Stephenson, 1975). The population means for each character were computed and were weighted by the $F - 1$ value (Adams, 1975b). At this stage 3 characters that had F values less than 1 were rejected, so that subsequent analysis was based on 88 standardised and weighted characters for each population.

A matrix of Manhattan Metric distances was calculated among the populations; this distance measures utilises the $d_1(j,k)$ formulation of Sneath & Sokal (1973). This matrix was subjected to a principal coordinates analysis (Gower, 1966) to produce an ordination of the populations. To further study the interrelations among the populations, the axes derived from the principal coordinates analysis were rotated according to the Varimax criterion (Veldman, 1967).

A similarity matrix was derived from the Manhattan Metric distance matrix and range-standardised so that the similarity values were in the range of 0–1. This similarity matrix was then used for a single-linkage clustering of the populations. The similarity matrix was also used to form overlapping clusters using a complete-linkage criterion for each cluster formed, at each of three different levels of similarity.

In an attempt to find significant correlations among environmental variables and the pattern of geographic variation, the projections of the populations on the first three ordination axes (both rotated and unrotated) were plotted against the environmental variables altitude, latitude, longitude, and distance from sea.

Several computer programs were written in Fortran IV to perform the above analyses. Programs CONTRS and DIFSYS perform the contour mapping and differential systematics, respectively; program SNK performs the analysis of variance and the SNK tests; all three are modifications of programs originally written by Adams (1969, 1970). Programs GOWER and GOWORD perform the principal coordinates analysis and are modifications of programs originally written by the Division of Computing Research, CSIRO (Williams, Dale, & Lance, 1971). Program VARS undertakes the Varimax rotation of ordination axes and is a modification of the subroutines written by Veldman (1967). Programs STRANS and WTSEL perform various character standardisations and weighting, respectively; program TAXDT computes various measures of taxonomic distance and, where necessary, similarity measures derived from the taxonomic distance measures. Program SLINK performs a single-linkage clustering on a similarity matrix.

Results

Of the original 91 characters, 33 showed both a significant F test and a significant SNK test (both at the 0.01 level). Each of these 33 characters was contour-mapped. These contour maps show the major, smoothed regional trends in population means for the character studied.

Each contour map may be considered as a separate univariate analysis. It is possible, however, to arrange subjectively the contour maps into five main groups on the basis of the pattern of variation exhibited. Of the 33 characters used, 29 fit into these five subjective groupings; the remaining 4 characters, all with lower (but still significant) F values, do not fit into any group or with one another. Examples of contour maps of characters in these five main groups are provided in Figs. 2.2–2.4. Under each contour map is given a summary of the SNK test for that character, to indicate the significance of the differences depicted on the contour map (Adams & Turner, 1970). In this summary, any two populations whose means are not underscored by the same line are significantly different for that character.

The most common pattern of geographic variation found is illustrated in Fig. 2.2. Eleven characters exhibit this pattern, including most of those with the highest F values. Therefore, this would appear to be the most significant pattern of variation shown by any individual character. Examples of characters exhibiting this pattern of variation are compounds 61 and 63A (Fig. 2.2). These show relatively high values for populations 3, 9, and 11; low values for populations 4, 5, 6, 7, 8, and 12; and low to intermediate values for populations 1, 2, and 10. Thus, this pattern shows a basic distinction between three of the coastal populations and the main inland populations, with populations 1, 2, and 10 forming a link between the two.

Another group of three characters, exemplified by compound 40, shows a moderately similar pattern of variation (Fig. 2.3). In these characters, high values are shown by coastal populations 9 and 11, intermediate values by the main inland populations, and low values by coastal population 10, which again links up to a certain extent with the adjacent inland populations.

A third group, of five characters, is exemplified by compound 39A (Fig. 2.3). These characters are again somewhat similar as regards the overall pattern of variation to those in the first two groups. Low values are associated with coastal populations 9, 10, and 11; intermedi-

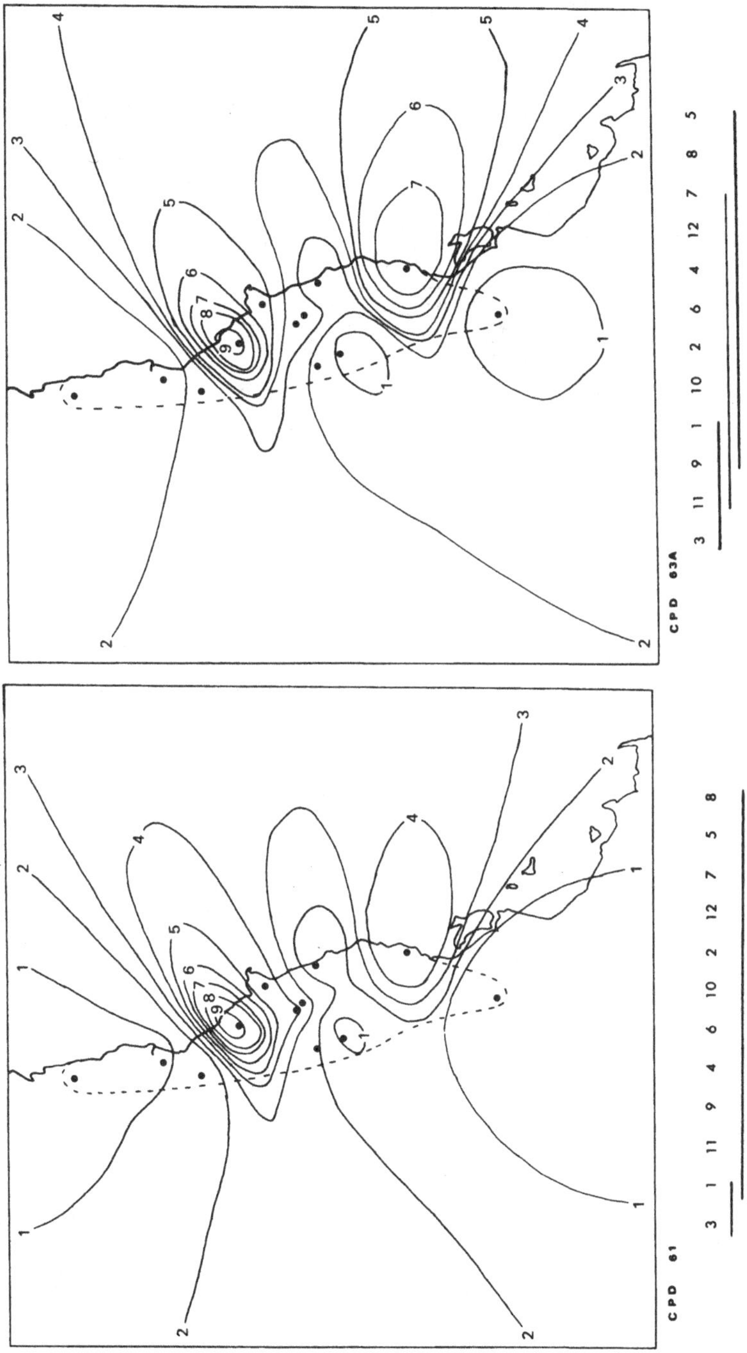

Fig. 2.2. Contour map (with summary of SNK test) for compound 61 (*left*) and compound 63A (*right*).

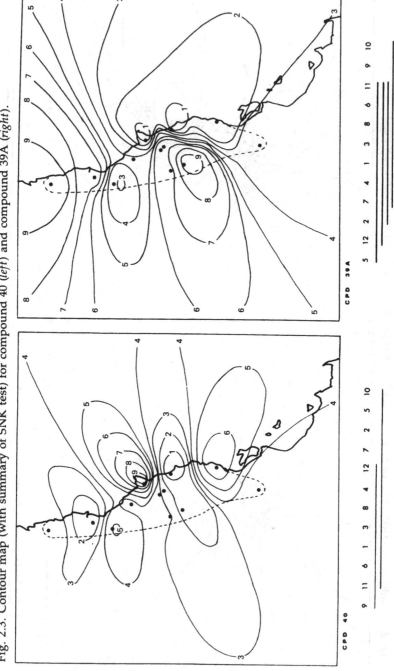

Fig. 2.3. Contour map (with summary of SNK test) for compound 40 (*left*) and compound 39A (*right*).

ate values with populations 3, 6, and 8; and intermediate to high values with the main inland populations, especially populations 5 and 12.

A fourth group, of two characters, is exemplified by compound 17 (Fig. 2.4). With these, high to intermediate values are shown by coastal populations 9, 10, and 11; low values by all the remaining populations.

The fifth group, of eight characters, is exemplified by compound 11 (Fig. 2.4). These characters show high values for population 6 and low to intermediate values for the remaining populations.

If it is possible to draw a general pattern from these individual contour maps, it would appear that the central coastal populations, especially populations 3, 9, and 11, are in the main distinct from the basically inland populations, especially populations 4, 5, 7, 8, and 12. Coastal population 10 links sometimes with the other coastal populations (as with compound 39A) and at other times with the adjacent inland populations (as with compound 63A). Population 3 usually groups with populations 9 and 11, but rarely shows relations with the inland populations, especially population 6 (as with compound 39A). The inland populations are usually fairly similar to one another, but often population 6 is differentiated (as with compound 11). Population 1 and, to a lesser extent, population 2 are sometimes intermediate between the inland populations and the coastal populations and often link with population 10 (as with compound 63A).

One approach to integrating the individual contour maps is to map a composite variable. In this case, the projections of the populations upon the first principal coordinates axis were used as the composite variable. As the first axis extracted accounts for 25% of the total variation in the 88 weighted characters, this contour map presumably represents a significant proportion of the total variation present among the populations. The contour map (Fig. 2. 5, left) shows close grouping of the coastal populations 3, 9, and 11. There is also a fairly close grouping of the main inland populations, with coastal population 10 and inland population 6 being somewhat intermediate.

Another approach to integrating the individual contour maps is the use of differential systematics. The composite differential formed from the 33 significant characters is shown in Fig. 2.5 (*right*). The characters were weighted by their F values to emphasise the more highly significant characters. High contour levels indicate regions of rapid change. Thus, the most rapid changes are found between the coastal populations 3, 9, 10, and 11 and their adjacent inland populations. In the case of population 10, the rate of change is less rapid than with the other three populations. Although to a lesser extent, there are also regions of change among the various coastal popula-

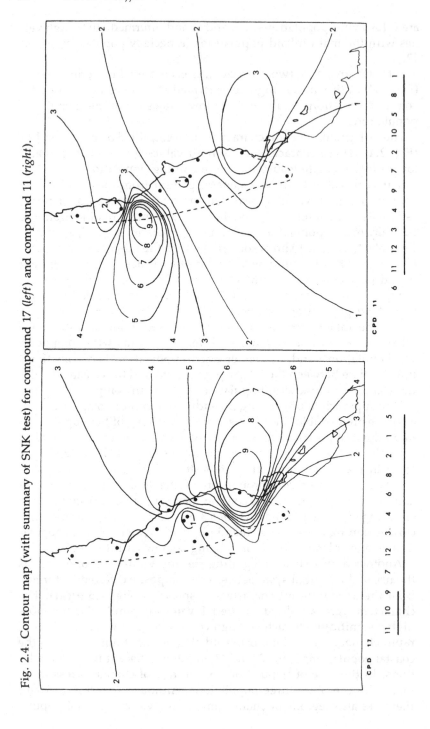

Fig. 2.4. Contour map (with summary of SNK test) for compound 17 (*left*) and compound 11 (*right*).

coastal populations 3, 9, and 11 are related to one another, but this group shows more internal divergences than does the group comprising the main inland populations. Populations 1 and 10 are not particularly closely related to one another or to the two groups of coastal and inland populations, respectively, but they appear to link these two groups to a certain extent.

A hypothesis to account for this pattern of variation may be formulated. The most closely related populations are in the central part of the range of the species, on the Atherton Tablelands. The highest mountains in northeast Queensland are on the eastern edge of these tablelands and are, in fact, between these central inland populations and the coastal populations. It is possible that *F. bourjotiana* survived the Pleistocene dry period (Kershaw, 1975) in a refugium in this general area. As the rainfall subsequently increased, *F. bourjotiana* migrated out from this refugium, both to the north and to the south, thus accounting for the present pattern of variation within the main inland populations.

F. bourjotiana presumably also migrated towards the coast to cover eventually the coastal lowlands. The divergence between the main coastal populations and the main inland populations may be due to higher selective pressures in the new coastal environment or to a longer period of disjunction. In a similar way, the marked variation among the various coastal populations may be due to selection or genetic drift occurring in the separated populations, or it may relate to there having been more than one refugium or more than one derivation of the coastal populations from inland populations.

It would be unwise at the moment to place too much weight on this hypothesis or to attempt to stretch the hypothesis much farther. It will prove of interest to compare the pattern of variation in *F. bourjotiana* with that in other rain forest trees currently under investigation, and it may at that time prove possible to formulate more detailed hypotheses to account for the pattern of geographic variation in the tropical rain forest trees of northeast Queensland.

Acknowledgments

I am extremely grateful to B. P. M. Hyland, G. C. Stocker, A. K. Irvine, K. Sanderson, and A. Dockrill, all of the Queensland Regional Station, CSIRO Division of Forest Research, for their encouragement and for their extensive help in the collection of the population samples. I am grateful also to R. P. Adams and to the CSIRO Division of Computing Research for copies of their respective computer programs, and to E. Shuwalow and A. Garcia for technical assistance. The study on variation and evolution in *Flindersia* is supported by a grant from the Australian Research Grants Committee.

General discussion

Nozeran: The variation you measured could be the result of the "genetic luggage" of the population, but it could also be the result of a population with the same genetic characteristics but adjusted, by a certain biochemical plasticity, to different environmental conditions. These two possibilities could be distinguished by the manner of sampling. Did you cultivate your population to exclude the possibility of adaptation?

Whiffin: No, these are natural populations that were sampled. It would not be feasible to cultivate these plants; they are large rain forest trees, and it would thus be at least a 50- or 60-year project. With regard to biochemical plasticity, there is no evidence that volatile oils show this kind of plasticity. As I mentioned, the evidence available, mainly in temperate species, indicates that the environmental effects on the volatile oil composition (as when one grows the plants under different conditions) is very small when compared with the variation among individuals and among populations in the field.

Baker: It seems to me that a rather important factor in the interpretation of these pictures would be knowledge of the reproductive biology of these trees. Can you tell us anything about the breeding systems or the seed-dispersal mechanisms?

Whiffin: Not much is known. They produce inflorescences of white flowers, which I suspect are insect pollinated. Seed dispersal is by wind, as the seed is winged, but generally the seed is not transported over very large distances. There is a tendency in *Flindersia bourjotiana* toward the replacement of many of the bisexual flowers in an inflorescence with functionally unisexual male flowers (Hartley, 1969), but little is known about the extent of outcrossing.

Janzen: The emphasis you have placed on the unchanging response of the resin chemistry to what you are calling environmental changes implies that the thing that actually challenges the plant (the thing these resins are adaptive in the context of) is something that doesn't change with respect to these environmental changes. The resins are there for some reason, in a selective natural history sense. I'm not going to worry about what that is at the moment. Now the question is: What are its characteristics if it does not change in response to the various kinds of physical and environmental changes that occur in its habitat? Why do you think this characteristic should be under such strong genetic control, as many other plant traits are not under such control?

Whiffin: Well, this is true of many chemical characters; very few chemical characters appear to show very much phenotypic variation. So you are really asking me to speculate on the function of volatile oils, and that is very difficult. No one has determined what their function is. There is some evidence that they may act as attractants or repellents to certain insects, but this is mostly speculation, particularly in the case of leaf volatile oils. It could be very difficult to explain why there are so many different compounds. In *Flindersia bourjotiana* there were over 90, and in other plant species there may be 150–200 different compounds. Why is there such a mixture of compounds? Are some of these compounds more effective than others? At present we don't know.

References

Adams, R. P. (1969). Chemosystematic and numerical studies in natural populations of *Juniperus*. Ph.D. dissertation, University of Texas.

– (1970). Contour mapping and differential systematics of geographic variation. *Syst. Zool.*, *19*, 385–90.

– (1972). Numerical analysis of some common errors in chemosystematics. *Brittonia*, *24*, 9–21.

– (1975a). Gene flow versus selection pressure and ancestral differentiation in the composition of species: Analysis of populational variation of *Juniperus ashei* Buch. using terpenoid data. *J. Mol. Evol.*, *5*, 177–85.

– (1975b). Statistical character weighting and similarity stability. *Brittonia*, *27*, 305–16.

Adams, R. P., & Turner, B. L. (1970). Chemosystematic and numerical studies of natural populations of *Juniperus ashei* Buch. *Taxon*, *19*, 728–51.

Ashton, P. S. (1969). Speciation among tropical forest trees: some deductions in the light of recent evidence. *Biol. J. Linn. Soc.*, *1*, 155–96.

Clifford, H. T., & Stephenson, W. (1975). *An Introduction to Numerical Classification*. New York: Academic Press.

Federov, A. A. (1966). The structure of the tropical rain forest and speciation in the humid tropics. *J. Ecol.*, *54*, 1–11.

Flake, R. H., von Rudloff, E., & Turner, B. L. (1973). Confirmation of a clinal pattern of chemical differentiation in *Juniperus virginiana* from terpenoid data obtained in successive years. In *Terpenoids: Structure, Biogenesis, and Distribution*, ed. V. C. Runeckles & T. J. Mabry, pp. 215–28. New York: Academic Press.

Gould, S. J., & Johnston, R. F. (1972). Geographic variation. *Annu. Rev. Ecol. Syst.*, *3*, 457–98.

Gower, J. C. (1966). Some distance properties of latent root and vector methods used in multivariate analysis. *Biometrika*, *53*, 325–38.

Hanover, J. W. (1966a). Genetics of terpenes. I. Gene control of monoterpene levels in *Pinus monticola* Dougl. *Heredity*, *21*, 73–84.

– (1966b). Environmental variation in the monoterpenes of *Pinus monticola* Dougl. *Phytochemistry*, *5*, 713–17.

Hardy, P. J. (1969). Extraction and concentration of volatiles from dilute aqueous and aqueous alcoholic solutions using trichlorofluoromethane. *J. Agric. Food Chem.*, *17*, 656–8.

Hartley, T. G. (1969). A revision of the genus *Flindersia* (Rutaceae). *J. Arnold Arbor.*, *50*, 481–526.

Hartley, T. G., & Hyland, B. P. M. (1975). Additional notes on the genus *Flindersia* (Rutaceae). *J. Arnold Arbor.*, *56*, 243–7.

Irving, R. S., & Adams, R. P. (1973). Genetic and biosynthetic relationships of monoterpenes. In *Terpenoids: Structure, Biogenesis, and Distribution*, ed. V. C. Runeckles & T. J. Mabry, pp. 187–214. New York: Academic Press.

Jardine, N., & Edmonds, J. M. (1974). The use of numerical methods to describe population differentiation. *New Phytol.*, *73*, 1259–77.

Jardine, N., & Sibson, R. (1971). *Mathematical Taxonomy*. London: Wiley.

Kershaw, A. P. (1975). Late Quaternary vegetation and climate in north eastern Australia. In *Quaternary Studies*, ed. R. P. Suggate & M. M. Cresswell, pp. 181–7. Wellington: Royal Society of New Zealand.

Loomis, W. D., & Croteau, R. (1973). Biochemistry and physiology of lower terpenoids. In *Terpenoids: Structure, Biogenesis, and Distribution*, ed.

V. C. Runeckles & T. J. Mabry, pp. 147–85. New York: Academic Press.

Mabry, T. J. (1973). The chemistry of geographical races. *Pure Appl. Chem.*, 34, 377–400.

Noy-Meir, I. (1973). Data transformations in ecological ordination. I. Some advantages of non-centering. *J. Ecol.*, 61, 329–41.

Powell, R. A., & Adams, R. P. (1973). Seasonal variation in the volatile terpenoids of *Juniperus scopulorum* (Cupressaceae). *Amer. J. Bot.*, 60, 1041–50.

Seal, H. L. (1964). *Multivariate Statistical Analysis for Biologists*. London: Methuen.

Sneath, P. H. A., & Sokal, R. R. (1973). *Numerical Taxonomy*. San Francisco: Freeman.

Sokal, R. R., & Rohlf, F. J. (1969). *Biometry*. San Francisco: Freeman.

Squillace, A. E., & Fisher, G. S. (1966). *Evidence of the Inheritance of Turpentine Composition in Slash Pine*. USDA Forest Service Research Paper NC-6, pp. 53–60.

Steel, R. G. D., & Torrie, J. H. (1960). *Principles and Procedures of Statistics*. New York: McGraw-Hill.

Turner, B. L. (1970). Molecular approaches to populational problems at the infraspecific level. In *Phytochemical Phylogeny*, ed. J. B. Harborne, pp. 187–205. New York: Academic Press.

Veldman, D. J. (1967). *Fortran Programming for the Behavioral Sciences*. New York: Holt, Rinehart and Winston.

von Rudloff, E. (1972). Seasonal variation in the composition of the volatile oils of the leaves, buds, and twigs of white spruce (*Picea glauca*). *Can. J. Bot.*, 50, 1595–603.

– (1975). Volatile leaf oil analysis in chemosystematic studies of North American conifers. *Biochem. Syst. Ecol.*, 2, 131–67.

Williams, W. T., Dale, M. B., & Lance, G. N. (1971). Two outstanding ordination problems. *Aust. J. Bot.*, 19, 251–8.

Zavarin, E., Critchfield, W. B., & Snajberk, K. (1969). Turpentine composition of *Pinus contorta* and *Pinus banksiana* hybrids and hybrid derivatives. *Can. J. Bot.*, 47, 1443–53.

Part II: Reproduction and demography

3

Chemical aspects of the pollination biology of woody plants in the tropics

HERBERT G. BAKER

Botany Department, University of California, Berkeley, California

In this chapter, I shall discuss only some pollinatory aspects of the flower biology of some tropical trees and lianes, leaving seed production and dispersal matters in the capable hands of Dr. Janzen. Floral features important in pollination biology are (1) the *attractants* that bring pollinatorily useful visitors to the flowers, (2) the *rewards* these visitors receive from the flowers, and (3) the *deterrents* that may prevent overexploitation of the floral resources by visitors.

The attractants hardly need elaboration here; color and its patterning, the shapes of flowers, their scents, and the combinations of these that constitute so-called nectar guides, are generally well known (e.g., Meeuse, 1961; Percival, 1965; Faegri & van der Pijl, 1971; Proctor & Yeo, 1973). The rewards to visitors are less well known and include energy-providing chemicals (Heinrich & Raven, 1972), body-building chemicals (Baker & Baker, 1975) and, sometimes, volatile chemicals that can be used for territory marking by male bees (Vogel, 1966) or even for attracting other male bees into leks (Dodson, 1975). The potential deterrents to the attentions of unwelcome visitors (or overenthusiastic legitimate visitors) may be mechanical or chemical.

We often see these three characters combined in syndromes because, of course, there must be coordination of the various aspects of the pollination process if it is to be repetitively successful. This has sometimes led to confusion among the categories, and indeed, it is possible for a reward to function as an attractant and vice versa. For example, the freely exposed nectar of a flower of *Tristania* (Myrtaceae) may glisten in sunshine and attract the attention of Australian nectar-loving birds. In subtropical South Africa, the black nectar of some of the sugar bushes of the genus *Melianthus* (Melianthaceae) is prominently displayed while held in the concave calyx against a red

57

background provided by the petals and, presumably, constitutes an attractant as well as a reward to the sunbirds that pollinate the flowers (Scott-Elliott, 1889).

Sometimes pollen, or, at least, the stamens containing pollen, is attractively colored and displayed, as in *Bixa orellana* of the Bixaceae. Pollen is the only reward for bees that visit this neotropical flower; there is no nectar. But in flowers, features usually of the corolla, less often of the calyx, provide most of the attraction, and the various syndromes of floral characters that attract various kinds of pollinators are well reviewed by Faegri & van der Pijl (1971), Baker & Hurd (1968), Proctor & Yeo (1973), and other authors. Therefore, we may concentrate on rewards, with some mention of deterrents. These rewards may be solid or liquid.

Solid rewards
Food bodies
The fleshy bracts of species of *Freycinetia* (Pandanaceae), growing in Southeast Asia, are said to be eaten by rodents and birds, notably bulbuls (Faegri & van der Pijl, 1971, pp. 141–2). In the neotropics, N. W. Uhl and H. E. Moore (pers. comm.) have shown that the fleshy sepals and petals of the staminate flowers of *Bactris major* (Palmae) are eaten by several kinds of beetles, especially the curculionid *Phyllotrox megalops*, which consume pollen as well as carrying it from staminate to pistillate flowers in this nectarless, monoecious species. Even in *Theobroma cacao* (Sterculiaceae), where the reward is meager (merely the staminodes that form an outer ring around the fertile stamens), it is sufficient for the pollinators, ceratopogonid midges of the genera *Forcipomyia* and *Lasiohelea* (Purseglove, 1968, p. 580).

Pollen
Much more frequently, however, the reward to a flower visitor is pollen produced in excess of the amount needed directly in pollination of stigmas. For example, among dicotyledonous trees, the flowers of *Cochlospermum vitifolium* (Cochlospermaceae) are visited in Central America by female anthophorid bees (G. W. Frankie, pers. comm.). These flowers produce no nectar, but the pollen-collecting bees are their pollinators.

Sometimes the excess pollen is displayed separately from the pollen that will be used in pollination. For example, there are short "feeding" stamens and longer "pollinating" stamens in *Tibouchina* spp. (Melastomataceae) and in *Cassia* and *Swartzia* (Caesalpiniaceae). Large euglossine bees are the pollinators of *Swartzia simplex* in Costa Rica (Harcombe & Riggins, 1968), and they fit well in the flowers;

while their mouthparts are dealing with the feeding anthers, the pollinating anthers make contact with the undersides of their thoraxes. There is some evidence of dimorphism in the pollen, too; that of the feeding anthers is about two-thirds the diameter of the grains in the pollinating anthers.

Pollen provides a flower visitor with chemicals that can be used in its own nutrition or that of its brood. Pollen grains contain a wide range of amino acids, both in the free state and in protein form, as well as lipids, carbohydrates, and, sometimes, various vitamins (literature review in Stanley & Linskens, 1975). My wife and I are just beginning a systematic analysis of pollen from species with different pollination systems to see if chemical composition correlates with pollinator type. In advance of our results, we can point to an interesting conclusion by Howell (1974), who has studied the pollen of some bat-pollinated species (whose pollen is ingested by the bats). She finds that the pollen, a major source of protein-building material for *Leptonycteris nivalis* (of the Microchiroptera), is particularly rich in protein and free amino acids, especially proline and tyrosine. She points out that proline constitutes more than 80% of collagen, which is an especially important constituent of tissues in the large wings and tail membranes of the bats. Tyrosine may function as a growth stimulant for the young bats, as it is well represented in the milk they drink from the mother's teats. Even if these details should be modified by future work, they suggest that the biochemistry of pollen will prove revealing of the adaptation of the grains to the nutritional needs of flower visitors. However, in this respect, we have much more information on the liquid rewards provided by flowers.

Liquid rewards
Stigmatic secretions

The liquid that exudes from the tissues of a stigma onto its surface is usually thought of as preventing drying out of the surface and providing a suitable medium for the germination of pollen grains (Baker, Baker, & Opler, 1973). These sticky secretions may also be smeared on the bodies of beetles crawling over the surface of the stigma, causing pollen grains knocked from the anthers to adhere to the beetles and be carried by them from flower to flower, as in *Luehea* spp. (Tiliaceae: Baker et al., 1973) and species of *Annona* (Annonaceae: Baker, unpubl.).

Undoubtedly, stigmatic secretions do perform these functions. Stigmatic exudates contain lipids and amino acids and may contain sugars, all in an aqueous medium (literature review in Baker et al., 1973); they do not dry up easily, and these chemicals may also serve as significant nutrients for the developing pollen tubes. But these chemicals may also serve in insect nutrition, and although different in composition from nectar that may be secreted in the same flower (Table 3.1), they may be no less nutritious. Even the tiny amounts provided by small flowers may serve to nourish flies and thrips.

Table 3.1. *Chemical constituents, particularly amino acids, in stigmatic secretion and nectar of flower of* Luehea candida *(Tiliaceae), La Pacifica, Guanacaste, Costa Rica* [a]

Constituent	Stigmatic secretion[b]	Nectar[c]
Alanine	3	3
Arginine	2	3
Asparagine		
Aspartic acid	1	1
Cysteine, cystine, cysteic acid	3	
Glutamic acid	2	1
Glutamine		
Glycine	1	1
Histidine		
Isoleucine	1	2
Leucine	1	2
Lysine	1	1
Methionine		
Phenylalanine		1
Proline	2	2
Serine	1	4
Threonine	2	2
Tryptophan		1
Tyrosine	2	1
Valine	2	2
β-Alanine	1	
Unknown 1	4	2
Unknown 2	2	

[a] 1 is least concentrated amino acid; 2–4 represent increasing levels of concentration. For methods of analysis, see Baker & Baker (1976).
[b] Also contains lipids, antioxidants, proteins, phenolics, and sugars.
[c] Also contains traces of lipids and sugars.

Nectars

Often much more abundant and, consequently, very important as rewards are the secretions of floral nectaries. Nectar is the floral reward to which we have given closest attention during the last 4 years.

Nectars are far from being the mere sugar water of popular belief (Baker & Baker, 1973a,b, 1975, 1976). Analyses by various workers (literature review in Baker & Baker, 1975) have demonstrated the presence in nectars of some or all of the following chemicals: sugars, proteins, amino acids, lipids, antioxidant organic acids, a variety of other nutritive organic substances in very small amounts, as well as alkaloids, glycosides, saponins, and phenolics, which may have a deterrent effect on some potential nectar feeders.

Amount of the nectar reward
Usually the free exposure of nectar is associated with the copious presentation of pollen as an alternative reward (for both are held to be features of more "primitive" flower types). On the other hand, those "more advanced" species that hide their nectar away, to be picked up only by specialized visitors, are usually also rather parsimonious in their pollen supply. This correlation has consequences for the chemical composition of nectar that will be discussed later.

The amount of the reward presented to a flower visitor is likely to be an important factor in determining the behavior of that visitor, as well as having important consequences for the breeding system of the plants (Heinrich & Raven, 1972). There must be enough energy income from the reward to exceed the energy expended by the visitor in finding it and picking it up. Consequently, the volume of nectar provided by a flower and its concentration of energy-giving chemicals also become important. These energy-giving chemicals are usually sugars, but we shall see later that they may include lipids.

Different flowers contribute differently in this matter of energy supply. Generally speaking, the volume of nectar produced is roughly proportional to the size of the flower (although there are some glaring exceptions: e.g., some large flowers, such as those of *Cochlospermum*, that produce no nectar). A *Bombax* flower may produce 200–300 μl of nectar each day and be attractive to fairly large birds as a result. By contrast, individual flowers of *Cordia dentata* or *Cordia panamensis* (Boraginaceae) contribute only a fraction of a microliter. Consequently, these *Cordia* trees have small pollinators, usually Coleoptera, Diptera, and some small Lepidoptera (Opler, Baker, & Frankie, 1975).

A compact inflorescence of small flowers can equal the output of a large single flower, sometimes with division of labor in the process. In the inflorescence of the West African savanna tree *Parkia clapper-toniana* (Mimosaceae), there are sterile flowers at the base of the inflorescence (which, actually, is uppermost because the inflorescence hangs down on a pendulous peduncle: Fig. 3.1). These flowers have the sole function of producing nectar, and this they do magnificently. Together, in one night they may produce as much as 5000 μl of nectar, which runs down and collects in a depressed ring above the ball of fertile flowers. This is amply sufficient to be attractive to large nectar-lapping bats of the genera *Epomophorus* and *Nanonyc-teris* (Baker & Harris, 1957; Fig. 3.2). The bat visitors transfer pollen from inflorescence to inflorescence on the fur of their necks and breasts, where they make contact with the fertile flowers. The visits

Fig. 3.1. Pendulous inflorescences of *Parkia clappertoniana* (Mimo-saceae). *From right to left:* unopened inflorescence, inflorescence in staminate phase, inflorescence in pistillate phase, inflorescence with half of fertile flowers removed to show manner in which they are borne. Sterile, nectar-producing flowers constitute cylindric portion of inflorescence; their nectar runs down into ring at junction of cylindric and spherical parts of inflorescence. (Photograph by H. G. Baker and B. J. Harris)

of the bats are important to *Parkia* because the fertile flowers within an inflorescence exhibit coordinated protandry (with each inflorescence in a staminate condition on the first night of anthesis and in a pistillate condition on the second night) as well as evidence of self-incompatibility. Trees of *P. clappertoniana* bring new inflorescences to anthesis continuously night after night for several weeks (Baker & Harris, 1957), assuring a food supply for their large visitors during that period.

The amount of the reward offered a flower visitor also depends on the concentration of sugar in the nectar. This varies considerably from species to species and shows a clear relation with the nature of the predominant pollinator. In a quick survey of Costa Rican forest trees in February 1974, we found nectar sugar concentrations as low as 11% in nocturnally open flowers of *Ochroma pyramidale* (Bombacaceae) and *Crataeva tapia* (Capparidaceae), both of which are pollinated by bats. At the other extreme, we found sugar concentrations as high as 75% in dry season flowers of *Andira inermis* (Fabaceae), visited by large bees.

Fig. 3.2. Nectar-lapping bat (*Epomophorus gambianus*) taking nectar from *Parkia clappertoniana*. Achimota, Ghana. (Photograph by H. G. Baker and B. J. Harris)

Table 3.2 shows the averages for sets of sugar concentrations in Costa Rican forest tree nectars, arranged according to predominant pollinator type. Even though the numbers of species involved are not large, it is apparent that bees are provided with, and can handle, rather viscous, syrupy nectar. (Incidentally, so can flies, which may dilute the nectar or even dissolve sugar crystals in liquid they regurgitate, following this by sucking up the resultant solution.)

Most Lepidoptera take up nectar through a long, narrow tube formed by fused maxillae; consequently, the nectar of the flowers they visit cannot be very viscous (i.e., concentrated). It is notable, in this connection, that a 40% sucrose solution is 4 times as viscous as a 20% solution, and a 60% solution has 28 times the viscosity of the 20% solution (Hodgman, Weast, & Selby, 1958). This limitation on the manageability of concentrated nectar is particularly severe in the case of flowers visited by hawkmoths, for these insects do not usually settle while feeding and take their nectar rather quickly. A similar limitation applies in the cases of the quick-feeding hummingbirds and bats of the neotropics (Baker, 1975). Naturally, exposed nectar tends to become stronger the longer it sits in an open flower, as a result of evaporation. With this in mind, Lynn Carpenter (pers. comm.) suggested that the tubular nature of most hummingbird flowers is particularly important in preventing nectar from evaporating to concentrations inappropriate to the foraging habits of hummingbirds.

Evaporation may be one cause of some differences we have found between the sugar concentrations of nectars from flowers produced, respectively, low down and high up on the same tree (Table 3.3). However, this cannot be the whole explanation of the tendency for

Table 3.2. *Mean sugar concentrations (in "sucrose equivalents," weight by total weight) in nectars of trees and lianes in lowland forests in Costa Rica, grouped according to their principal pollinators; determinations made in February 1974 (dry season)*

Pollinator	No. of species examined	Sugar concentration (%)	Range (%)
Short-tongued bees	5	46	25–75
Long-tongued bees	15	46	24–75
Butterflies	20	29	16–40
Settling moths	2	41	39–42
Hawkmoths	11	24	17–40
Bats	5	17	11–26
Hummingbirds	29	21	12–29

nectar to increase in concentration with height on the tree, for the samples described in this table were all of freshly produced nectar. Perhaps more active photosynthesis in the upper part of the tree provides a greater supply of sugar to the nectaries located there. Whatever the cause, the effect may be one of the factors responsible for the tendency of bees of various species to forage at characteristically different heights in the trees, a phenomenon being investigated in Costa Rica by Gordon W. Frankie.

The sugar concentration percentages given here represent "sucrose equivalents," estimated by using a suitably calibrated refractometer and assuming that all the sugar present is sucrose. Actually, it usually is not; sometimes in the nectary or subsequently in the nectar, sucrose is hydrolyzed to a greater or lesser extent to glucose and fructose. This process does not, in itself, greatly change the refractive index of the nectar, but there is some inaccuracy involved in estimating sucrose equivalents if the two hexose sugars are not in equal proportions in the nectar as secreted, and there may also be slight inaccuracy introduced if other sugars are present (e.g., maltose, melibiose, or melezitose). Consequently, it is useful to have actual chemical analyses of the sugars present.

Sugars in nectar
Thus far, we have analyzed the sugars in nectars from 68 species of Costa Rican lowland forest trees and lianes. The great majority of these contain sucrose, glucose, and fructose together in significant amounts. Only 2 species, *Anemopaegma orbiculata* (Bignoniaceae)

Table 3.3. *Mean sugar concentrations in nectars taken from flowers borne at different heights above ground in five lowland forest trees in Guanacaste Province, Costa Rica; samples taken in dry season*

Flower	Height above ground (m)	Sugar content (%)
Andira inermis	11	40
	2.5	25
Cordia alliodora	10	25
	5	15
Dalbergia retusa	High	20
	Low	13
Tabebuia neochrysantha	High	20
	Low	10
Tabebuia rosea	High	35
	Low	35

and *Carica papaya* (Caricaceae), show pure sucrose nectar; sucrose is missing from the nectars of *Cordia dentata* (Boraginaceae) and *Stigmaphyllon lindenianum* (Malpighiaceae). Too much significance should not be given to the pure sucrose result for *Carica papaya* because the related *Carica dolichaula* (*Jacaratia dolichaula*) showed the following proportions: sucrose, 0.54; glucose, 0.32; and fructose, 0.13 (Baker, 1976).

Where multiple determinations have been made on the same species, they have shown generally constant proportions for each of the sugars (as also found by Percival, 1961). An interesting exception to this, however, is provided by some nectars of staminate and pistillate flowers of some dioecious species (Table 3.4). Although "staminate" and "pistillate" nectars agree in *Coccoloba padiformis* (Polygonaceae), they differ radically in *Trichilia colimana* (Meliaceae), *Triplaris americana* (Polygonaceae), and *Simarouba glauca* (Simaroubaceae). The biologic significance of such differences has yet to be elucidated.

Table 3.5 summarizes the sugar compositions of nectar for Costa Rican trees and lianes, arranged on the basis of predominant pollinator type. Although the data are, as yet, rather skimpy, it looks as if flowers pollinated by butterflies, moths, and hummingbirds tend to have sucrose dominance (a condition noted for tubular flowers by Percival, 1961), whereas the nectar of bat-pollinated flowers tends to be dominated by hexose sugars. Bee-pollinated flowers show no

Table 3.4. *Proportions (by weight) of sugar components of nectars from flowers on staminate and pistillate plants of four dioecious tree species in Costa Rica*

Plant	Sucrose	Glucose	Fructose	Melezitose
Coccoloba padiformis				
Staminate	0.28	0.33	0.39	
Pistillate	0.28	0.32	0.40	
Trichilia colimana				
Staminate	0.45	0.50	0.05	
Pistillate	0.39	0.27	0.33	
Triplaris americana				
Staminate	0.66	0.12	0.15	0.07
Pistillate	0.07	0.64	0.29	
Simarouba glauca				
Staminate	0.03	0.47	0.50	
Pistillate	0.16	0.24	0.60	

clear picture. Obviously, more data must be acquired before firm conclusions can be drawn.

Any correlations that stand up to further testing may relate to the taste preference of the animals concerned, for there can be little doubt that these sugars have different tastes and, if the pun can be forgiven, so do the various flower visitors. Very few species produce nectars with comparable proportions of the various sugars in comparable overall concentrations. Consequently, even if nothing but sugars were to contribute to the tastes of nectars, very few species would produce nectars with exactly the same taste to a discriminating visitor. We shall see later that there *are* other substances that contribute to giving the nectar of each species a separate taste identity. These include amino acids, proteins, lipids, organic acids of various sorts, vitamins, alkaloids, glycosides, saponins, and various phenolic substances.

Amino acids in nectar

Since 1972, my wife and I have been giving close attention to the occurrence of amino acids in nectar (Baker & Baker, 1973a,b, 1975, 1976). We have observed the amino acid concentrations in nectars from over 1200 species from many parts of the world. We have made analyses of the amino acid complements of over 300 species.

Amino acids are always present in nectars, although the amounts are small by comparison with the sugars. About 4 mg/ml appears to be an upper level of concentration. However, the amino acids occur in amounts that may be significant nutritionally to the flower visitor either in protein building (for repair or reproduction) or as gustatory stimuli (Baker & Baker, 1973a,b, 1975).

Table 3.5. *Numbers of floral nectars of tropical trees and lianes in Costa Rican lowland forests with more than half their sugar (by weight) as sucrose or hexoses, arranged by type of predominant pollinator*

Pollinator	No. of species sampled	No. with nectar dominated by	
		Sucrose	Hexoses
Short-tongued bees	12	5	7
Long-tongued bees	14	6	8
Butterflies	17	10	7
Moths	10	7	3
Hummingbirds	8	7	1
Bats	7	1	6

It is notable that the concentration of amino acids in nectar tends to be greater if nectar is the only (or the chief) source of protein-building material for the usual flower visitors than it is if the visitors have an abundant alternative (Table 3.6). In Costa Rica, settling moths, butterflies, and wasps (with the exception of some social wasps) depend upon nectar for their own nourishment, and the nectars of the flowers they visit have relatively high amino acid concentrations. The flower-visiting bats in Costa Rica make use of fruit juices and pollen as sources of protein-building materials and consume some insects (Baker, 1973; Heithaus, Opler, & Baker, 1974; Heithaus, Fleming, & Opler 1975); the nectars they take are weak in amino acids.

The hummingbirds, especially females at times of reproduction, are avid insect eaters. Flowers could not possibly provide them with a significant alternative supply of protein-building materials, and they do not.

A detailed consideration of the relation between nectar amino acid concentrations and pollinator behavior in Costa Rica will be published in collaboration with Paul A. Opler (who collected most of the nectar samples and made many observations on pollinator behavior). He and Gordon W. Frankie have also been involved in studies of nectar amino acid concentration in relation to the height at which flowers are produced above the ground. The results of these studies

Table 3.6. *Means of total amino acid concentrations in floral nectars of plants of all life forms growing in lowland forests of Costa Rica, grouped according to type of predominant pollinator[a]*

Pollinator	No. of determinations	Amino acids (μM/L)
Settling moths	51	1150
Wasps	32	875
Butterflies	84	820
Short-tongued bees	76	650
Flies	22	585
Hawkmoths	34	450
Long-tongued bees	49	398
Bats	15	275
Hummingbirds	55	255

[a] Amino acid estimations made colorimetrically; for method see Baker & Baker (1973b, 1975). Almost all samples collected by P.A. Opler.

will be published elsewhere; at present, it appears that, unlike sugars in nectar, which tend to be more concentrated higher up in a tree than at lower heights above the ground, the amino acid concentrations are lower at higher elevations. The significance of such a trend is not clear at present, but, again, the result may be connected with the height preference for foraging by various flower visitors.

Nectars contain from 2 to 24 amino acids, according to the species concerned. Table 3.7 shows the percentage occurrences of the various amino acids. Some are found more frequently than others,

Table 3.7. *Percentages of nectars examined containing recognized protein-building amino acids and one or more nonprotein amino acids*

Amino acid	First 283 species (%)[a]	69 tropical trees and lianes (%)[b]
Alanine	97	100
Arginine	100	99
Asparagine	11	4
Aspartic acid	40	46
Cysteine, cystine, cysteic acid	41	49
Glutamic acid	67	75
Glutamine	33	31
Glycine	81	88
Histidine	19	14
Isoleucine	75	78
Leucine	71	72
Lysine	42	36
Methionine	20	28
Phenylalanine	10	14
Proline	82	91
Serine	92	90
Threonine	89	84
Tryptophan	37	41
Tyrosine	36	49
Valine	67	68
Nonprotein	36	55

[a]Results from first 283 species (of all life forms) examined from all sources.
[b]Results from 69 tropical tree and liane species from Costa Rican lowland forests.

but the 69 species of tropical trees and lianes from Costa Rican low-land forests whose nectars we have analyzed have a frequency of oc-currence of the various protein-building amino acids comparable to those found in the 283 species that were the first to be analyzed from anywhere in the world. Therefore, it does not seem likely that there is anything strikingly unusual about the nutritive qualities of tropi-cal tree and liane nectars as a whole. (We shall consider later the rather high proportion of tropical tree and liane nectars that contain nonprotein amino acids.)

Arginine, alanine, serine, threonine, and proline are the most commonly occurring nectar amino acids, and frequently they are the most abundant in a nectar. On the other hand, the aromatic amino acids and those with sulfur in the molecule are rather uncommon, as is histidine. No single nectar contains all the protein-building amino acids – at least, at the level that can be detected by the sensitive dansylation and polyamide thin-layer chromotography methods we apply (Baker & Baker, 1976) – and only one species' nectar so far has been found to contain all the so-called essential amino acids (i.e., *Erythrina breviflora;* see Table 3.8). However, it must be remembered that the great majority of flower visitors in the tropics take nectar from more than one plant species, and for many, there are possibil-ities of diet supplementation from pollen or other sources.

The amino acid complements of nectars are remarkably constant from plant to plant in or among populations (data from temperate species to be published elsewhere; see also Table 3.8), even when the "sexes" of the flowers are different. However, in making analyses, one must be careful to compare only floral nectars; extrafloral nectar has a different composition from floral nectar even when both are produced on the same plant. (A paper on this subject is in prepara-tion in collaboration with P. A. Opler.) A difference in composition between floral and extrafloral nectars is not surprising when one con-siders that they are feeding different constellations of animals: pol-linators at the flowers and, often, ants at the extrafloral nectaries.

Apart from their potentially nutritional function, nectar amino acids, with their differing presences, proportions, and concentra-tions, can also modify the taste of the nectars that contain them. Even to the human palate, some amino acids are sweet, some are bitter, and some are sour, and Shiraishi & Kuwabara (1970) have shown that some insect chemoreceptors react differentially to these groups of amino acids (although, unfortunately, the flies they tested were not flower visitors). So, in combination with the sugars, the amino acid complement may contribute to determining the taste of nectar to a discriminating flower visitor.

Table 3.8. *Amino acid complements from floral nectars of individual plants of five species of trees from tropical America*[a]

Amino acid	Erythrina breviflora[b]				Coccoloba padiformis[c]				Simarouba glauca[d]		Triplaris americana[d]		Randia sp. III[d]	
	1	2	3	4	♀1	♀2	♀3	♀4	♂	♀	♂	♀	♂	♀
Alanine	2	2	3	2	3	3	3	3	3	3	2	2	2	2
Arginine	3	3	3	3	3	4	3	3	2	2	4	3	4	4
Aspartic acid	1	1	1	1	2	1	1	1	1	1	1	1	1	1
Asparagine									2	1				
Cysteine, cystine, cysteic acid	1	1	1	1	1	1	1	1	1	1	1	1	1	1
Glutamic acid	1	1	1	1	2	1	1	1	2	1	2	1	1	1
Glutamine														
Glycine	1	1	1	1	1	2	2	2	3	3	1	1	1	1
Histidine	1	1	1	1										
Isoleucine	3	3	3	2	1	1	1	1	2	2	1	1	1	1
Leucine	3	3	3	2	1	1	1	1	3	3	3	2	1	1
Lysine	2	2	3	2	1	1	1	2						
Methionine	2	1	1	1					2	2			2	2
Phenylalanine	1	1	1	1										
Proline	3	3	2	3	3	4	2	3	2	1	2	3	1	1
Serine	2	3	2	2	3	4	3	3	2	2	4	3	4	4
Threonine	4	4	4	3	3	1	1	1	3	3	2	2	1	1
Tryptophan	1	1	1	1					1	1	1	1	1	1
Tyrosine	2	2	2	2	1	1	1	1	3	2	3	2	1	2
Valine	2	2	2	2	1	1	1	1	2	1	2	1	1	2
Unknown 1	2	2	2	2										
Unknown 2	2	2	2	2										
Unknown 3					2	2	2	2						
Unknown 4									2	2				
Unknown 5									2	2				
Unknown 6											2	2		
β-Alanine									1	1				
Homoserine									1	1				

[a] 1 is least concentrated amino acid; 2–4 represent increasing levels of concentration.
[b] Samples collected from four separate localities in Mexico by R. W. Cruden.
[c] Samples collected from Costa Rica by G. W. Frankie.
[d] Samples collected from Costa Rica by P. A. Opler.

Lipids in nectar

All nectars contain sugars and amino acids, but this does not exhaust the list of potential contents. Lipids have long been known to be present in and on pollen grains, and they constitute part of the reward to a pollen gatherer. However, 7 years ago, Vogel (1969) pointed out that a number of plants, mostly growing in South America, produce drops of oil that are collected by anthophorinine bees. The glands that secrete these oils he calls "elaiophors." In subsequent papers, Vogel (1971, 1974) has developed these observations, with the Malpighiaceae being the only family of tropical trees, lianes, and vines on which he has made observations.

Vogel (1974) finds that female bees, particularly of the genus *Centris*, collect the oil from elaiophors on the sepals of flowers of neotropical Malpighiaceae, using their forelegs to gather the drops and transferring them to their hindlegs, where the oil is trapped on the hairs. Then these bees convey the oil to their nests where, along with pollen, it provides nourishment for their larvae. Vogel's photographs illustrate the collection process elegantly.

Oil-producing flowers, in Vogel's experience, usually have a yellow corolla. He claims that they produce no nectar and that the production of one liquid or the other represents *alternative* strategies for rewarding visitors.

However, 3 years ago, we reported lipid-containing nectar, a milky, aqueous liquid that also contains sugars, amino acids, and other substances, in *Jacaranda ovalifolia* (Bignoniaceae), as well as in other, unrelated taxa (Baker & Baker, 1973b). Later, we indicated that about one-third of 220 species of nectars we tested for the presence of unsaturated lipids gave positive results (Baker & Baker, 1975). At that time, 40 families were represented among the species whose nectar contained lipoid substances, and the range of their occurrence has increased subsequently.

In lowland Costa Rican forests, lipid-containing nectars seem to be found most frequently in the trees (Table 3.9), where *Centris* bees often forage. Among the trees and lianes, nectar lipids have been found particularly frequently in the Caesalpiniaceae and Bignoniaceae, without any obvious tendency for yellow flowers to predominate.

Sometimes, nectaries and elaiophors are not spatially separated and their secretions are intimately mixed. In species of *Catalpa* (Bignoniaceae) there is a slight separation. Here, the aqueous nectar secretion (from the nectary that forms a disk beneath the superior ovary) begins slightly earlier than the release of oil from the stalked multicellular glands that occur very close to the nectary on the inside of the corolla and the lower parts of the ovary wall. Some lipids are also produced by the subepidermal cells of the nectary itself. Consequently, the nectar is at first clear and then becomes milky. Ultimately, even the multicellular glands become detached and fall into the nectar.

This discovery that lipids are common constituents of nectar has led us to look again at some members of the Malpighiaceae. Vogel (1974) claims that almost all New World species have elaiophors only (producing no reward except oil), whereas Old World species have normal nectaries. However, we find that smears onto chromatography paper of the secretions of the sepal glands of *Byrsonima crassifolia, Hiraea obovata,* and *Stigmaphyllon lindenianum* (all Costa Rican representatives of the Malpighiaceae) contain water-soluble substances, including sugars and amino acids (Table 3.10) as well as lipids. As long as 15 years ago, Percival (1961) reported the presence of sugars (sucrose, fructose, and glucose) in the nectar of *Stigmaphyllon* (as *Stigmatophyllum*) *ciliatum.*

Even *Centris* bees need these other chemicals, for their own nutrition if not for that of their larvae, and *Centris* bees of both sexes in Costa Rica are by no means restricted in their visits to malpighiaceous trees and lianes. They also visit other plants in whose flowers lipids may or may not be present as constituents of normal nectar. Consequently, we believe that the secretion of nectar and of oil are not *alternative* strategies, but that a balance is usually struck between the proportions of these two secretions and that, in the case of the neotropical Malpighiaceae, the balance has been swung usually far over to the oil side under the influence of bees with a predilection for using oil for larval nutrition.

Table 3.9. *Percentages of floral nectars from lowland Costa Rican plants (arranged by life form) that contain lipids*[a]

Type of plant	No. of species[b]	% with lipids
Herbs	64	23.4
Shrubs	70	22.9
Trees	73	41.1
Lianes, vines	47	25.5
Epiphytes	7	14.3

[a]Lipid content identified by osmic acid test (Baker & Baker, 1975).
[b]Almost all samples collected by P. A. Opler.

Our own demonstration that oil is often a constituent of nectar simplifies enormously the evolutionary problem of accounting for the presence of elaiophors in unrelated families and in distantly related neotropical genera of Malpighiaceae, but not in their paleotropical representatives. Vogel claims (1974, p. 538) that "the immediate ancestors of oil flowers were pollen flowers or autogamous at least; they did not possess typical nectaries. In both insects and plants, oil-bound interaction evolved polyphyletically, even within related subgroups." *Now,* the assumption in Vogel's first sentence is unnecessary, and the assumption in the second is easier to accept.

Table 3.10. *Amino acid complements of nectars derived from sepal elaiophors of three members of Malpighiaceae in Guanacaste Province, Costa Rica* [a]

Amino acid	*Byrsonima crassifolia*	*Stigmaphyllon lindenianum*	*Hiraea obovata*
Alanine	3	1	4
Arginine	1		1
Asparagine			
Aspartic acid			
Cysteine, cystine, cysteic acid	1	3	2
Glutamic acid	1		2
Glutamine			
Glycine	2	2	1
Histidine			
Isoleucine			2
Leucine		2	1
Lysine			
Methionine			
Phenylalanine			
Proline	1	5	4
Serine			
Threonine			1
Tryptophan			
Tyrosine	1	3	2
Valine			
Unknown 1			2
Unknown 2			2

[a] 1 is least concentrated amino acid; 2–4 represent increasing levels of concentration.

Antioxidants in nectar
Other substances potentially important in the nutrition of flower visitors are present in many of the nectars of tropical trees and lianes.
For example, many nectars contain antioxidants, most notably ascor-
bic acid (vitamin C), and these are particularly often present in lipid-
containing nectars, where they may prevent rancidity (Baker &
Baker, 1975). Also, they may affect the taste of the nectar.

Potential deterrents in nectar
Still other, scarcely nutritious substances can be found in nectar.
These may have an effect upon the nectar's taste and possibly a significant effect upon the nervous system of a flower visitor. They are
the alkaloids, glycosides, and phenolic substances we have shown to
be present in a minority of flowers (Baker & Baker, 1975), as well as
nonprotein amino acids. Possibly these chemicals have a protective
function against robbers as they do in some seeds (Rehr, Bell, Janzen, & Feeny, 1973).

   We have seen (Table 3.7) that nonprotein amino acids are present
(up to as many as four at the same time) in the nectars of more than
half the tropical trees and lianes examined. By analogy with the role
such acids appear to play in chemically protecting seeds of tropical
plants from predation, we may suppose they play a similar role in
nectar. In this respect, it may be significant that the proportion of
nectars containing at least one nonprotein amino acid is higher for
the tropical tree and liane sample than for the worldwide group of
283 species. Table 3.11 shows, in addition, that nonprotein amino
acids are more common in the nectars of tropical tree and liane
flowers that are pollinated by Hymenoptera than in those that are
visited by Lepidoptera, bats, or birds. This matter deserves further
study.

   If nonprotein amino acids, alkaloids, glycosides, and phenolics
reduce the list of flower visitors, they should be particularly likely to
be found in habitats where the greatest variety of potential flower
visitors occurs. The extra frequency with which tropical tree and
liane nectars contain nonprotein amino acids fits this theory. Table
3.12 shows the proportions of species whose nectars we have found
to contain alkaloids and phenolics over a transect from Costa Rica
through lowland California and up into the mountains to the alpine
tundra of the Colorado Rockies. There is a gradual decline in the
frequency of occurrence of alkaloids from the lowland tropics to the
Colorado tundra, and the same is true for phenolics. At the tundra
end of the transect the range of potential flower visitors is greatly

Table 3.11. *Frequencies of occurrence of nectars containing one or more nonprotein amino acids in lowland tropical tree and liane species in Costa Rica, grouped according to type of predominant pollinator*

Pollinator	No. of species[a]	Species with nonprotein amino acids	
		No.	%
Overall	66	36	55
Wasps	8	6	75
Short-tongued bees	14	10	71
Long-tongued bees	12	8	67
Butterflies	10	4	40
Moths	17	9	53
Hummingbirds	12	5	42
Bats	9	3	33
Hymenoptera	34	24	71
Lepidoptera	27	13	48
Vertebrates	21	8	38

[a]Almost all samples collected by P. A. Opler.

Table 3.12. *Frequencies of occurrence of nectars containing alkaloids and phenolics in plants of all life forms on transect from lowland forest in Costa Rica to alpine tundra of Colorado*

Location	Alkaloids		Phenolics	
	No. of species examined	% with alkaloids	No. of species examined	% with phenolics
Costa Rica: lowland forest[a]	198	12.2	164	50.6
California: below 8000 ft[b]	275	6.9	248	31.0
California: above 8000 ft[b]	67	7.4	54	29.6
Colorado: subalpine forest[b]	27	7.4	31	29.0
Colorado: alpine tundra[b]	26	0.0	31	19.4

[a]Samples collected by P. A. Opler.
[b]Samples collected by I. Baker and H. G. Baker.

reduced, and the pressure to deter inappropriate visitors is correspondingly small.

Inorganic ions, such as potassium, may be present in nectars in quantities sufficient to discourage bee visits, as Waller (1973) has shown in onion flowers, but there is, as yet, no evidence from the tropics on this subject. In addition, we have yet to obtain evidence about the ranges of pH values shown by tropical as opposed to temperate flower nectars; in both these instances there are potentials for influencing the tastes of nectars.

Deception

If the syndrome of adaptations that attracts flower visitors is present without the provision of a reward, this constitutes deception. Deception is well known in pollination biology, particularly in relation to insects with poorly developed discriminatory powers (e.g., Faegri & van der Pijl, 1971; Proctor & Yeo, 1973).

Partial deception certainly exists among tropical trees in those cases where staminate flowers of a monoecious or dioecious species provide a reward (pollen or nectar or both) while the pistillate flowers do not. The pistillate flowers must be visited "by mistake" by pollinators anticipating the same reward they get from staminate flowers (Baker, 1976). An example of such mistake pollination is provided by the species of *Carica* (Caricaceae), where the staminate flowers offer both pollen and nectar for a wide array of visitors (birds, bees, flies, butterflies, and moths), but the pistillate flowers provide no reward and are visited fleetingly, in crepuscular conditions, by sphingid moths that spend proportionately much more time at the rewarding staminate flowers (Baker, 1976). Other cases of partial deception involve monoecious species, such as *Stelechocarpus burakol* (Annonaceae), where pollen is the only reward to a flower visitor and is, of course, contributed only by the staminate flowers.

Clear cases of *total* deception, where no reward is provided by any flower, are, at best, extremely rare among tropical trees. Possibly, the energy outlay for a flowering episode here is so considerable that a small saving in not providing any reward to pollinators might be offset by the risk that the pollinators would not be deceived. Clearly, we need more information than is currently available on the energy costs of providing rewards.

The suggestion has been made (Gentry, 1974) that some vines of the Bignoniaceae with hermaphrodite flowers, such as species of *Cydista, Clytostoma,* and *Phryganocydia,* practice deception, offering no nectar reward but mimicking other species that do. These vines flower repeatedly in short bursts, and the assumption is made that they take advantage of naive bees that are learning the hard way which flowers are nectar sources and which are not. But until such bees are actually seen foraging for the nonexistent nectar at these flowers, this remains an assumption; the bees may be rewarded with pollen, or the flowers may be self-pollinated.

A recommendation

The necessity for speculation rather than deduction from hard facts, particularly in relation to visitor deception, emphasizes that even when we know something about the pollination requirements of an individual tree or liane, we know very little about the interactions between individuals of the same and different species of flowering plants in acquiring the services of pollinators. The need for quantitative information of this sort linking the biotic components of tropical ecosystems is outstanding.

In order that some progress may be made toward remedying such deficiencies, I suggest that, in future, no autecologic study, let alone any synecologic investigation, should be considered satisfactorily accomplished if it does not include some information on the pollination biology, breeding systems, seed dispersal mechanisms, seed germination requirements, and seedling establishment peculiarities of the plants concerned. The roles of animals at all stages of the reproductive processes of the plants need to be elucidated.

This reproductive information is more difficult to acquire than details of the vegetative morphology of plants, but it is no less important because it is concerned with the *fitness* of the plants: the extent to which they contribute, generation by generation, to the plant life of an area. Fortunately, a growing number of students of tropical biology are carrying out multidisciplinary and interdisciplinary studies that will throw much light on the mysteries, and solid advances may be expected in the next few years.

Acknowledgments

Most of the nectar samples utilized in this work were collected by Paul A. Opler, Gordon W. Frankie, Robert W. Cruden, and Spencer C. H. Barrett. The laboratory tests and measurements were devised and carried out by Irene Baker. The fieldwork was made possible by grants from the National Science Foundation (GB-25592; GB-25592 A#2; BMS 73 01619 A01), administered by the Organization for Tropical Studies. I am extremely grateful to all these persons and organizations.

General discussion

Givnish: Dr. Baker, one of the patterns you described included a decrease in the amount of amino acids in flower nectar with increasing height in the forest. Could this possibly be due to the fact that the pollinators at greater height, having alternative sources of protein and amino acids, may not require them in nectar? Could you relate this to your statement that hummingbird- and bat-pollinated flowers are relatively weak in sugar solutions by determining whether the increase in sucrose content and decrease in amino acids with height occur in the same plants?

Baker: I didn't give actual results for amino acids, but our studies have concerned the lower levels and the upper levels of the same trees. The inverse relation with height also holds when species are grouped according to their occurrence at different levels in the forest, and it further holds when herbaceous, shrubby, and tree species that are members of the same family are compared: There is a decrease in the amino acid concentrations as you go from the herbaceous layer to the tree layer. This seems to be something that is inherent in flowering plants, and I think it is related to more than just the poliinators. A suggestion made in discussion outside this room is that it might be related to the fact that the nitrogen used in the manufacture of amino acids comes from the roots, and there may be, therefore, some siphoning off of the materials for making amino acids in the lower branches of the trees. Thus, analyses of phloem and xylem sap at different levels would be of interest.

Bawa: You said there are no apparent differences between male and female flowers with regard to the complements of amino acids. What about the amounts of amino acid; do you have any data on that?

Baker: There is no obvious distinction between staminate and pistillate flowers as far as either concentration or actual complements of amino acids are concerned. This is strikingly different from the picture with sugars, where the nectars sometimes varied considerably in constitution from staminate to pistillate flowers.

Lee: In the case of diurnal fluctuations in output of nectaries, were there ever instances where you found a higher output of fluid from the nectaries (with higher concentrations) during the day as against the normal sort of sequence, which one would suspect is much greater output during the night – both in terms of quantity of solution and altered concentrations of sugars?

Baker: The data we have for flowers that produce nectar both day and night are largely of temperate origin, and they show that the same flower will produce a weaker nectar at night than during the daytime. I haven't found any exceptions to that picture yet. Most flowers produce nectar during the day rather than at night, except those that are pollinated by bats or moths.

Policansky: Is the fact that there is more sugar in the nectar at the top of the tree possibly related to the fact that there are fewer amino acids or a lower concentration from some chemical point of view?

Baker: I'd have to defer to a real biochemist to answer that question, but I have a suspicion that the two things are positively rather than negatively correlated in most cases. Normally, when you get a nectar that is stronger in sugar, you get a nectar that is stronger in amino acids. It is only in the cases of the height relation that the correlations are reversed.

Janzen: I can't resist noting that an underlying assumption of the suggestion given to Dr. Baker – the question asked by Dr. Lee and many other questions related to nectar – is that a nectary is a sort of a leak in the plant and, therefore, the content of what is leaked is a reflection of the milieu in the general vicinity. I argue that the burden of proof of this assumption lies with those who believe it, because there is a large body of circumstantial evidence and also of reason that supports the idea that a nectary is, in fact, an adaptive structure of the plant. Therefore, the contents of its output should have been molded by various selective processes in the past. If someone believes that it is to some degree a leak, I would like to see data supporting the idea.

Zimmermann: Nectaries are certainly not mere leaks, as the ample literature on fine structure of secretory tissues and physiology of secretion indicates (see Lüttge & Schnepf, 1976). On the other hand, the nature of nectar will certainly reflect the nature of its source, the phloem. There are different kinds of nectaries. Some of them are actively secreting. They can be cut off the plant and floated on a sugar solution, from which they will selectively pump sugars (e.g., Frey-Wyssling et al., 1954). But there are also passive nectaries.

Dr. Baker, you described the situation from the point of view of the insects and nectar-collecting animals. A plant physiologist studies nectar secretion from the point of view of the plant, regardless of what insects are attracted.

Baker: I am aware that nectaries vary in construction and probably in the methods of release of the nectar (which is an active process), and it will be most interesting to see if any correlations can be found between nectary construction and functioning, on the one hand, and nectar constitution, on the other. I have already pointed to differences in constitution of floral and extrafloral nectars from the same plant.

I have been aware that, even as late as the 1960s, there were plant physiologists who regarded nectar as an "excretion" by the plant and for whom the utilization of nectar by animals, especially insects, was of minor interest. By the opposite token, entomologists think that we look at things too much from the plant point of view. If *both* criticize me, I feel I must have struck something like an appropriate balance.

References

Baker, H. G. (1973). Evolutionary relationships between flowering plants and animals in American and African forests. In *Tropical Forest Ecosystems in Africa and South America: A Comparative Review*, ed. B. J. Meggers, E. S. Ayensu, & W. D. Duckworth, pp. 145–59 (Chap. 11). Washington, D.C.: Smithsonian Institution Press.

– (1975). Sugar concentrations in nectars from hummingbird flowers. *Biotropica, 7*, 37–41.

– (1976). "Mistake" pollination as a reproductive system with special reference to the Caricaceae. In *Variation, Breeding and Conservation of Tropical Forest Trees*, ed. J. Burley & B. T. Styles, pp. 161–9. London: Academic Press.

Baker, H. G., & Baker, I. (1973a). Amino acids in nectar and their evolutionary significance. *Nature Lond., 241*, 543–5.

– (1973b). Some autecological aspects of the evolution of nectar-producing

flowers, particularly amino acid production in nectar. In *Taxonomy and Ecology*, ed. V. H. Heywood, pp. 243–64 (Chap. 12). London: Academic Press.
- (1975). Nectar-constitution and pollinator-plant coevolution. In *Coevolution of Animals and Plants*, ed. L. E. Gilbert & P. H. Raven, pp. 100–40. Austin: University of Texas Press.
Baker, H. G., Baker, I., & Opler, P. A. (1973). Stigmatic exudates and pollination. In *Pollination and Dispersal*, ed. N. B. M. Brantjes, pp. 47–60. Nijmegen: University of Nijmegen.
Baker, H. G., & Harris, B. J. (1957). The pollination of *Parkia* by bats and its attendant evolutionary problems. *Evolution, 11*, 449–60.
Baker, H. G., & Hurd, P. D. (1968). Intrafloral ecology. *Annu. Rev. Entomol.*, *13*, 385–414.
Baker, I., & Baker, H. G. (1976). Analyses of amino acids in flower nectars of hybrids and their parents, with phylogenetic implications. *New Phytol.*, *76*, 87–98.
Dodson, C. H. (1975). Coevolution of orchids and bees. In *Coevolution of Animals and Plants*, ed. L. E. Gilbert & P. H. Raven, pp. 41–9. Austin: University of Texas Press.
Faegri, K., & van der Pijl, L. (1971). *The Principles of Pollination Ecology*, 2nd ed. Oxford: Pergamon Press.
Frey-Wyssling, A., Zimmermann, M., & Maurizio, A. (1954). Ueber den enzymatischen Zuckerumbau in Nektarien. *Experientia, 10*, 491.
Gentry, A. H. (1974). Coevolutionary patterns in Central American Bignoniaceae. *Ann. Missouri Bot. Gard., 61*, 728–59.
Harcombe, P., & Riggins, R. (1968). Observations on the pollination biology of *Swartzia simplex* (Swartz) Spreng. Research project, O.T.S. Advanced Course in Reproductive Biology of Tropical Plants (mimeographed). San José, Costa Rica: Organization for Tropical Studies.
Heinrich, B., & Raven, P. H. (1972). Energetics and pollination ecology. *Science, Wash., 176*, 597–602.
Heithaus, E. R., Fleming, T. H., & Opler, P. A. (1975). Foraging patterns and resource utilization in seven species of bats in a seasonal tropical forest. *Ecology, 56*, 841–54.
Heithaus, E. R., Opler, P. A., & Baker, H. G. (1974). Bat activity and pollination of *Bauhinia pauletia:* Plant-pollinator coevolution. *Ecology, 55*, 412–9.
Hodgman, C. D., Weast, R. C., & Selby, S. M. (1958). *Handbook of Chemistry and Physics*, 40th ed., p. 2168. Cleveland: Chemical Rubber.
Howell, D. J. (1974). Bats and pollen: Physiological aspects of the syndrome of chiropterophily. *Comp. Biochem. Physiol., 48A*, 263–76.
Lüttge, U., & Schnepf, E. (1976). Elimination processes by glands: Organic substances. In *Transport in Plants IIB, Tissues and Organs*, ed. U. Lüttge & M. G. Pitman, pp. 244–77. Heidelberg: Springer-Verlag.
Meeuse, B. J. D. (1961). *The Story of Pollination*. New York: Ronald Press.
Opler, P. A., Baker, H. G., & Frankie, G. W. (1975). Reproductive biology of some Costa Rican *Cordia* species (Boraginaceae). *Biotropica, 7*, 234–47.
Percival, M. S. (1961). Types of nectar in angiosperms. *New Phytol., 60*, 235–81.
- (1965). *Floral Biology*. Oxford: Pergamon Press.
Proctor, M., & Yeo, P. (1973). *The Pollination of Flowers*. London: Collins.
Purseglove, J. W. (1968). *Tropical Crops: Dicotyledons*, vol. 2. London: Longmans.

Rehr, S. S., Bell, E. A., Janzen, D. H., & Feeny, P. P. (1973). Insecticidal amino acids in legume seeds. *Biochem. Systematics, 1*, 63–7.

Scott-Elliott, G. F. (1889). Ornithophilous flowers in South Africa, *Ann. Bot. Lond., 4*, 265–80.

Shiraishi, A., & Kuwabara, M. (1970). The effect of amino acids on the labellar hair chemosensory cells of the fly. *J. Gen. Physiol., 56*, 768–82.

Stanley, R. G., & Linskens, H. F. (1975). *Pollen: Biology, Biochemistry, Management.* Berlin: Springer-Verlag.

Vogel, S. (1966). Scent organs of orchid flowers and their relation to insect pollination. In *Proceedings, Fifth World Orchid Conference,* Long Beach, California, pp. 253–9.

– (1969). Flowers offering fatty oil instead of nectar. In *Abstracts, XI International Botanical Congress, Seattle,* p. 229.

– (1971). Oelproduzierende Blumen, die durch ölsammelnde Bienen bestäubt werden. *Naturwissenschaften, 58,* 58.

– (1974). Oelblumen und ölsammelnde Bienen. *Tropische Subtropische Pflanzenwelt, 7,* 285–547.

Waller, G. D. (1973). Chemical differences between nectar of onion and competing plant species and probable effects upon attractiveness to pollinators. Ph.D. thesis, Utah State University.

4

Seeding patterns of tropical trees

DANIEL H. JANZEN
Department of Biology, University of Pennsylvania,
Philadelphia, Pennsylvania

Individual tropical trees display seeding (fruiting) patterns ranging from nearly continuous throughout the year (e.g., *Ficus* spp. on the small Caribbean island of San Andres: Ramirez, 1970) to intracohort synchrony at 120-year intervals (e.g., the bamboo, *Phyllostachys bambusoides:* Janzen, 1976). Between these extremes lies a multitude of patterns of seed production. A prominent division among tropical tree species is whether an individual bears seeds every year or skips years between large seed crops. My goal is to explore the possible adaptive significance of some of these patterns, with special emphasis on the role animals may have had in their evolution. The question is: What determines how long a tropical tree should wait between reproductive events?

Even when well documented (rarely the case), the interpretation of tropical tree seeding and flowering patterns is no easy task. A tree that flowers but does not set fruit is not necessarily abnormal or behaving in a maladaptive manner; the "sterile" flowering tree is likely to be simply acting as a male that year. Abortion of flowers and fruits does not necessarily result from a failure in pollination, but may be the outcome of choice of parentage by the female genome (Janzen, 1977) or the discarding of flowers produced only for pollen or for pollinator attraction. The adaptive significance of a particular timing of seed production can relate, at the least, to pollinator activity, fruit and seed developmental rates (adjusted and absolute), dispersal agent activity, seed predator behavior, resource allocation options within the plant, and germination demands by the seedlings. All these things may in turn be related to the direct impact and cueing effects of changes in the physical environment. Even a change in the physical environment can be subject to natural selection; natural selection may drive one tree species to respond to a dry spell as a seeding cue, another to respond to the same dry spell as a traumatic event and abort its seeds, and a third to show no response. Finally, it is technologically very difficult to know in any one year if a particular tree that does not flower and/or seed so behaves because it did not receive an appropriate cue, did not have enough reserves to allow it to respond to a cue, had its flowers destroyed by "bad weather" or herbivores, did not receive appropriate pollen, or is involved in some internal counting scheme that is insensitive to external events.

For this to be a definitive essay, I would need clear answers to at least the following questions about a series of different tropical tree species in the same habitat:

1. How does the size of the seed crop within an individual crown vary with the number of years between seed crops?
2. How does the intensity of seed predation vary with the size of a seed crop and the duration of its availability on an individual tree crown?
3. What are the consequences of synchronization of an individual tree's seed crop with those of neighbors (conspecifics and otherwise) that share seed predators?
4. How does skipping 1, 2, 3, . . . *n* years between an individual's seed crops affect its seed predators?

For those species of trees whose seeds are dispersed by animals (i.e., satiable and competitively distractible dispersal agents: McKey, 1975) the same four questions are important with respect to dispersal agents. Dispersal agents are of utmost importance with respect to the conspicuous conflict between producing enough seed to satiate seed predators (or seedling predators at a later date) and producing so much seed that dispersal agents do not remove it from the vicinity of the parent tree. It is not the purpose of this chapter, however, to delve deeply into the interaction between dispersal agents and seed crops.

The above four questions have not been answered for any tropical flora, though I am currently attempting to provide them for the deciduous lowland tropical forest of Parque Nacional Santa Rosa, in northern Guanacaste Province, Costa Rica.

Therefore, I beg the indulgence of the reader in allowing me to rely heavily on as yet incompletely documented information from my own studies in the dry Pacific lowlands of Costa Rica. This area includes the northern end of Puntarenas Province and most of Guanacaste Province. At one time it was clothed in largely deciduous forest sparsely crossed by evergreen riparian forest along the rivers from the adjacent volcano slopes to the east. At present the general study area is primarily pasture and croplands. There are small forest patches of a few hectares, scattered fencerow and pasture trees, and one large patch of forest in the extreme northwest corner, which contains Parque Nacional Santa Rosa. Since the early 1960s, the general study area has been the site of a number of ecologic studies by investigators associated to some degree with the Organization for Tropical Studies and largely funded by the U.S. National Science Foundation (e.g., Daubenmire, 1972; Wilson & Janzen, 1972; Frankie, Baker, & Opler, 1974; Bawa, 1974; Fleming, 1974; Heithaus, Opler, & Baker, 1974; Heithaus, Fleming, & Opler, 1975; Janzen, 1967, 1970a,b, 1971a,b,c, 1972, 1973, 1975a,b,c; Janzen & Wilson, 1974; Frankie & Baker, 1974; Turner, 1975; Opler, Baker, & Frankie, 1975). This chapter is not intended to be a definitive review of the tropical sexual phenologic literature, but rather an essay about the problems and rewards of such studies.

Methodology
Understanding the seeding patterns of tropical trees is fraught with methodologic difficulties that need special mention.

Gathering and recording data
Statements such as "this was a good seeding year for species X" have been useful in the past in alerting us to the possible existence of seeding and flowering pattern differences among species, but have largely outlived their usefulness. We now know that such differences exist. We need unbroken records of many years' duration of flowering and seeding by many *marked* individuals in the same population. We need such records for several tens or hundreds of individuals of one species under a multitude of circumstances, rather than such records on a few individuals of many species.

Individuals must be tagged and their locations mapped in some unambiguous way. All phenologists have had the experience of discovering that an adjacent unmarked tree is the desired species as well, when it suddenly flowers 3 years after the marked one did. Unbroken records of data will be taken by many different people working sequentially and must be accurately kept. It would be a tragedy if what was believed to be tree 96 was gradually replaced in the records by a juvenile maturing next to it, or if tree 37 was stolen by poachers and replaced in the records by a nearby individual of similar size.

Recognizing flowering and fruiting is difficult in many species unless specimens are watched closely or unless the flowering and fruiting structures persist unambiguously through the census period. The natural history of the species for many types of individuals (e.g., subadults, competitively disadvantaged, diseased) must be well known. One cannot assume that flowering means fruiting or that a large flower crop means a large fruit crop. Long-term records are probably easiest to obtain for species whose inflorescences or infructescences persist for many months, and these species should probably receive priority in setting up and recording where a variety of only partly experienced or interested individuals will be recording the data over the years (such as in a park or field station).

It is imperative that the records be to some degree quantitative. If only one branch flowers in the crown that year, that is what should be recorded; let the data analyst worry about how to compare this with a half-sized tree that flowered over the entire crown. Flower and fruit crops should be numerically approximated wherever possible. A crop of 100 fruits may be extremely heavy for a small tree, yet a trivial event for a large one of the same species. Therefore, it is also important that the size and apparent competitive circumstance of the crown of each recorded individual be noted at supra-annual intervals. Other tree size parameters such as diameter at breast height (DBH) and height should be recorded, but are probably of less direct importance.

Age and size of trees

It is conspicuous that subadult tropical trees sometimes have seeding patterns different from those of adult trees. For example, *Hymenaea courbaril* (in the general study area) with a DBH between 6 and 14 in. often flower, but do not produce more than 10 fruits (in the rare cases where fruits are produced at all). Field data from Parque Nacional Santa Rosa suggest that it is commonplace for hermaphroditic species of trees to act only as males when subadult in size. I suspect that as *H. courbaril* attains adult competitive status in the canopy, the intervals between large seed crops become gradually shorter. The transition from a no-fruit subadult to a fully developed adult with large fruit crops at regular long intervals is gradual rather than abrupt. In some varieties of mango (*Mangifera indica*), the young trees produce annual small crops of fruit, but as they age, begin to skip years between larger fruit crops (Singh, 1960).

Subadult trees are often smaller and, therefore, more easily seen by the investigator than adults, and thus flowering and seeding records taken from them may be unrepresentative of the breeding population. Furthermore, sexually mature subadult trees are more common along edges, roadsides, etc., and, being easily seen, tend to produce subjective impressions of reproductive activity out of proportion to their abundance in the population as a whole.

Injured trees

There may be strong selection for a lethally wounded or diseased tree to expend its remaining reserves on a final flower or seed crop. For example, I found that Mexican *Acacia cornigera* subject to very heavy defoliation (owing to the removal of their protective ant colony) often produce a flower crop, and on rare occasions, a few seeds, before dying (unpubl. field notes, Veracruz and Oaxaca, 1963–4). A large adult *Pterocarpus rohrii*, broken off at its stump by wind, dropped its leaves and went into full flower 2 months after the usual flowering time and even set several dozen apparently normal fruits (May 8–14, 1976; Parque Nacional Santa Rosa). Such injured plants often flower and fruit out of phase with conspecifics and are thus handy sources of sexual material for plant collectors; though such specimens are undoubtedly useful to taxonomists, they occur with sufficiently high frequency in herbaria as to render most herbarium material useless for the determination of seeding patterns and phenology.

However, in addition to the lethally injured individuals, trees may go through disease and competitive bouts where they are temporarily weak. Such trees often flower but set no seed, and are thus acting as males, though they may be morphologically hermaphroditic. Presumably, they are thus reproducing, but at a lower cost (and probably success) per reproductive season. Combined with subadult trees that also are acting as males, injured or competitively disadvantaged trees can occur in such numbers as to give the illusion of a "bad year for seed" almost every year.

Location of trees

It is commonplace for tropical trees in forest conditions (i.e., under severe intercrown competition) to display different seeding patterns from those of trees left in the open after forest clearing or of trees grown in the open. The differences are of several sorts. For example, *Hymenaea courbaril* trees in the Guanacaste lowlands growing in full sun generally bear large seed crops more frequently than do those in forest. Fully insolated *H. courbaril* also seem to display a larger range of seed crop sizes, less complete (accurate?) suppression of fruit production in an off year, and greater variation in the period between large seed crops than do their counterparts in nearby forest. Phenologic records taken from plantation trees in the tropics may in like manner be unrepresentative. Insolated trees are, of course, a kind of experiment that is useful in understanding the regulatory mechanisms of periodic seeding but, unless forest trees are used for comparison, tells us next to nothing about seeding patterns.

Flowering vs. fruiting

It is a common mistake to equate flowering with fruiting in tropical phenologic records. In the older bamboo literature (e.g., Janzen, 1976), flowering and fruiting records are hopelessly confused, rendering much of the "data" useless. There are two problems.

First, different species of trees have vastly different periods between the flowering event and the fruiting event. In Costa Rican deciduous forest, many trees delay fruit maturation for 6 months to a year, with the result that flowers from this year's season are often on the tree at the same time as the fruits from last year's season (e.g., *Enterolobium cyclocarpum*, *Pithecellobium saman*). A tree may, therefore, set no fruit in a given flowering season, the apparent full complement of fruit belonging to a previous season. The strong interspecific differences in fruit maturation time also mean that flowering times are worthless for determining the month of fruit development unless fruit development time is also known. Fruiting synchrony (or lack thereof) is by no means a mandatory physiologic consequence of flowering synchrony (or lack thereof).

Second, a morphologically hermaphroditic tree that flowers but does not fruit has not failed in any sense. In many tree individuals, the entire or partial adaptive significance of flowering is the pollination of other trees, and thus the number of fruit set is no measure of success of the sexual event. Furthermore, the adaptive goal in pollination is not pollen flow, but gene flow. A plant may settle for second-rate pollen to produce a full-size crop of fruit, yet not have been as efficiently or successfully pollinated as another tree with the same fruit/flower ratio or fruit crop size that obtained pollen of a higher genetic quality (Janzen, 1977). Even abortion of part of the seeds in a multiseeded fruit may be the result of something other than pollination failure (i.e., may be a physiologic device to increase the fruit/seed ratio within the fruit and thereby increase the reward to the dispersal agent).

Small flower crops

Small flower crops commonly occur on a large crown, or a normal-sized flower crop appears on one branch only. Such flowerings (and associated fruitings) are variously ignored, viewed as abnormal, or stressed, in accordance with the interests and perceptiveness of the investigator. Such crops are not necessarily failures of the plant's physiologic processes, but may well be part of some adaptive pattern of seed production. If an individual tree has a large seed or flower crop at one time of year and a small seed or flower crop at another, this may be a highly adaptive pattern of seed allocation or seed parentage during the year. On the other hand, there are many cases where the branches of a tropical tree become nonsynchronous and produce flowers or fruits at a time or in quantities such that they will be neither pollinated nor dispersed.

Individuals vs. population

It is particularly difficult to establish population seed patterns from data on a single tree, because different individuals in a population normally behave differently. Latin binomials do not seed; individuals do. As most of man's interest in seeding patterns has been concentrated on harvesting the seed, he has generally ignored the question of *which* trees produce the seed. However, to understand the adaptive value of seeding patterns, it is imperative that we obtain information about individuals and thereupon build a population view.

An individual in fruit does not mean a population in fruit (or flower). Long-term records of individual trees and their neighbors are needed. When it is a seeding year for a given species, commonly only a fraction of the individuals in the population set seed. Furthermore, when the next seed year comes around, the same individual trees may not necessarily produce the next big population-level seed crop. If someone says that a species of tree has two seeding peaks per year, does this mean the same individuals or different ones? The answer is vital for understanding the evolution and ecology of seeding patterns.

Geographic variation

Seed production by a species of tropical tree is subject to a good deal of between-habitat variation, often inconsistent from year to year. It is common for conspecific populations only a few miles apart to be nonsynchronous within the year. Careful documentation of the actual site of the trees in seed or flower becomes imperative for the data to have any meaning. A report that "flowering occurred all over the Googong hills" is a frustratingly useless piece of information.

Terminology
Predator satiation

This phenomenon occurs when more seeds (or seedlings) are produced at one time and place than the seed predators that arrive can eat (or oviposit on and thereby cause to be eaten). If a species of animal can prey on the seeds of a species of tree, if at least one such animal gets to the seed crop, and if some seeds survive, then there was predator satiation. As in the case of *Eucalyptus regnans*, described by Evans (1976), satiation of seed and seedling predators may occur only in certain years or by certain individuals. It is probable that with many adult individuals of many species of trees, for many, if not all, of their reproductive years, the set of organisms feeding on ovules, seeds, and young seedlings is not satiated, and, therefore, the trees do not reproduce.

Escape by seed predator satiation may be conceptually distinct from (1) escape from seed predators by seeding at a season when the seed predators are not or cannot be present, (2) escape by having the seeds or seed crops so far apart in space or time that they are not found by seed predators, and (3) escape by having the individual seed so well protected chemically or morphologically that it cannot be eaten by a potential seed predator. However, all these escape mechanisms may be operating simultaneously within one tree's seed crop and involve the same organisms. All these escape mechanisms are manipulated by the tree directly through the pattern of appearance of seeds (pattern of seed maturation) and indirectly through the pattern of removal of seeds by dispersal agents (at a cost of fruits or seeds).

In the simplest case of seed predator satiation, the number of seed predators to arrive at an individual seed crop is independent of the size of the seed crop and its rate of removal by dispersal agents. Here, the larger the crop and the faster it is removed by dispersers, the greater the percent of seeds that survive. However, such a pattern can be easily complicated within the tree population; certain individuals may lose all their seeds while others satiate the predators by differential attractiveness of very small and very large seed crops to seed predators and dispersal agents. Furthermore, there are seed predators that selectively seek very large seed crops (e.g., nomadic passenger pigeons, jungle fowl, Malaysian bearded pigs). Under the threat of such animals, the tree may produce the most surviving seeds by having either a seed crop of a size just below the level of interest of these animals or a huge crop that satiates the nomadic as well as the local arrivals.

Predator satiation probably comes about most commonly through selection for synchronization of conspecifics through some environmental cue perceived in common, through selection for physiologic mechanisms that reduce the spread of the seeding time around the mode for individual crowns and populations, and through selection for more rapid removal of seeds by dispersal agents. The most spectacular tropical examples are the Southeast Asian dipterocarps (Janzen, 1974), bamboos, and *Strobilanthes* (Janzen, 1976); however, on a smaller scale, the timing and size of virtually every individual tree's seed and seedling crop are probably involved in some degree of predator satiation.

Mast fruiting

This refers to the phenomenon whereby the population(s) of trees in an area are synchronized on a supra-annual schedule, so that a tree in fruit is likely to be fruiting in the same year as its conspecifics. Mast fruiting may be achieved either by response to an external cue (e.g., Dipterocarpaceae, beeches, oaks) or by an internal calendar that counts from the year of germination (e.g., bamboo, *Strobilanthes*). In various older studies, mast fruiting has been occasionally referred to as "gregarious fruiting" (or "gregarious flowering"), but I avoid this term because it implies some sort of conscious response by the trees to each other: mast fruiting trees do not fruit in response to each other (unless there is some as yet undiscovered pheromone system in operation), but rather in response to a cue or to a personal clock. They are kept on schedule, not by reacting to each other, but by removal of the deviants by natural selection. The term "mast fruiting" derives from the common name for beech and oak mast, the large numbers of beech or oak seeds found on the forest floor in a mast year in extratropical deciduous forests. It also applies to the supra-annual synchronized production of large conifer seed crops at northern latitudes. It seems that virtually all the large expanses of extratropical conifers and hardwoods display mast fruiting at the habitat level, and all the evidence indicates that this is the result of coevolution with seed predators (Janzen, 1971b).

The time between mast crops may be referred to as the "intermast period" and, of course, is the subject of natural selection just as is the size of the seed crop produced in a mast crop.

Iteroparous and semelparous

An "iteroparous organism" is one that has more than one distinct reproductive bout during its life span; a "semelparous organism" is one that does all its reproduction in one bout (Cole, 1954). These terms are widely used by zoologists, but botanists tend to replace them with "polycarpic" and "monocarpic."

Designation of a perennial plant as semelparous requires some caution. Many individuals of *Agave*, for example, are not strictly semelparous because they produce plantlets on the inflorescences capable of vegetative extension of the adult. In one sense such species of *Agave* (and apparently semelparous bamboo) are in fact iteroparous plants displaying extreme developmental polymorphism, a polymorphism that is probably selected for through the advantage of converting almost all reserves into seed (or bulbil) production at long intervals.

Annual fruiting

Perhaps the most uncomplicated fruiting pattern displayed by individual tropical trees is for each healthy adult female or hermaphrodite in the population to bear a relatively similar-sized crop each year during some kind of fruiting season. Year-to-year variation in the size of an individual's fruit crop is then largely a function of the vicissitudes of weather, action of herbivores on vegetative parts, and pollinator activity. Annual fruiting, as it seems to be a response to a seasonal cue, requires minimal storage of resources (e.g., minimal sapwood parenchyma), minimal decisions about allocation of resources, minimal evaluation of the amount of resources stored, and no ability to count years. The important adaptive modification is in the rates of development in the flower-to-seed progression. These rates are the mechanism for achieving the most adaptive distribution of seed maturation times within the tree crown. They are also the mechanism whereby the seeding distribution in one crown is adjusted to bear the best temporal relation to the seasonal cycle and to the seeding distributions of neighboring trees (through the medium of seasonal timing). However, even with something as simple as annual fruiting, the tree has to arrive at an optimal distribution of resources among the conflicting demands of vegetative growth, insurance, seed size, seed number, flower number, pollen quantity, abortion option, etc.

Examples

Well-documented examples of annually fruiting species of tropical trees are surprisingly rare. A population of supra-annually asynchronous biennial bearers can easily give the illusion of being a species with annually bearing individuals. In heavily disturbed forest, many biennially bearing individuals freed from crown competition display abnormal annual fruiting, making the deception worse. In the phenologic study by Frankie et al. (1974) of the Guanacaste lowlands, trees are often noted to flower, but no information is given about whether a fruit crop followed from all or any of those individuals.

Detailed records of individuals will probably show that healthy female or hermaphrodite adults of at least the following species of trees (or woody vines) in the Guanacaste lowlands will bear fruits in most years: *Acacia collinsii, Albizia caribaea, Anacardium excelsum, Apeiba tibourbou, Bauhinia pauletia, Bauhinia ungulata, Bombacopsis* quinata, Bursera simaruba, Caesalpinia eriostachys, Calycophyllum candidissimum, Cecropia peltata, Chlorophora tinctoria, Cochlospermum vitifolium, Cordia alliodora, Enterolobium cyclocarpum, Genipa caruto, Gliricidia sepium, Godmania aesculifolia, Guazuma ulmifolia, Hemiangium excelsum, Jacquinia pungens, Lonchocarpus costaricensis, Lonchocarpus minimiflorus, Luehea candida, Muntingia calabura, Ochroma pyramidale, Parkinsonia aculeata, Pithecellobium saman, Plumeria rubra, Sterculia apetala, Swietenia humilis, Tabebuia neochrysantha, Tabebuia rosea, Thevetia ovata, Triplaris americana. Opler et al. (1975) state that all individuals of Cordia alliodora in the Guanacaste lowlands flower each year, but do not give the detailed basis for this statement. I offer this qualified and only subjectively documented list merely as substantiation of the existence of species with annually fruiting individuals, and I would not be surprised to discover that a number of the species listed above in fact have individuals that do not fruit on an annual basis.

There are apparent documentations of species with annually seeding individuals in the literature, but one should view all of them with healthy skepticism. Phillips (1927) states that the South African Platylophus trifoliatus and Cunonia capensis flower every year (the implication is that individuals do, but no detailed documentation is given). However, he also points out that more than 90% of the seeds seem to be inviable, and one wonders if his observations were in a normal year. He says of Olinia cymosa that "flowers are produced very irregularly. Some trees bear [what?] every year, others every second, third, or fourth year, but full fruit-crops occur every third or fourth year only." The only information that can be extracted from the cryptic statement is that some individuals skip years between flower(?) crops. Medway's (1972) records are much better, but deal almost entirely with only one member of a population. At least one can say with some assurance that populations of the following Malaysian species may contain some individuals that fruit annually or more often: Xylopia stenopetala, Bhesa robusta, Erythroxylum cuneatum, Diospyros maingayi, Litsea rostrata, Parkia speciosa, Ficus ruginervia, and Ficus sumatrana. However, even these individuals occasionally skipped a year. The eight species listed above constitute only 18% of the species chosen for Medway's study without apparent previous knowledge of their fruiting patterns. A single old tree of Acacia nilotica in Sudan bore fruit for 4 consecutive years (Khan, 1970).

From this I suspect that when tropical forest sexual phenology is better understood, tree species with individuals that bear a fruit crop every year will definitely be in the minority. The primary cause may well be the advantage of bearing a fruit crop larger than can be produced with only 1 year's photosynthate, and the primary driving force may be interactions with dispersal agents, pollinators, and seed predators.

Interactions with animals
No predispersal predation

There appear to be two somewhat different selection regimens associated with annual fruiting by individuals in the tropics. There are species in which predispersal seed predation does not occur (or is trivial). In this case, the variance of the seeding distribution within an individual's crown, and the temporal placement of that distribution with respect to neighbors, are largely selected for by the activities of the dispersal agents and pollinators and the demands of postdispersal seed predation. Here, intrapopulation seeding synchrony may even be a totally accidental outcome of selection to hit a physical environmental window, such as having the seeds on the ground at the beginning of the rainy season. With such trees, vertebrate dispersal agents may select for very attenuated seeding distributions by individuals and for strong asychrony among individuals. Potential candidates in the Guanacaste lowlands are *Muntingia calabura, Cecropia peltata,* and *Trema micrantha.* Some of the species of Trinidadian Melastomataceae discussed by Snow (1965) probably fall into this category when they grow in communities less rich in melastome species. As Snow's study suggests, however, the dispersal agents may also select for quite low variance of the seeding distributions of the individuals and within populations of such species.

If there are no predispersal predators, and if dispersal is by inanimate forces, the picture is somewhat different. As McKey (1975) has emphasized, such trees are not likely to compete for dispersal agents. The wind is unlikely to be satiated. The seeding distributions of individuals may thus be only a by-product of demands for timing of when the seed should hit the ground. For example, in my study area *Ochroma pyramidale, Cochlospermum vitifolium,* and *Bombacopsis quinata* appear to be species with annually seeding individuals, wind-dispersed seeds, and no predispersal seed predation. However, once on the ground, the seeds are fed on by birds, rodents, and sucking bugs (Lygaeidae, Pyrrhocoridae). The seeds germinate when the rains come, and thus there would appear to be strong selection for timing that places the seeds on the ground before the rains arrive and minimizes the amount of time seeds spend on the ground during the dry season. Such selection should produce strong intrapopulation and intracrown seeding synchrony, but only as a side effect of selection for traits unrelated to other individuals.

Predispersal seed predation

There are species whose annually seeding individuals sustain substantial predispersal seed predation. Their seeding distributions and their relations to those of neighbors should be determined by predator satiation and escape in time and space provided by dispersal agents, as well as by the various factors mentioned in the previous paragraphs. The seeding distributions of such trees may take many forms, but frequently the individual seeding distributions are quite narrow. This means that the period during which the developing fruits or seeds are susceptible to predation is minimized. Likewise, and presumably for the same reason, synchronization of fruit and seed maturation within the population is well developed. Finally, the population seeding peaks of the different species are at a variety of places during the total season in which seeding could potentially occur. Presumably the location of peaks within that season is strongly influenced by the unfavorableness of that timing for the seed predators. Interspecific synchronization might be developed by generalist seed predators in habitats containing such species of trees, but in the Guanacaste lowlands the major insect seed predators are so host-specific that satiation of their predation capacity generally does not take place at the interspecific level, and thus they will not select for interspecific synchronization. A few examples of trees in this category in the Guanacaste lowlands are *Guazuma ulmifolia, Albizia caribaea, Cordia alliodora, Bauhinia pauletia,* and *Pithecellobium saman.*

How do seed predators respond to annually fruiting individuals? As far as the generalists are concerned (mostly vertebrates), the story is probably uncomplicated. When a herd of peccaries encounters a *Guazuma ulmifolia* fruit crop on the ground (laid out by the tree to be dispersed), it eats the fruit. That the same tree had a seed crop the year before probably means nothing specific to that herd of peccaries. If a mutant individual *G. ulmifolia* appeared that produced a crop every other year, it would probably not lower the total amount of food available in the herd's foraging range enough to lower the overall seed predation on that individual tree through reduction in the size of the peccary herd. This is not to say, of course, that if the population or individuals of *G. ulmifolia* were on a 2-year fruiting cycle the peccary herd would not suffer. It would, because biennial bearing would lead to a more intense fluctuation of the peccaries' food – a fluctuation they might not be able to overcome through nomadism and fat storage, which, even if effective, costs something.

Annual seeding *may* be of great importance to specialist seed predators. If the general habitat of the host tree is favorable to waiting out the period between seed crops, there can be a resident population of seed predators that simply wait in the vicinity of the parent tree, genetically "knowing" that there will be a new seed crop at their birthplace next year. For the seed predators to disperse from their birthplace would be adaptive only in the context of avoiding competition for seeds too heavily attacked and in the context of locating unexploited seed crops (e.g., new reproductive individual adults and adults that suffered local extinction of their seed predators).

The omnipresence and predictability of an annual seed crop on each reproductive female or hermaphrodite of the tree species should result in the maximum possible densities (and seed predation) of specialist seed predators that the rest of the environment will support (e.g., parasites, dry seasons, intertree distances). However, there will be cases where the physical environment is harsh enough during part of the year so that the beetles have a higher fitness by migrating to a local moist creekbed, for example, in which to pass the end of the dry season. Here, then, there will be no local population of specialist seed predators associated with each individual parent tree. The seed crop must be relocated each year by nomadic behavior. The ineffectiveness built into such annual relocation may be the primary reason for the persistence of annually fruiting trees that are attacked by predispersal specialist seed predators. The other likely reason such a tree can survive is that even if the seed predators do not have to leave the vicinity of the parent tree, there is predation by generalist carnivores on the free-living adult specialists (beetles, moths, wasps) while they wait the 8–11 months between seed crops.

An example: Pithecellobium saman

An example of an annually fruiting tree and a brief description of its interaction with its primary predispersal specialist seed predator should now be given. *Pithecellobium saman* is a large, wild, leguminous tree native to the Guanacaste lowlands (as well as to much of the drier neotropics). It bears scattered pink inflorescences over its huge crown from March to May (last half of the dry season; the details of timing vary from habitat to habitat). Nearly all adults flower each year; not all set seed each year, but most do insofar as detailed information is available. *P. saman* sets 1–15 pods per inflorescence-bearing branch. These pods remain 1–3 cm in length and 2–4 mm wide through September–October (and sometimes later), and then during the last one-third to one-fourth of the rainy season expand rapidly to full size (10–20 cm long, 1–2 cm wide). During the period of fruit dormancy, the number of pods is reduced by abortion and fruit damage to about 1–4 per branch. By November–December most trees bear full-sized pods, and these ripen and fall by February–

March. Again, the variation in seed maturation timing is related to habitat.

Once the pods are full-sized and the seed approximately full-sized, *Merobruchus columbinus* bruchid beetles appear and glue single eggs to the pod surface. The larvae bore into the seed through the green pod wall. A maximum of one bruchid develops per seed, but many may enter (others apparently die of cannibalism). At 1–2 months from the time of oviposition, an adult completes the exit hole through the side of the seed that was started by the bruchid larva. The adult then cuts a further exit hole through the side of the nearly ripe seedpod, and leaves the tree. The newly emerged beetles do not reoviposit on mature pods either in vitro or in the field.

At the time of pod fall and beetle emergence, the dry season is well advanced, and vegetation below the adult *P. saman* is quite dry. Sweeping this vegetation with an insect net does not yield adult *M. columbinus*, but, of course, they could be in some more secretive locality near the parent. However, sweeping nearby riparian vegetation or shady places beneath individual evergreen trees does produce adult *M. columbinus* during daylight hours. Presumably, they are whiling away the days waiting for next year's *P. saman* seed crop to mature. I have not yet found them feeding at flowers during the late dry season or the following wet months. However, I suspect they take nectar and pollen from flowers, like other species of bruchids. They do not pass another generation on some alternate host, as I have made collections and rearings from all other species of seeds in the area that are large enough to harbor *M. columbinus*. Incidentally, a large number of other legumes in the area have seeds quite large enough to support a *M. columbinus* larva (e.g., *Acacia farnesiana,** *Acacia collinsii,** *Acacia cornigera,** *Hymenaea courbaril, Phaseolus lunatus,** *Mucuna pruriens, Bauhinia ungulata,** *Lonchocarpus costaricensis,** *Caesalpinia eriostachys, Caesalpinia coriari,** *Caesalpinia conzatii, Gliricidia sepium, Dioclea megacarpa,** *Enterolobium cyclocarpum, Pithecellobium dulce,** *Prosopis juliflora,** *Parkinsonia aculeata,** *Cassia grandis,** *Schizolobium parahybum, Andira inermis;* those marked with an asterisk are attacked by their own species of bruchid).

It thus appears that *M. columbinus* adults relocate the fruiting tree anew each year, though it should be added that many *P. saman* trees grow in the same riparian vegetation that harbors *M. columbinus* adults, and for them the location of the parent tree and the resting habitat are the same. Incomplete results show that samples of 54 individual crops give an average of about 46% of the seeds killed by *M. columbinus* with certainty (range, 12.4%–100%; assay based on bruchid exit holes and on finding dead beetles in the seeds through the use of x-rays). In addition, up to half again as many of the seeds

may be aborted, many of which may be seeds killed by the bruchid before the seed was independent of the parent, with resulting
termination of nutrient flow by the parent and death of both the
bruchid and the seed.

A large, healthy *P. saman* frequently produces 10,000–20,000 pods
in a single crop. Assuming an average of about 10 filled seeds per
pod (this figure varies strongly among trees), a large tree may easily
generate 92,000 adult beetles, half of which are females. I have no
good estimates of how many females will reappear at the tree
10 months later, but would guess from the number of eggs laid that it
is about 4000. Thus, there is at least an 87% mortality of the adult
female beetles between seed crops. This particular picture is uncomplicated by bruchid parasites; unless there is an as yet undiscovered
egg parasite, *M. columbinus* has virtually no parasites (based on thousands of rearings).

Once the adults have emerged and the fruits have fallen, in an undisturbed community the fruits are chewed and swallowed along
with the hard seeds by large animals (peccaries, tapir, deer) and thus
carried off, or the outside sweet pulp is chewed off by small vertebrates, who also carried off some pods (squirrels, agoutis, small terrestrial rodents). The fruits are avidly eaten by cattle, who pass the
seeds intact. In contemporary, largely vertebrate-free habitats, the
pods are often left lying under many trees. On rare occasions, *Stator
limbatus* enter these pods through the *M. columbinus* exit holes. This
tiny bruchid then oviposits on undamaged seeds. However, the inside of a *P. saman* pod is not easily traversed and thus they do not
take a high fraction of the seeds remaining (in contrast to the case
with *Zabrotes interstitialis* in *Cassia grandis* pods: Janzen, 1971c). *S.
limbatus* is, for a bruchid, amazingly generalized, and the larvae can
feed on at least 6 of the 150 or more species of legume seeds in the
Guanacaste lowlands (Janzen, 1975a).

If we assume for the moment that *M. columbinus* can undergo no
evolutionary change, there are several ways that *P. saman* might
change and thus lower the percent seed mortality, yet remain a population of annually fruiting individuals.

1. The rate of development and maturation of the fruit could increase. The bruchid would thus still be inside the seed when the
newly fallen fruit is eaten entire by a large frugivore. The bruchids in
certain African *Acacia* seeds eaten by ungulates do not survive (Lamprey, Halevy, & Makacha, 1974), and this suggests that the bruchids
in the *P. saman* seeds also might not survive the passage through the
animal's gut. Such mortality could favor a fast mutant of *P. saman* in
two ways. First, if the bruchid larvae are very small, they may be

killed before the seed is consumed and thus some crippled but living seedling *P. saman* might be produced. Second, this bruchid mortality could lower the density of bruchids in the vicinity of *P. saman* at subsequent seed crops *if* there is any degree of local population formation by *M. columbinus* around the parent tree that produced them. However, if the population of *M. columbinus* is distributed in the general habitat independently of the sizes of the clutches of bruchids generated by each *P. saman* crown, then the fast mutant *P. saman* would not be favored. (If the fast mutant were favored for some other reason unrelated to the bruchids, and thus over time the bruchid population as a whole came to have a higher death rate of adults from mammalian fruit consumption, then seed predation by *M. columbinus* would fortuitously decline).

2. The synchronization of ripening within a tree's seed crop could increase, thereby shortening the period that the ripening pods are actually available for oviposition. This will work only to the degree that the beetles are time-limited in their choice of pods of appropriate age. The physiologic system to produce ever more synchronized fruit ripening will probably be bought at an ever increasing cost per unit of synchrony.

3. The time of fruit maturation could change to a season more unfavorable to the bruchids. For example, if pods were to mature at the end of the dry season, the newly emerged bruchids would have first to pass the 6-month rainy season and then 3–4 months of the dry season before they could oviposit. This might be more detrimental to the bruchids than the current sequence of passing 3–4 months of the dry season followed by 6 months of rainy season. It might also be more detrimental because the bruchids would have to locate and be active in the tree crowns during the hottest, windiest, and driest time of year. Again, to the degree that there are no local populations of *M. columbinus* associated with each *P. saman*, such a change probably could not occur through favoring individual mutants through reduced seed predation, but would have to be the by-product of other selective pressures. A single mutant that matures fruit late in the dry season would at present have a huge population of newly emerged adult *M. columbinus* to contend with; these beetles would probably produce a second generation of bruchids on the crop of the hapless mutant.

It is evident that each of the above-suggested possible changes in *P. saman* would cost the tree something in resources. The current seeding pattern of *P. saman* probably involves a compromise between each of the suggested changes and the opposite end of the gradient. For example, the behavior of keeping the pods very small until shortly before maturation has at least the adaptive value of preventing several generations of bruchids within a given crop. The intrapopulation synchronization of *P. saman* fruiting may in large part by driven by the possibility that individuals with late-maturing pods are subjected to a second generation of bruchids derived from those that emerge from the earliest-maturing pods on the same or different trees. In short, even if mammals select very strongly for attenuated fruiting distributions in *P. saman*, I doubt they will appear.

Repeated within-season fruiting
There are a few reports of tropical trees that bear fruit more than once a year. Medway (1972) describes an individual of *Bhesa robusta* as bearing fruit several times per year in Malayan rain forest; in the same forest individuals of two species of scrambling figs (*Ficus ruginervia* and *Ficus sumatrana*) produced successive crops of figs at a rate of about twice a year. Such successive fruiting seems likely to persist only where the seeds are very free from seed predators and where there is strong selection for frequent fruiting or asynchronous fruiting within the year, as where short-lived pollinators have to be able to find available flowers shortly after emergence, as in *Ficus* (Ramirez, 1970). Flowering records can be deceptive with respect to this phenomenon. An individual of *Genipa caruto* apparently may flower several times in 1 year in the Guanacaste lowlands (Frankie et al., 1974), but to the best of my knowledge individuals still bear only one mature seed crop (11 months later). In South Africa, *Virgilia capensis* individuals may flower two or three times a year, but only one of the flowerings produces a seed crop (Phillips, 1926); presumably, the hermaphroditic tree is acting as a male during the other flowerings.

Unsynchronized supra-annual fruiting
General considerations
There are many ways an annually fruiting individual may come to skip its seed crop. It may be damaged by weather or herbivores and thus not have enough reserves to seed; it may lose its flowers to weather or herbivores; insufficient pollen of the appropriate type may arrive; or it may be a mutant that, for example, requires higher than normal amounts of stored reserves to initiate fruit formation (or at times, flower formation). Whatever the cause of skipping a year's seeding, there should be several important immediate effects.

First, I expect that in the subsequent seeding year there will be more reserves with a consequent exceptionally large fruit crop (however, the tree may use these reserves for life-support functions other than seed production). How much larger depends on several variables. What does it cost to store these reserves? The answer is un-

known, as are the answers to the following questions: Are structures
and transport systems present that can store these reserves? Is the
consequent larger fruit crop produced from a normal-sized flower
crop and thus at a lower flower cost per fruit? Are pollinators available to provide enough acceptable pollen (assuming the tree is an
obligate outcrosser) for the larger fruit crop? There are almost no data
on actually how much larger such a fruit crop is. Biennially bearing
varieties of mangos (*Mangifera indica*) have crops that are more than
twice the size of annually bearing varieties (Singh, 1960); suggesting
that it may be more economical to make a few large seed crops than
many small ones. However, since biennial (or greater) bearing is the
normal state for wild mango trees, the annually bearing horticultural
varieties may have rather badly disrupted physiologies, and this
may be why they set so little fruit.

Second, if the seed crop is larger (as expected) in the year after a
skipped fruiting, there should be more effective satiation of seed
predators. However, whether this occurs will again depend on several variables. Do the dispersal agents differentially respond to the
larger crop or does the larger crop simply satiate the dispersal agents,
thereby leaving many of the larger number of seeds below the parent
tree to be taken by seed predators? Of course, if such an enlarged
pool of seeds below the parent tree absorbs an increased fraction of a
finite number of seed predators in the general vicinity of the parent
tree, such an accumulation may be functional in reducing the seed
predation on dispersed seeds. Does the same number of seed predators arrive at the seed crop as before, or is the larger crop so large that
it is found by or attracts a larger number of seed predators as well?
This larger number of seed predators may appear because a larger
location cue is produced (odor, color) and because certain species of
seed predators are interested only in very large seed crops.

Third, the skipped fruiting year probably excludes the presence of
an established local population of specialist seed predators waiting at
the site when the larger seed crop appears. If more than one year is
skipped between crops, it is extremely unlikely that a seed predator
will wait at the tree for 2 years or more for a fruiting year to occur.

As a mutant that skips years between fruit crops begins to spread
through a population, several attendant changes are likely to occur.
There will be strong selection for physiological self-evaluating de-
vices to appear within the plant so that it will fruit only after it has attained a certain level of reserves for sexual reproduction or only after
a certain number of years have passed. Counting years alone (and
not stored reserves) should be the mechanism only in habitats where
the rate of accumulation of reserves by adults is quite predictable.
There should be selection for the tree to wait long enough so that its
reserves are large enough to alleviate the problems mentioned in the

previous three points; such a delay in seed production will have the disadvantage of lengthening the periods during which the mutant is not contributing seeds to the habitat. This disadvantage may be partly offset by producing dormant seeds and seedlings that live many years while waiting for a gap in the canopy. The effective density of the mutant's species will decline (i.e., the number of trees acting as females per year will decline), and this should change the competition for pollinators and select for visitation by pollinators that are effective over longer distances. Decrease in effective density should also intensify selection for specialized pollination systems, such as monoecism and dioecism. By storing resources for later sexual reproduction, the tree may make itself a more desirable food item for wood- and twig-eating insects; on the other hand, it now has more reserves to use in case of vegetative emergencies (e.g., disease, attack, encroachment by another tree crown, root elongation in case of drought). Perhaps of greatest importance, skipping years between seed crops should lead to selection for traits that lead to sychronization of fruiting years with other conspecifics (or other neighbors that share seed predators).

Leaving these hypothetical considerations aside, we know very little of actual asynchronous supra-annual fruiting by individual tropical trees. Even when some information is available, it is generally unrecorded whether a skipped year is simply a response to a bad weather event, whether it is the individual or the population that is doing the skipping of potential fruit crops, and whether animal interactions with the trees are any different than if the tree fruited annually. Frankie et al. (1974) reported that in the Guanacaste lowlands some individuals of *Dalbergia retusa* and *Piscidia carthagenensis* that fruited in 1969 did not flower in 1970, whereas others of the same species did. In 1971, the 1969-fruiting individuals flowered again, and then in 1972 did not. They also list other species as not flowering every year, but do not note if it is the individual or the population that is doing the skipping or to what degree flowering equals fruiting. Medway (1972) lists a number of species of Malayan rain forest trees that skip years between fruiting, but as only one to a few trees were under observation for each species, there is no way of knowing to what degree sexual reproduction was asynchronous within the population (with the exception of the Dipterocarpaceae, discussed later).

An example: Cassia grandis

I have kept records to date on only one species of tree that displays unambiguous asynchronous supra-annual fruiting. That is the legume tree *Cassia grandis* (see later section for reference to *Hy-*

menaea courbaril). It is a native (wild) tree in riparian vegetation in the moister parts of the dry Pacific lowlands of Costa Rica (northern Puntarenas Province, near Las Juntas de Abangares). In this area, large, reproducing *C. grandis* are moderately common in roadsides, creek banks, pastures, and small woodlots.

When an adult *C. grandis* has other tree crowns adjacent to it (somewhat comparable to a forest situation), its flowering and fruiting pattern is quite clear. It makes a crop of up to several hundreds of thousands of pink-orange flowers when largely or entirely leafless during the middle of the dry season (February–March). Some 2% of these flowers set thin green pods (6–11 cm long, 2–4 mm wide) that remain this size or slightly larger through a wave of moth larvae attack and various selective abortions by the parent plant. Then in the last half of the rainy season (beginning about September), they enlarge rapidly to as much as 80 cm long and 4–5 cm in diameter. At this time a very large pod crop is 400 pods (far less than 1% of the flowers in the crop that produced them). These pods mature in the dry season (January to May), and most fall at that time (a few may hang on the tree for as long as a year later, making a census difficult). When large animals are present, the large pods are chewed by rodents and broken open by deer and peccaries in search of the sweet molasseslike pulp around the seeds. The hard seeds pass intact through large mammals.

In the year of pod maturation, there are few or no flowers. The following year, the tree produces another large flower crop, and the cycle repeats itself. About half the trees are in flower each year, so the breeding population size is about half that which might be recorded by a forester. However, when a *C. grandis* is isolated in a pasture or roadside, major branch sections of the crown often get out of phase with the remainder of the crown. One part of the tree crown will contain mature fruit and another part will be covered with flowers in the same dry season, giving the false illusion that the tree fruits and flowers in the same season. Presumably this lack of synchrony in the crown occurs in a severe year when only part of the crown has enough reserves to set fruit. I am currently setting up defloration experiments in a deliberate effort to asynchronize parts of crowns that are in synchrony, and vice versa. The phenomenon is reminiscent of the efforts by apple growers to turn biennial bearers into annual bearers.

There are three predispersal seed predators on *C. grandis* seeds in addition to the moth larvae that attack the small green pods. In the central portion of the tree's range in the Pacific lowlands of Cost Rica, *Megasennius muricatus* bruchids lay 10–50 eggs singly on the surface of the immature but full-sized pods about 1–2 months before the pods are ripe enough for the dispersal agents to be interested in them. The first instar larvae bore through the pod wall, locate a seed, drill through the somewhat soft seed coat, and complete development inside. They pupate there, and the adult then emerges by cutting an exit hole through the wall of the indehiscent pod about the time the pod is ripe enough for dispersal agents to be interested in it. As near as I can determine, these adults then disappear until next year's seed crop is nearly ripe. They do not have a second generation in the seeds of another species, but if there happens to be a *C. grandis* with somewhat slowly maturing pods nearby, there seems no reason why there should not be a second generation of bruchids on it.

In the center of the distribution of *C. grandis* in my study area, and throughout the rest of its range in Costa Rica, *Pygiopachymerus lineola* bruchids treat the pods in the same manner as *M. muricatus*. They lay clusters of two to eight eggs at regular intervals on the pod, take as many as 50%–60% of the seeds in an average-size crop of several hundred pods, cut exit holes through the pod margin rather than the pod side as does *M. muricatus*, and definitely do continue to oviposit on pods in the tree's crown as long as there are pods available of the right age (Janzen, 1971c). However, once these adults have emerged and the pods are mature, *P. lineola* adults leave for some nearby creekbed (where they can be taken with a sweep net) to pass the remainder of the dry season. They have no other seed hosts in the study area.

Finally, if dispersal agents have not yet removed the pods, a moth lays its eggs in the exit hole of *M. muricatus* or *P. lineola*, and the moth larvae eat out the sweet, juicy pulp inside the pod while it hangs on the tree or lies on a dry piece of ground. As the seeds are cleaned by these moth larvae, the very small adult of the bruchid *Zabrotes interstitialis* enters the larger bruchids' exit holes and glues its eggs directly to the seeds. It oviposits on both intact and damaged seeds, as the larvae of the two larger bruchids commonly do not eat all the seed contents. *Z. interstitialis* normally kills all the viable seeds in a *C. grandis* pod if the dispersal agents do not remove it. This bruchid will recycle in the tree's seed crop as long as there are seeds available on which to oviposit.

In summary, there are two bruchids that kill a potentially smaller number of seeds as the *C. grandis* crop gets larger. However, if the crop is so large that the dispersal agents are slow in removing it, there is a third bruchid that will kill the remaining seeds. I suspect that the two bruchids are a primary driving force for the behavior of making a large crop every 2 years rather than half this size crop every year. On the other hand, the presence of *Z. interstitialis* should place a ceiling on the size of the crop (and thus the number of years skipped between crops), as there probably was a limit to the number of *C. grandis* pods the original vertebrate dispersal guilds would absorb.

I should reemphasize that the seed predators are not the only selective force operating on the behavior of skipping 1 year or more between seed crops by individuals in an unsynchronized manner. First, there is the possibility of economy of scale, in that the flower cost per acceptably pollinated fruit produced may decline as the number of fruits to be set per flower crop rises. Looked at another way, only half the flower crop every year may not bring in enough high-quality pollen to make even half a fruit crop, whereas a full flower crop every other year may bring in more pollen than is needed. Second, the effect on seedling recruitment of skipping years between seed crops depends in great part on the seed and seedling dormancy behaviors. If seeds are just added every several years to a large pool of dormant juveniles, out of which a few initiate a competitive attempt each year, then skipping years may have little effect on the number of competing juveniles entering the community. Alternately, if the seeds produced from each crop germinate that rainy season, skipping years may mean many years when an individual parent is making no attempt to compete for the available gaps in the canopy. Third, there is no information on how skipping years influences the ultimate longevity of the tree. It might well be that a tree that skips a year between seed crops lives longer than the annually seeding morph because it has more tactical options for the uses of its reserves and thus overall has a higher lifetime reproductive potential.

Mast fruiting (synchronized supra-annual fruiting)

There are two rather distinct kinds of synchronized supra-annual fruiting, both referred to as "mast fruiting" for brevity. On the one hand, there may be a cohort of trees all of the same age that after having grown vegetatively for a number of years, flower, seed, and die in the same year or other short period; bamboos and *Strobilanthes* (niloo) are the two best tropical examples of such semelparous behavior (Janzen, 1976). On the other hand, there may be an iteroparous mast fruiting population whose members all respond to the same weather cue and are thus synchronized; the subadult trees use the same weather cue and thus became synchronized with the adult population. Extratropical forests are dominated by such tree species (e.g., beech, oak, conifers, hickories, hazelnuts; Janzen, 1971d), but the only spectacular example in the tropics occurs in the Southeast Asian dipterocarp forests, where a severe dry spell seems to be the cue for many species in the habitat to fruit synchronously (Burgess, 1972; Janzen, 1974). Other, less obvious tropical examples are now coming to light, but are poorly investigated.

It is my hypothesis that in both the above kinds of mast fruiting, seed (and seedling) predation is the driving selective force. The seed predators determine the seeding distributions of the population and thereby select strongly for synchrony and for large seed crops by individuals. Such reproductive behavior may be viewed simply as a form of predator satiation achieved by the individual through seeding at a time when its neighbors are also doing so.

The advantages of outcrossing can also select for supra-annual flowering synchrony within the species (but not among them). However, most species that display mast fruiting are so common that even if their flowering were spread over many years, it seems unlikely that flowering individuals would be so far apart that pollination failure would be the entire driving force for supra-annual synchronization.

Mast fruiting: iteroparous

Leaving aside Dipterocarpaceae for the moment, there are a few hints of mast fruiting iteroparous tree species in various parts of the tropics. In Parque Nacional Santa Rosa, nearly all large individuals of *Ateleia herbert-smithii* (chaperno) bore flowers in 1971, and about half subsequently bore fruit in 1972. In 1974 again nearly all the population flowered, and about half again bore fruit in 1975. There was no flowering observed in the park in the intervening years or in the 1975–6 potential flowering season. In the Guanacaste lowlands, *Andira inermis* (almendra) appears to have a mast year every other year (Table 4.1), though less than 1% of the trees are out of phase. Opler et al. (1975) state that in the Guanacaste lowlands "all individuals of Costa Rican *Cordia gerascanthos* come into flower synchronously for two weeks on alternate years" (1972 and 1974); they also fruited in 1976 (pers. obs. of several hundred large adults in Parque Nacional Santa Rosa). However, no information is given by Opler et al. (1975) about how many trees of what size are involved in their records. Phillips (1927) noted several species of South African trees that have what he called flowering seasons at 2-, 3-, or 4-year intervals. D. McKey, working in Cameroon rain forest in the Edea-Douala area, and T. Struhsaker, working in Kenya rain forest near Fort Portal (pers. comm.), have encountered a number of species of trees that are skipping at least 1 year between synchronized fruit crops. *Triplochiton scleroxylon* produces flowers and fruit "in abundance only in occasional years" in Nigerian rain forest; "the periodic-

Table 4.1. *Flowering (x) records for* Andira inermis *in Guanacaste Province, Costa Rica.*

Tree	Year 1972	1973	1974	1975	1976	Tree	Year 1972	1973	1974	1975	1976
2	x		x		x	40	x				x
3	x		x		Dying	41	x				x
4	x		x		x	42	x				x
5	x		x		x	43	x				x
6	x		x		x	44	x				x
7	x		x		x	47	x		x		x
8	x				x	49	x		x		x
9	x				x	50	x		x		x
10	x				x	53	x				x
11	x				x	54	x		x		x
12	x				x	55	x		x		x
14	x				x	56	x				x
15	x		x		x	57	x				x
16	x				x	58	x		x		x
17	x		x		x	59	x		x		x
18	x				x	63	x		x		x
19	x		x		x	65	x				x
20	x				x	73	x		x		x
21	x				x	74	x		x		x
22	x				x	75	x		x		x
23	x				x	77	x		x		x
24	x				x	80	x		x		x
25	x		x		x	81	x		x		x
26	x				x	82	x		x		x
27	x				x	83	x				x
28	x					84	x				x
29	x				x	85	x		x		x
31	x				x	86	x		x		x
32	x				x	87	x		x		x
33	x				x	97	x		x		x
34	x				x	98	x		x		x
35	x					99	x		x		x
36	x				x	100A	NR	NR	NR	x	
37	x				x	100	NR		NR	x	
38	x					101	x	NR	x		x
39	x										

NR, no record.

ity is irregular but is generally between three and ten years" (Lowe, 1968).

The Guanacaste trees with apparently very regular intermast periods pose a most perplexing problem. No environmental cue occurs at invariant 2-year intervals. Of course, a tree can count 2-year periods, but how does it align its counting with that of the other members of the population? Pheromones seem a possible way, but these have never been recorded for plants. The only other possibility is that an occasional very severe season stops every tree from fruiting, with the result that in the following year all fruit and a 2-year calendar maintains the synchrony after that. Such catastrophic weather events would have to occur rather frequently for the subadult trees to become synchronized as they begin sexual activity.

An example: Hymenaea courbaril
Observations

The behavior of *Hymenaea courbaril* (guapinol) in the Guanacaste lowlands appears to be intermediate between that of *Andira inermis* and *Ateleia herbert-smithii*, which have very good synchrony, and that of *Cassia grandis*, which is unsynchronized.

The reproductive behavior of *H. courbaril* and how it relates to seed predators requires detailed examination. My observations suggest that an isolated individual adult in a forest will have the following reproductive behavior. Sometime between the middle of February and the middle of May, depending on the local region, it will begin to open a few hermaphroditic flowers per inflorescence each night. There may be several thousand inflorescences in such a crown. For about 1 month, it will do this and will be visited each night by nectar-seeking bats. It appears to be obligatorily outcrossed (Bawa, 1974), but I say that with caution for reasons that will become obvious. It will usually abort all flowers a few days to several weeks after they have opened. These aborted, opened flowers have slightly enlarged tiny fruit. As many as 50% of the flowers will drop as unopened buds, and these normally contain curculionid weevil larvae (*Anthonomus* spp.). Either no pods will be matured, or on occasion there will be a single pod here and there in the crown. Such flowering will be repeated for several years, and then one year, instead of aborting all the flowers, the tree will set a crop of 100–500 pods, depending on the size of the tree and the crown's exposure. The pods will be expanded to full size (10–20 cm long, 5–10 cm wide, 2–4 cm thick) within 1–2 months of flowering, and then hang on the tree in this state until December to March, when they mature (harden) and drop. Following this event, sometimes there are no flowers in the first year after fruiting, but in general there will be another run of flower-only years. During these years the tree is effectively a male tree, and the flower crop size and dynamics should be subject to the normal selection pressures associated with competition among males for mates in any organism.

Young, diseased, or shaded trees may act as male trees indefinitely. When a healthy tree in a forest has its competitors removed by lumbering, it sometimes begins to bear a fruit crop almost every year. Some of these fruit crops reach the phenomenal size of 3000 pods. There is a good deal of variation in the population concerning how individuals respond to such apparent elimination of competitors. I assume that the increased seeding frequency of insolated trees is the simple consequence of their being able to store resources more rapidly than in a forest, so that almost every year their counting mechanism tells them that it is time to seed.

In the Guanacaste lowlands, *H. courbaril* has two major predispersal seed predators (omitting the predators on flower buds). *Rhinochenus stigma* females (Curculionidae) arrive at the tree sometime during the last half of the rainy season and lay clutches of about six eggs in a green pod. It appears that they do this late enough in the life of the pod so that the copious resin in the pod wall has begun to set and thus does not well out and drown the eggs or first instar larvae as they drill further through the pod wall. The larvae burrow inward until they find one of the large seeds, and as many as five larvae may go through development inside one seed. There is an average of about 4.5 seeds per pod (range, 1–15) and *R. stigma* kills 1 to many. It rarely kills all of them, however. The larvae pupate in the seed, and the adults emerge to walk about the inside of the pod and feed on the pulp and seed fragments. They do not kill seeds missed by the larvae (the seed coat is too hard for them to penetrate). They cannot escape from the indehiscent seedpods on their own, but when an agouti (*Dasyprocta*) chews open the pod, the adults quickly flee (or are eaten). *R. stigma* is becoming locally extinct in my study area because the agoutis are disappearing owing to habitat destruction.

At about the same time *R. stigma* is ovipositing in the pods, females of *Rhinochenus transversalis* are laying single eggs in the pods. The larva mines through the pod pulp, taking a notch out of each seed. It thus kills or severely damages each seed in the pod. However, it eats some seeds nearly entirely; probably these are seeds that have *R. stigma* larvae in them. I postulate that such behavior is adaptive because the *R. transversalis* larva pupates in the pod cavity and would be eaten by the adults of *R. stigma* emerging from the seeds. In tens of thousands of pods opened, I have never found a *R. stigma* and *R. transversalis* adult in the same pod. *R. transversalis* is not dependent on an agouti to open the pod, because the larva cuts an exit hole nearly all the way through the pod wall before it pupates. The newly emerged adult completes the job, often before the pod falls from the tree.

Both *R. stigma* and *R. transversalis* leave the tree after leaving the pod. I have been unable to locate them in the interim between seed crops, but I have no reason to believe they sit at their parent tree for 3–6 years waiting for the next seed crop to appear. It should be noted that they have no other host plants in the habitat. *R. stigma* can live at least a year in vitro with no water and only dried seed contents on which to feed. When a tree is freed of competition and thus starts to fruit every year, the intensity of seed predation does not conspicuously build up on its successive crops. I interpret this to mean that the beetle has no way of being aware of this event, and every year spreads out over the habitat to relocate trees anew at the correct time. Of course, if most of the *H. courbaril* population becomes exposed trees that display nearly annual fruiting, a mutant beetle that simply stays with its parent tree may be strongly selected for.

I am not yet prepared to state with absolute firmness that the bigger the seed crop in a *H. courbaril* crown the lower the percent seed mortality caused by the two *Rhinochenus*. However, the beetles kill 30%–80% of the seeds in most crowns, and it appears that bigger seed crops have a lower percent seed mortality than smaller ones. Another 10 years of data on the same trees are needed before more positive statements are possible.

However, once the seeds have escaped the weevils, the story is only half told. When the pods hit the ground, they are opened by agoutis and the pulp eaten. In addition, some of the seeds are eaten (preyed on). The agoutis bury the remainder just as squirrels bury acorns (except that *H. courbaril* seeds are not de-embryonated by agoutis). If the seed crop is very small, consisting of less than 50 pods, nearly all seeds appear to be consumed as the pods are opened over 1–2 months, and that is the end of the seed crop. The seeds that are buried are dug up and eaten throughout the remainder of the year. Some germinate when the rains come. If the agouti (or other vertebrate) finds them when little more than the cotyledons have appeared above ground, it simply eats off the cotyledons and thus preys on the seedling. If the crop is only medium-sized, it is doubtful that any seeds survive the agouti predation to the seedling stage. If the crop is very large, some may survive. In one recent case during the 1975 fruiting season in Parque Nacional Santa Rosa, an input of approximately 1000 beetle-free pods from three adult trees resulted in only 23 seedlings (though more may appear in the 1976 rainy season from the site). At least three agoutis have territories and foraging ranges overlapping the site. I have watched an adult wild agouti eat five *H. courbaril* seeds in 2 hours. If each animal consumes 10 seeds per day, such a seed crop would survive a maximum of only 150 days.

In summary then, it appears that *H. courbaril* has to store resources for a number of years to get a flower and seed crop that is large enough to survive the flower bud predation and predispersal seed predation by weevils and the postdispersal seed predation by agoutis. The agoutis are sufficiently generalist and sufficiently intelligent that there might well be a point of diminishing returns in this game. If the seed crop is so large that the seeds missed by weevils satiate the dispersal (burial) abilities and interests of the agoutis, it may simply attract (or allow) another agouti into the area and thereby raise the effective size of the predator guild that has to be satiated. If the crop is very large, yet not so large as to be left unburied, the agoutis may respond simply by having more offspring or getting fatter. The trick is to engineer the size of the seed crop so that the number of pods surviving the weevils is enough to cause maximal seed burial and minimal attraction of new agoutis to the area, coupled with minimal reproduction by the resident agoutis. The agoutis can probably store fat and reproduce only at a certain maximum rate, and they are territorial. These two facts mean that up to a point more seeds can be dumped on a site without an immediate increase in the number of agoutis present. The game may also be played by modifying the nutrient content of the fruit, but as yet I have no information on this for *H. courbaril*.

Interpretation
With this background in mind, a start can be made at interpreting the population-level fruiting pattern of *H. courbaril*. It is obvious from the data in Tables 4.2–4.4 that the population of *H. courbaril* within a small area is not perfectly synchronized on a supra-annual cycle (all the trees described in Tables 4.2–4.4 are within an area 6 km by 10 km and in similar terrain). First, most adults flower in most years. If fruiting synchrony were perfect at long intervals (as in *Andira* and *Ateleia*), there would be strong selection against even male flowering in nonfruiting years. Second, within the 60 km^2 (at least) that the nectivorous bats range over, there are some *H. courbaril* that bear a large fruit crop in any year. Third, if we decide to designate 1971 and 1974 as mast years, there are trees even within my marked series that do not conform; tree B13 is only 14 m from B15 (Bosque Humido Group) and trees 371 and 376 are about 50 m apart (Rio Guapote Group). Furthermore, some Rio Guapote Group trees that produced large crops in 1971 did not do so in 1974 (e.g., numbers 305, 348, 351, 377, 396) and vice versa (e.g., numbers 307, 312, 314,

387, 389, 390). Fourth, there is, however, a distinct fruiting peak in 1974 for all three groups and there probably was one in 1971 if the Rio Guapote Group is representative of the other two groups. Fifth, the Sendero Natural Group and the Bosque Humido Group seem to be more tightly synchronized than the Rio Guapote Group.

An important determinant of whether a *H. courbaril* tree will set a large fruit crop is the amount of reserves it has stored; once they reach a certain level, the tree fruits heavily. If they have not reached that high level when a cueing weather event occurs, the tree does not

Table 4.2. *Flowering and fruiting parameters of* Hymenaea courbaril (*Sendero Natural Group*) *in Parque Nacional Santa Rosa, Guanacaste Province, Costa Rica*

Tree	DBH (cm)	Year			
		1972	1973	1974	1975
D31	38.5	+		1	
D35	40.6			6	2
D47	16.8			+	
D96	39.6	1	+	+	1
D111	34.4	60	+	300	1
D139	10.3				
D144	16.1	27	+	200	29
D146	14.2		+	50	1
D148	14.3	+	1	110	1
D155	14.8	1	+	+	15
D157	28.6	+	+	76	2
D160	16.4		+	15	+
D161	21.9		+	25	+
D172	17.8	2	4	15	+
D176	27.5	+	+	100	1
D196	25.0	+	+	15	+
D229	23.7	1	+	300	46
D230	25.8	+	+	15	+
D264	20.8	1	+	+	9
D267	26.8	+	+	200	+
D312	23.4		+	15	+
Total trees in flower		14	17	20	18
Total pods produced		93	5	1443	108

+, flowered; number, number of mature fruit produced; italic number, normal large fruit crop for that tree.

Table 4.3. *Flowering and fruiting parameters of* Hymenaea courbaril (*Bosque Humido Group*) *in Parque Nacional Santa Rosa, Guanacaste Province, Costa Rica*

Tree	DBH (cm)	Year 1973	1974	1975
B1	14.0	+	*200*	+
B2	12.8	+	11	5
B3	10.3		+	+
B4	15.0	+	*100*	22
B5	15.7		+	+
B6	11.3		+	+
B7	8.9		+	+
B8	10.6		12	2
B9	25.3	+	*200*	3
B10	12.2		+	+
B11	14.5	+	18	+
B12	13.1	+	+	+
B13	24.5	+	*300*	7
B14	15.7			+
B15	35.7	+	+	*100*
B16	27.0	+	50	1
B17	28.0	+	*200*	1
B18	12.0		75	+
B19	14.5		46	1
B20	15.0	+	+	
B21	21.6	+	+	+
B22	17.5	+	+	
B23	19.2		+	+
B24	21.4	+	+	+
B25	19.1	+	12	3
B26	23.8	+	+	+
B27	28.2	+	*200*	1
B28	18.2	+	1	+
B29	28.5	+	*1000*	600
Total trees in flower		19	28	27
Total pods		0	2425	746

+, flowered; number, number of mature fruit produced; italic number, normal large fruit crop for that tree.

Table 4.4 *Flowering and fruiting parameters of* Hymenaea courbaril *(Rio Guapote Group) in Parque Nacional Santa Rosa, Guanacaste Province, Costa Rica*

Tree	DBH (cm)	Year 1971	1972	1973	1974	1975
300	17.1	+		+	+	
301	13.2	+			+	+
303	18.2	+		+	4	
304	16.5	17		+	+	+
305	24.8	100	256	1	2	1
306	8.1		+			
307	14.0	6	85	+	68	4
309	5.6					
310	8.3					
312	13.5	2	+	+	100	+
313	11.2		3			+
314	21.2	+	1	+	200	+
315	23.5	100	25	+	58	+
316	11.0+9.7	32	65	4	46	45
317	19.8	1	+	+	+	+
318	24.2	+	+	+	+	+
319	19.1	52	36	+	8	+
320	5.5	+	+	26	+	41
321	19.2	10	2		+	+
322	9.1+8.8					+
323	8.0					
325	7.7	+		+	+	5
326	6.2			+		
327	5.0					
328	7.0	+	+			+
329	9.7				+	
330	6.0					
331	4.8					
332	14.3					+
333	12.2	+		+	+	+
334	16.6	+		+		6
335	15.5	2		+	+	+
336	15.1	+	+	120	+	1
337	7.2			4	+	+
338	20.9		+		+	
339	11.5	+	+	19	+	+
340	18.7				+	
341	28.5	4	+		+	
342	21.5	+	+		+	

Table 4.4 (*cont.*)

Tree	DBH (cm)	Year				
		1971	1972	1973	1974	1975
343	24.8	+	6	+	11	
344	29.3	+	+	+	+	
345	23.8	+				
346	16.2	+			+	+
347	14.6	+	+	+	+	+
348	12.8	*250*		+	11	1
349	20.0	75			30	+
350	19.0	+	+	+	30	
351	20.8	*100*			+	+
352	11.8 + 5.1				+	+
354	16.1			+	2	+
355	8.8	10			+	
356	12.3	40			+	
357	10.6	10			+	
358	6.8 + 9.8				+	
361	17.1	10			+	
362	12.2	+	+	+	+	
363	23.9	+	+	+	+	+
364	13.3		+		+	
365	23.3	1	+	+	35	+
366	18.7	+	+	+	+	+
367	23.5	+	+	+	+	+
368	8.5					
369a	20.5	+	7	+	+	+
369b	26.4	+	+		2	
370	21.2	+	+	+	+	+
371	29.6	*100*	+	+	*200*	+
371A	10.7				1	+
372	22.0	50	+		+	+
373	21.1	40			4	+
374	20.8	+	+		+	+
375	21.4	+		+	+	+
376	38.0	30	+	*100*	37	+
377	19.7	*100*	+	+	+	+
378	19.8	+	+	+	3	+
379	15.9	+	+	+	+	+
380	19.2	+	+	+	2	+
382	20.6	+			65	+
383	14.2	+	+			

Table 4.4 (*cont.*)

Tree	DBH (cm)	Year 1971	1972	1973	1974	1975
384	29.8	+	+	+	4	+
385	14.2	+		+	+	+
386	29.8	*100*	20		*100*	+
387	26.3	+	+	4	*100*	+
389	26.4	25	+	+	*100*	+
390	24.5	+		+	*300*	+
390A	25.1	+	+	2	1	+
392	25.5	+		+	1	+
393	25.5	+	2	19	14	+
394	20.2	+	+	2	+	+
395	13.2 + 16.4	+	+	+	5	+
396	27.1	*100*	+	+	5	+
397	17.2	+		+	1	+
398	14.5		+		+	+
399	21.6	+	+	+	3	+
400	16.3	+		+	5	+
Total trees in flower		72	52	56	79	65
Total pods produced		1367	508	301	1538	104

+, flowered; number, number of mature fruits produced; italic number, normal large fruit crop for that tree.

respond to the cue with an incomplete fruit crop. However, the large number of small seed crops in the 1974 mast year in the Sendero Natural Group contradicts this generalization. A tree may also produce a large fruit crop in 2 consecutive years, presumably because it had built up reserves to such a level that one large fruit crop only partly exhausted them. Whatever the weather cue used, it is weak enough so that trees spread for 6 km along the banks of a seasonally dry river (Rio Guapote Group) do not perceive it uniformly. When the trees are tightly bunched in a small area (Sendero Naturalea Group, Bosque Humido Group), they respond in a more coordinated manner. Whereas a year such as 1973 would appear to be disastrous for the host-specific *Rhinochenus* weevils, and not all that pleasant for agoutis, neither set of animals is eliminated from the area. Agoutis have other foods they can eat (though perhaps not reproduce on), and *Rhinochenus* can (and did) migrate back into the area of marked trees from *H. courbaril* that fruited in 1973 in other areas a few kilometers away. The synchrony would have to be much tighter within the *H. courbaril* population before *Rhinochenus* could be eliminated from the area.

Southward from the Parque through the Guanacaste lowlands the picture changes, but unfortunately there are two changing variables. As one moves south, the annual rainfall increases slightly and the degree of habitat destruction increases greatly. The latter change means that the surviving adult *H. courbaril* are under less competition than those to the north in the Parque. The trees in the Rio Estanque Group (Table 4.5) illustrate the effect of one or both of these changes on *H. courbaril*. A periodicity of fruiting as seen in the Parque is evident, but with a quite different timing (1970, 1972, and 1974 being mast years). It is noteworthy that whereas 1971 was apparently a mast year in the Rio Guapote Group in the Parque, it was

Table 4.5. *Flowering and fruiting parameters of* Hymenaea courbaril *(Rio Estanque Group) in Bagaces area, Guanacaste Province, Costa Rica*

Tree	DBH (cm)	Year					
		1970	1971	1972	1973	1974	1975
94	41.3	*319*	+	47	+	21	
210	21.6	*300*	+	*500*	22	*310*	
211	23.9	1	+	41	+	+	
212	11.4	+	+	35	4	+	+
213	10.5	2			+	22	+
214	10.5				+	+	
215	29.0	40	30	20	*101*	*150*	+
216	20.5	*250*	+	*175*	60	75	10
218	24.9	*300*	+	*100*	300	50	+
219	24.3	30	+	12	3	*100*	+
220	30.1	*1000*	2	300	60	*200*	108
221	12.1	2	+	5	83	23	
222	21.9	*800*	+	12	+	*200*	7
223	30.3	30	+	*500*	18	*300*	
224	7.7						
225	33.8	*400*	+	2	1	20	+
226	29.1	*1000*	+	4	40	250	+
Total trees in flower		15	14	14	16	16	10
Total pods		4474	32	1753	692	1721	125

+, flowered; number, number of mature fruit produced; italic number, normal large fruit crop for that tree.

not a mast year in the Rio Estanque Group, and the reverse can be said for 1972; these two sites are about 50 km apart.

Further south, in the vicinity of Cañas, the fruiting records of five huge and heavily insolated trees are even more confusing with respect to synchronization (Table 4.6). These trees are all within 4 km of each other. Only 1973 could be identified as a year in which all trees responded with a heavy fruit crop, and heavy fruit crops are variously scattered among the other years. Whereas 1971 and 1974 may be viewed as the lowest years, I remind the reader that these were the mast years in Parque Nacional Santa Rosa, which is about 85 km to the northwest. Tree 4 perhaps shows best the effects of complete insolation. Its large pod crops were produced in different parts of the crown, and its 3-year runs of large crops are the longest I have recorded in the Costa Rican lowlands. For all practical purposes, tree 4 had no pods in 1975 and it did not even flower in 1976.

An example: Dipterocarpaceae

As a final example, the Southeast Asian rain forest Dipterocarpaceae are probably the most spectacular (Janzen, 1974). The most important features are the following:

1. Examining forestry records from 1925 to 1970 in rain forest Malaya, Burgess (1972) concluded that "most dipterocarp tree species flower gregariously at intervals of two to five years." It appears that in Sarawak and Brunei there are even dipterocarp forests with inter-mast periods as long as 9–11 years (Janzen, 1974). In dipterocarp forests, members of this family tend to constitute 50%–100% of the

Table 4.6. *Flowering and fruiting parameters of* Hymenaea courbaril *in Cañas area, Guanacaste Province, Costa Rica*

Tree	DBH (cm)	Year							
		1968	1969	1970	1971	1972	1973	1974	1975
16	39.9	+	*1000*	30	10	33	*1000*	35	17
4	42.6	*500*	*500*	*500*	125	*500*	*400*	200	31
1	34.7	+	*700*	10	2	100	*800*	46	800
17	26.8	+	+	10	+	150	*500*	6	+
18	27.2	+	+	40	+	25	*600*	8	300
Total pods produced		500	2200	590	137	808	3300	295	1148

+, flowered; number, number of mature fruit produced; italic number, normal large fruit crop for that tree.

canopy member individuals. Any given mast year involves many to most of the species, and thus it is evident that they are all using the same cue to determine in which year to fruit.

2. There can be very heavy seed predation by insects on dipterocarp seed crops, and these insects are amazingly generalist among the dipterocarps (Daljeet-Singh, 1974). For example, the weevil *Alcidodes dipterocarpi* has been reared from the seeds of nine species of dipterocarps, and *Nanophyes shoreae* from seeds of seven dipterocarps. This is similar to the case with insects that feed on the seeds in extratropical oak and conifer mast crops (Janzen, 1971d). Such broad host specificity is in striking contrast to the extreme host specificity shown by insect seed predators in most tropical forests. Although a detailed comparative study is yet to be done, there is much circumstantial evidence to support the notion that the general edibility of dipterocarp seeds is much greater than that of tropical forest seeds in general. There is conflicting opinion about the degree to which mammals prey on dipterocarp seeds, but certainly pigs and humans do (Janzen, 1974; Whitmore, 1975). Medway's (1972) statement that Malayan rain forest vertebrates were not preying on dipterocarp seeds in his observation plot should be considered against the disappearance of the really big seed predators that used to roam that forest (indigenous people, elephants, rhinos, tapirs, pigs, forest bovids). Medway offers no observations about how he knows small rodents did not eat the seeds.

3. A dry spell or year is obviously a major part of the cue to which dipterocarps respond, and this in turn results in synchronization at the population and habitat level (Burgess, 1972; Janzen, 1974). In the current (1976) mast year in Malaya, there are at least 50 rain forest species of dipterocarps flowering and fruiting. This year follows an exceptionally dry year, and this year is the first dipterocarp mast year since 1968 (F. S. P. Ng, P. S. Ashton, pers. comm.). Burgess (1972) and other authors are bothered by the fact that a dipterocarp general flowering does not follow every dry year, but this is to be expected if the trees have to accumulate a certain amount of reserves before they are primed to respond to the dry weather cue. Burgess (1972) points out that as many as half the individuals of a dipterocarp species in a stand may not flower in a general flowering year; I assume these are simply individuals with low reserves, individuals that will respond in the next mast year. The presence of these individuals supports the idea that a dipterocarp individual's seed crop must be of a certain minimum size before it is worth producing in the first place.

4. Wood (1956) noted that "despite the disparity in flowering times in most of the genera, the tendency was for the fruits of those that flowered late to develop faster than those that flowered earlier,

so that except in a few species of *Vatica* the fruits mostly fell between mid-August and mid-October." Assuming that such a pattern is general, I interpret it as the result of strong selection by seed predators for interspecific synchronization of seed fall within the year, and strong selection by competition for pollinators for asynchronization of interspecific flowering times (Janzen, 1974). I expect the flowering times of the different dipterocarp species to be rather evenly distributed between the time of the flowering cue and the time of habitat-wide synchronized fruit fall. I would expect the flowering-to-fruiting time to be well buffered from outside influences owing to the importance of hitting the "fruit drop window," and thus even arboretum and botanic garden data could be used for this parameter. Flowering-to-fruiting times such as are recorded by Ng & Loh (1974) for precisely located dipterocarp species should be used to examine this hypothesis, bearing in mind that there are likely to be local variations in this parameter within species.

5. Owing to the high proportion of dipterocarp individuals in many Southeast Asian rain forests, intra- and interspecific mast fruiting should have a severe effect on the animal community, as it means that much of the habitat-wide fruit and seed food is arriving as large pulses at long intervals; this should serve to drive the overall animal density downward, select for extreme generalists, and select for highly nomadic species. Also, at least on some of the very poor soil sites in Sarawak and Brunei, and on faunally impoverished islands, the overall animal density should be lower initially. The combination of these two factors should make mast fruiting a sufficiently good form of escape from seed predators so that tree community structure is largely set by competitive interactions among the incredibly large numbers (Chim & On, 1973) of seedlings and saplings found in dipterocarp forest following mast years. I thus expect clump distributions of adult trees of each species, with the clumps often located on those microsites on which the particular species of dipterocarp is the superior competitor; such an expectation is not at variance with Ashton's (1969) detailed descriptions of such forests or with the frequent reference in Malaysian forestry literature to nearby pure stands of this or that species of dipterocarp. Seeds of rain forest dipterocarps are likely to be selected for short-distance dispersal (so as to remain on the microsite of the parent), but to have properties such that in rare circumstances they are spread widely (in search of new and, therefore, unoccupied patches of the favored microsite); their poorly developed wind dispersal fits this expectation very well.

6. I expect the plants to be only moderately outcrossed, largely through facultative inbreeding. Three pressures are selecting for only moderate outcrossing. First, there is probably a shortage of pollina-

tors at the times of the infrequent flowering (much of the habitat's flower resources are badly pulsed in a dipterocarp forest). Assuming an individual tree has enough reserves for a large seed crop, I expect it to use foreign pollen only to the degree that is advantageous or available and to supplement this through polyembryony, apomixis, actual self-pollination, etc. In short, selfing may be viewed as a mechanism for matching the reserves for seed production with the kinds and quantities of foreign pollen available and desired. Second, the distance from one clump to another conspecific clump may be long and unlikely to be crossed by the kinds of generalist pollinators that should be available when the infrequent flowerings occur. Third, a finely tuned microsite competitive specialist may have a good deal to lose by exposing its genome to genetic variance derived from outcrossing. At least, it may have more to lose than does a more edaphically generalist species in a forest where the distribution of individuals within species is less clumped and the outcome more of the activities of seed predators and dispersal agents than of intraspecific competition (Janzen, 1970b). Clumped plants seem to be exceptionally well protected chemically at all vegetative levels, from the trunk to the leaves (Janzen, 1974). The consequence is that outcrossing may be less important to dipterocarps in the biochemical battle with herbivores than it is to plants that rely in greater part on chemical and behavioral evolutionary versatility to stay in the game.

It is my opinion that all the above facts and interpretations add up to a description of the outcome of the interaction between animals and trees over millions of years, operating at levels from the individual to the habitat. I wonder what would happen if we could let the Guanacaste lowlands remain undisturbed for enough more millions of years? Would it eventually end up in this state, with, for example, an adaptive radiation from the evergreen *Andira inermis* and *Hymenaea courbaril* as the starting point? Or is there something about the combination of soils, weather, starting germ plasm, and topography that produces the dipterocarp forest result in Southeast Asian rain forests and a different system in Central America. I opt for the latter hypothesis, but if I stretch my imagination even more than I have in the previous pages, the former hypothesis does not seem altogether ridiculous.

In the context of these comments, the contrast between the dipterocarps in Malayan rain forest and those in the much drier semideciduous forest of northern Malaya is instructive. P. S. Ashton informs me that in the northern, drier areas there are no mast years; instead, large seed crops may be found in any year on some or all individuals of some species of dipterocarps. We could let such a system go for an infinite number of years and it probably would not change into a mast year system like that operating in dipterocarp rain forests, for the following reasons. First, the more seasonal is the rainfall in a tropical habitat, the more difficult it is to imagine a weather cue that will be perceived by most of the members of a population, to say nothing of several different populations. A supra-annual dry spell is very noticeable in a rain forest, but a wet spell in a dry forest is much less obvious over a large area because of its confusion with the beginnings of the rainy season, because there are always local wet areas in dry forest, and because soils are water reservoirs with long retention times. Second, the more seasonal a habitat, the more severely it should depress the seed predator populations between seed crops; thus, the selective advantage of seed crop synchronization on a supra-annual basis is reduced. Third, as incipient habitat-wide masting begins to develop in a rain forest, I suspect it will have a greater depressant impact on the generalist seed predators than in a dry forest. This is because many generalist seed predators can feed on other plant parts and these are generally more available during the rainy season in dry forest (vegetative parts of deciduous plants) than they are in a rain forest.

Mast fruiting; semelparous
Bamboos
The interaction of bamboo and *Strobilanthes* (niloo) with their seed predator has been reviewed recently (Janzen, 1976). Only the essential conclusions are related here. In its shortest and most generalized form, a representative non-rain forest bamboo life cycle in the wild is roughly as follows. After growing by rhizome and shoot production for a species-specific period of 3–120 years, nearly all the members of one species in one area (and in one habitat) produce wind-pollinated flowers, set large quantities of seed, and die. This seed germinates immediately or when the first rains (or spring) appear, but is preyed upon heavily by local animals (e.g., rodents, small birds), highly nomadic animals (e.g., elephants, bovids, jungle fowl, columbids, pigs), and apparently the offspring of both groups. This seed predation is proportionately heaviest on the tails of the seeding distribution, thus maintaining the synchrony. The new cohort of surviving seedlings, largely derived from seed in the center of the seeding distribution of the parental cohort, then grows vegetatively for the same length of time as did its parents and repeats the process. The timing of seeding is set by an internal physiologic calendar, rather than an external weather cue, as in iteroparous mast seeding trees, and is, therefore, extremely well buffered against out-

side perturbations (e.g., transplantation to quite different habitats and parts of the world).

The system is found throughout the subtropics and tropics of the world, occurring as far north as the mountains of Japan and river bottoms of Indiana (United States) and as far south as Argentina and Chile. At least 200 species of bamboo are involved, and all commercially important species of bamboo appear to be semelparous. Semelparous bamboos are virtually nonexistent on the equator, and the two species of African transequatorial bamboos have (apparently) very poorly synchronized cohorts. Most of the species known to be semelparous are in the Indian-Asian tropics, but I suspect that most of the neotropical bamboos will be found to be semelparous. Poor development of the strategy in the wet equatorial tropics may be a reflection of competition ecology between bamboo and other woody plants, but as there are nonmasting species of bamboo in the equatorial tropics, its absence may be due to the difficulty of counting years by day-length cycles very near to the equator.

Strobilanthes (Acanthaceae)

In India, Sri Lanka, and countries immediately to the east, the mast seeding bamboos are generally sympatric with many species of mast seeding semelparous *Strobilanthes* (sensu lato). These small shrubs or trees have extremely well-defined cohorts and intermast periods. For example, one cohort of *Strobilanthes kunthianus* flowered in the Nilgiris and Palnis hills of southern India every twelve years from 1838 to 1970 (Janzen, 1976). Their small seeds are eaten by the same birds and small animals that eat bamboo seeds, and it is my hypothesis that these animals maintain the cohort synchrony by relatively heavier predation on the tails of the seeding distribution than the central part. *Strobilanthes* are apparently pollinated by large numbers of highly nomadic honeybee colonies (*Apis* spp.). It should be noted that pollination difficulties for both bamboo and *Strobilanthes* plants that are far out of phase with the remainder of the cohort should also select for intracohort flowering synchrony, irrespective of the activity of seed predators.

Conclusion

In this chapter I have described a rough progression from tree species whose individuals bear seed every year through those with asynchronous and then synchronous skipping of potential fruiting years. A final step, which seems to have been taken in a major way only by bamboo and *Strobilanthes*, is synchronous skipping of potential fruiting years accompanied by total exhaustion of reserves for sexual reproduction and reliance on an internal calendar for synchronization. This final step could have its ancestry either in an iteroparous mast seeding species or in the synchronization of a population of a perennial semelparous species. I favor the latter origin.

I hypothesize that seed (and seedling) predation has been the major driving force for all these deviations from simple annual fruiting by every adult individual in the population. Furthermore, I suspect that seed predation (as well as dispersal and pollination interactions with animals) has been a major selective force in setting the timing and variance of the seeding distribution of individual trees within the year in which they fruit.

An alternative hypothesis suggests that the entire progression has been driven by economies of scale (it is cheaper to make $2x$ seeds with one flowering event than to make x seeds twice with two flowering events). In this system, pollination would be seen as the primary selective force for synchronization within and between years. Such a system does not account for multispecies assemblage synchronization of seed production (e.g., dipterocarps) and in fact should select against it through competition for dispersal agents and pollinators. Economies of scale are unlikely to be the cause of supra-annual seeding of trees like *Hymenaea courbaril*, which bear a flower crop nearly every year and thus have already invested in flowers nearly every year. There are also other difficulties, but I will let the reader imagine them by applying this alternative hypothesis to the various detailed parts of a seed predation hypothesis.

Acknowledgments

This study was supported by NSF BMS75-14268 and inspired by the Fourth Cabot Foundation Symposium. The manuscript was initiated while I was associated with the Department of Ecology and Evolutionary Biology, Division of Biology, University of Michigan, Ann Arbor, Michigan 48109. It has profited from discussion with P. S. Ashton, T. C. Whitmore, P. W. Richards, B. Williamson, J. L. Harper, H. G. Baker, G. W. Frankie, C. M. Pond, P. A. Opler, D. E. Wilson, J. H. Vandermeer, D. A. Boucher, and others.

General discussion

Whitmore: Dr. Janzen mentioned that *Hymenaea courbaril* is a species that flowers roughly every second year. It stores up carbohydrates in the trunk, starch was suggested, in order to build up an energy source with which it can later flower and fruit. Has anyone measured these storage products, or is the suggestion hypothetical?

Janzen: It has been measured, for beech trees in Europe starting about 1880 and was measured progressively more frequently up through about 1930 before interest was lost in the subject. I've measured it for *Hymenaea,* but not for other tropical trees, so when I said that the starch content builds up in the sapwood, that was based on actual data.

Policansky: Dr. Janzen, you showed us small, green pods that remain on the tree for 4 months before they mature; did they suffer loss by predation or other effects?

Janzen: No, there is a 100% survivorship for those pods.

Policansky: Why?

Janzen: This is a good question. The tannin content on a dry weight basis is about 40%. Comparable values occur in *Terminalia* fruits, although not as hydrolyzable tannin, and in the bark of certain mangrove trees. I suspect the fruits are low in nutrient content; the soluble carbohydrate content is almost zero.

Tomlinson: As an extension to this question, what is the biologic significance of the long period of fruit immaturity?

Janzen: I am assuming that in some sort of natural selection sense, the rainy season (during which the fruit remains immature) is an inappropriate time for the fruits to mature, with regard to the time of either seed germination or seed dispersal. Nevertheless, the seed has to survive this period. It could do it as a small but inactive fruit, which is what happens (e.g., *Enterolobium cyclocarpum*). Otherwise, it could remain on the tree as a fully expanded fruit that photosynthesizes, but then it would have to be well protected chemically. This strategy is adapted by *Hymenaea* trees. I neglected to mention that *Hymenaea* fruits also remain for 10 months on the tree, but expand to full size immediately after flowering. Again there is little fruit loss during the waiting period, probably because there is a very copious production of resin, which may deter squirrels and parrots.

Givnish: I have a general question about the predation hypothesis, or why plants should synchronize their reproductive efforts, because one has cause to wonder how *Cassia grandis* survives at all. A factor in addition to the ones already considered is that many seed predators are also seed dispersers. In certain instances, consumption of the seeds by the dispersal agents completely destroys them; in others, the dispersal agents do not damage the seeds and may actually facilitate their germination. Among plants whose seeds are destroyed by the dispersal agent – as squirrels destroy acorns, for example – one might expect to find cyclic fluctuations in seed output and animal populations. In the north temperate zone, one does find mast fruiting among nut-bearing trees as a general phenomenon. Among plants that are characteristically dispersed by birds, where the fruit tends to be a beak-sized berry with an indigestible hard seed(s) inside, I know of no documented case of mast fruiting. This is to be expected because, for one thing, it is to the plants' advantage to maintain a large population of benign dispersal agents, and for another, the nature of the dispersal agents – here birds – is migratory, so that synchrony would have to develop throughout the whole area to be effective. In the case of *C. grandis,* if the ultimate dispersal agent does not digest the fruit, then might not this be an important selective force maintaining its apparently unadaptive strategy?

Janzen: The answer to such a question would be very extended. For the moment I will say that the lack of synchrony among individuals is probably related to the behavior of dispersal agents in the area.

Ashton: First, how do you explain the fact that in the north of Malaya, where there are not geographically isolating boundaries but where there is a change in climate with an increase in seasonality over a very short distance, the several *Dipterocarpus* species that transgress that climatic boundary flower annually north of the boundary, whereas they flower intermittently, in the way you described, south of it. I should point out, nevertheless, that the individual trees in the population north of the boundary where I have seen this annual flowering do not flower every year. Second, from my observation, but without quantitative information, which we hope to have very shortly, individual trees (rather than the whole population) north of this boundary produced not less but more fruit each year than the individual trees that fruit only at long intervals south of the boundary.

Janzen: The first question I would relate to the problem that an unseasonal habitat makes it hard to establish a synchronizing cue in a tropical system that is effective over a large area. On the other hand, a severe dry season in a rain forest is a precise cue and will synchronize flowering extensively. In a habitat where there is a severe dry season of variable length and intensity each year, it is hard to establish a cue that is distinguishable by all members of the population. I would argue that this is one of the primary reasons (but not the only reason) why community-wide synchrony does not appear in deciduous forests. In most of the seasonally dry tropics, there is not community-wide synchrony by a response cue. Bamboos do it by counting internally, not by using an external cue.

The second question deals not so much with my hypothesis as with the situation of a dipterocarp tree, which has to store enough to make a big enough seed crop to satiate all predators. I would suggest that an individual dipterocarp north of this line is, in fact, capable of making more photosynthate per year for seeds than is one that lives in the rain forest.

Ashton: Or alternatively that your hypothesis is not correct.

Janzen: I don't see how it is related to my hypothesis. The question deals with the amount of energy an individual dipterocarp in the rain forest converts into seeds. I would argue that in a dry forest area, the tree either has more energy to convert or it has made an internal allocation decision to put more energy into seeding; the rain forest area has less energy to convert per year or it has made an internal allocation decision to put less energy into seeding. Perhaps rain forest trees have shorter life spans.

Ng: I was interested in Dr. Janzen's description of *Pithecellobium saman* because it has evolved quite a different biologic cycle in Malaysia. Yours is a 1-year tree, shedding its leaves, flowering, and fruiting in a 12-month cycle; ours is a 6-month tree, giving two cycles per year. The fruits do not persist in the small phase, but just keep on growing from fertilization to maturity. After 6 months, fruit is ripe and then a new cycle begins.

Janzen: What is the seed source for the Malayan *P. saman?*

Ng: I don't think anybody knows.

Borchert: Are these geographic variations within the same species?

Janzen: It is introduced in Malaya.

Borchert: If the Malayan tree were reintroduced to Central America, would it revert to the New World cycle?

Janzen: The Malayan source could be from South America, or Panama, or Mexico; i.e., very diverse.

Borchert: If this different behavior is common then I cannot see how the growth cycle can be adaptive; it rather must be controlled by an environmental cause.

References

Ashton, P. S. (1969). Speciation among tropical forest trees: Some deductions in the light of recent evidence. *Biol. J. Linn. Soc. Lond.*, *1*, 155–96.

Bawa, K. S. (1974). Breeding systems of tree species of a lowland tropical community and their evolutionary significance. *Evolution, 28*, 85–92.

Burgess, P. F. (1972). Studies on the regeneration of the hill forests of the Malay Peninsula. *Malayan Forester, 35*, 103–25.

Chim, L. T., & On, W. F. (1973). Density, recruitment, mortality and growth of dipterocarp seedlings in virgin and logged over forests in Sabah. *Malayan Forester, 36*, 3–15.

Cole, L. C. (1954). The population consequences of life history phenomena. *Quart. Rev. Biol., 29*, 103–37.

Daljeet-Singh, K. (1974). Seed pests of some dipterocarps. *Malayan Forester, 37*, 24–36.

Daubenmire, R. (1972). Phenology and other characteristics of tropical semideciduous forest in northwestern Costa Rica. *J. Ecol., 60*, 147–70.

Evans, G. C. (1976). A sack of uncut diamonds: The study of ecosystems and the future resources of mankind. *J. Ecol., 64*, 1–40.

Fleming, T. H. (1974). Population ecology of two species of Costa Rican heteromyid rodents. *Ecology, 55*, 493–510.

Frankie, G. W., & Baker, H. G. (1974). The importance of pollinator behavior in the reproductive biology of tropical trees. *An. Inst. Biol. Univ. Nac. Autón. México, 45* (Ser. Bot.), 1–10.

Frankie, G. W., Baker, H. G., & Opler, P. A. (1974). Comparative phenological studies of trees in tropical wet and dry forests in the lowlands of Costa Rica. *J. Ecol., 62*, 881–919.

Heithaus, E. R., Fleming, T. H., & Opler, P. A. (1975). Foraging patterns and resource utilization in seven species of bats in a seasonal tropical forest. *Ecology, 56*, 841–54.

Heithaus, E. R., Opler, P. A., & Baker, H. G. (1974). Bat activity and pollination of *Bauhinia pauletia:* Plant-pollinator coevolution. *Ecology, 55*, 412–19.

Janzen, D. H. (1967). Synchronization of sexual reproduction of trees with the dry season in Central America. *Evolution, 21*, 620–37.

– (1970a). *Jacquinia pungens*, a heliophile from the understory of tropical deciduous forest. *Biotropica, 2*, 112–19.

– (1970b). Herbivores and the number of tree species in tropical forests. *Amer. Nat., 104*, 501–28.

– (1971a). Escape of juvenile *Dioclea megacarpa* (Leguminosae) vines from predators in a deciduous tropical forest. *Amer. Nat., 105*, 97–112.

– (1971b). The fate of *Scheelea rostrata* fruits beneath the parent tree: Predispersal attack by bruchids. *Principes, 15*, 89–101.

– (1971c). Escape of *Cassia grandis* L. beans from predators in time and space. *Ecology, 52*, 964–79.

– (1971d). Seed predation by animals. *Annu. Rev. Ecol. Syst., 2*, 465–92.

– (1972). Escape in space by *Sterculia apetala* seeds from the bug *Dysdercus fasciatus* in a Costa Rican deciduous forest. *Ecology, 53*, 350–61.

– (1973). Sweep samples of tropical foliage insects: Effects of seasons, vegetation types, elevation, time of day, and insularity. *Ecology, 54*, 687–708.

– (1974). Tropical blackwater rivers, animals, and mast fruiting by the Dipterocarpaceae. *Biotropica, 6*, 69–103.

– (1975a). Interactions of seeds and their insect predators/parasitoids in a

128 *Daniel H. Janzen*

tropical deciduous forest. In *Evolutionary Strategies of Parasitic Insects and Mites*, ed. P. W. Price, pp. 154–86. New York: Plenum Press.

– (1975b). Behavior of *Hymenaea courbaril* when its predispersal seed predator is absent. *Science, 189,* 145–7.

– (1975c). Intra- and inter-habitat variations in *Guazuma ulmifolia* (Sterculiaceae) seed predation by *Amblycerus cistelinus* (Bruchidae) in Costa Rica. *Ecology, 56,* 1009–13.

– (1976). Why bamboos wait so long to flower. *Annu. Rev. Ecol. Syst., 7,* 347–91.

– (1977). A note on optimal mate selection by plants. *Amer. Nat., 111,* 365–71.

Janzen, D. H., & Wilson, D. E. (1974). The cost of being dormant in the tropics. *Biotropica, 6,* 260–2.

Khan, M. A. W. (1970). Phenology of *Acacia nilotica* and *Eucalyptus microtheca* at Wad Medani (Sudan). *Ind. For., 96,* 226–48.

Lamprey, H. F., Halevy, G., & Makacha, S. (1974). Interactions between *Acacia*, bruchid seed beetles and large herbivores. *E. Afr. Wild. J., 12,* 81–5.

Lowe, R. G. (1968). Periodicity of a tropical rain forest tree *Triplochiton scleroxylon* K. Schum. *Commonwealth For. Rev., 47,* 150–63.

McKey, D. (1975). The ecology of coevolved seed dispersal systems. In *Coevolution of Animals and Plants*, ed. L. E. Gilbert & P. H. Raven, pp. 159–91. Austin: University of Texas Press.

Medway, L. (1972). Phenology of a tropical rain forest in Malaya. *Biol. J. Linn. Soc. Lond., 4,* 117–46.

Ng, F. S. P., & Loh, H. S. (1974). Flowering-to-fruiting periods of Malaysian trees. *Malayan Forester, 37,* 127–32.

Opler, P. A., Baker, H. G., & Frankie, G. W. (1975). Reproductive biology of some Costa Rican *Cordia* species (Boraginaceae). *Biotropica, 7,* 234–47.

Phillips, J. F. V. (1926). *Virgillia capensis* Lamk. ("keurboom"): A contribution to its ecology and sylviculture. *S. Afr. J. Sci., 23,* 435–54.

Phillips, J. (1927). Mortality in the flowers, fruits and young regeneration of trees in the Knysna forests of South Africa. *Ecology, 8,* 435–44.

Ramirez, W. B. (1970). Host specificity of fig wasps (Agaonidae). *Evolution, 24,* 680–91.

Singh, L. B. (1960). *The Mango.* London: Leonard Hill.

Snow, D. W. (1965). A possible selective factor in the evolution of fruiting seasons in tropical forest. *Oikos, 15,* 274–81.

Turner, D. C. (1975). *The Vampire Bat.* Baltimore: Johns Hopkins University Press.

Whitmore, T. C. (1975). *Tropical Rain Forests of the Far East.* Oxford: Clarendon Press.

Wilson, D. E., & Janzen, D. H. (1972). Predation on *Scheelea* palm seeds by bruchid beetles: Seed density and distance from the parent palm. *Ecology, 53,* 954–9.

Wood, G. H. S. (1956). The dipterocarp flowering season in North Borneo, 1955. *Malayan Forester, 19,* 193–201.

5

Strategies of establishment in Malayan forest trees

F. S. P. NG

Forest Research Institute, Kepong, Selangor, Malaysia

The Malay Peninsula or Malaya lies between latitudes 1.25° N and 6.50° N. The highest and lowest temperatures ever recorded in the lowlands were 103° F (39.4° C) and 60° F (15.6° C), respectively (Wyatt-Smith, 1963). In the highlands the lowest temperature recorded was 36° F (2.2° C). There are regional differences in rainfall, but on the whole there is enough rain to ensure evergreen conditions throughout the country throughout the year. The natural vegetation is tropical rain forest containing about 2500 species of trees (Corner, 1940).

The study of establishment strategies in Malayan forests has been carried out almost entirely by members of the Forest Department, although the terms "recruitment," "regeneration," "enrichment," and "rehabilitation" would be more familiar to foresters than the rather more abstract term "establishment strategy."

My own studies on establishment are centred in the nursery of the Forest Research Institute, Kepong, in which I am raising seedlings for the purpose of writing a manual for seedling identification (Ng, 1973, 1975). This gives me an opportunity to study the behaviour of a wide range of species with minumum interruption. In the forest, it is difficult to keep seedlings under observation for long periods because of the very high mortality rates under natural conditions. Seedlings tend to be eaten, buried under litter, washed away by rain, trampled by animals, starved by lack of light or water, etc. In the nursery the seeds are sown, as soon as possible after collection under a light cover of soil in a loamy mixture. The boxes are kept under light shade, watered twice daily, and protected from the larger predators by wire netting. Under such standard conditions, the intrinsic properties of the seedlings in germination and establishment are revealed with minimum external interference.

So far, about 300 species have been germinated from freshly collected seeds. Excluding nonindigenous species, species that have not yet been identified, and species that have not yet completed germination, we have data for about 200 species, listed in Table 5.1. This represents about 8% of the indigenous tree flora. Although this is not a random sample in the statistical sense, it is, nevertheless, not deliberately biased, because our policy has been to test every species for which seed was available during the past few years. Whether the conclusions based on this sample are applicable to the Malayan tree flora as a whole remains to be seen. My seedling project will probably be completed when a total of 400–500 species has been tested. Hence, at this halfway mark, it is appropriate to analyse the data in hand in order to formulate a theoretic framework for verification at the end of the project.

In observing the behaviour of seedlings, three areas have come to my attention in which contrasting establishment strategies appear to have evolved. These are (1) germination rate, (2) initial seedling morphology, and (3) initial seedling size. The consequences of the different strategies must remain largely a matter for speculation until experiments of a more critical nature are carried out. I propose to attempt to define the different strategies and to work out the percentage of species involved in each strategy. In this way, I hope to differentiate strategies that are within the mainstream of evolution in tropical rain forest from those strategies that belong to lesser streams.

Germination rates
Rapid germination
The seeds of most dominant trees in Malayan forests germinate within a few days of shedding, and after a few weeks, any seeds that have not germinated are found to be dead. This brief period of viability has been noticed by all who have ever worked in Malayan forests because its result is that the forest floor is densely carpeted by seedlings beneath and around the mother trees shortly after heavy fruiting. This rapid germination of all viable seeds within a short time after seed fall is obviously a well-defined, but puzzling, strategy of establishment. Superficially, it appears ineffective because the seedlings compete directly with each other and with the parent.

However, if the majority of species have adopted this strategy, it must have been within the mainstream of evolution in the humid tropical environment. Our data indicate that the majority have indeed adopted this strategy. I have defined "rapid germination" as total germination of all viable seeds within 12 weeks of seed fall. The period 12 weeks was chosen because all dipterocarps, with only one known exception in Malaya, would fall within the category so defined. As the Dipterocarpaceae is the dominant family in Malayan forests, it would serve as an eminently suitable biologic standard for comparing performance of other trees. I have also been influenced by practical considerations of forest management. Species that germinate totally within 12 weeks form a distinct category worthy of recognition by silviculturists, seed technologists, and nurserymen. They are most vulnerable to extinction by logging because the forest floor does not hold a reservoir of their seeds. The need for methods of seed storage for such species is becoming increasingly urgent. At the same time, such species simplify nursery management because their rapid rate of turnover ensures that space and labour are not tied up for uneconomically long periods.

If we accept 12 weeks as the boundary, then 118 of the 180 species for which germination data are available fall within this category. This represents 65% of our sample and includes nearly all emergent and main-storey species so far tested. I am not aware of any further experiments in the Malayan region designed specifically to demonstrate what the selective advantage of rapid germination might be.

However, it has been observed by Burgess (1975) that in natural forest, seeds of *Shorea curtisii* are heavily attacked by ants, which eat the cotyledons as soon as they are exposed in germination. In one experiment, where pregerminated seeds were sown in four quadrants in the forest, not a single seedling succeeded in establishing itself. Those that were not destroyed by ants were killed by crickets or possibly rats. Burgess suggested that a high concentration of seedlings is necessary to ensure the survival of a few, but he did not suggest how exactly a high concentration would help. There are at least two possibilities. One is that the survivors are most likely to be found in areas of highest seedling density because in such areas there may be more food than the predators can consume. The trouble with this hypothesis is that the area of greatest seedling density is directly under the parent tree, so that subsequent development of the surviviors would have to be at the expense of the parent, or vice versa. At best, such a strategy, if it operates in this manner, would lead to a one-for-one replacement on the spot; but there is a lot of evidence, particularly from agriculture, that it is difficult to maintain the same species on the same ground through two or more rotations. The other possibility is that the surviors are most likely to be found at the edge of the dispersal area, where seedling density is lowest, and when there is at the same time a high concentration of food around the parent tree to which predators would be attracted. Experimental support of this hypothesis can be found in a study by Wilson & Janzen (1972) in which *Scheelea* palm seeds were found to suffer a higher intensity of predation if they were clustered than if they were isolated.

Table 5.1. *Germination rate, initial morphology, and seed size of 210 species of Malayan trees*

| Species | Sample size[a] | Germination[b] | | | | Initial morphology | | | Seed size class[c] | Comment[d] |
| | | Prolonged | | | Epigeal | Semi-hypogeal | Hypogeal | Durian | | |
		Rapid	Delayed	Inter-mediate						
Actinidiaceae										
Saurauia roxburghii	292	2–4			+				1	
Alangiaceae										
Alangium ebenaceum	50			4–14	+				4	
Anacardiaceae										
Bouea macrophylla	18	3–4						+	4	A
Dracontomelum mangiferum	16	5–7			+				4	
Annonaceae										
Cyathocalyx pruniferus	78	3–8			+				3	
Mezzettia leptopoda	7		12–20					+	5	
Monocarpia marginalis	39		18–36		+				4	
Polyalthia glauca	18	6–7						+	3	
Apocynaceae										
Alstonia angustiloba	97	2–8			+				2	
Dyera costulata	16			7–18	+				4	
Hunteria zeylanica	17	2–6			+				3	

Araliaceae						
Arthrophyllum ovalifolium	81	4-14	+			2
Bignoniaceae						
Oroxylum indicum	33	3-4	+			2
Bombacaceae						
Bombax valetonii			+			3
Durio carinatus					+	5
D. griffithii	13	1-3			+	3
D. zibethinus	10	2-4			+	6
Burseraceae						
Canarium littorale	31	6-12	+			6
C. megalanthum			+			6
C. patentinervium			+			5
C. pseudo-sumatranum	22	5-12	+			5
C. reniforme			+			5
Dacryodes costata	14	4-6	+			3
D. kingii	6	6-12	+			4
D. rostrata	7	3-6	+			4
Santiria laevigata	30	4-10	+			3
S. oblongifolia	17	4-8	+			3
S. rubiginosa	17	7-16	+	+		2
Scutinanthe brunnea			+			4
Triomma malaccensis	6	2-3	+			3
Casuarinaceae						
Casuarina equisetifolia	69	2-11	+			1

Table 5.1 (*cont.*)

| Species | Sample size[a] | Germination[b] | | | Initial morphology | | | | Seed size class[c] | Comment[d] |
| | | Rapid | Prolonged | | Epigeal | Semi-hypogeal | Hypogeal | Durian | | |
			Delayed	Inter-mediate						
Celastraceae										
Kokoona reflexa									3	B
Bhesa robusta	24		2–6					+	3	
Combretaceae										
Terminalia catappa	26		6–12		+				6	
T. phellocarpa	22		3–6		+				4	
T. subspathulata					+				3	
Cornaceae										
Mastixia pentandra	93			6–18	+				3	
Dilleniaceae										
Dillenia aff. grandifolia	92			11–52	+				2	
D. ovata					+				2	
D. suffructicosa	23		5–7		+				2	
D. sumatrana					+				2	
Dipterocarpaceae										
Balanocarpus heimii	43		2–7		+				6	
Dryobalanops aromatica	21		1–2		+				3	

134

Species	N	Range					No.	
D. oblongifolia	17	2–3				+	3	
Dipterocarpus baudii	7	3–4				+	4	
D. oblongifolius	100	2–4				+	3	
Hopea dyeri	136	1–5		+			2	
H. helferi	50	2		+			2	
H. nervosa	9	3–6		+			3	
H. nutans	39	2–3		+			2	
H. odorata	96	1–3		+			2	
H. subalata	54	2–9		+			3	
Parashorea densiflora	62		4–30	+			4	C
Shorea assamica	10	2–4		+			3	
S. leprosula	39	1–5		+			3	
S. maxima	25	1–3		+			6	
S. ovalis	85	3–5		+			4	
S. parvifolia	26	1–4		+			3	
S. platyclados	80	1–3		+			3	
S. resinosa				+			4	
S. singkawang	15	2–4		+			6	
S. sumatrana	98	2–3		+			3	
S. talura	105	3–7			+		3	
Vatica lowii	73	3–10		+			3	
V. stapfiana	86	3–6				+	5	B
V. wallichii	29	4–5				+	5	
Ebenaceae								
Diospyros argentea	5	3–5		+			6	
D. confertiflora				+			3	
D. diepenhorstii				+			6	
D. ismailii	45		3–36	+			4	

135

Table 5.1 (cont.)

Species	Sample size[a]	Germination[b]				Initial morphology				Seed size class[c]	Comment[d]
			Prolonged								
		Rapid	Delayed	Inter-mediate	Epigeal	Semi-hypogeal	Hypogeal	Durian			
Ebenaceae (cont.)											
D. maingayi	15			5–26				+	4	
D. pendula	63			10–18			+		4	
D. pilosanthera	47	4–6			+				3	
D. sumatrana	17	2–5			+				3	
D. trengganuensis					+				3	
Elaeocarpaceae											
Elaeocarpus floribundus	5		24–36		+				4	
E. petiolatus	25			11–16	+				3	
E. stipularis	11			10–20	+				3	
Ericaceae											
Gaultheria leucocarpa	184	3–9			+				1	
Erythroxylaceae											
Erythroxylum kochumenii					+				3	
Euphorbiaceae											
Antidesma neurocarpum	42			7–14	+				2	
Baccaurea griffithii	57	2–3			+				4	

Species								
B. motleyana	50	2–4			+		4	
Cheilosa malayana	7	3–4			+		4	
Elateriospermum tapos	68	3–6			+		5	
Fahrenheitia pendula	47	2–7			+		3	
Glochidion obscurum	64			4–15	+		2	
G. sericeum	51			4–38	+		1	
Macaranga tanarius					+		2	
Trigonistemon elegantissimus	30	1–2			+		5	
Fagaceae								
Lithocarpus encleisacarpus						+	4	
L. ewyckii	23	6–19				+	4	
Flacourtiaceae								
Hydnocarpus woodii	12		23–124		+		5	D
Guttiferae								
Garcinia cataractalis						+	3	E
G. nigrolineata						+	4	E
G. opaca	16		13–26	8–13		+	4	E
G. parvifolia	10					+	3	E
Mesua ferrea	18	2–3				+	4	
Hamamelidaceae								
Maingaya malayana					+		3	
Icacinaceae								
Platea latifolia	22	7–14			+		5	
Lauraceae								
Litsea castanea	73	7–18				+	4	B

Table 5.1 (*cont.*)

Species	Sample size[a]	Germination[b]			Initial morphology				Seed size class[c]	Comment[d]
		Rapid	Delayed	Prolonged Intermediate	Epigeal	Semi-hypogeal	Hypogeal	Durian		
Lecythidaceae										
Barringtonia macrostachya	10		47–151				+		7	E
Careya arborea							+		6	E
Leeaceae										
Leea indica	11	3–9			+				2	
Leguminosae										
Adenanthera bicolor	12			5–44	+				2	
A. pavonina	85			3–18	+				2	
Crudia curtisii	23			8–30		+			6	
Dialium maingayi	8			1–70	+				3	
Fordia lanceolata						+			3	
Intsia palembanica	17			2–90	+				5	
Koompassia malaccensis	28	2–3			+				6	
Milletia atropurpurea	22			11–48			+		6	B
Ormosia venosa	10	2–7				+			3	
Parkia javanica	45			1–104	+				4	
P. speciosa	5	2			+				4	
Peltophorum pterocarpum	54			1–24	+				3	

Pithecellobium clypearia	30	1–2				+		2
P. ellipticum	56	2–3				+		4
P. pahangense	48	2–3				+		3
Pterocarpus indicus	48			2–16	+			3
Sindora echinocalyx	11		40–158		+			3
Linaceae								
Ixonanthes icosandra	5	5–7			+			2
Loganiaceae								
Fagraea fragrans	64	2–4			+			1
Lythraceae								
Lagerstroemia speciosa	124	2–7			+			2
Magnoliaceae								
Aromadendron elegans	99	3–12			+			2
Meliaceae								
Amoora malaccensis	17	3–7				+		6
Chisocheton macrothyrsus	49	5–11				+		4
Dysoxylum angustifolium	50	3–8		9–34		+		3
D. arborescens	30		16–35			+		3
D. cauliflorum	82	4–7				+		3
Lansium domesticum	13	3–5				+		3
Sandoricum koetjape	39				+			4
Walsura neuroides	17			9–24			+	3
Moraceae								
Artocarpus elasticus	87	3–8					+	3
A. gomezianus	24	6–9					+	2
A. integer	5	5–6				+		5

139

Table 5.1 (cont.)

Species	Sample size[a]	Germination[b]							Seed size class[c]	Comment[d]
		Prolonged			Initial morphology					
		Rapid	Delayed	Inter-mediate	Epigeal	Semi-hypogeal	Hypogeal	Durian		
Moraceae (cont.)										
A. lanceifolius	44	3–8					+		4	
A. lowii							+			
Ficus benjamina	46			2–14	+				1	
Parartocarpus bracteatus						+			4	
Myristicaceae										
Horsfieldia brachiata	18			10–22			+		4	
Knema curtisii	18	9–10					+		4	
K. furfuracea	50			6–14			+		3	
K. laurina	100	5–7					+		4	
K. scortechinii	10	8					+		3	
K. stenophylla							+		2	
Myristica crassa	7			7–15			+		5	
M. malaccensis	38			7–19			+		6	
Myrtaceae										
Eugenia claviflora	100	3–5			+				3	
E. graemeandersoniae	44			3–15		+			3	

Species								
E. grandis	55	3–10			+		3	
E. microcalyx	92	5–11			+		2	
E. operculata	99	1–4		+			2	
E. polyantha	39	3–12		+			2	
E. sp. 44	53		3–22		+		3	
E. sp. 66	19		5–22		+		4	
Leptospermum flavescens	38	3–5		+			1	
Tristania merguensis	87	2–3		+			2	
Olacaceae								
Scorodocarpus borneensis						+	6	
Strombosia javanica	80		10–27				5	
Passifloraceae								
Paropsia vareciformis	76	2–5		+			2	
Podocarpaceae								
Podocarpus neriifolius	90	3–10		+			3	
Polygalaceae								
Xanthophyllum griffithii	100	2–6					3	
X. obscurum	25	2–5			+		4	
Rhizophoraceae								
Anisophyllea corneri	10		5–24			+	7	E
A. disticha	17					+	3	E
A. grandis	22		7–40			+	7	E
A. griffithii		32–151					5	E
Pellacalyx saccardianus							1	
Rhizophora mucronata	56	6–12		+		+	5	

Table 5.1 (cont.)

Species	Sample size[a]	Germination[b]				Initial morphology				Seed size class[c]	Comment[d]
			Prolonged				Semi-				
		Rapid	Delayed	Intermediate	Epigeal	hypogeal	Hypogeal	Durian			
Rosaceae											
Prunus arborea	5		14-22				+		2	
Rubiaceae											
Anthocephalus cadamba					+				1	
Gardenia carinata	40	4-7			+				1	
Morinda citrifolia	276			5-30	+				2	
Randia densiflora					+				1	
R. scortechinii	77	6-11			+				1	
Rutaceae											
Euodia glabra	6		18-22		+				1	
Micromelum minutum	62	2-3			+				2	
Murraya paniculata	11	2-3			+				2	
Sapindaceae											
Arytera littoralis	93	3-7					+		3	
Grossonephelis penangensis	22			3-18			+		6	
Erioglossum edule	98	2-3					+		2	
Guioa pleuropteris	40	3-5			+				2	
Harpullia confusa	91	3-8				+			2	

Species									
Nephelium glabrum	78	2–6			1–13	+		3	B
N. lappaceum	22	1–3				+		4	B
N. malaiense	36	1–5				+		3	
Pometia pinnata	168		+					3	
Sapotaceae									
Madhuca utilis	29	4–12	+		4–16			5	
Mimusops elengi	17	2–5	+					3	
Payena lucida	26	4–6	+					4	
Planchonella glabra	24		+					4	
P. maingayi	29		+		10–26			3	
Sarcospermataceae									
Sarcosperma paniculatum	15				5–16		+	3	
Simarubaceae									
Irvingia malayana	35	5–10	+					6	
Sterculiaceae									
Heritiera simplicifolia	20	1–5	+					3	
Pterocymbium javanicum	105	1–2	+					3	
Pterospermum javanicum	149	1–3	+					3	
Pterygota alata	82		+		2–20			3	
Sterculia foetida	55	2–3	+					4	
S. parviflora	29	1–4	+					3	
Tetrameristaceae									
Tetramerista glabra	6		+	13–36				3	
Theaceae									
Adinandra acuminata	117	3–12	+					2	
Schima wallichii	96	2–3	+					2	

Table 5.1 (cont.)

Species	Sample size[a]	Germination[b]			Initial morphology				Seed size class[c]	Comment[d]
			Prolonged							
		Rapid	Delayed	Intermediate	Epigeal	Semi-hypogeal	Hypogeal	Durian		
Tiliaceae										
Grewia blattaefolia	36			11–36	+				3	
G. laurifolia	28		6–10		+				3	
Trigoniaceae										
Trigoniastrum hypoleucum	74		3–7		+				4	
Verbenaceae										
Teijsmanniodendron pteropodum	10			5–22					5	
Vitex pubescens	15		6–7					+	3	
Violaceae										
Rinorea anguifera	37		1–5		+				2	
Total	209	118 (65%)	12 (7%)	50 (28%)	134 (64%)	21 (10%)	39 (18%)	16 (8%)		

[a]Number of viable seeds in test sample (i.e., number of seedlings successfully germinated).
[b]Weeks in which first and last seedlings appeared in samples of at least five viable seeds.
[c]1, < 0.3 cm long; 2, 0.3 – 1.0 cm; 3, 1.0 – 2.0 cm; 4, 2.0 – 3.0 cm; 5, 3.0 – 4.0 cm; 6, 4.0 – 5.0 cm; 7, 6.0 – 8.0 cm.
[d]A, cotyledons slightly exposed; B, cotyledons partially exposed; C, seed coat adherent to cotyledons; D, cotyledons normally exposed, sometimes not; E, cotyledons apparently absent.

144

The predominance of rapid germination as a reproductive strategy must, in any case, force us to reject the prevalent idea that it is wasteful or ineffective. What seems more likely is that it effectively attracts and concentrates predators to the centre of the area of seed-fall, creating a massive diversion and sacrificing the vast majority of seedlings in order that the few dispersed furthest may have a better chance of survival (see Chap. 4).

Prolonged germination

Having defined rapid germination, we are left with species that begin to germinate after at least 12 weeks of dormancy and those that begin to germinate before 12 weeks but do not achieve total germination within 12 weeks. Those that are dormant for at least 12 weeks may be described as exhibiting "delayed germination." Only 7% of our sample fall into this group, which, although quite artificial and arbitrary in definition, serves a purpose because it is useful, in tropical forestry, to know which seeds can be easily stored for a significant period. The remaining species are listed as showing "intermediate germination" and constitute 28% of our sample.

The species showing delayed and intermediate germination grouped together make up 35% of our sample and may be described as exhibiting "prolonged germination." I think this combined grouping is ecologically meaningful. Species that exhibit prolonged germination tend to produce seedlings a few at a time over an extended period, and this seems to be a definite strategy. Predator pressure and infraspecific competition are reduced. There is a long-lasting reservoir of viable seeds on the ground. While they are on the ground, they can be moved by rain and animals, so the dispersal time is effectively lengthened and the ultimate dispersal area enlarged.

In the forests of Malaya the difference between the two contrasting strategies is pronounced. Species with rapid germination are made conspicuous by their carpets of seedlings after fruiting. Species with prolonged germination do not produce such carpets, but have seeds that lie around for a long time; if the seeds are big enough (e.g., in *Barringtonia, Anisophyllea, Intsia, Sindora*), one can observe and pick up viable seeds at virtually any time of the year.

Dormancy as such does not appear to be important in an ever-moist ever-warm environment. What seems more important is differential dormancy, whereby individual seeds of the same crop exhibit different degrees of dormancy, so that a small proportion germinates at a time over an extended period. Nevertheless, despite the obvious advantages, prolonged germination, which is the result of differential dormancy, is a minority phenomenon; this suggests that somehow it is an inferior strategy for trees of tropical rain forests.

It is also worth noting that leguminous seeds do not always exhibit dormancy, as is often assumed. A distinction must be made between soft-seeded and hard-seeded legumes (Ng, 1974). Soft-seeded legumes can be cut through with a sharp razor blade, and these germinate rapidly. The hard-seeded ones are difficult to cut, and these generally exhibit prolonged germination. The difference in hardness is not just in the seed coat but also in the embryo and albumen and is reflected in marked differences in moisture content of the seed as a whole. The genera *Pithecellobium* and *Koompassia* appear to be wholly soft-seeded. The genus *Parkia* contains both soft-seeded (e.g., *P. speciosa*) and hard-seeded (e.g., *P. javanica*) species.

Initial seedling morphology

The initial morphology of seedlings has traditionally been described by the two contrasting terms: "epigeal" (above ground) and "hypogeal" (below ground), with reference to the position of the cotyledons at the stage when the seedling axis is fully straightened. As the lifting up of the cotyledons is obviously the function of the hypocotyl, I think the emphasis should have been focussed on the behaviour of the hypocotyl rather than on behaviour of the cotyledons.

Duke (1965) proposed that the terms epigeal and hypogeal should be replaced by the terms "phanerocotylar" (cotyledons exposed) and "cryptocotylar" (cotyledons hidden). I agree that it is important to distinguish between exposed and hidden cotyledons, but it was unfortunate that Duke made phanerocotylar synonymous with epigeal and cryptocotylar with hypogeal. There is room for both sets of terms if one set is applied solely to the hypocotyl and the other to the cotyledons. Our work with Malayan seedlings shows that the behaviour of the hypocotyl is independent of the behaviour of the cotyledons, so that four combinations result, as illustrated in Fig. 5.1. However, because it is impossible to amend established usage without causing confusion, I have, for the present purpose continued to use the terms epigeal and hypogeal and have added two more: "semihypogeal" and "durian."

Epigeal germination

Epigeal germination is that in which a hypocotyl is developed and the cotyledons are exposed (Fig. 5.2). An impressive 64% of our sample exhibit this condition. The epigeal condition allows the cotyledons to function as photosynthetic organs. Hence, the seedling can become independent of stored food resources at a very early stage. It is probably significant that all species with seeds smaller than 3 mm (measured along the longest axis excluding wings) exhibit the epigeal condition. In general, the frequency of epigeal germination declines with seed size (Table 5.1).

Nevertheless, some epigeal species have not exploited this photosynthetic advantage. In *Diospyros argentea* and *Diospyros diepenhorstii* in Malaya and several species of *Diospyros* in Ceylon (Wright, 1904) and India (Troup, 1921), the cotyledons emerge and are shed almost immediately afterwards without ever turning green. However, the first two leaves are precociously developed as if to compensate for the early loss of the cotyledons.

Semihypogeal and hypogeal germination

The semihypogeal condition occurs when the hypocotyl is undeveloped and the cotyledons are exposed (Figs. 5.3, 5.4). The exposed cotyledons lie on or in the ground, usually with the broken seed coat still adhering more or less loosely to their outer surfaces. This is evidently not a good position for photosynthesis because green cotyledons at ground level are rarely encountered. But with the weight of the cotyledons fully supported by the ground, conditions are specially favourable for the cotyledons to be developed into bulky storage organs. All semihypogeal seedlings have bulky cotyledons.

Fig. 5.1. Initial seedling morphology in four types of germination: (A) epigeal, (B) hypogeal, (C) semihypogeal, (D) durian. Pie chart shows relative percentage of each type in survey made (see Table 5.1).

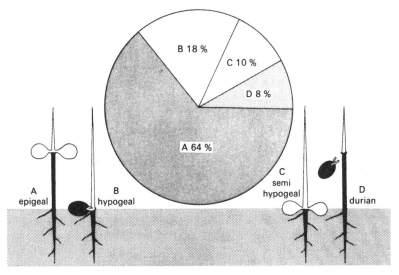

148   F. S. P. Ng

Many species exhibiting the semihypogeal condition have previously been described as hypogeal. For example, Burger (1972) did this with *Aglaia eusideroxylon* and *Dysoxylum gaudichaudianum*. However, he correctly described *Pithecellobium jiringa* as "semi-hypogeous". The semihypogeal condition merges into the hypogeal condition, in which the hypocotyl is undeveloped and the cotyle-

Fig. 5.2. Epigeal germination: *Pittosporum ferrugineum* (Pittosporaceae).

dons are hidden within the seed coat (Fig. 5.5). I use the term "merge" because many species are intermediate in the sense that their cotyledons are partially exposed. I have, in Table 5.1, placed the intermediates in the hypogeal class purely as a matter of convenience. The resting of the seed body on the ground in the hypogeal condition also favours the development of bulky storage tissue, but

Fig. 5.3. Semihypogeal germination: *Amoora* sp. (Meliaceae).

in this case, the endosperm may develop this bulk instead of the cotyledons. The storage function may also be passed on to a specialised hypocotyl, as in *Garcinia, Barringtonia* (Fig. 5.6) and *Anisophyllea* in which cotyledons and endosperm are absent or undeveloped. In such cases, the hypocotyl is contained within the seed coat and does not appear above ground. There are some other peculiarities

Fig. 5.4. Semihypogeal germination: *Pithecellobium ellipticum* (Mimosaceae).

about this "storage hypocotyl." It does not grow during germination, does not become erect, and in *Garcinia* and *Anisophyllea* is eventually abscissed. I have lumped these examples into the hypogeal grouping because a true hypocotyl, in the behavioural sense, is lacking.

The semihypogeal condition is exhibited by 10% of our sample

Fig. 5.5. Hypogeal germination: *Scorodocarpus borneensis* (Olacaceae).

and the hypogeal condition by 18%. Together, they make up 28%, representing a strategy in which the photosynthetic potential of the cotyledons is restricted or sacrificed altogether. The ability of the seed to store food reserves is enhanced. As if to emphasise their abundance of stored reserves, the semihypogeal and hypogeal seedlings often produce scale leaves before foliage leaves. The smallest

Fig. 5.6. Hypogeal germination: *Barringtonia macrostachya* (Lecythidaceae).

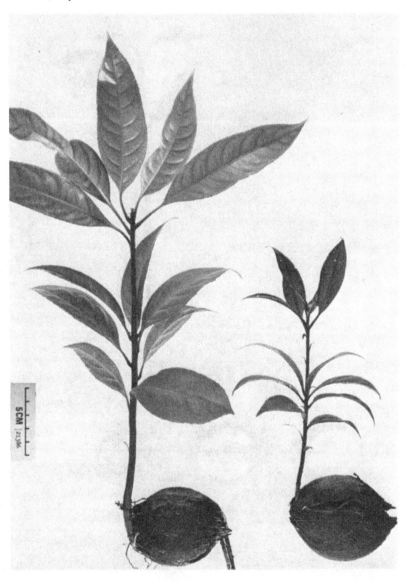

seeds found in this combined grouping are in *Eugenia microcalyx;* they were 4–5 mm long. All seeds 6 cm or longer are in the hypogeal or semihypogeal group.

Jackson (1968, 1974) described "cryptogeal germination," in which the hypocotyl is undeveloped, the cotyledons exposed or hidden, and the embryonic axis (plumule, cotyledonary node, and radicle) pushed into the ground by the elongating cotyledonary stalks. As a result, the shoot arises from below the ground even if the seed is germinated on the surface. As a hypocotyl is not developed, cryptogeal germination may be regarded as a special case of hypogeal or semihypogeal germination, depending on whether the cotyledons are hidden or exposed. I have come across cryptogeal germination in Malaya only in the palms *Eugeissona tristis* and some species of *Livistona.* Whitmore (1973) suggests that this is a device to get the seedling of *Eugeissona* into the ground through the litter layer. Jackson suggests that cryptogeal germination is an adaptation that enables young seedlings to survive bush fires in the savanna lands of Africa. In the context of tropical rain forest, where the role of fire in plant evolution may be discounted, Whitmore's suggestion is an attractive one.

Durian germination

The durian condition occurs when the hypocotyl is developed and the cotyledons are hidden. It occurs mainly in the genera *Durio, Rhizophora,* and *Dipterocarpus,* which are too important in the tropics to be ignored. Burger (1972) dealt with the durian (*Durio zibethinus*) in a curious way that suggests he was aware of the difficulties in terminology, but was not prepared to propose amendments. He described the cotyledons as hypogeous and the hypocotyl as epigeous. In fact, the cotyledons, with the seed coat around them, tend to be lifted above the ground (Fig. 5.7).

Only 8% of species in our sample exhibit the durian condition, which appears to be absent from nontropical regions. This suggests that it is selectively at a disadvantage outside the tropics and perhaps within the tropics as well. In its most basic manifestation, the developing hypocotyl lifts the seed body well above the ground before shedding it, with the cotyledons still enclosed (Fig. 5.8). Hence, the photosynthetic potential of the cotyledons is lost, while at the same time, energy is spent in lifting a body for no obvious advantage. The impression that the basic durian condition is selectively negative or at best neutral is strengthened by the fact that many species in this group have compensatory strategies to get around their basic problem.

In *Durio* spp., as also in *Polyalthia glauca* and *Diospyros maingayi,* the cotyledons are abscissed and discarded with the seed body before the hypocotyl loop straightens. In *Dipterocarpus* and some spe-

154   *F. S. P. Ng*

cies of *Vatica* (Fig. 5.9) the lifting of the seed body is avoided by the development of long cotyledonary stalks connecting the seed body in the ground to the elevated cotyledonary node. In *Rhizophora* the question of lifting does not even arise. The hypocotyl develops viviparously, and the seedling frees itself by leaving its cotyledons behind on the tree, within the seed and fruit body.

Fig. 5.7. Durian germination: *Durio zibethinus* (Lecythidaceae).

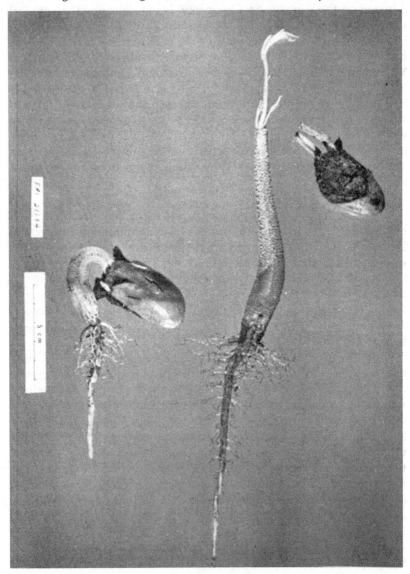

Hence, the durian condition is distinguished more by its variations than by its basic form, which suggests that the basic form is merely tolerated in the tropical rain forest, if not actually selected against.

Initial seedling size

It has been mentioned in passing that seed size has some influence on whether seedlings are or are not epigeal. Seed size probably also determines the intial size of the seedling. In the concept of initial size, we are concerned with the size the seedling attains as a result of stored reserves, not what it builds up by its own effort. The concept is difficult to define exactly, but it is readily grasped when one compares the minute seedlings of a small-seeded family such as Rubiaceae with those of a large-seeded family such as Dipterocarpaceae at the same stages of early development.

Fig. 5.8. Durian germination: *Strombosia javanica* (Olacaceae).

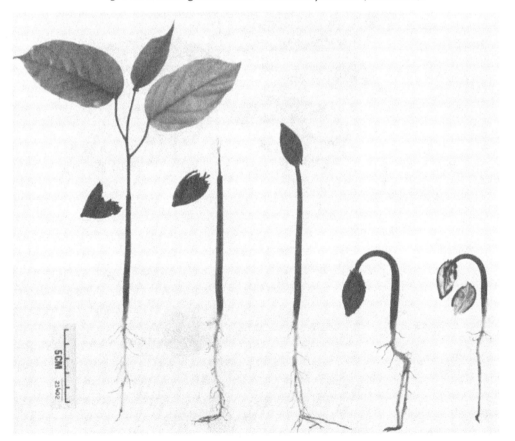

Our experience in the nursery suggests that initial seedling size makes quite a difference in establishment. Seedlings from seeds 3 mm in diameter or smaller are more susceptible to fungus if overwatered and are easily desiccated if underwatered. On the other hand, they can find a foothold in the tiniest crevices, probably because their roots are finer. In the forest, they are easily smothered by litter.

Fig. 5.9. Durian germination: *Vatica wallichii* (Dipterocarpaceae).

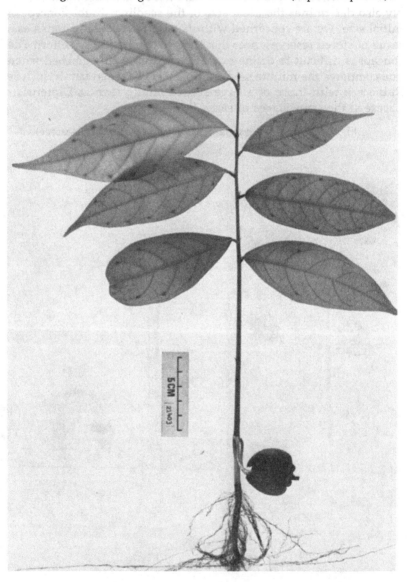

Big seedlings are generally hardier and easier to manage, but in nature they seem to have difficulty anchoring themselves to the ground. One sometimes comes across dipterocarp seedlings in natural forest that are unable to stand erect because they have not been able to become rooted properly. This is particularly so in areas where the soil is compacted.

However, the most important differences between big and small seedlings may be in the amount of food reserves they contain, which determines their ability to persist under dense shade. A good example of the role of stored reserves in the life of seedlings is provided by *Parashorea densiflora,* the seed of which measures about 25 mm by 25 mm and produces a correspondingly large seedling. It has what appears to be a mycorrhizal problem under nursery conditions. The seedling develops a number of scale leaves and then dies. Rarely one foliage leaf may be produced but to no avail. What is remarkable is that the seedlings do not die until after 6 months. A small seedling could not possibly survive as long on its own reserves. I have also come across albino seedlings in various species, and my impression is that those from smaller seeds die first.

A big seedling, therefore, is better equipped to wait in the shade for a gap to appear in the canopy. It is less easily overwhelmed by fungus while waiting, despite the dark and humid conditions on the forest floor. On the other hand, mobility of the species is reduced because the seeds that produce big seedlings have to be fewer, larger, heavier, and consequently less widely dispersed. Hence, there must be a limit to size beyond which the disadvantages would outweigh the advantages. In order to discover the optimum size, one should first be able to quantify initial seedling size; but this is easier said than done. The best measure would probably be the dry weight of the embryo and endosperm in a good seed, but this is obviously difficult to determine. As a very crude and ready measure, I have used the length (longest axis) of the seed, excluding wings, to indicate the size of the seed and, therefore, of the seedling. The "seed" is true seed in the botanical sense if it can be separated intact from the fruit. If not, the stone is measured if it can be separated intact from the rest of the fruit. Failing both, the whole fruit is measured, but excluding wings. The results are given in Table 5.1 and summarised in Table 5.2.

The most common size class is class 3 (1.0–2.0 cm), exhibited by 74 species (35%). Seeds of 1 cm or longer are found in 157 species (75%); this supports the common belief that big seeds predominate in tropical forests.

Conclusions

The mainstream of evolution in the tropical rain forest of Malaya appears to favour large seeds and seedlings (75% of species have seeds 1 cm or longer) and germination that is rapid (65%) and epigeal (64%).

Large seedlings are able to survive longer under dense shade and apparently are less susceptible to fungal attack, probably an important factor under conditions of low light intensity and high humidity. At the same time, large seedlings have thick taproots and need looser soil in order to anchor themselves. This is no problem in primary forest. The seeds that produce large seedlings are large; hence, they are produced in smaller numbers and are less widely dispersed. We have to adopt the view that efficiency in dispersal is not necessarily the same as efficiency in species survival. In climax tropical rain forest, which covered the whole of the Malayan tropics until comparatively recent times, the ability to survive under shade in humid conditions must have been of major importance in evolution. Species mobility appears to be of secondary importance, because in nature, the opportunities for mobile species are limited to relatively small gaps of sporadic occurrence (Chaps. 23, 27). In other parts of the world, periodic fire, grazing by large herds of herbivores, and extremes of cold, heat, and dryness recreate at regular intervals large tracts of open land for colonisation. The opportunities for pioneering species are enhanced. In the tropical rain forest the opportunities have been so limited that pioneers elsewhere are here reduced to the

Table 5.2. *Frequency distribution of species and initial seedling morphology according to seed length*

Size class	Definition (length in cm)	No. of species	Species with epigeal germination	
			No.	%
1	<0.3	13	13	100
2	0.3–1.0	39	31	79
3	1.0–2.0	74	48	65
4	2.0–3.0	43	23	53
5	3.0–4.0	19	9	47
6	4.0–6.0	18	10	55
7	6.0–8.0	3	0	0
Total		209		

status of biological nomads, to use a term coined by van Steenis (1958).

There are, of course, different degrees of nomadism. Table 5.1 lists 23 species that tend to be found in clearings and at the forest fringes. These are undoubtedly nomads. The species are *Adenanthera bicolor, Adenanthera pavonina, Adinandra acuminata, Alstonia angustiloba, Anthocephalus cadamba, Arthrophyllum ovalifolium, Casuarina equisetifolia, Dillenia suffructicosa, Euodia glabra, Fagraea fragrans, Gaultheria leucocarpa, Guioa pleuropteris, Ixonanthes icosandra, Leea indica, Macaranga tanarius, Morinda citrifolia, Oroxylum indicum, Paropsia vareciformis, Pellacalyx saccardianus, Pithecellobium clypearia, Randia scortechinii, Saurauia roxburghii,* and *Schima wallichii.*

It is instructive that all have seeds shorter than 1 cm, and all except *Pithecellobium clypearia* exhibit epigeal germination. The apparent advantages of small seeds are that they are produced in larger numbers, are dispersed more widely, and can become rooted in harder ground. In the shade they are easily starved and overwhelmed by fungus, but in the open they can photosynthesise and grow immediately. Their cotyledons are photosynthetic, which gives them a faster start. Rapid germination is probably a strategy whereby predators can be attracted to places of high seedling density (i.e., to the centre of the dispersal area), so that the few seedlings that happen to be dispersed further may have a better chance of escape. Prolonged germination or seed dormancy is often thought to be specially associated with nomadic or pioneer species. Our data do not support this idea. Of the 23 undoubted nomads, only 6 (25%) exhibit prolonged germination, compared with 35% for the entire sample. Data on one (*Anthocephalus cadamba*) were not available. As Whitmore (1975) stressed, there is a need to distinguish between the germination of dormant seeds and the rapid seeding-in of species that produce seed continuously and that are widely dispersed. We conclude that dormancy is not particularly characteristic of pioneer species in the humid tropics. Early reproductive maturity, high frequency of seeding (or continuous seeding), small seed size, epigeal germination with photosynthetic cotyledons, a fine root system, large numbers of seeds, and wide dispersal are probably far more important.

Acknowledgments

I would like to thank all the staff of the Forest Research Institute who helped in collecting seeds or raising the seedlings, and in particular Mat Asri bin Ngah Sanah, who kept the germination records, and Isaac S.Y. Ho, who took the photographs.

General discussion

Ashton: Our student in Malaysia, H. T. Chan, has quantitative information from *Shorea ovalis* relevant to your prediction that the survival rate of dipterocarp seeds and seedlings is higher, further away from the tree. Another student, S. K. Yap, has a similar confirmation for an understory species, *Xerospermum intermedium.*

In regard to these proportions of "gap phase," or secondary forest, species that have short and long dormancy, how did you test this dormancy? Are you talking about delayed germination or dormancy, and how do you investigate it?

Ng: I am talking about delayed germination, established by checking a seed population every week until no further germination takes place. After a long period without appearance of shoots, I look for what is left in the ground to see if all are rotten. If I find any seed still in good condition, I wait for another few months.

Ashton: Does the experiment not depend very much on the condition under which seeds are grown? If you put them all into shade, you might find them dormant over a long period. If, on the other hand, you grow them under light conditions, you are not really demonstrating that they don't have dormancy.

Ng: Yes, germination can depend on other conditions, whether the seed has been eaten by birds or lies on the forest floor and is desiccated before the rains come. I am merely trying to establish a base line with constant methods and carry out a broad survey. Species may subsequently be studied in more detail.

Givnish: Dr. Ng, you hypothesize that the tendency toward larger seed sizes in the humid tropics is due to the facts that the seeds have to penetrate a large amount of litter and have to grow up in the shade. It was my impression that we don't really know much about the conditions under which the seeds of the trees presently in the canopy or at various lower levels develop.

Ng: Quite a number of forest plots have been made to study what the foresters call "recruitment," but few seedlings persist for more than a year or two, unless artificial clearing of the older trees is practiced. I don't know how one can do such experiments on the ground without manipulating the canopy.

Baker: The germination method you mention that includes the seed within the seed coat, but by extension of the cotyledonary stalks pushes the embryo down into the ground, is also represented in temperate floras. In the Cucurbitaceae, the genus *Marah* shows this (Schlising, 1969), and it has the apparent function of putting the embryo below ground in a circumstance where there might be aridity very soon after germination takes place. A very large tuber develops below ground, for which reason the plant is called "man root." There is also some evidence that the same germination procedure takes place in some of the temperate oak trees of the genus *Quercus,* where there is an extension of the cotyledonary stalk and burying of the embryo.

In Costa Rica we looked at seed size in a number of forest trees and other life forms and related this to the germination methods. We found the same correlation between small seeds and epigeal germination and large seeds and hypogeal germination that you have reported. Of additional interest is the finding that the large seeds tend to be the only ones that store starch; they may have oil as well. But in the small seeds the more compact energy

source, oil, is the one selected for. Have you any observation on the food reserves of your plants?

Ng: No, we have made no seed analyses.

Whitmore: Some of them are very oil-rich – e.g., the illipe nuts (*Shorea* spp.),
which have very big seeds.

Baker: Do they have starch as well?

Whitmore: I don't know.

Janzen: The tropical oak in lowland Costa Rica, *Quercus oleoides*, has a self-
burying acorn as well, which has long been regarded as an antifire device;
the same device occurs in *Quercus virginiana* in the eastern United States.
But, at least in Costa Rica, as soon as the acorn germinates and the underground part starts to develop, a moth lays an egg on the acorn and the
larva produced begins to eat the inside of the acorn. There is a resulting race
between the acorn's transfer of nutrients into the underground tuber and the
rate of seed consumption. If growth is slow, the acorn loses and no tuber is
formed; if growth is fast, the tuber is formed and further development is
possible. I argue that fire does not have to be the only driving force for this
kind of self-burial.

Tomlinson: Similarly, the explanation of why palms exclusively have hypogeal germination and why many of them have some biologic adaptation
for burying the plumule is simply because they lack the ability to make secondary growth. Therefore, the seedling axis is obconical due to a phase of
"establishment growth" (Tomlinson & Zimmermann, 1966). If the axis is not
buried, the seedling has to have some other mechanical device to support its
trunk – usually in the form of stilt roots. Otherwise, if it can develop well
below the soil surface, it can produce the broad root-bearing surface below
soil level. So the correlation there is readily explained (see Chap. 11).

References

Burger, D. (1972). *Seedlings of Some Tropical Trees and Shrubs Mainly of S.E.
Asia.* Wageningen: Pudoc.

Burgess, P. F. (1975). *Silviculture in the Hill Forests of the Malay Peninsula.*
Research Pamphlet No. 66. Kuala Lumpur: Forest Department.

Corner, E. J. H. (1940). *Wayside Trees of Malaya.* Singapore: Government
Printer.

Duke, J. A. (1965). Keys for the identification of seedlings of some prominent woody species in eight forest types in Puerto Rico. *Ann. Missouri Bot.
Gard., 52,* 314–50.

Jackson, G. (1968). Notes on West African vegetation. III. The seedling morphology of *Butyrospermum paradoxum. J. W. Afr. Sci. Assoc., 13,* 215.

– (1974). Cryptogeal germination and other seedling adaptations to the
burning of vegetation in savanna regions: The origin of the pyrophytic
habit. *New Phytol., 73,* 771–80.

Ng, F. S. P. (1973). Germination of fresh seeds of Malaysian trees. *Malayan
Forester, 36,* 54–65.

– (1974). Seeds for reforestation: A strategy for sustained supply of indigenous species. *Malayan Forester, 37,* 271–7.

– (1975). The fruits, seeds and seedlings of Malayan trees. I–XI. *Malayan
Forester, 38,* 33–99.

Schlising, R. A. (1969). Seedling morphology in *Marah* (Cucurbitaceae) related to the Californian Mediterranean climate. *Amer. J. Bot.*, *56*, 552–61.

van Steenis, C. C. G. J. (1958). Rejuvenation as a factor for judging the status of vegetation types: The biological nomad theory. In *Proceedings of the Symposium on Humid Tropics Vegetation, Kandy*. Paris: UNESCO.

Tomlinson, P. B., & Zimmermann, M. H. (1966). Anatomy of the palm *Rhapis excelsa*. III. Juvenile phase. *J. Arnold Arbor. 47*, 301–12.

Troup, R. S. (1921). *The Silviculture of Indian Trees*. Oxford: Clarendon Press.

Whitmore, T. C. (1973). *Palms of Malaya*. London: Oxford University Press.

– (1975). *Tropical Rain Forests of the Far East*. Oxford: Clarendon Press.

Wilson, D. E., & Janzen, D. H. (1972). Predation on *Scheelea* palm seeds by brucid beetles: Seed density and distance from the parent tree. *Ecology*, *53*, 954–9.

Wright, H. (1904). The genus *Diospyros* in Ceylon. *Ann. Roy. Bot. Gard. Peradeniya*, *2*, 1–78.

Wyatt-Smith, J. (1963). *Manual of Malayan Silviculture for Inland Forests*. Malayan Forest Record No. 23. Kuala Lumpur: Forest Department.

6

Studies on the demography
of tropical trees

JOSÉ SARUKHÁN

Departamento de Botánica, Instituto de Biología,
Universidad Nacional Autónoma de México

The study of population dynamics, population genetics, environmental physiology, and natural selection is, for many biologists, basic to the understanding of natural systems and their components. The different traits of organisms are determined by processes of natural selection, and accordingly, it is necessary to ask questions about selective forces and the adaptive significance of those traits to understand the evolutionary paths the organisms may be following (Orians et al., 1974).

The action of selective processes can best be studied in the context of populations. Phenotypic or genotypic structure, competition, growth and death rates, etc. are all aspects of the biology of species that must be approached at the population level.

One of the several ways in which populations can be studied to elucidate biologic questions of evolutionary relevance is through an actuarial or demographic treatment. Demographic studies, developed originally for the investigation of human populations, soon constituted the backbone of population theory regarding animals. The contrasting dearth of demographic studies of plants has been repeatedly stressed (Sarukhán & Harper, 1973; Harper & White, 1974). Fortunately, awareness of the need for these studies and of their inherent potential seems to have grown considerably during the last 5 years, resulting in several pioneer demographic studies. A complete review of investigations relating to plant demography and ancillary aspects has been provided by Harper & White (1974), and it should be consulted for much basic information on the subject.

Of recent studies, only two refer to tropical plants. Hartshorn's (1972) analysis of the life cycles of *Pentaclethra macroloba* and *Stryphnodendron excelsum* constitutes the first formal actuarial treatment of populations of tropical trees. More recently, Van Valen (1975) published a detailed account of the life history of *Euterpe globosa*, a tall canopy species characteristic of rain forests in Puerto Rico, based largely on data previously gathered by Bannister (1970). No other studies on the demography of tropical species are known to the author.

As actuarial and ancillary data about numerical fluctuations in plant populations have begun to form a coherent body of information, it has become clear that a theoretic structure of plant demography cannot be built uncritically on information originating from animal demography (Harper & White, 1974). Plant plasticity, for example, precludes consideration of age as necessarily the pivotal character to which relate all other population and individual attributes, as happens with most animal populations. The concept of an individual itself as a population of discrete (structural) units departs from the idea of the unity of the animal individual. This "populational" organization of plants emphasizes the very different ways in which individual plants and individual animals (except sedentary or highly socially integrated animals) react to environmental stimuli.

Plant population demography offers ample opportunities for research and will certainly yield unsuspected information (as it already has) on, for example, the ways plant numbers are regulated, the possible selective pathways for different environmental conditions, and the fundamentally important plant-animal relation. This chapter presents an overview of the topics that are being investigated, using as examples several tropical arboreal species in Mexico.

The project on comparative demography of trees

During mid-1974, a project on the comparative demography of trees of Mexico (both temperate and tropical) was started in the Botany Department of the Institute of Biology, University of Mexico, under the direction of the author. The aims of this long-term study are various. In the long run, it is expected to discover whether general patterns of population regulation exist for species of widely different biologies growing in strikingly different environments and, if so, which factors are responsible for such patterns and in what ways they act on plant populations. It is expected to provide insight into possible selective pressures and the response of plants to them. Over a shorter term, it is expected to produce a better understanding of the biology of tropical species, the mechanisms of numerical fluxes, the factors that cause them, the nature of the biotic interactions affecting the populations under study, and the role played by the selected species within their communities.

The sites selected for pilot study represent three interesting extremes of the range of forest communities in Mexico: (1) a site in the high evergreen tropical forest (*"selva alta perennifolia"* in the system of Miranda & Hernández, 1963) of the lowlands of the Gulf of Mexico, representing the most mesic tropical conditions occurring in Mexico; (2) the low deciduous tropical forest (*"selva baja caducifolia"* of Miranda & Hernández) on the lowest slopes of the Pacific Sierra Madre Occidental, which represents the driest extreme of communities of species of neotropical origin; and (3) a single-species conifer forest of *Pinus hartwegii* at 3400 m above sea level (timber line at 4000 m) on the slopes of the Iztaccihuatl volcano.

All the basic actuarial and other information is being obtained similarly in the three sites, so as to provide a basis for comparisons. Different groups are responsible for studying each site, although a free flow of staff among the sites is encouraged. In this way, another aim of the project is being achieved: the training of students in field projects that have a continuing and planned trajectory. This chapter refers to studies carried on at the two first sites only. A brief description of the sites and the selected species is given first; results are then presented and discussed.

High evergreen tropical forest

The site for this vegetation type is located in the general area of Los Tuxtlas, in the state of Veracruz, in a biologic reserve belonging to the Institute of Biology, University of Mexico. The forest in this area is a typical species-rich high evergreen forest; canopy trees have heights of 30 m or more, and there are about three arboreal strata: the lower from 5 to 12 m, the intermediate from 13 to 19 or 24 m, and the higher from 20 or 25 to 35 or 40 m (Pennington & Sarukhán, 1968). Several of the more common canopy trees are *Nectandra ambigens, Lonchocarpus cruentus, Cynometra retusa, Vochysia hondurensis, Pseudolmedia oxyphyllaria,* and *Brosimum alicastrum.* The forest grows on soils of volcanic origin on a somewhat hilly terrain.

Annual rainfall is about 4.5 m, and no month has less than 60 mm of rain; May is the month with least precipitation. Average yearly temperature is around 24° C, the coolest month being January (17° C average temperature). *"Nortes"* (cool, northerly winds, bringing heavy rain), resulting from the displacement of cold continental masses of air to the Gulf of Mexico, produce low temperatures from December to February and account for some 15% of the yearly precipitation.

It is interesting that this locality constitutes the northernmost distribution of the high evergreen tropical forest on the American continent.

At this site, two species were chosen to represent contrasting habits: *Nectandra ambigens,* a tree typical of the forest canopy and one of the most common species, and *Astrocaryum mexicanum*, a palm typical of the lower stratum of the forest and undoubtedly the most abundant plant in it; it is the species that gives to the interior of the forest the characteristic appearance a visitor perceives from the ground. By selecting these two abundant species, we hoped to compare a species adapted to the relatively variable macroclimatic conditions affecting the canopy of the forest (e.g., high light intensity, lower relative humidity, severe effects of wind) with a species adapted to the relatively more even microclimatic conditions within the forest. An interesting addition is the period in which both species share the same physical environment during the germination and growth of saplings and younger trees of *N. ambigens*.

Both species possess several practical advantages from the viewpoint of demographic studies. Their seeds, seedlings, and young individuals are characteristic and easy to identify and handle; individuals, at least in some of the life cycle stages, are plentiful and widely distributed in the forest. But undoubtedly, *A. mexicanum* is the better experimental subject. Among other characteristics, its size range (maximum height about 6 m), the neat organization of its growth form, its discrete methods of flowering, and the possibility of determining the age of individual trees make it an ideal species for demographic study.

The species flower and fruit at different times during the year. The flowering peak for *A. mexicanum* occurs rather abruptly (3 weeks after the first inflorescences open, only 10% remain closed) in mid-April. Isolated inflorescences open before or after this peak. Fruits start forming by mid-April and are fully mature and begin to drop by the end of August or mid-September.

Flowering in *N. ambigens* starts in mid-May and lasts for 5–7 weeks. Fruits mature and drop to the ground in October. Leaves are renewed just before flowering begins and attain a deep crimson color that, together with the showy pink blooms, makes this species easy to spot in the forest canopy.

Permanent sites for the study of both species have been established, representing a variety of conditions. For *A. mexicanum*, populations of three different densities were selected, each with a separate replicate site. Each site is 600 m², and within each all individuals of the palm, from newly germinated seedlings to mature plants, have been tagged and their locations registered on a grid. Equally, all individuals of other species with a trunk ⩾3.3 cm DBH have been registered, with their heights and crown covers, and located on the grid. Sites for detailed seedling observations (one-third of the total area) have been carefully delimited to avoid trampling. Densities (seedlings to mature plants) for *A. mexicanum* range from 950 to 4350 per hectare.

Owing to the characteristics of the *N. ambigens* populations, to which reference will be made later in this chapter, attention has been concentrated on the numerically abundant seedling phases. Randomly chosen permanent sites, 1 m², have been established, 16 representing situations of high canopy tree density and 25 representing situations of low canopy tree density. Two populations of marked seedlings are being observed: those germinated in 1974 and earlier and the cohort of seedlings germinated in 1975. Seedling densities observed range from 3 to 124.

Low deciduous forest

The general locality of this vegetation type is the lower Pacific slopes of the Sierra Madre Occidental in the state of Jalisco. The site is located in another biologic reserve belonging to the Institute of Biology, University of Mexico, near the village of Chamela, Jalisco.

The soils of the region are very sandy, derived from highly metamorphosed granites. They have rather low water-holding capacity and tend to be shallow on hilly ground. Annual rainfall of about 900 mm, a clearly marked 6-month dry season, and a mean yearly temperature of 25° C, make this an area of seasonal vegetation. It constitutes what may be considered the driest extreme along a moisture gradient that starts with the evergreen tropical forest.

This is a species-rich community: There are 64 species of trees represented by individuals with trunks 3.3 cm or more DBH, belonging to 27 families in a 1-ha sample plot. The average height of the canopy is between 11 and 13 m, and all species except one or two heliophiles (e.g., *Jacquinia* and *Coccoloba*) become leafless for 3–8 months in the dry season. The forest offers a striking contrast during the rainy season, when every species rushes into vigorous vegetative growth.

Some of the most abundant species in the canopy of the forest are *Caesalpinia eriostachys*, *Cordia elaeagnoides*, *Lysiloma divaricata*, *Neea* sp., *Recchia mexicana*, *Pterocarpus acapulcensis*, *Acacia acatlensis*, and *Ruprechtia fusca*.

The species selected for study was *Cordia elaeagnoides*, one of the clearly dominant species of the canopy. It is a tree with an average height of 11 m (maximum height observed in the area is 15 m). *C. elaeagnoides* flowers profusely during late November and December, the time at which it also starts shedding its leaves. Fruits are normally one-seeded nutlets, which retain all the dried up floral parts that serve then as dispersal structures in the wind. The tree is leafless

168   José Sarukhán

from January to the end of May or early June, and during the first rains vigorous new leaf growth occurs. Water availability is clearly a determining factor in how quickly and for how long leaves are shed. Individuals growing near houses or in areas where they receive more water than is available from rain keep their leaves much longer than trees in the forest.

Normal density of C. *elaeagnoides* is about 120 individuals 3.3 cm DBH or more per hectare. Seeds are easy to spot on the ground, either with or without dispersing structures. However, seedlings have been difficult to identify properly.

Fig. 6.1. Population flux model for *Astrocaryum mexicanum* based on averages of data presented in Table 6.1. for all permanent sites studied from May 1975 to March 1976. Arrows show direction of flux in time, starting from flower stage. Box heights are log scale of number of individuals at each stage in life cycle. Bold figures in boxes are real numbers of plants, flowers, or fruits at each stage. Figures between boxes are survival probabilities from one stage to next in horizontal flow and contribution of one category of plants to another in vertical flow. Figures under bottom row of boxes show years spent in each category. Roman numerals above boxes of immature and mature categories are height classes as follows: I, 1–50 cm; II, 51–100 cm; III, 101–150 cm; etc. *Inset* shows survivorship curve for average population of same species. See text for further explanation.

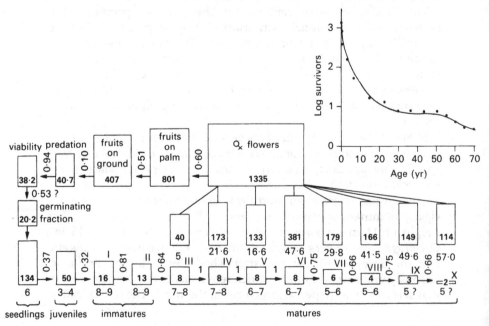

Two sites of contrasting soil conditions have been selected: one on the slopes of a hill with rather shallow soil that supports a low vegetation (6–7 m high) and one on flat ground with much deeper soil and a taller vegetation (10–11 m). In the first case, a 1-ha permanent site has been established, and in the second a 0.42-ha site. Procedures for registering both C. *elaeagnoides* individuals and the accompanying vegetation are the same as those described for A. *mexicanum*.

General results
Because some of the studies started in mid-1974 and others in mid-1975, most observations are incomplete, and results so far obtained can be dealt with only in a tentative way. However, some sets of data, particularly concerning A. *mexicanum*, lend themselves well to a preliminary analysis, and it is largely on data obtained from this species that the following discussion is based.

It is convenient to introduce the discussion of results with a model of population flux in A. *mexicanum* as a guide both to the data obtained on this species and to the different aspects of population biology dealt with so far in A. *mexicanum*, N. *ambigens*, and C. *elaeagnoides*. Clearly, it is at this stage that comments and opinions about approaches adopted, methods followed, and results obtained will be greatly appreciated.

A model of numerical flux in the population
Using the data obtained from the permanent study sites, it is possible to construct a vertical life table for A. *mexicanum*. Such a life table is represented in Fig. 6.1 as a population flux model. It shows an average population of the six permanent, 600-m^2 sites (Table 6.1); all the data included, except number of female flowers per inflores-

Table 6.1. *Numbers of plants of* Astrocaryum mexicanum *in permanent sites* ($600\ m^2$) *and their reproductive behavior*

Site	Seed-lings	Juve-niles	Imma-tures	Matures (a)	Total (b)	Fruited 1975 No.	% (a)	% (b)	Flowered 1976 No.	% 1975
A	261	65	36	65	427	34	52	8	37	76.5
AA	111	78	28	64	281	31	48	11	39	93.5
B	150	60	46	42	298	18	43	6	24	77.8
BB	174	52	19	75	320	26	35	8	31	69.3
C	54	25	11	25	115	12	48	10	13	58.3
CC	54	18	28	21	121	9	43	7	12	66.7

cence, were obtained at the sites from the end of May 1975 to March 1976. The model is clearly a tentative one, as not even a full yearly cycle has been covered. However, there seems to be a fairly strong internal consistency in the figures, as all the data derive from completely independent observations, except the germination value from seeds in the soil, which is estimated from the number of seedlings observed to emerge between October and December 1975. Population sizes are represented at scale, using the log of the number of individuals as the height of each box. Inside each box, the actual number of plants, flowers, or fruits is represented. Figures between boxes in the horizontal direction of flux are transition values representing the fraction of individuals that survives from one age class to the next; in the vertical direction of flux, figures between boxes denote the contribution of one or several categories of plants to another (e.g., the number of fruits produced per plant in the different mature categories or the transition from fruits to seedlings through a given germination value). The time (in years) for an individual to pass through each age class is shown under each box.

General population survivorship

The lack of a horizontal time scale obscures the survivorship curve implicit in the different logarithmic heights of the boxes. Such a survivorship curve is represented in the inset graph in Fig. 6.1, a clear Deevey type III curve, which shows a severe decrease in plant numbers at the early stages of life, particularly through fruit predation and seedling mortality, and a phase of rather low and constant mortality between the ages of 10 and 45 years, after which mortality increases abruptly. This abrupt drop in survivorship could be an effect of the very small numbers of plants involved at these ages in the populations. A strikingly similar curve for *Euterpe globosa* (another tropical forest palm) has recently been reported by Van Valen (1975) based on information obtained earlier by Bannister (1970) in Puerto Rico.

Depending very much on the amount of information involved, such survivorship curves as the one obtained for *A. mexicanum* could also be interpreted as formed by two or three exponential lines, each corresponding to the mortality rate characteristic of a stage in the life cycle of the organisms, and this may be what happens in many cases.

General survivorship curves for *C. elaeagnoides* and *N. ambigens* cannot yet be assembled.

Seedling survivorship

Although detailed actuarial studies on plant populations are still scarce, all observations referring to seedling survival point to this stage as the most important one from the viewpoint of numerical changes, once the seed-seedling transition has been accomplished. Equally, all data confirm that the younger the seedling, the greater the mortality risk. A detailed account of survivorship is not yet available for a given cohort of *A. mexicanum* seedlings, but the statement concerning mortality risk applies to a mixture of some five or six generations of seedlings present in the forest floor, until they reach the juvenile stage. For specific cohorts, survivorship may well vary from the average 0.37 value in the model, owing to year-to-year variations of the environment, but the data seem to support the idea that this could be a good estimate of the average conditions for different sites and years. It will be possible to confirm or modify this value when the first year's observation of the 1975 cohort of seedlings is completed in November-December 1976 and data for separate successive cohorts are obtained.

Observations on two different populations of seedlings of *N. ambigens* (one produced in 1974 and earlier and a second produced in 1975) show survivorship rates that could tentatively be interpreted as exponential, with an average monthly survivorship probability of 0.834. If this value is approximately correct, a cohort would have a mean life of 4 months, and all seedlings belonging to it would have disappeared from the forest soil in a little more than a year. Actually, of the total number of seedlings (1488) registered in March 1976, 25% are those germinated prior to 1975 and the rest belong to the last cohort of 1975, so clearly either the average survivorship value for each month is somewhat higher than 0.834 or risks of mortality are not wholly exponential. However, these data are not too different from those generated by observations carried out in forests of Puerto Rico, where mean lives of 6 months were found for all seedlings in the forest soil (Smith, 1970). Clearly, an understanding of mortality processes in *N. ambigens* will increase as data are obtained during the next few years.

Juvenile, immature, and mature stages and age structure
The transition of an *A. mexicanum* individual from the juvenile stage (a plant without above-surface trunk and possessing between 6 and 10 leaves of an average length of 75 cm) to the immature stage seems also to be particularly hazardous: Over two-thirds of the juveniles cannot survive to the youngest mature stage. This juvenile-mature transition is also a stage in which, as will be discussed later, a remarkable change in biomass allocation begins to take place. A similar situation seems to occur with young plants of *Euterpe globosa* (Van Valen, 1975): Mortality is highest in the seedling stage and next highest during the juvenile-mature transition. This latter stage could be a time of delicate metabolic balance in plants that have been "frozen" in immaturity owing to suppression by both dominant conspe-

cific individuals and other dominant components of the forest. It is possible to speculate that if a termination of such freezing does not occur in time (e.g., in the form of new gaps in the canopy or death of suppressing individuals) the delicate balance in plants that cannot grow actively but, nevertheless, have to keep alive could be tipped over the negative slope and so mean the death of the plants. Evidence for the suspended animation of saplings or young trees in the lower strata of tropical forests is ample (Richards, 1966), but there are no studies of the significance of this state on the chances of survival of the individuals.

Survivorship between the immature stages in the model seems high – in fact, quite similar to that among the mature categories. However, mortality is apparently higher between the last immature (nonreproducing) stage and the first mature (reproducing) stage. Whether this occurs as the result of the extra effort needed for the production of flowers and fruits, which causes a number of weak individuals to die in the attempt, is not clear. Further observations of the tagged individuals in the permanent sites may answer this question.

The chance for survival from one life class to another in the case of mature individuals is both high and nearly constant from age 25 to 55 years and probably represents a situation in which death risks are low, not concentrated in any specific age, and probably accidental. Such accidental causes of death are mainly catastrophic events, like the fall of large canopy trees or their large limbs, particularly during seasons of high winds, which in the study area occur mainly between September and January, either as cyclones or as *nortes*. Events of this nature have been observed during the study period and their effects on *A. mexicanum* individuals registered.

The characteristic growth form of palms provides an excellent opportunity for assessing the individual's age. Each senescent leaf leaves a fairly neat scar, bordered by two rows of sharp, flattened spines, when it becomes detached from the trunk. Such scars can also be seen in the underground stems of most seedlings and juvenile plants. If the rate of leaf production over time is determined, it becomes possible to establish with fair accuracy the age of an individual. Leaves of randomly chosen individuals of *A. mexicanum*, representing all height classes present in the permanent sites, were tagged to assess their rates of formation and death. Although a full year has not yet been covered in the observations, it is now possible to give a first approximation of the rate at which plants of different height classes are growing. Table 6.2 shows the number of leaves estimated to be produced in a 12-month period for different height categories, the average number of scars present for each category, the number of living leaves forming the crown of the plants, and consequently the ages attained by plants at such categories. The age of an individual will, therefore, be the time required to form a number of scars, using the height-specific leaf-formation rate, plus the number of living leaves. The duration of the 0.5-m interval categories in the population model have been calculated from the data on 1-m intervals from Table 6.2. Interscar distances seem to be rather constant; consequently, total number of scars (and thus age) becomes a function of the height of the trunk (Fig. 6.2) in a strong correlation ($r = 0.93312$, $P < 0.5\%$).

The resulting age distribution of individuals of *A. mexicanum* (Fig. 6.3) shows a very mature, stable structure: Over 50% of all individuals are seedlings between 0.5 and 6 years of age, 19% are juveniles, and all other categories possess well under 10% of the total individuals in a decreasing fashion as age increases.

Certainly, *A. mexicanum* (as is true of most other palms) constitutes an exception among tropical trees in the ease with which age can be defined for individuals. Attempts at establishing ages of individuals of *Cordia elaeagnoides* have faced several problems, despite the potential presence of indicators of differential growth in the wood as a result of the striking seasonality of the environment in which this species grows. A study is being carried out to explore the wood anatomy in detail and to solve the problem of obtaining sample cores from the extremely hard wood; observations on girth increment by means of microdendrometers and the seasonal scarification of cambium are already in process in the permanent sites in the field. The distribution of breast height diameters of 115 individuals (≥ 3.3 cm DBH) in a 1-ha permanent plot is "normal," which suggests a stable

Table 6.2. *Approximate growth rates and age determination of individuals of* Astrocaryum mexicanum

Trunk height (m)	Estimated no. of leaves in 1 yr	Average no. of scars in trunk	No. of leaves in crown	Total years taken to grow (+ 6 yr as seedling)
Juveniles	2.25	8	8.0	9–10
Immatures				
0–1	2.54	50	11.0	25–28
Matures				
1–2	3.05	95	12.0	39–44
2–3	3.24	137	13.5	51–58
3–4	3.91	180	13.5	61–70

population, but there is no evidence that age and DBH are directly correlated in this species.

A wholly different situation exists in the case of *Nectandra ambigens*. Despite the fact that the tree is among the most abundant forest canopy species and possesses large numbers of seedlings in the forest soil, there is a remarkable absence of saplings and young individuals. The species is represented in the form of trees only by the tall canopy individuals. In the study area the ratio of seedlings to canopy trees is between 100:1 and 200:1. It seems that only catastrophic events will provide the conditions for the development of young trees.

Reproduction

Demographic data on reproduction of tropical trees are either very scarce or rather difficult to find. Interest in the reproduction of tropical trees has increased noticeably in the last few years (e.g.,

Fig. 6.2. Relation between trunk height and number of scars present on it in 104 *Astrocaryum mexicanum* plants. $r = 0.93312$; $p < 0.5\%$.

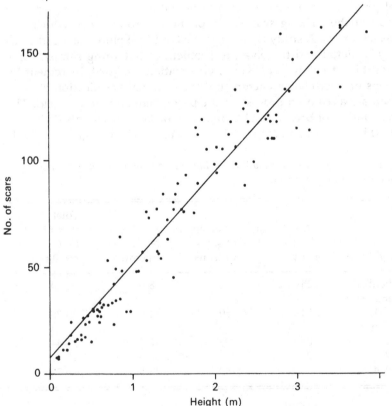

Bawa, 1974; Bawa & Opler, 1974; Frankie & Baker, 1974; Opler, Baker, & Frankie, 1975), but the references cited have more relevance to breeding systems, plant-pollinator relations, and the genetics or taxonomy of the species studied. With very few exceptions (e.g., Ng, 1966, for Malaysian trees, and Van Valen, 1975, for *Euterpe globosa*), information on age at first reproduction, reproductive schedules, age-specific reproduction, etc., is largely ignored for tropical trees with the possible exception of some cultivated species, like mango (Singh, 1960).

The preliminary data presented in Fig. 6.1 show that first success-ful reproduction takes place for *A. mexicanum* at between 32 and 36 years of age; from then on, the reproductive contribution of the age classes increases to nearly 10 times the initial at about 51–58 years, decreases steadily after this age, but is maintained through the life of the older individuals. This species fits well with Harper & White's (1974) general relation between age at first reproduction and func-tional life span for numerous plant species.

Age-specific reproductive values can be tentatively estimated on the basis of the transition values from the mature plant categories to the "fruits on tree" category. An extra category has been introduced in the model (Fig. 6.1) to estimate the loss of reproductive potential by undeveloped or unfertilized female flowers, which is about 40%. If these values are correct, the older age categories possess the high-

Fig. 6.3. Age distribution of average *Astrocaryum mexicanum* popu-lation of 259 individuals. Ages determined on basis of number of scars left by leaves on trunk and leaf-formation rate.

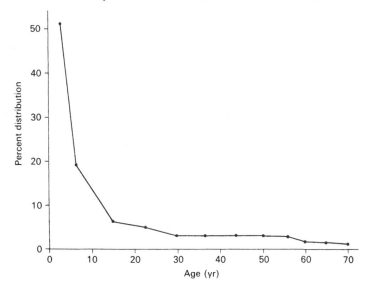

est reproductive contribution per individual. This tends to support the view that mortality risks at old ages in *A. mexicanum*, at least under the conditions of the study, may be due more to accidental causes than to senescence of the individuals.

The plants that fruit in a year (see Table 6.1) are a fairly constant proportion of both the mature (average, 45% in 1975) and the total population (average, 7% in 1975). In all sites, more individuals flowered in 1976 than fruited in 1975; this may suggest certain loss of inflorescences due to various factors; already approximately 6% of inflorescences produced in 1976 have been recorded as gnawed by squirrels and, therefore, lost. Up to the last visit to the sites in March 1976, between 58% and 93% of the individuals that fruited in 1975 had produced at least one inflorescence.

Of the 22 individuals that, in an average of all sites, fruited in 1975, exactly half produced one inflorescence, 7 produced two, 3 produced three, and 1 produced four. The situation was repeated in 1976: Of 24 individuals that flowered (on average), 10 produced one inflorescence, 8 produced two, 4 produced three, and 2 produced four. The average number of fruits per infructescence was 21, with a minimum of 4 and a maximum of 43 (SD 28%). An average reproductive plant produced 35 fruits, the minimum fruits per fruiting tree being 4 and the maximum 112 (SD 58%). The greater variability of fruits per tree than of fruits per infructescence is due to a varying number of infructescences per plant and suggests that a plant would produce inflorescences in excess of one only if there is enough energy available to ensure a good chance of producing a more or less constant number of mature fruits.

Assessing reproductive activities in species like *Cordia elaeagnoides* or *Nectandra ambigens* is far less tidy and practicable. The profuse branching of the trees and their height impose the need for cumbersome estimations and roundabout procedures to produce a reasonable idea of the reproductive characteristics of the trees. At present, nearly 100 randomly placed, 0.25-m^2 mesh traps are being used to estimate the number of fruits produced by trees of *C. elaeagnoides* and the time of their production. For nearly a year, monthly or fortnightly collections have been made, and the material is currently being processed. The traps are being used also to study the propagule flora dispersed by wind and its distribution in time. This will be followed by a comparative morphologic study of the propagules, which should give some insight into their germination requirements. Lateral observations on direct counts of parts of the trees will be made to estimate the number of fruits produced per plant in *C. elaeagnoides*. A similar procedure is planned to start for *N. ambigens* in August 1976.

Seed predation

Perhaps one of the most intricate biotic relations is that between animals and plants at the level of seed predators (Janzen, 1971), carefully evolved from chemical and behavioral interrelations. It constitutes, in many species belonging to virtually all families of plants, a major factor in control of population size.

Predation represents the single most important cause of death in populations of *A. mexicanum*. Fruit predation in the palms and on the ground together reduced the seed population from 801 to 40.7, a 95% mortality (Fig. 6.1).

Observations to assess the effect of predators on survival of *A. mexicanum* seeds were carried out during the 1975 fruiting season (September). Although preliminary, these observations give insight into the importance of several vertebrate predators in the fate of seeds. Predation of *A. mexicanum* fruits can take place when the fruits are still attached and on the ground after they have fallen. An arboreal squirrel (*Sciurus aureogaster*) is apparently responsible for an average 50% loss of fruits. The squirrel avoids the numerous spines on the trunk and underneath the leaves by climbing from a nearby branch to the top of the palm crown, and sliding on the upper sides of the leaves (which are unarmed) until these bend low under the weight of the animal. In this position, the squirrel can reach the hanging infructescences without touching either the trunk or the basal parts of petioles, which are heavily armed with spines. The fruits are easily dislodged from the rachis of the infructescence and taken away by the squirrel. There are suggestions that fruits are taken to caches, although this has not been observed directly. Several *A. mexicanum* palms have been observed growing on bifurcations between large limbs of trees at heights of 10–15 m; this suggests the location of old caches.

By positioning traps under marked infructescences, it was possible to assess the number of fruits that would escape squirrel predation and fall to the ground. Traps were installed in 40 randomly chosen individuals, and the number of fruits were counted at the beginning of the observation and 1 month later (both those remaining on the rachis and those trapped). Results were highly variable, ranging from 0 to 100% predation. Analyzing chances of escaping predation as a function of the number of fruits in the infructescences suggests that infructescences with few fruits have a greater percentage of their fruits stolen than those with large numbers of fruits. Another source of variation in the data was the density of *A. mexicanum* palms in the permanent sites: Those sites with higher palm densities showed, in general, lower levels of predation than sites with low palm density.

Predation on the ground can be attributed to several species of rodents, but it is not known at present which species and in what proportions are most responsible for the very high levels of seed mortality. Introduction of three densities of fruits (4, 20, and 40 per square meter) showed again that, despite a high overall predation, fruits had a greater probability (0.32) of escaping predation in sites of high density of plams than in sites of lowest densities (0.01).

Whether palm densities in the sites and fruit densities in infructescences or the environmental characteristics of the sites are responsible for the differential mortality risks apparent in the first observations on predator activity remains unclear and will be the object of special attention during the following fruiting seasons.

Extremely high seed losses due to predation are also characteristic of *Cordia elaeagnoides* in the low deciduous forest. Chances for a seed to escape predation varied in the different experimental sites from 0.13 to 0.02.

Dynamics of seed populations in the soil

Seeds that have escaped predation on the tree become, after their detachment from the parental plant, part of the soil seed bank for varying periods. Their exit from the seed bank can only result from (1) germination, (2) removal by predation, and (3) death due to senescence or pathogens. Permanence in the bank and, therefore, much of the strategy for maximizing the number of successful individuals in the populations, are achieved through a finely selected dormancy behavior. The detailed mechanics of numerical changes of seed populations under different environmental conditions have been little studied for species of tropical forests. The few studies existing on seed populations in the soil refer mostly to floristic accounts of seed banks (e.g., Symington, 1933; Keay, 1960; Bell, 1970; Kellman, 1970; Guevara & Gómez-Pompa, 1972).

Following an approach used to study temperate grassland species (Sarukhán, 1974), experiments were started with seeds of *Cordia elaeagnoides* in sites with deep and shallow soils with the slope of the terrain as the factor determining the depth of the soil. Known numbers of seed were introduced in fixed sites in a random fashion and then harvested at 2-month intervals during the year after their introduction.

Two series of experiments were started, the first in March 1974 and the second in March 1975. For both series, the seed bank was reduced to 3% at the second month after the introduction of seed into the soil and was under 1% after the fifth month. All the losses from the seed bank were due to predation, as the maximum germination recorded in the introduced lots of seeds was 1.4%. A rodent, *Liomys pictus pictus*, seems to be responsible for a considerable proportion of the total predation. Other species, like *Oryzomys palustris*, could also play an important role in inflicting losses to the seed bank of *C. elaeagnoides*. A new series of experiments is planned to assess more adequately the effect of ground predators.

It is to be noted that only about 15% of the nutlets produced by *C. elaeagnoides* is viable. One wonders whether such behavior is a mechanism to save some seed from predation, provided, of course, the cost of producing an empty nutlet is lower for the plant than that of producing a seeded nutlet.

Observations on allometric relations and energy budgets

As a complement to the demographic observations, a number of allometric relations are being established in some of the species, and energy budgets and the allocation of dry matter to different purposes are being determined.

A. mexicanum has lent itself best to these studies. It is now feasible to relate certain parameters (e.g., length of spathe or length of the leaf blade) to total number of male and female flowers, dry matter of the inflorescence, and leaf area and dry matter of the whole leaf.

It is possible now to acquire a very close estimate of leaf area indices for each plant and to relate these to nearest neighbor data and age of the plants in a search for competitive interactions in the population.

Equally, it will be possible to estimate from the yearly biomass production the proportion allocated to reproductive and vegetative ends, comparing energy allocation strategies for plants of different stands, ages, density conditions, reproductive behavior, etc.

Observations on a number of plants of different ages, from seedlings to reproductive mature individuals, show an interesting pattern of dry matter distribution on the individuals (Fig. 6.4). Although data on newly germinated seedling were not available for the construction of Fig. 6.4, it seems that at this stage nearly a third of the biomass of the new plant, not including the weight of the fruit, is represented by roots and the rest by leaves. The proportion represented by leaves is maintained at a relatively constant value up to the age of 10 years, but the proportion of roots decreases to 10% at the same age, a value that is maintained thereafter.

The proportion represented by leaves decreases steadily from 75% at 10 years to 40% at about 50 years. It is probable as a consequence of this decrease in the proportion of leaf biomass in the plants that a budgetary imbalance of senile plants may be caused. This aspect is now under consideration in the field observations.

The main cause of the change in the proportion of leaf biomass is the active growth of the trunk after the age of 100 years, increasing from about 5% at this age to 35% at 50 years. Spines represent a constant 2%–3% of the plant's weight, and reproductive structures (remnants of old infructescences, actual fresh infructescences, and all the floral buds present in the plant) represent only 3% of the mature plant.

Clearly, to determine adequately allocation to reproductive and vegetative purposes, total yearly growth should be considered. The information available on allometric relations and on dry weights of parts as inflorescences, fruits, leaves, trunk, etc., allows such analyses.

Conclusions

In summarizing the information contained in the flux model (Fig. 6.1), it is convenient to refer briefly to two points. First, very important numerical changes are evident during the life cycle of *A. mexicanum*, particularly at the seed and seedling levels and through the juvenile phase. However, it is impossible to state at present whether any kind of selection and/or regulation on *A. mexicanum* is

Fig. 6.4. Approximate patterns of standing biomass distribution at different ages in *Astrocaryum mexicanum*, based on data from 13 individuals.

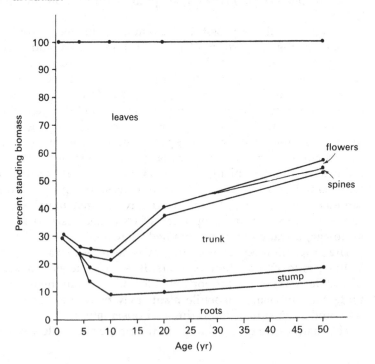

taking place at those stages. Extended observations of cohorts of seedlings, a more complete overview of predator-plant relations, and a study of the intra- and interpopulation diversity will be necessary for an understanding of selective processes acting on the population. Equally, studies on the effects of density on the vegetative and reproductive behavior of juvenile and mature plants will throw light on the ways population numbers are being regulated.

A second point for consideration is that the information, as presented in the flux model, is amenable to treatment in mathematical models. This is within the plans of the project; there are already precedents for such mathematical modeling for plant populations (Hartshorn, 1972; Sarukhan & Gadgil, 1974). However, it is premature to attempt such a project on data about *A. mexicanum*. Such models will allow a more rigid description of population dynamics and a prediction of population changes resulting from the manipulation of population parameters, an exercise that will provide valuable insight into general selective and regulating processes in the populations.

Surely a promising and stimulating future awaits the field of plant demography as an increasing body of information permits the development of a proper conceptual frame for studying the biology of plant populations. The perspective of continued research projects in the field of population biology of tropical trees in itself should constitute an optimistic format for the development of this important field of biology.

Acknowledgments

Data presented in an introductory and general form in this chapter have been and are being obtained with the close collaboration of the following persons: Daniel Piñero and Pilar Alberdi, concerned with all the studies on *Astrocaryum mexicanum* and *Nectandra ambigens*; Luis A. Pérez-Jiménez on the demography and general biology of *Cordia elaeagnoides*; and Fernando Guevara on the seed population dynamics of *C. elaeagnoides*. The research project is being carried out with a grant from the Consejo Nacional de Ciencia y Tecnología (CONACyT, PNIET proyecto No. 931, México).

General discussion

Givnish: It impresses me that plant demographers dealing with long-lived plants or plants with long-lived ramets are faced by extremely difficult problems. If one studied *Sequoiadendron gigantea* or *Sequoia sempervirens* in the same way one studies *Nectandra*, one could find that no seedlings "ever" become adults, at least in single populations studied over relatively short periods. One might have to wait for a glaciation to clear a terminal moraine to see a massive reestablishment of *S. gigantea* or for some lesser catastrophe in coastal California to see a new cohort of *S. sempervirens* emerge. Another problem on a different scale is that of forest herbs – one can often find, for example, 10 or more species growing in an area of 2–3 m². One wonders in studying their demography (1) whether they react to subtle differences in the environment and are coexisting on that basis; (2) whether, although they are established, they only flower or reproduce depending on the sorts of conditions they encounter over a long span of years (i.e., wait for favorable changes in the canopy); (3) whether their seeds have differing abilities to germinate in years of different conditions; or (4) whether there are micro-topographic differences to which the seeds and adults respond differentially. The problem in studying *Nectandra* or any long-lived plant is that one does not know whether the conditions under which a plant is found today are similar to those existing when it germinated or whether the conditions under which one finds seedlings today will support adults in the future. Would it not be appropriate to study the demography of long-lived plants in a somewhat different way than, say, the way one studies *Ranunculus?* That is, would it not be better to look at the distribution of an organism's life stages among various environmental patches than to study the details of seed and plant demography in one place? For example, one might look at gap phase and mature phase forest and see where seedlings are and whether they are found as saplings in succeeding stages.

Sarukhán: I am not sure what answer you want. If you mean in your question that the research is difficult to do, the answer is yes! I think the only way we can understand numerical fluxes in plant populations and get to know their causes will be to follow more or less closely some of the existing examples (e.g., the research done on *Ranunculus* spp. that you mentioned). Much depends on having what one may call the "luck" of finding the right species, like the buttercups or *Astrocaryum mexicanum*, which are extremely "cooperative" and allow a number of technical shortcuts. Demographic studies of plants have been neglected until very recently for much the same reasons you give, and largely because of this, plant population theory has been underdeveloped in relation to animal population theory. It is high time we become more courageous and produce as detailed and as complete demographic information as possible, although it may take many years to build up a good body of information and theory of plant populations. I think one important lateral contribution of the work on *Ranunculus* is that it proved that detailed demographic studies, even with the complication of vegetative reproduction, are quite feasible. The longevity of plants may or may not be a problem. *Nectandra ambigens* is not such a long-lived species and there is evidence that many frequent catastrophic events in the forest are pushing turnover of individuals pretty tightly. We have been able to produce a tentative population model for *A. mexicanum* in only 10 months. In a proportionate way, we will be able to do something similar for *N. ambigens,* and I am sure we will be able to find a number of shortcuts along the way. In addition, if one has good collaborators, as I happen to have, the task is less cumbersome and fair progress is possible.

I agree that in order to gain a fuller understanding of the role of a species in a community and of the ways the plant reacts to all different conditions found in the community one should look also at the "macrodynamics" level. The way in which we have selected the permanent sites and their size provide us with this "macrovision" to an extent. One has to be clear also that we are in the process building up the fine details of populations dynamics; after that we will move on to consider more general situations.

References

Bannister, B. A. (1970). Ecological life cycle of *Euterpe globosa* Gaertn. In *A Tropical Rain Forest: A Study of Irradiation and Ecology at El Verde, Puerto Rico*, ed. H. T. Odum & R. F. Pigeon, pp. B-299–314. Oak Ridge, Tenn.: U.S. Atomic Energy Commission.

Bawa, K. S. (1974). Breeding systems of tree species of a lowland tropical community. *Evolution, 28,* 85–92.

Bawa, K. S., & Opler, P. A. (1974). Dioecism in tropical forest trees. *Evolution, 29,* 167–79.

Bell, C. R. (1970). Seed distribution and germination experiment. In *A Tropical Rain Forest: A Study of Irradiation and Ecology at El Verde, Puerto Rico*, ed. H. T. Odum & R. F. Pigeon, pp. D-177–182. Oak Ridge, Tenn.: U.S. Atomic Energy Commission.

Frankie, G. W., & Baker, H. G. (1974). The importance of pollinator behavior in the reproductive biology of tropical trees. *An. Inst. Biol. Univ. Nac. Autón. México, 45 (Ser. Bot.),* 1–10.

Guevara, S., & Gómez-Pompa, A. (1972). Seeds from surface soils in a tropical region of Veracruz, Mexico. *J. Arnold Arbor., 53,* 312–35.

Harper, J. L., & White, J. (1974). The demography of plants. *Annu. Rev. Ecol. Syst., 5,* 419–63.

Hartshorn, G. S. (1972). Ecological life history and population dynamics of *Pentaclethra macroloba,* a tropical wet forest dominant, and *Stryphnodendron excelsum,* an occasional associate. Ph.D. thesis, University of Washington.

Janzen, D. H. (1971). Seed predation by animals. *Annu. Rev. Ecol. Syst., 2,* 465–92.

Keay, R. W. J. (1960). Seeds in forest soil. *Nigerian Forest. Inf. Bull. N.S., 4.*

Kellman, M. C. (1970). The viable seed content of some forest soils in coastal British Columbia. *Can. J. Bot., 48,* 1383–5.

Miranda, F., & Hernández, X. E. (1963). Los tipos de vegetación de México y su clasificación. *Bol. Soc. Bot. México, 28,* 29–179.

Ng, F. S. P. (1966). Age at first flowering in Dipterocarps. *Malayan Forester, 29,* 290–5.

Opler, P. A., Baker, H. G., & Frankie, G. W. (1975). Reproductive biology of some Costa Rican *Cordia* species (Boraginaceae). *Biotropica, 7,* 234–47.

Orians, G., Apple, J. L., Billings, R., Fournier, L., Gilbert, L., McNab, B., Sarukhán, J., Smith, N., Stiles, G., & Yuill, T. (1974). Tropical population biology. In *Fragile Ecosystems,* eds. E. G. Farnworth & F. B. Golley, pp. 5–65. New York: Springer-Verlag.

Pennington, T. D., & Sarukhán, J. (1968). *Arboles Tropicales de México.* Instituto Nacional de Investigaciones Forestales, S.A.G. México. 413 pp.

Richards, P. W. (1966). *The Tropical Rain Forest.* Cambridge: Cambridge University Press.

Sarukhán, J. (1974). Studies on plant demography: *Ranunculus repens* L., *R. bulbosus* L., and *R. acris* L. II. Reproductive strategies and seed population dynamics. *J. Ecol., 62,* 151–77.

Sarukhán, J., & Gadgil, M. (1974). Studies on plant demography: *Ranunculus repens* L., *R. bulbosus* L. and *R. acris* L. III. A mathematical model incorporating multiple modes of reproduction. *J. Ecol., 62,* 921–36.

Sarukhán, J., & Harper, J. L. (1973). Studies on plant demography: *Ranun-*

culus repens L., *R. bulbosus* L. and *R. acris* L. I. Population flux and survivorship. *J. Ecol.*, *61*, 675–716.

Singh, L. B. (1960). *The Mango*. London: Leonard Hill.

Smith, R. F. (1970). The vegetation structure of a Puerto Rican rain forest before and after short-term gamma irradiation. In *A Tropical Rain Forest: A Study of Irradiation and Ecology at El Verde, Puerto Rico*, ed. H. T. Odum & R. F. Pigeon, pp. D-103–40. Oak Ridge, Tenn.: U.S. Atomic Energy Commission.

Symington, C. F. (1933). The study of secondary growth on rain forest sites in Malaya. *Malayan Forester*, *2*, 107–17.

Van Valen, L. (1975). Life, death, and energy of a tree. *Biotropica*, *7*, 259–69.

Part III: Architecture and construction

7

Branching and axis differentiation in tropical trees

P. B. TOMLINSON
Harvard University, Harvard Forest,
Petersham, Massachusetts, and
Fairchild Tropical Garden, Miami, Florida

A tree is a photosynthetic device consisting of assimilating units (leaves) arranged on one or more woody, self-supporting axes. Leaves are the products of primary meristems, and the number of leaf-bearing units (shoots) in most trees is increased by the proliferation of the original seed-borne or plumular shoot meristem. This increase is determined by the genetic makeup according to a precise pattern. This implies that there is a precise organisation to the tree, and this assumption has been justified and explored fully by Hallé & Oldeman (1970) by means of their concept of architecture, which has permitted a typologic categorisation of growth models. Their system is followed in this discussion because it allows a great deal of information to be summarised concisely (Fig. 7.1). A further extension of their scheme allows one to recognise that a tree may adjust itself to environmental circumstances by modification of the existing architecture either when meristems not normally active as determined by genetic organisation (i.e., "reserve" meristems) are brought into play, or when some division of meristem labor, which exists in many trees, is altered from strict conformity to the model. This response to environmental change is embodied in the concept of reiteration developed by Oldeman (1974) as a logical extension of the recognition of organised tree architecture (see also Chap. 23). This chapter considers the way in which meristems are proliferated in trees in relation to both architectural and reiterative patterns and illustrates how division of meristem labor may contribute to a highly organised and adaptive pattern of growth.

This examination of various methods of branching in trees is an extension of an earlier discussion of the subject (Tomlinson & Gill, 1973), but with the addition of many newer ideas (see Acknowledgments) reflecting how much our understanding of tree organisa-

tion has advanced in recent years. The approach is entirely morphologic, but analysis at this level is necessary, especially as it is
expected that a standardised nomenclature can be developed. However, this standardisation can come only if it is expressed in terms of
the full range of diversity shown by woody plants, and this means
emphasis must be given to tropical examples.

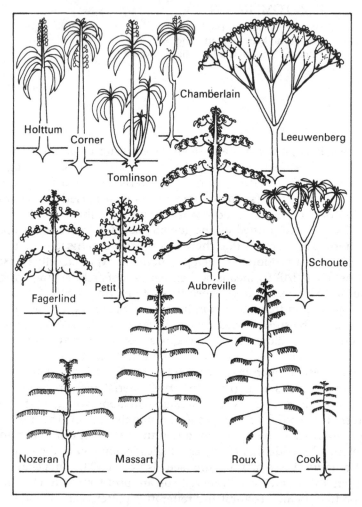

Fig. 7.1. Architectural models of tropical trees (after Hallé &
Oldeman, 1970; Oldeman, 1974) with patronymic terminology employed by them. Root system is stylised. Shoot system, represented in one plane, indicates presence, position, and frequency of
branches; branch orientation; and inflorescence position. Individual models refer to most frequently expressed condition within a

In this account, only the branching of vegetative parts is considered; some plants that are unbranched in the vegetative state show profuse branching at reproductive maturity. The hapaxanthic *Corypha* palm provides a striking example (Tomlinson & Soderholm, 1975). Because *Corypha* is monocarpic, this example further allows a precise quantitative statement about fluctuation in meristem number

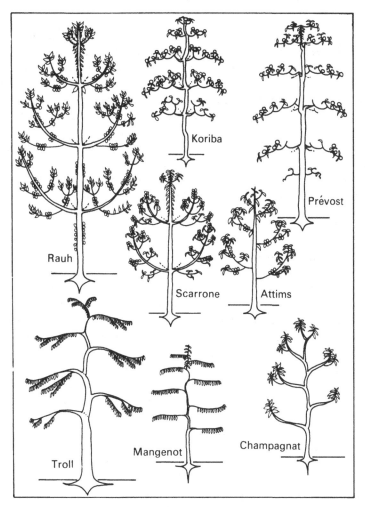

spectrum of possibilities; thus system is neither all-inclusive nor complete. It should be emphasised that architecture is a dynamic concept and models are best represented by a series of drawings illustrating developmental stages, as in original publications, which should be consulted. For present purposes simplistic drawings are used.

from generation to generation, as a calculation indicates that about 250,000 seeds are produced by a single palm, the ultimate products of a single vegetative meristem. Wastage via seed predation and seedling mortality is colossal, but only a single seed-borne vegetative meristem from any one palm need survive to reproductive maturity to ensure a stable population.

Unbranched trees

Single-stemmed palms, of which the coconut is a familiar example, remain vegetatively unbranched because they lack lateral vegetative meristems completely. Lateral meristems are entirely determinate and reproductive as inflorescences. No method has yet been devised for inducing a lateral meristem to switch from a reproductive to vegetative state, although the economic benefits of this could be considerable. A number of dicotyledons, of which *Carica papaya* is a familiar example, are, architecturally, unbranched and belong to the same (Corner's) model. However, foliage leaves here do subtend lateral vegetative meristems, which, although strongly inhibited, may develop under appropriate conditions, usually the result of damage. This branching constitutes a reiteration of the model.

Branched trees
Dichotomy (terminal branching)

The simplest mechanism for proliferation of a meristem is by its equal division into two daughter meristems of initially equal growth potential. We will term this "dichotomy," setting semantic obstacles aside, although it is appreciated that other terms exist (e.g., terminal branching: Bugnon, 1971).

Dichotomy in existing seed plants is known for a few palms, of which *Hyphaene* species provide the most striking example. Existence of a dichotomy is difficult to demonstrate in large organisms, but the circumstantial evidence may be compelling, as in *Nypa* (Tomlinson, 1971). Recent work, as summarised by Fisher (1974, 1976), shows that the process is not uncommon in some other families of monocotyledons, and it is now known for about 10 species. A recent example in the dicotyledons is provided by Boke (1976) for two species of *Mammillaria* (Cactaceae).

Obviously such examples should be distinguished from the bifurcation that is common in woody plants as the result of lateral branching below a terminal meristem that has aborted, become parenchymatous, or become converted into a terminal inflorescence or flower (Chap. 9; Tomlinson, Zimmermann, & Simpson, 1970). Truly dichotomous branching implies continuous growth activity of the apical meristem as it bifurcates, as emphasised by Bugnon (1971) and recently documented for *Flagellaria indica* (Fig. 7.2) by Tomlinson & Posluszny (1977).

Such examples of dichotomy in angiosperms are few and specialised (although an evolutionist or morphologist may find them intriguing), but they are significant because they raise an important

Fig. 7.2. *Flagellaria indica* L. (material from Rigo Road, East of Port Moresby, Papua New Guinea). Stages in dichotomy of vegetative apex of aerial lianescent shoot, from different axes fixed in FAA. A. Single apex with leaf primordia P_1 and P_2 evident, showing distichous phyllotaxis. B. Apex expanded and clearly divided into two identical growth centers; P_1 has not quite completed its basal encircling growth. C. Later stage. D. Late stage of dichotomy, each daughter shoot producing its first leaf primordium, showing that they are not mirror images. (Photographs by Usher Posluszny)

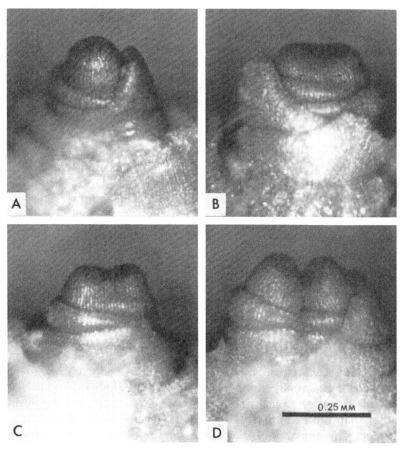

morphogenetic and adaptive question. Reconstruction of fossil trees suggests that equal dichotomy was the basic method of meristem proliferation in several now extinct groups (e.g., Lepidodendrales). The evidence is, of course, circumstantial, as we have no direct information about the growth of fossil trees, but the existence of dichotomy is not seriously doubted. Why, therefore, should dichotomy have been superseded by the almost ubiquitous lateral branching of modern gymnospermous and angiospermous trees? – especially when the few rare examples in angiosperms demonstrate that there is no morphogenetic restriction that excludes the possibility of this method of branching in seed plants. The answer seems to be that the lateral (axillary) branching of gymnosperms and angiosperms permits an elaborate organisation of shoot form, and the precise position of a lateral meristem – in the axil of a leaf – allows a precise control over the expression of that lateral meristem. This control can determine whether or not the meristem initially grows out; whether or not the meristem is differentiated as a vegetative branch, a short shoot, a flower, or a lateral inflorescence; and whether or not the branch is orthotropic. One architectural model is sufficient to account for trees with dichotomous branching (Schoute's model), whereas at least 20 can be recognised in branched trees (Fig. 7.1). Subsequent discussion is exclusively of axillary branching and considers how this is related to overall tree organisation.

Axillary branching
    Axillary branching involves the development at each node of one or more lateral meristems from the terminal meristem, which may or may not continue its activity. Bugnon (1971) suggested that the essential morphogenetic feature of lateral branching is the establishment of a new growth center that does not involve all the histogenetic regions of the parent meristem. Normally, however, superficial morphologic features are sufficient to allow one to distinguish lateral from other types of branching. In doubtful cases, use may be made of the existence of a distinct "shell zone" (Shah & Patel, 1972) at the base of the lateral meristem. Histologic study shows that the lateral meristem is otherwise always subordinate to the terminal meristem, at least initially, although this may change subsequently when secondary effects come into play.
    Attention must be drawn to the frequent existence of more than one lateral meristem in the axil of a single leaf. Sometimes these meristems, essentially occupying a single site, differ in growth potential. In many tropical trees serial or multiple buds are the norm rather than the exception, though their development remains little studied. The anatomy of their nodes merits detailed investigation.

The complex physiological interrelations among multiple lateral meristems are evident when one considers that dormant and active buds may exist within the same leaf axil, as in *Rhizophora;* that vegetative and reproductive branches may occupy virtually the same site, as in *Calyptranthes* (Myrtaceae); and that determinate branches (spines) and indeterminate branches may be subtended by the same leaf, as in *Ximenia* (Olacaceae).

Syllepsis and prolepsis

Given the simplest case (i.e., the existence of a single axillary meristem), two alternative developmental possibilities exist: The meristem may or may not develop contemporaneously with the terminal meristem that produces it. With the first possibility, branch and parent axis are chronologically of the same age; with the second possibility, a lateral meristem undergoes a period of rest (dormancy or correlative inhibition), after which it may extend to form a lateral shoot, chronologically younger than the supporting axis. In most temperate trees a lateral branch is always 1 year younger than its parent. Unfortunately, this is sometimes regarded as a "norm" for tree growth, and a case where branch and parent axis are contemporaneous is regarded as an exception, indicated by reference to "precocious branching." In fact, neither one is "standard"; they represent, in the simplest situation, two alternative possibilities.

The neutral terms "syllepsis" and "prolepsis," originally used by Späth (1912) in his discussion of lammas shoots, have been adopted, with redefinition, by Tomlinson & Gill (1973) to refer to these two developmental possibilities. Fortunately, in the architecture of most trees the two types of branching can be distinguished by fairly consistent morphologic features, which probably apply in about 95% of examples: Syllepsis produces a branch axis (sylleptic shoot) that lacks basal bud scales (or their persistent scars), has a long basal internode (hypopodium), and has little or no transition in leaf morphology and size at the first few nodes (Fig. 7.3A); prolepsis produces a branch axis (proleptic shoot) that has basal bud scales (i.e., at least one or a pair of reduced prophylls or their persistent scars), initially congested nodes (no hypopodium), and a gradual transition in leaf morphology and size at the first few nodes, usually from bud scales to foliage leaves (Fig. 7.3B).

The etymology of these words is somewhat involved, and they exist in the modern forester's vocabulary, but without the precise meaning with which they can be invested if we extend Späth's terminology and adopt a cosmopolitan view of tree growth.

Of interest are those examples in which a sylleptic and proleptic shoot may originate successively from the same node. This is possible, as in coffee, where there are serial buds, of which one develops by syllepsis, with the possibility of being followed at a subsequent time by another that has been dormant.

The quite consistent distinction between these two types of branching, which is very evident when a wide range of woody plants is examined, has remained obscure because syllepsis is a predominantly tropical phenomenon. For example, it is known as an architectural feature in few genera native to New England, although examples are found in *Clethra, Cornus, Liriodendron,* and *Sassafras.*

Fig. 7.3. Most common morphologic features by which proleptic branches (B) can be distinguished from sylleptic branches (A). For explanations, see text.

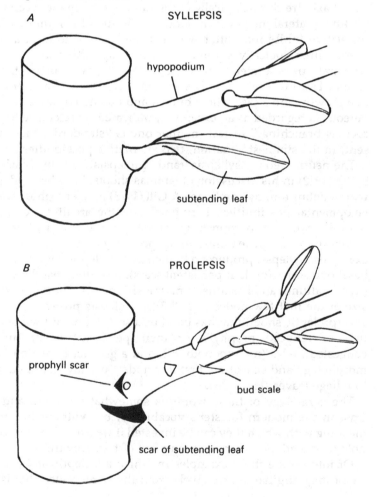

Further south in the eastern United States it becomes more common, so that the predominantly tropical tree flora of southern Florida includes some 40 out of 120 examples. Why syllepsis should be non-adaptive in a temperate flora is not obvious.

Periodicity of branching

The development of lateral meristems, whether by syllepsis or prolepsis, bears a relation to the periodicity of growth activity of the parent terminal meristem. Where growth of the terminal meristem is *continuous* (the ever-growing state: Koriba, 1958) branching itself may be either continuous or intermittent (diffuse). Normally this branching pattern is established only after an initial period of seedling development in which the axis is unbranched. Subsequently, with continuous branching each node subtends a branch (two or more with decussate or whorled phyllotaxis). Coffee provides an example, and the condition characterises Roux's and Cook's models (Fig. 7.1). Continuous branching always involves syllepsis; no example of proleptic architecture is known in the above two models.

Intermittent branching involves the production of one or more tiers of branches at irregular (i.e., unpredictable) intervals, as the development of lateral meristems into branches (usually by syllepsis) seems correlated with the vigour of the terminal meristem. Examples can be found in a number of mangrove and associated genera (*Avicennia, Bruguiera, Ceriops, Rhizophora*), and this architecture distinguishes Attims's model. Where the level of vigour (specifically, rate of growth) of the shoot is high, branching may become continuous (e.g., in *Conocarpus, Laguncularia, Lumnitzera*, all Combretaceae). Gill & Tomlinson (1971) suggested that in *Rhizophora mangle* intermittent branching is a consequence of a simple endogenous feedback mechanism that regulates shoot vigour but is affected by seasonal fluctuations in climate. Consequently, *Rhizophora* shows intermittent branching in both seasonal and nonseasonal climates.

Where growth of a shoot is *rhythmic* or episodic (i.e., with regular alternation of growth and rest), branching is also rhythmic and is clearly correlated with the activity of the terminal meristem. The branches then are precisely positioned on the shoot relative to the articulations on the shoot that may appear by virtue of the presence of bud scale scars. Most temperate trees show rhythmic branching, with the distal meristems on each increment developing the major architectural features of the tree. Branch development is proleptic. However, many tropical trees with rhythmic growth of the trunk develop a tier of sylleptic branches towards the *end* of each flush of trunk growth. This method of rhythmic branching is most common in examples of Aubréville's model, of which *Bombax* and *Terminalia* species provide familiar examples (see Chap. 13). The characteristic physiognomy of these trees is flat-topped, as the leader is unbranched for only a brief period, unlike temperate trees, in which the leader is unbranched for almost a year. The distinction that can be made between intermittent and rhythmic branching on the basis of the few examples studied in detail (Hallé & Martin, 1968; Gill & Tomlinson, 1969, 1971) is likely to be refined as more trees are investigated. The greatest need is for well-worked-out examples.

Morphology of branch expression
We can deal very briefly with the familiar morphologic distinction between "monopodial" and "sympodial" branching. In monopodial branching, lateral branch meristems are produced (continuously, intermittently, or rhythmically) by a permanent terminal meristem. In sympodial branching, lateral branch meristems successively function for a limited period as a terminal shoot and are successively evicted; there is no permanent terminal meristem.

This distinction is precise and should not be used for superficial features of habit. Many trees have a single trunk (monocaulous) that is morphologically a sympodium (e.g., in Chamberlain's model, Fig. 7.1). Tall single-trunked forest trees can have sympodial construction of that trunk (e.g., *Alstonia boonei:* Hallé & Oldeman, 1970, p. 50, Prévost's model). For gross physiognomy of trees, the terms "excurrent" and "decurrent" are appropriate and useful (Zimmermann & Brown, 1971, p. 130). An excurrent tree has a pronounced trunk and a narrow, cone-shaped crown; a decurrent tree has a poorly defined trunk and a spreading crown.

Sympodial branch systems in tropical trees show a wide variety of expression; in the more specialised examples, each *unit of the sympodium* (i.e., portion of the axis produced by the limited activity of a given meristem) may be short, with a fixed number of parts. This modular construction is considered in detail in Chap. 9. The ultimate modular reduction is to a single foliage leaf with a terminal inflorescence, as in *Leptaulus daphnoides* (Icacinaceae), as reported by Hallé & Oldeman (1970), and in many *Piper* species.

It is useful to follow the suggestion of Koriba (1958) and contrast two main types of sympodial growth: substitution and apposition. Sympodial growth by substitution occurs when the terminal meristem either aborts or becomes reproductive (flower or inflorescence) and makes no further contribution to the vegetative architecture of the tree. Substitution of this kind is, of course, common in temperate trees, although where there is seasonal abortion of terminal buds careful morphologic analysis may be needed to demonstrate that the continuing unit arises from the ultimate lateral meristem of the parent unit (i.e., is a pseudoterminal bud resulting in pseudomono-podial branching).

Sympodial growth by apposition occurs when both terminal and lateral meristems continue to function, the terminal meristem of each unit being evicted into a subordinate position (usually an erect short shoot) and extension growth of the axis being continued by a vigorous lateral, which in its turn eventually becomes abruptly erected (e.g., Fig. 7.7A). This is the familiar method of branching within the horizontal tiers of many tropical "pagoda trees" (*Terminalia* branching: see Chap. 13). The adaptive feature of this type of branch tier probably lies in the fact that meristems are long active as leaf-producing rosettes, and lateral expansion of the whole tier is produced by rapidly growing units of extension; proliferation that fills in the available space occurs when one original meristem is apposed by two daughter laterals. An essential feature of this arrangement is that apposition shoots are produced by syllepsis, although meristems commonly exhibit rhythmic growth. It is important to appreciate that the branch system of such a tree functions as a complex of interrelated meristems. Hallé & Oldeman (1970) include such examples in Aubréville's model, where it is precisely expressed, but the tendency towards *Terminalia* branching is found in the outer shoots of branch systems of trees belonging to other models (e.g., *Rhizophora* in Attims's model; *Bumelia* spp. in Rauh's model), indicative of a generalised physiologic response.

Plagiotropy vs. orthotrophy

The recognition of distinct branch tiers or branch complexes leads naturally to a discussion of the essential difference that exists in trees between trunk and branch axes, as it is largely upon this differentiation that the existence of the elaborate organisation, made possible by axillary branching, depends. Many trees show a clear differentiation between "orthotropic" and "plagiotropic" shoots, which may be distinguished by a combination of morphologic features and physiologic responses (tropisms). An orthotropic shoot has an erect orientation (negatively geotropic), radial symmetry, and phyllotaxis, most commonly decussate or spiral. A plagiotropic shoot has a horizontal or oblique orientation (more or less diageotropic) and dorsiventral symmetry either by virtue of a distichous phyllotaxis or, if spiral or decussate, by secondary leaf orientation (petiolar pulvini or twisting of internodes; Fig. 7.4).

In many trees orthotropic shoots correspond to trunk axes and plagiotropic shoots to branch axes, although we have seen that a branch tier may be determined by a complex of interacting meristems (branch complex) rather than by the behaviour of a single meristem. In *Terminalia* branching each sympodial unit is initially plagiotropic and finally orthotropic.

The two categories, in their extreme expression, represent opposite ends of a spectrum of possibilities, and many intermediates exist, depending on the degree to which growth correlations exist. For purposes of discussion, two basically contrasted methods of tree organisation may be recognised: one in which all axes are *inherently* orthotropic, but in which plagiotropy is imposed by apical control; the other in which there exists a *primary differentiation* of shoot meristems, in which each primary meristem of the shoot from its inception shows a preferred symmetry and orientation, so that a marked distinction between orthotropic and plagiotropic shoots may exist within a single individual. This may be described as meristem differentiation.

The situation is best explained by examples. In many trees that conform to Rauh's model (Fig. 7.1) axes are all orthotropic, although normally only one evident orthotropic shoot functions as the leader, exerting a controlling influence on lateral shoots that show some degree of plagiotropy. The essentially orthotropic nature of a lateral meristem can be demonstrated when the apical controlling influence is lost by damage to the leader: An existing lateral then immediately substitutes for the lost leader. In *Pinus* initial growth of lateral shoots is erect, a horizontal position being adopted as control is exerted. With progressive development of a plagiotropic branch complex in

Fig. 7.4. Distal part of plagiotropic shoot (*Psidium guajava*) with decussate phyllotaxis, leaves secondarily oriented into a single plane. Flowers and hence fruits are axillary.

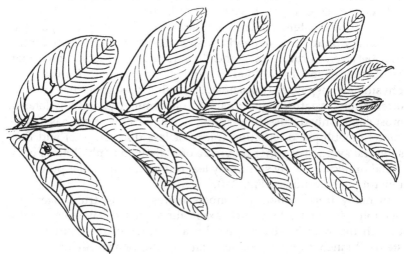

such a shoot system, a reversion to the orthotropic state becomes more difficult, because the control is exerted by a plurality of associated meristems. The advantage of this organisation is clear. Potentially any axis can substitute for a damaged leader in "repairing" the tree. This, no doubt, accounts for the common existence of this model in both temperate and tropical trees. We can consider this change in shoot orientation to represent a dedifferentiation or phase reversion by its meristem.

The architecture of the tree may be such that no controlling leader is developed, as in trees made up of a series of short equivalent orthotropic modules, and the trunk axis may be represented only by a determinate seedling axis, to the height of its first branching (e.g., Leeuwenberg's model, Fig. 7.1; see Chap. 9).

Where there is a marked primary differentiation of meristems – some orthotropic, others plagiotropic – the latter are frequently very stable (i.e., the plagiotropic state is not easily lost by dedifferentiation). Coffee provides an example, as the single orthotropic trunk axis produces branches continuously (by syllepsis), the branches plagiotropic, with pronounced secondary leaf orientation so that the primary decussate leaf arrangement still results in a flattened shoot. Further evidence of meristem differentiation is indicated by the infrequency of further vegetative branches on these plagiotropic shoots and the restriction of flowers to them.

A higher degree of organisation may be exemplified by many species of *Trema* (Ulmaceae). Here differentiation among primary meristems is even more pronounced, as the orthotropic leader has *spiral* phyllotaxis, continuous branching by syllepsis, and does not produce flowers, whereas the plagiotropic laterals have *distichous* phyllotaxis and produce flowers but infrequent vegetative branches (Fig. 7.5). In such plagiotropic shoots dedifferentiation (i.e., reversion from plagiotropy to orthotropy) is difficult. The high degree of trunk-branch differentiation shown by *Araucaria* accounts for its symmetry (Chap. 10), and in the classic experiments of Vöchting (1904) on *Araucaria*, rooted cuttings retained their diageotropic response indefinitely.

Cocoa (Nozeran's model, Fig. 7.1) exhibits a high degree of meristem differentiation and organisation, as each orthotropic (trunk) axis – with spiral phyllotaxis – is of limited growth, terminating in a tier of spirally arranged plagiotropic branches that originate by syllepsis shortly before the parent meristem aborts. Further growth in height above this branch tier is provided by a new orthotropic meristem that develops a proleptic branch at a node below the previous plagiotropic whorl. Experiments show that meristems capable of producing further orthotropic trunks are present at lower nodes

Fig. 7.5. Features of plagiotropy and orthotropy (based on *Trema floridana*). *Lower figure.* Perspective view of erect, orthotropic shoot with spirally arranged leaves (leaf arrangement shown in *inset, top right*). Plagiotropic shoots develop in axil of each leaf on erect shoot; each shoot has distichous phyllotaxis (*inset, top left*) and remains unbranched but produces axillary flower clusters. Inset figures drawn from microtome sections, simplified by omission of indumentum. Petiole or midrib with midvein shown solid black; stipules hatched.

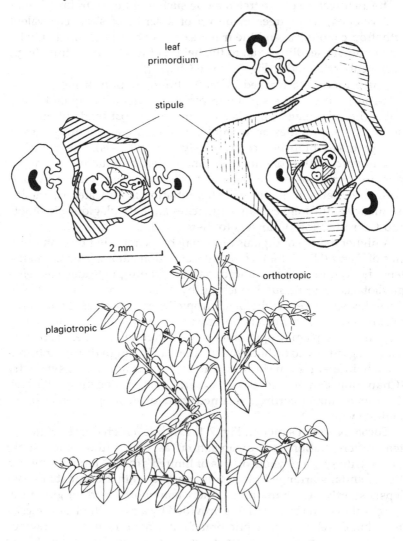

leaf primordium

stipule

2 mm

orthotropic

plagiotropic

(Greathouse & Laetsch, 1969). The morphogenetic situation is complex, as the tree retains a precise architecture without the release of unnecessary erect renewal shoots that would otherwise produce an interlacing network of competing branch tiers. Part of this morphogenetic control undoubtedly comes from the correlation among prolepsis (involving a period of meristem dormancy), orthotropy (spiral phyllotaxis), syllepsis (no meristem dormancy), and plagiotropy (distichous phyllotaxis). This correlation between the length of rest and subsequent symmetry of shoot seems common in tropical trees and needs emphasis, as it may depend on contrasted methods of anatomic attachment resulting from contrasting developmental procedures.

Mixed axes
The existence of orthotropic axes in which plagiotropy is eventually induced by apical control (as in Rauh's model) and the change from plagiotropy to orthotropy in the units that make up the branch complex in *Terminalia* branching demonstrate that the "degree of differentiation" of a meristem may be changed – either by external influences or by modification of internal correlations. Of special interest are those woody plants in which there is an inherent change of expression within a single meristem producing axes of architectural significance. Such axes are described by Hallé & Oldeman (1970) as "mixed." Among the several possibilities the most distinctive are those axes in which an initial orthotropic phase shows a pronounced distal curvature, the site of a future branch complex. Plants made up of such axes are recognised in the definition of Mangenot's model (Fig. 7.1) by Hallé & Oldeman (1970). Because growth in height is quite limited, tall trees are not formed and examples are few. The change in differentiation level that produces the change in orientation remains an interesting morphogenetic response. In *Vaccinium corymbosum* (Ericaceae), a temperate example from the eastern United States, the meristem phase change occurs within a single season, but does not involve a change in phyllotaxis, which is always spiral. In *Dicranolepis persei* (Thymeleaceae), described by Hallé & Oldeman (1970, p. 122), there is a change in phyllotaxis between the proximal orthotropic segment with spirally arranged leaves and the distal plagiotropic segment with distichously arranged leaves. Growth in height is possible only by continued superposition of successive units.

Secondary axis differentiation
Though axis orientation is initially a property of the primary shoot meristem and its interaction with other meristems, secondary changes may have important consequences, mostly mediated via the

influence of reaction wood (Zimmermann & Brown, 1971, pp. 98–105). In angiosperms tension wood developed on the lower stem surface may play a role in establishing the initially oblique or even horizontal orientation of an otherwise orthotropic lateral; subsequently, tension wood developed on the upper side of the stem may prevent the branch from bending below the horizontal under its weight. The example of *Terminalia catappa* (Chap. 13) illustrates this well. There is plenty of evidence to show that reaction wood is a cause, not a consequence, of branch position (e.g., Zimmermann, Wardrop, & Tomlinson, 1968). In gymnosperms, with compression wood, the position and effects of reaction wood tend to be reserved.

Of special interest are those tropical trees in which tension wood raises one out of a number of equivalent, but competing, lateral axes inserted at the same level so that one of them eventually becomes one unit of a sympodial trunk axis. *Hura crepitans* (Euphorbiaceae) illustrates the process well (Fig. 7.6). Each axis is determinate, terminating in an inflorescence (or its aborted vestige) immediately below which develops a group of two to four (usually three) branches. All are initially the same diameter and have the same oblique orientation. By presumed competition for nutrition, one becomes more vigorous, thickens more rapidly, and develops tension wood on the upper surface until a vertical orientation is achieved. The less successful modules remain as obliquely horizontal branches. Despite this sympodial construction, *Hura crepitans* makes a tall forest tree, to 50 m. However, it is not an isolated example; the same architecture exists in a wide variety of unrelated trees (Koriba's model, Fig. 7.1). These all provide examples of secondary determination of axis orientation.

A similar process may determine architecture in those many trees described by Hallé & Oldeman (1970) as Troll's model (Fig. 7.1), in which axes are essentially plagiotropic, growth in height being achieved by their successive superposition. According to F. Hallé (pers. comm.), secondary erection of axes, after leaf fall, is here a probable feature of growth, though evidence for it still remains to be established.

Sequential branching and reiteration

The concept of architecture, or genetically expressed growth form, provides the syntax by means of which we are beginning to understand tree construction. It is of fundamental significance because it allows us to recognise that two factors are involved in the normal expression of tree growth. The first is its inherent architecture expressed when we grow a tree under optimal conditions, as in a nursery. However, trees grow in forests or other natural communi-

Fig. 7.6. Branch reorientation induced by secondary growth in *Hura crepitans* (Euphorbiaceae). *A.* Young shoot from above at level of branching immediately below aborted terminal inflorescence, represented by its scar in crotch of fork. *B.* Similar level in older shoot, main axis recognisable as a trunk by its greater thickness and now almost erect orientation, lateral branches still inclined and much thinner. *C, D.* Section of trunk axis shown in *B* stained with chlorazol black to distinguish reaction fibers of tension wood. *C* represents upper sector (much reaction wood); *D*, lower sector (little reaction wood).

ties where conditions are rarely optimal, and the second factor is the response the tree's genetic pattern permits when it is environmentally disturbed. Essentially these two factors may be said to be the genotypic and phenotypic responses of the tree, but it must be appreciated that the responsiveness of the tree to environmental change is itself under genetic control.

The second process involves the concept of "reiteration" of the tree model that exists as a logical extension of the architectural concept (Oldeman, 1974). It is explained in an ecologic context in Chapter 23, but is introduced at this stage because it permits us to distinguish two types of tree branching within an architectural context:

1. Branching determined in direct conformity with the tree's genetic model (i.e., "sequential branching," in which branches are produced in a regular and often highly predictable sequence, determined solely by the unmodified architecture). This branching is recognised largely by qualitative criteria, but its existence is quite clear in trees with a high degree of symmetry that conform precisely to the model, as in some *Araucaria* species (e.g., *A. heterophylla*).

2. Branching outside the model (i.e., "reiterative branching," in which meristems not brought into play in the original architecture of the tree are induced to elaborate new shoots, usually seen as additional trunk axes and branch complexes). Where the tree is symmetric and highly organised, this type of reiteration is conspicuous, as it produces new miniature tree models inserted on the existing framework. But a moment's reflection will indicate that it must occur in less symmetric species. Such familiar features as epicormic branching, suckering, proliferation by branch and stump sprouts exemplify this process. Such branching is inevitably by prolepsis, as it comes from dormant meristems. In trees in which sequential branching is entirely by syllepsis (e.g., those with continuous branching, as in Roux's and Cook's models), the existence of a proleptic shoot immediately indicates reiteration.

However, not all trees possess reserve meristems, and some reserve meristems are short-lived (e.g., 2–3 years in *Rhizophora mangle:* Gill & Tomlinson, 1971). In such instances a more subtle process of reiteration is possible and is dependent on the process of dedifferentiation (or phase change) when a plagiotropic shoot reverts to orthotropy. In *Rhizophora* sylleptic laterals, which in the architecture of the intact tree are subject to strongly induced plagiotropy (Fig. 7.7A), will remain orthotropic and substitute for a damaged leader. Forked trunks (rare in *Rhizophora*) may be initiated in this way; normally one axis becomes dominant. Scrubby mangrove on unfavourable sites consists of small tress made up of numerous reduced orthotropic axes, no one capable of inducing apical control over the others (Fig. 7.7B).

Clearly these processes of reiteration permit a high degree of flexibility of tree organisation under environmental stress; the degree of flexibility is then in itself an important ecologic parameter. Proleptic reiterated branches permit repairing of damaged trees. Reiteration by dedifferentiation may permit an initially plagiotropic branch, which becomes orthotropic, to exploit fully a gap in the canopy that opens near it. Both processes can occur within a single tree. Their significance is indicated by recent observations on *Agathis australis*

Fig. 7.7. Features of architecture and reiteration in *Rhizophora mangle* (Rhizophoraceae). A. Distal portion of lateral branch, illustrating plagiotropy by apposition growth, a characteristic feature of *Terminalia* branching. This represents branch construction in unstressed trees. B. Entire dwarf plant with leaves removed (aerial roots stippled) from scrubby mangrove, representing a highly stressed tree. All axes remain orthotropic, and no dominant trunk is developed.

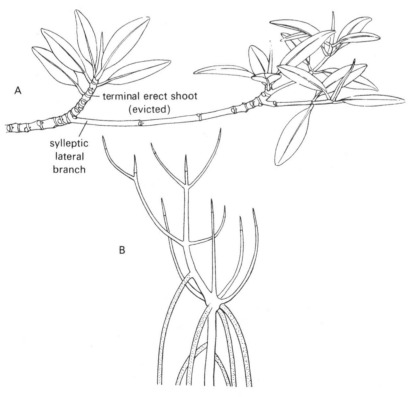

(Araucariaceae) in New Zealand. This represents Massart's model, but with some instability of the plagiotropic laterals. In a plantation of 90 young trees of varying size (the tallest 16.5 m high), 43.3% still grew in exact conformity with their model, and 56.7% showed some degree of reiteration, either by forking below a damaged leader (24.5%) or by dedifferentiation so that the ends of horizontal branches had turned erect (32.2%). It could be estimated that all trees taller than 20 m would show some degree of reiteration, and this accounts for the rather broad crown on the massive trunk, which characterises adult trees of this species in the forest. This might be said to be the "shape" of the tree, but the shape of the juvenile tree, determined by model conformity, is obviously narrowly conical.

Reiteration of both sorts, but particularly that which initiates proleptic orthotropic shoots, is normally responsible for the larger limbs of older trees; morphologically these are not branches but trunk axes. This is one method by which the excurrent habit of trees is produced. Herein might lie an explanation of the nonsynchronous behaviour of different parts of an individual in reference to flowering or flushing, a tropical phenomenon frequently commented upon. The "individual" is probably made up of several reiterated models, not necessarily developmentally in phase with each other.

A distinction between sequential and reiterative branching is important in an appreciation of the adaptive nature of tree organisation. Growth models represent efficiency in tree organisation determined by natural selection. Equally the reiterative plasticity may be an important medium on which the processes of natural selection are at work. In fact, it seems that in many temperate trees it is the dominant force, related to the marked environmental stress to which they are subjected. In the tropics, stress (at least, climatic stress) is less pronounced, reiterative processes may be minimised, and the architecture of the tree is not only more diversified but more readily recognised. This is why the recognition of the twin processes (architecture and reiteration) that determine tree form have awaited an analysis of tropical tree form.

Acknowledgments

The substance and ideas in this talk owe much to discussions with Francis Hallé and Roelof Oldeman, and I apologise if their concepts are distorted or misrepresented in my presentation. Earlier discussions with Macolm Gill have been quite catalytic. Support from the Cabot Foundation and Atkins Fund of Harvard University has permitted access to the tropical vegetation types that are the appropriate and ultimate source of inspiration. Recent visits to the South Pacific have been made possible by grants from the National Geographic Society, Washington, D.C.

References

Boke, N. (1976). Dichotomous branching in *Mammillaria* (Cactaceae). *Amer. J. Bot., 63*, 1380–4.

Bugnon, F. (1971). Essai pour une caractérisation des types fondamentaux de ramification chez les végétaux. *Mém. Soc. Bot. Fr., 1971,* 79–85.

Fisher, J. B. (1974). Axillary and dichotomous branching in the palm *Chamaedorea. Amer. J. Bot., 61,* 1046–56.

– (1976). Development of dichotomous branching and axillary buds in *Strelitzia* (Monocotyledoneae). *Can. J. Bot., 54,* 578–92.

Gill, A. M., & Tomlinson, P. B. (1969). Studies on the growth of red mangrove (*Rhizophora mangle* L.). 1. Habit and general morphology. *Biotropica, 1,* 1–9.

– (1971). Studies on the growth of red mangrove (*Rhizophora mangle* L.). 3. Phenology of the shoot. *Biotropica, 3,* 109–24.

Greathouse, D. C., & Laetsch, W. M. (1969). Structure and development of the dimorphic branch system of *Theobroma cacao. Amer. J. Bot., 56,* 1143–51.

Hallé, F., & Martin, R. (1968). Etude de la croissance rythmique chez l'Hévéa (*Hevea brasiliensis* Müll.-Arg. Euphorbiacées-Crotonoïdées). *Adansonia, 8,* 475–503.

Hallé, F., & Oldeman, R. A. A. (1970). *Essai sur l'architecture et la dynamique de croissance des arbres tropicaux.* Paris: Masson.

Koriba, K. (1958). On the periodicity of tree-growth in the tropics. *Gard. Bull. Singapore, 17,* 11–81.

Oldeman, R. A. A. (1974). L'architecture de la forêt guyanaise. *Mém. O.R.S.T.O.M.,* no. 73, 1–204.

Shah, J. J., & Patel, J. D. (1972). The shell zone: Its differentiaton and probable function in some dicotyledons. *Amer. J. Bot., 59,* 683–90.

Späth, H. L. (1912). *Der Johannistrieb.* Berlin: Parey.

Tomlinson, P. B. (1971). The shoot apex and its dichotomous branching in the *Nypa* palm. *Ann. Bot. Lond., 35,* 865–79.

Tomlinson, P. B., & Gill, A. M. (1973). Growth habits of tropical trees: Some guiding principles. In *Tropical Forest Ecosystems in Africa and South America: A comparative review,* ed. B. J. Meggers, E. S. Ayensu, & W. D. Duckworth, pp. 129–43. Washington, D.C.: Smithsonian Institution Press.

Tomlinson, P. B., & Posluszny, U. (1977). Features of dichotomizing apices in *Flagellaria indica* (Monocotyledones). *Amer. J. Bot., 64,* 1057–65.

Tomlinson, P. B., & Soderholm, P. K. (1975). The flowering and fruiting of *Corypha elata* in South Florida. *Principes, 19,* 83–99.

Tomlinson, P. B., Zimmermann, M. H., & Simpson, P. G. (1970). Dichotomous and pseudodichotomous branching of monocotyledonous trees. *Phytomorphology, 20,* 28–32.

Vöchting, H. (1904). Ueber die Regeneration der *Araucaria excelsa. Jb. Wiss. Bot., 40,* 144–55.

Zimmermann, M. H., & Brown, C. L. (1971). *Trees: Structure and Function.* New York: Springer-Verlag.

Zimmermann, M. H., Wardrop, A. B., & Tomlinson, P. B. (1968). Tension wood in aerial roots of *Ficus benjamina* L. *Wood Sci. Tech., 2,* 95–104.

8

Architectural variation at the specific level in tropical trees

FRANCIS HALLÉ

Institut de Botanique, Montpellier, France

My main research is a genetic approach to plant form in its entirety aimed at improving our understanding of the inheritance of plant architecture. It has been established that vegetative architecture is a constant and stable characteristic of the plant (Hallé, 1971, 1974; Hallé & Oldeman, 1975). Of course, ecologic conditions often involve some changes of form, but these are only quantitative, the basic architecture remaining unaltered.

For instance, the commonly cultivated tree *Muntingia calabura* L. (Tiliaceae, tropical America) has quite a different appearance depending upon whether it is growing in forest shade or in open secondary vegetation. Although the physiognomy shows a contrast between a tall, narrow-crowned (Fig. 8.1*A*) and a low, broad-crowned (Fig. 8.1*B*) individual, the architectural model itself is free from ecologic variations, at least within certain limits (see Chap. 9). As a general rule, therefore, we may consider the architecture as a constant character, valuable for specific identification; indeed, the architectural model of a species has been quoted in several recent identifications (Hallé, 1970, 1973; Mabberley, 1973, 1974, 1975). It is obviously one of our aims in this symposium to bridge the gap between taxonomy, the study of plant parts, and developmental morphology of the whole plant.

Nevertheless, this general rule of architectural constancy at the specific level makes it difficult to explain the polymorphism that occurs in most families and even in some genera. For instance, as far as I know, one can find 8 models in the Moraceae, 10 in the Sterculiaceae, and 10 within the single genus *Euphorbia* L. (Cremers, 1975). The existence of this polymorphism in the higher taxa inevitably implies a mechanism of architectural variation at the specific level, and this gives rise to the question of the nature of such a mechanism.

210   *Francis Hallé*

Of course, the stability of the architecture as a hereditary character is compatible with a certain degree of variation; this kind of situation is well known in genetics. If tropical trees are living systems, and I am pretty sure they are, their architecture must permit some variability.

Architectural modification
    Observation demonstrates that a specific qualitative variation of the vegetative architecture does exist and, as a first step toward understanding it, I present some examples, divided into three groups: intraspecific variation correlated with sex, ecologically imposed variation, and variation due to mutation.

Intraspecific variation correlated with sex
    In *Cycas circinalis* L. (Cycadaceae, Southeast Asia) the intraspecific variation is correlated with sex; male and female plants belong to different architectural models, as essentially shown by Chamberlain (1911). The female *Cycas* tree conforms to Corner's, the male to Chamberlain's model (Fig. 8.2). In the former, the trunk is a monopodium, with ovulate sporophylls borne laterally; in the latter, the trunk is a sympodium, each module ending in a male cone.

Ecologically imposed variation
    In several trees recently recorded by Kahn (1975) – *Euphorbia mellifera* Ait. (Euphorbiaceae, West Africa); *Arbutus unedo* L. (Ericaceae, Mediterranean region), *Mangifera indica* L. (Anacardiaceae, Southeast Asia), *Isertia coccinea* (Aubl.) Gmel. (Rubiaceae,

Fig. 8.1. Two different forms of *Muntingia calabura* determined by ecologic circumstances, architecture remaining unchanged. *A*. Tree (arrow) in dense forest. *B*. Open-grown tree.

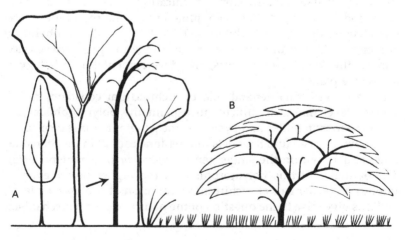

Guianas) – the vegetative architecture is variable and obviously correlated with the environment, most closely with the amount of incident light. In full sun, these trees conform to Leeuwenberg's
model; in the shade, to Scarrone's model (Fig. 8.3). The difference is
established by the continued activity of the apical meristem to form a
distinct trunk in the latter, and by its early abortion in the former.

Fig. 8.2. Architectural variation correlated with sex in *Cycas cir-
cinalis. A.* Female plant (Corner's model). *B.* Male plant (Chamberlain's model).

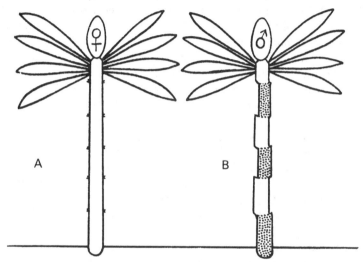

Fig. 8.3. Architecture influenced by environmental conditions. *A.*
Open-grown tree, corresponding to Leeuwenberg's model. *B.* Tree
of same species growing in shade, corresponding to Scarrone's
model. For further explanation, see text.

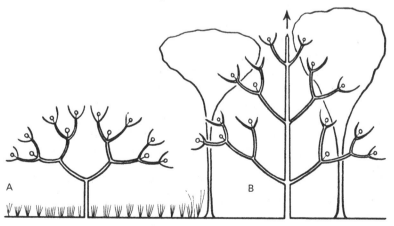

Under these circumstances we have a contrast between fully and partially modular construction (see Chap. 9). The amplitude of this variation is narrow, and the only conclusion we can draw is that there is a close developmental relation between these two patterns.

Variation due to mutation

Much more interesting is the third group, where are gathered what I propose to call "architectural mutations." The term "mutation" seems to be accurate because there is a sudden appearance of the modified architecture, which, at least in some cases, is heritable. Most of the architectural mutations described below concern tropical crop trees. Observations on cultivated plants are important, as they enable us to study wide monospecific and homogeneous plant populations, in which mutants are easily detected. The following examples illustrate the concept of architectural mutation (Figs. 8.4, 8.5).

Leeuwenberg's model is the normal architecture of cassava (*Manihot esculenta* Crantz; Euphorbiaceae, South America), as shown in Fig. 8.4A. The mutation "pétiolule" belonging to Chamberlain's model (Fig. 8.4B) was discovered by Cours (1951).

The normal form of the travelers' tree (*Ravenala madagascariensis* Gmel.; Strelitziaceae, Madagascar) is Tomlinson's model, although it is often seen as Corner's model owing to artificial removal of the basal suckers (Fig. 8.4C). The mutation "apical sexuality," corresponding to Holttum's model, was found in Lae, Papua New Guinea, in 1972 (Fig. 8.4D).

The opposite situation is also known, where Holttum's model is the normal form, as in maize and in tobacco (Fig. 8.4D). Several mutations (indeterminate growth) with lateral sexuality have been described in both plants: by Singleton (1946) in maize and by Jones (1921) in tobacco (Fig. 8.4C).

Rauh's model is the normal architecture of *Pinus* and of rubber (*Hevea brasiliensis* Müll.-Arg.; Euphorbiaceae, South America), as shown in Fig. 8.4E. Kozlowski & Greathouse (1970) described the mutation "foxtail" in *Pinus caribaea* Maorelet (Pinaceae, Central America), without branching, without annual rings in the wood, but fertile. Male cones have been recorded on trees growing in Kepong, Malaysia (Fig. 8.4F). The same mutation, but sterile, appears in rubber, and is known as "lampbrush." It has been demonstrated by Hallé & Martin (1968) that the lampbrush mutation can be produced experimentally by removing most of the foliar surface when the leaves are young. Both foxtail and lampbrush belong to Corner's model.

In *Pinus pinaster* Ait. subsp. *maritima* H. del Vill. (Pinaceae, Mediterranean region; Rauh's model), a mutation called "piboteau," with

Fig. 8.4. Architectural mutations. *A. Manihot esculenta,* normal form, conforming to Leeuwenberg's model. *B.* Pétiolule mutation conforming to Chamberlain's model. *C. Ravenala madagascariensis,* shown here without basal suckers and conforming to Corner's model. *D.* Mutant form with apparent terminal inflorescence, observed in Botanic Garden, Lae, Papua New Guinea. *E.* Rauh's model, illustrative of the architecture of many trees. *F.* Lampbrush and foxtail mutations of *Hevea* and *Pinus* spp., a modification of Rauh's model characterized by lack of branching.

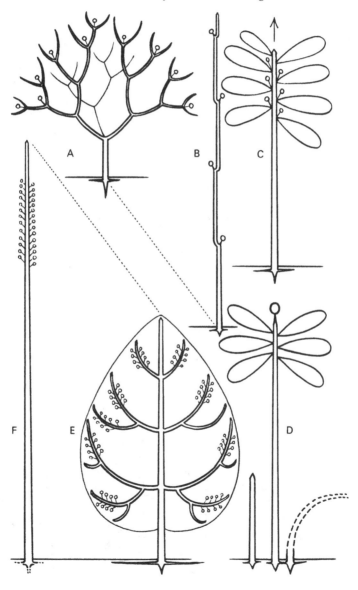

Fig. 8.5. Architectural mutations. *A*. Piboteau mutation in *Pinus pinaster*, which because of its sympodial trunk growth represents Koriba's model rather than original Rauh's model (cf. Fig. 8.4*E*). *B*. Roux's model, normal architecture of coffee. *C*. Plagiotropic mutation of coffee that corresponds to Troll's model. *D*. Petit's model, found in certain Malvales. *E*. Little- or unbranched mutation conforming to Corner's model. *F*. Abortive terminal mutation of cotton. *G*. Similar mutation in *Abroma augusta*.

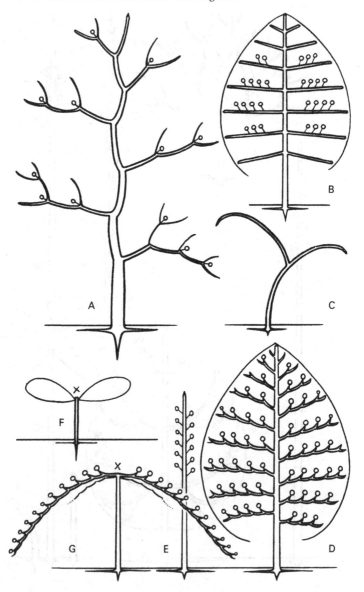

apical female sexuality, was recently described by Dupuy & Guédès (1975). This *Pinus piboteau* belongs to Koriba's model (Fig. 8.5*A*).

Roux's model is the normal architecture of coffee (*Coffea arabica* L.; Rubiaceae, Ethiopia) as well as of *Paliurus australis* Gaertn. (Rhamnaceae, Mediterranean region; Fig. 8.5*B*). The plagiotropic mutation, corresponding to Troll's model, was discovered by Carvalho & Antunes Filho (1952) in coffee grown from seeds (Fig. 8.5*C*). It is well known that a plagiotropic coffee is easily obtained by propagation of branches of the coffee tree. In *Paliurus australis* Gaertn., a sterile plagiotropic mutation has been experimentally produced by cultivation of etiolated plants by Roux (1968).

The model of Petit (Fig. 8.5*D*) is normal for several members of the Malvales, such as jute (*Corchorus capsularis* L., Tiliaceae, pantropical), cotton (e.g., *Gossypium hirsutum* L., Malvaceae, America), and *Abroma augusta* (L.) L.f. (Sterculiaceae, Southeast Asia). A short branch or unbranched mutation was described by Patel, Ghose, & Sanyal (1945) in *Corchorus,* and by Kearney (1930) in cotton. Both mutations belong to Corner's model (Fig. 8.5*E*). Another mutation, "abortive terminal" (Fig. 8.5*F*), was found in cotton and carefully studied by Quisenberry & Kohel (1971); nearly the same mutation exists in *Abroma,* and was found in the Frère Gillet Botanical Garden, Kisantu, Zaire, in 1969 (Fig. 8.5*G*). The abortive terminal mutation occurs in many herbaceous plants such as *Mollugo, Gisekia, Glinus, Solanum, Euphorbia,* and *Boerhaavia.* It may have played an important part in the process of miniaturization (Hallé & Oldeman, 1975) or evolution from trees to herbs.

In Table 8.1 are listed most of the examples of architectural mutations known at the present.

The profound heterogeneity of these examples must be stressed. A number of internal or environmental conditions may give rise to these mutations:

1. Profoundly modified ecologic or experimental environment (apical sexuality in *Ravenala,* plagiotropy in *Paliurus*)
2. Surgery or chemical mutagen experiment (apical sexuality in *Impatiens,* dichotomous branching in maize)
3. Viruses or mycoplasmas (forma monstrosa in *Opuntia*)

The genetic significance of these architectural mutations is also variable. In a few examples, the genetic constitution of the species is not modified, and the mutation is either not heritable (lampbrush in *Hevea*) or its heritability is not clearly demonstrated (foxtail in *Pinus*). In some cases, the mutation is genetic and involves a single nuclear gene (indeterminate growth in maize, short branch in cotton, lack of branching in jute and sunflower). In these cases, the mutant form of

216   *Francis Hallé*

Table 8.1. *Known architectural mutants*

Species	Normal model	Architecture and kind of mutation
Nicotiana tabacum L. Jones (1921)	Holttum	Corner Indeterminate growth
Zea mays L. Reeves & Stansel (1940) Singleton (1946)	Holttum	Corner Indeterminate growth
Zea mays L. Mouli (1970)	Holttum	Schoute Dichotomous branching
Hyphaene ventricosa Kirk. Lewalle (1968)	Corner	Schoute Dichotomous branching
Ravenala madagascariensis Gmel. Pers. obs. (1972)	Corner	Holttum Apical sexuality
Elaeis guineensis Jacq. Henry (1948) Henry & Scheidecker (1953)	Corner	Attims Vivipary
Opuntia subulata (Muchl.) Eng. Nozeran & Neville (1963)	Corner	Attims[a] Forma monstrosa
Helianthus annuus L. Hockett & Knowles (1970)	Leeuwenberg	Holttum Without branching
Impatiens balsamina L. Nanda & Purohit (1966)	Corner[a]	Leeuwenberg[a] Apical sexuality
Corchorus capsularis L. Patel, Ghose, & Sanyal (1945)	Petit[a]	Corner Without branching
Gossypium hirsutum L. Kearney (1930)	Petit	Corner Short branch
Gossypium hirsutum L. Quisenberry & Kohel (1971)	Petit	[b] Abortive terminal
Abroma augusta (L.) L.f. Pers. obs. (1969)	Petit	[b] Abortive terminal
Pinus pinaster Ait. Dupuy & Guédès (1969, 1975)	Rauh	Koriba Piboteau
Abies alba Mill Guinier (1935)	Massart	Corner Without branching
Pinus caribaea Maorelet Kozlowski & Greathouse (1970) Kummerov (1962)	Rauh	Corner Foxtail

Table 8.1 (*cont.*)

Species	Normal model	Architecture and kind of mutation
Hevea brasiliensis Müll.-Arg. Hallé & Martin (1968)	Rauh	Corner Lampbrush
Manihot esculenta Crantz Cours (1951)	Leeuwenberg	Chamberlain Pétiolule
Mangifera indica L. Scarrone (1969)	Scarrone	Leeuwenberg[a] Apical sexuality
Coffea arabica L. Carvalho & Antunes Filho (1952)	Roux	Troll[a] Plagiotropy
Paliurus australis Gaertn. Roux (1968)	Roux	Troll Plagiotropy
Theobroma cacao L. Soria (1961)	Nozeran	Corner Neoteny

[a] Architecture incompletely known.
[b] Architecture unknown.

the gene is recessive, and the inheritance is a very simple Mendelian one. On the other hand, the abortive terminal mutation in cotton, studied by Quisenberry & Kohel (1971), is not under the control of nuclear genes and has proved to be cytoplasmically inherited through the maternal parent. This result must be compared with those of Demarly (1974), who obtained, in *Lactuca* plants of equal genotypic constitution, several forms of branching (distal branching, basal branching, no branching). In Demarly's experiment, the different plants originated from the same tissue culture, and the equality of the genetic constitutions was easily demonstrated by crossing. The results in cotton and lettuce suggest that cytoplasmic heredity plays an important part in the genetic control of the vegetative architecture.

Conclusions

In spite of the considerable diversity of architectural mutations, this survey leads to a preliminary conclusion that seems to be of importance. In most cases, the form of the mutant is not arbitrary; it belongs to a known model, one which is normal in other species. It was recently assumed by Meyen (1973) and Stone (1975) that plant morphology is a closed system and that species must conform, over

218   *Francis Hallé*

and over again, to a few basic designs. Some experimental results
(Nanda & Purohit, 1966; Roux, 1968; Hallé & Martin, 1968; Mouli,
1970) now support the concept of plant architecture as a closed system. Stevens (1974) suggests that the concept of a closed system may
apply to all sorts of morphologic patterns, irrespective of whether
they are living or inanimate.

Future research on tree architecture and its genetic background
will depend upon knowledge of other examples of architectural mutations, and I would appreciate receiving information on the subject
based on either field observation or published report. The possibility
of crossing an architectural mutation with its normal form provides a
method for genetic analysis of vegetative architecture and leads ultimately to the causal phylogeny of tree models.

General discussion

Tomlinson: The mutation of *Ravenala* is interesting because the mutation is
in the direction of the genus *Phenakospermum*, of the same family. *Phenakospermum* is distinctive in the family Strelitziaceae in that it propagates by
basal stoloniferous suckers, and there is a very interesting correlation between the sexuality of the axis and the method of vegetative growth. Would
you speculate that *Ravenala* that did not have basal suckers would be poorly
adapted and would not survive?

Hallé: I don't know, because all the apical flowering *Ravenala* I saw were in
cultivation and suckers were removed for propagation purposes.

Tomlinson: Is there a possibility that this manipulation was the stimulus for
the flowering?

Hallé: No, I don't think so.

Janzen: I would like to know what is the sex determination mechanism in
cycads? You showed male and female plants. Does that start out as a 50 : 50
sex ratio? Or is sexuality determined later in the life of the plant, which then
remains either a male or female?

Hallé: If I remember correctly, I found many more males than females. I don't
know if the genetic mechanics of sex determination are known in this particular *Cycas* species.

Alvim: I would like to ask Dr. Hallé about the change in the architecture of
Theobroma cacao (Nozeran's model). In some areas the plagiotropic branches
are almost orthotropic, although they retain the distichous phyllotaxis; they
are very thick and vigorous. I wonder how would you interpret this? Is it a
mutation?

Hallé: Obviously I cannot answer without seeing the plant, but I would be
very interested to know the mechanism of growth of the trunk. Is it the same
sympodial mechanism of normal cocoa, or does a single meristem continue
to grow?

Alvim: The first orthotropic shoots produce a crown and plagiotropic
branches. Subsequently, another upright, very thick axis is developed from
one of the plagiotropic branches. This retains the same phyllotaxis of the
plagiotropic branch, but lacks a plagiotropic orientation. They are essentially orthotropic and very vigorous. I shall try to propagate or graft it to see

if it retains this character. It is certainly associated with some environmental stimulus because it is found only in a few areas where the conditions must be different; but only 1% or 2% of the plants show the change.

Oldeman: The axes you describe are a reiteration of the model. If this is true, they represent rejuvenation in the tree and you might well be able to make very good cuttings of them. They are related to increased energy levels in the environment because reiteration is always opportunistic.

Borchert: In relation to Dr. Hallé's and Dr. Alvim's statements, can one always assume that the observed morphologic variation is genotypic and not phenotypic? In your own example of the transition of *Hevea* from rhythmic to continuous growth and vice versa, you have expressed within the same plant two models, so obviously within one plant of the same genotype both behavioral forms are possible. Likewise, foxtailing is never observed in the temperate zone, yet many temperate zone *Pinus* when grown in the tropics (e.g., Colombia) for lumber production, grow foxtails. Some of the California pines do it consistently; this is one reason they have not been successfully used for timber production in the tropics. So again we must have with the same genotypic basis different phenotypic expressions. Can one easily refer to all of these as genetic mutations?

Hallé: I agree with you completely. This is why I put the term "mutation" in quotation marks.

Whitmore: In a young field of rubber planted in Malaya, some of them foxtail and others do not, yet all are the same clone.

Givnish: Have you looked at a few families that have a diversity in architectural models and considered how there may be an adaptive radiation within the family accordingly?

Hallé: This is a very important question, but the answer is not easy. For instance, in the boundary between forest and savanna, you find mainly an architecture with high potential for vegetative propagation. Otherwise, there is no clear evidence of a relation between architecture and ecology.

Oldeman: I do not quite agree. Certainly the relation between architecture and ecology is rather loose. On the other hand, when a tree conforms to the model throughout its whole life, it generally grows in an environment where energy is fairly constant. Such trees are uncommon in the temperate region; nor does one find such trees in the "mature" rain forest canopy, because most of the canopy trees have grown up from the shade into full light – i.e., they have experienced an appreciable change of energy offered during their lifetimes.

References

Carvalho, A., & Antunes Filho, H. (1952). Novas observacoes sobre o dimorfismo dos ramos em *Coffea arabica* L. *Bragantia, 12*, 81–4.
Chamberlain, C. J. (1911). The adult cycad trunk. *Bot. Gaz., 52*(2), 81–104.
Cours, G. (1951). Le manioc à Madagascar. *Mém. Inst. Sci. Madagascar B, 3*, 203–400.
Cremers, G. (1975). Sur la présence de dix modèles d'architecture végétative chez les Euphorbes malgaches. *C.R. Acad. Sci. Paris, 281*, 1578–8.
Demarly, Y. (1974). Quelques données nouvelles en amélioration des plantes. *Agron. Trop., 24*, 9.
Dupuy, P., & Guédès, M. (1969). Homologies entre rameaux courts, rameaux longs et rameaux reproducteurs chez le pin maritime (*Pinus pinaster*

Ait. ssp. *maritima* H. del Vill.): Notion de métamorphose chez les gymnospermes. *C. R. Sci. Soc. Biol., 163*(11), 2426–9.

– (1975). Une précieuse forme du pin maritime: *Pinus pinaster* Ait. subsp. *atlantica* H. del Vill. cv. "piboteau." *Bull. Soc. Bot. Fr., 122*, 163–72.

Guinier, P. (1935). Une curieuse mutation chez le sapin (*Abies alba*): Les sapins sans branches. *C. R. Sci. Soc. Biol., 119*, 863–4.

Hallé, F. (1971). Architecture and growth of tropical trees exemplified by the Euphorbiaceae. *Biotropica, 3*, 56–62.

– (1974). Architecture of trees in the rain forest of Morobe District, New Guinea. *Biotropica, 6*, 43–50.

Hallé, F., & Martin, R. (1968). Etude de la croissance rythmique chez l'Hévéa (*Hevea brasiliensis* Müll.-Arg. Euphorbiaćee-Crotonoïdée). *Adansonia, 8*, 475–503.

Hallé, F., & Oldeman, R. A. A. (1975). *Essay on the Architecture and Dynamics of Growth of Tropical Trees*. Tr. B. C. Stone. Kuala Lumpur: Penerbit Universiti Malaya.

Hallé, N. (1971). *Crioceras dipladeniiflorus* (Stapf.) K. Schumann. Apôcynacée du Gabon et du Congo. *Adansonia, 11*, 301–8.

– (1973). *Captaincookia*, genre nouveau monotypique néo-calédonien de Rubiacéae-Ixoreae. *Adansonia, 13*, 195–202.

Henry, P. (1948). Un *Elaeis* remarquable: Le palmier à huile vivipare. *Rev. Bot. Appl., 311, 312*, 422–7.

Henry, P., & Scheidecker, D. (1953). Nouvelle contribution à l'étude des *Elaeis* vivipares. *Oléagineux, 10*, 681–8.

Hockett, E. A., & Knowles, P. F. (1970). Inheritance of branching in sunflowers, *Helianthus annuus* L. *Crop Sci., 10*, 432–6.

Jones, D. F. (1921). The indeterminate growth factor in tobacco and its effects upon development. *Genetics, 6*, 433–44.

Kahn, S. (1975). Remarques sur l'architecture végétative dans ses rapports avec la systématique et la biogéographie. *D.E.A. Biol. Vég.*, Montpellier. 33 pp.

Kearney, T. H. (1930). Short branch: Another character of cotton showing monohybrid inheritance. *J. Agric. Res., 41*, 379–87.

Kozlowski, T. T., & Greathouse, T. E. (1970). Shoot growth and form of pines in the tropics. *Unasylva, 24*, 2–10.

Kummerov. J. (1962). Ueber Wachstumsanomalien bei *Pinus radiata* unter tropischen Bedingungen. *Ber. Dtsch. Bot. Ges., 75*, 37–40.

Lewalle, J. (1968). A note on *Hyphaene ventricosa*. *Principes, 12*, 104–5.

Mabberley, D. J. (1973). Evolution in the giant groundsels. *Kew Bull., 28*, 61–96.

– (1974). The pachycaul *Lobelia* of Africa and St. Helena. *Kew Bull., 29*, 535–84.

– (1975). The pachycaul *Senecio* species of St. Helena, "Cacalia paterna" and "Cacalia materna." *Kew Bull., 30*, 413–20.

Meyen, S. V. (1973). Plant morphology in its nomothetical aspects. *Bot. Rev., 39*, 205–60.

Mouli, C. (1970). Mutagen-induced dichotomous branching in maize. *J. Hered., 61*, 150.

Nanda, K. K., & Purohit, A. N. (1966). Experimental induction of apical flowering in indeterminate plant *Impatiens balsamina* L. *Naturwissenschaften, 54*, 230.

Nozeran, R., & Neville, P. (1963). Nature "virale" de certaines transformations observées chez des Cactacées. *Nat. Monsp., 15*, 109–14.

Patel, J. S., Ghose, R. L. M., & Sanyal, A. (1945). The genetics of *Corchorus* (jute). III. The inheritance of corolla colour, branching habit, stipule character and seed coat colour. *Ind. J. Genet. Plant Breeding, 4*, 75–9.

Quisenberry, J. E., & Kohel, R. J. (1971). Abortive terminal, a cytoplasmically inherited character in cotton, *Gossypium hirsutum* L. *Crop Sci., 11,* 128–9.

Reeves, R. G., & Stansel, R. H. (1940). Uncontrolled vegetative development in maize and teosinte. *Amer. J. Bot., 27,* 27–30.

Roux, J. (1968). Sur le comportement des axes aériens chez quelques plantes à rameaux végétatifs polymorphes: Le concept de rameau plagiotrope. *Ann. Sci. Nat. (Bot.), 12,* 109–255.

Scarrone, F. (1969). Recherches sur les rythmes de croissance du manguier et de quelques végétaux ligneux malagasy. Thesis, Université de Clermont-Ferrand. 2 vols.

Singleton, W. R. (1946). Inheritance of indeterminate growth in maize. *J. Hered., 37,* 61–4.

Soria, V. J. (1961). A note on cacao plant which flowered at the age of three months. *Cacao, 6,* 11–12.

Stevens, P. S. (1974). *Patterns in Nature.* Boston: Atlantic Monthly Press.

Stone, B. C. (1975). Translator's preface. In *Essay on the Architecture and Dynamics of Growth of Tropical Trees,* by F. Hallé, & R. A. A. Oldeman. Kuala Lumpur: Penerbit Universiti Malaya.

9

Modular construction and its distribution in tropical woody plants

MARIE-FRANÇOISE PRÉVOST

O.R.S.T.O.M., Abidjan, Ivory Coast

The definition of module (a free translation of the French *article*) has been given by Prévost (1967). The word seems to have a descriptive value, and we can define "modules" as simple morphogenetic shoot units of determinate growth, constant in their expression, derived one from the other by a sympodial mechanism, the resulting sympodium being linear, branched in one plane, or branched in three dimensions. The important fact in this definition is the limitation of the apical activity of modules and consequently the sympodial construction that results from their proliferation. Determinate growth of each module may result from different causes:

1. Formation of a terminal flower or an inflorescence, as in cassava, *Manihot esculenta* Crantz (Euphorbiaceae: Médard, 1973); castor bean, *Ricinus communis* L. (Euphorbiaceae); and frangipani, *Plumeria* spp. (Apocynaceae)
2. Formation of a terminal tendril, as in the woody climber *Landolphia dulcis* Pichon (Apocynaceae: Cremers, 1975)
3. Formation of a spine, as in *Carissa macrocarpa* A.DC. (Apocynaceae: Brunaud, 1970)
4. Simple parenchymatization of the original apical meristem, as is common in many trees in the Apocynaceae during an initial vegetative phase of growth (Corner, 1952; Prévost, 1967, 1972; Hallé & Oldeman, 1970; Brunaud, 1971).

Modules may be orthotropic or plagiotropic in their orientation and morphology as an expression of endogenous activity, without obvious relation to external (ecologic) factors. In this chapter, only

Present address: ORSTOM, B. P. 165, 97301 Cayenne Cedex, French Guiana

223

branched trees are considered (see Hallé & Oldeman, 1970), although it is recognized that the same construction can occur in trees with a single sympodial trunk that is apparently unbranched (Chamberlain's model).

Types of modular construction

The question asked here is: What is the place of modular construction within the architectural models recognized by Hallé & Oldeman (1970)? Some trees exhibit this construction in both trunk and branches; we can call them "fully modular models." Others show the construction only in the trunk or only in the branches; they are "partially modular."

Fully modular models

Three models are included in this category: Leeuwenberg's, Koriba's, and Prévost's.

In Leeuwenberg's model the axes are all equivalent and orthotropic, as in cassava and frangipani, as already noted. The model is common in the Araliaceae (e.g., many species of *Schefflera:* see also Chap. 12). The epicotyl (i.e., the first module, the one produced by the plumular meristem) can exceed 10 m, as in *Didymopanax morototoni* Decne. & Planch. (Araliaceae), from northern South America, and *Anthocleista nobilis* G. Don. (Loganiaceae), from West Africa. This is important in making a tall tree possible.

In Koriba's and Prévost's models two types of axes are differentiated: (1) orthotropic modules, which collectively and successively give rise to the trunk, and (2) plagiotropic modules, which collectively give rise to branches. The mechanism of trunk formation differs in these two models. Essentially it is formed by secondary erection of successive orthotropic modules in Koriba's model, but by primary orthotropic orientation in Prévost's model. However, the topic is not dealt with further here (see Hallé & Oldeman, 1970, pp. 46–57; Chap. 7).

The American balsa tree, *Ochroma lagopus* Swartz (Bombacaceae), and the African abale, *Combretodendron africanum* (Welw.) Exell (Lecythidaceae), are good examples of Koriba's model. The latter, a tall forest tree, demonstrates that modular trunk growth can produce tall single-trunked trees. The same is true of Prévost's model, which may be exemplified by the emien, *Alstonia boonei* De Wild. (Apocynaceae), reaching a height of 50 m in the West African rain forest.

Partially modular models

In Nozeran's model the trunk is developed by a succession of modules; a familiar example is cocoa, *Theobroma cacao* L. (Sterculiaceae), native to the Amazon region. In this model the branches are plagiotropic, but not modular, as they are monopodial. The situation is reversed in Fagerlind's and Scarrone's models: The branches, but not the trunk, are modular. In Fagerlind's model the branches are plagiotropic; this model is not common among familiar trees, but is shown by some Rubiaceae (e.g., *Randia fitzalani* F. Müll.; see Fagerlind, 1943). In Scarrone's model the branches are orthotropic, as in the familiar mango, *Mangifera indica* L. (Anacardiaceae).

Taxonomic distribution of modular construction
Family level
The distribution of modular construction in some tropical families has been studied by Kahn (1975). His findings for the Apocynaceae, Euphorbiaceae, and Rubiaceae are summarized in Table 9.1.

Table 9.1. *Distribution of modular construction in Apocynaceae, Euphorbiaceae, and Rubiaceae*

Family	Model	No. of species
Apocynaceae	Leeuwenberg[a]	17
	Koriba[a]	7
	Prévost[a]	8
	Scarrone[b]	6
	Nozeran[b]	1
	Rauh[c]	2
	Other[c]	c.9
	Total	50
Euphorbiaceae	Leeuwenberg[a]	6
	Koriba[a]	8
	Prévost[a]	2
	Scarrone[b]	1
	Nozeran[b]	4
	Rauh[c]	15
	Total	36
Rubiaceae	Leeuwenberg[a]	3
	Koriba[a]	1
	Prévost[a]	0
	Fagerlind[b]	5
	Scarrone[b]	1
	Nozeran[b]	0
	Total	10

a, fully modular models; b, partially modular models; c, nonmodular models.

The architectural model is known for about 50 species of Apocynaceae, and on the basis of the distribution shown in Table 9.1, it is clear that this can be described as a modular family; 64% of known species (32 out of 50) represent a fully modular architecture, and only 22% (11 out of 50) belong to wholly nonmodular models. This leaves a small proportion (14%) of partially modular examples.

The distribution of the 36 species of Euphorbiaceae whose architecture is known shows that the family is appreciably modular, with 44% (16 out of 36) fully modular species, but only 14% (5 out of 36) partially modular species. Note, however, the abundance of species that conform to Rauh's model (42%, 15 out of 36), which is a nonmodular type, exemplified in the family by *Hevea brasiliensis* Müll.-Arg. Rauh's model is of interest in comparison with Leeuwenberg's because in both, aerial shoots are equivalent and orthotropic. However, in the former construction is monopodial, with the growth segments developed successively as "units of extension," if one may so translate the term *"unité de croissance"* used by Hallé & Martin (1968) in their study of this species. In Leeuwenberg's model the units are modules, and construction is sympodial. Scarrone's model is intermediate, as we shall see later.

The Rubiaceae is not a modular family. Of its 45 species with known architecture, only 10 (22%) are modular. Of these, only 4 species (9%) are fully modular.

Generic and specific levels
We can speak of a genus as polymorphic (Hallé, 1969) when it includes a variety of species exhibiting different architectural models. *Alstonia* (Apocynaceae) possesses examples of the three fully modular models: *A. sericea* Bl., from Malaysia (Leeuwenberg); *A. macrophylla* Wall., from Malaysia (Koriba); and *A. boonei* De Wild., from tropical Africa (Prévost). All eight species of *Alstonia* whose architecture is known belong to one of these three models (Corner, 1952; Prévost, 1967; Hallé & Oldeman, 1970; Veillon, 1976). The genus *Cordia* (Boraginaceae) is somewhat comparable: *C. curassavica* Lam. and *C. laevifrons* Johnst. display Prévost's model and *C. alliodora* follows Fagerlind's model. All these species are from Africa, although the genus is widespread. The genus *Euphorbia* (Euphorbiaceae) is the most diverse, with examples representing 11 models. In Madagascar alone, Cremers (1975) has observed 10 of them.

At the specific level, the question may be asked whether in the development of a species two distinct architectural models can be accommodated within the same genetic framework, the variation being determined by ecologic factors. In *Fagara rhoifolia* (Lam.) Engl. (Rutaceae), for example, Hallé & Oldeman (1970) have described ontogenetic changes that allow this species to be assigned initially to

Scarrone's, but subsequently to Koriba's, model. When the tree is young the trunk grows rhythmically, producing periodically pseudowhorls of orthotropic branch complexes with a modular construction. When the tree reaches a height of about 8 m, the apical meristem of the trunk loses its vigor and is substituted for by one or more relay axes that have a modular construction, but continue growth in height. In *Pseudopanax crassifolium* (Araliaceae) a similar loss of dominance has been recorded by Philipson (Chap. 12), with a change from Scarrone's to Leeuwenberg's model. The phenomenon may be common in the family, as Veillon (1976) found it in three other genera (*Dizygotheca, Myodocarpus,* and *Tieghemopanax*).

In the avocado, *Persea gratissima* (Lauraceae), Aubert & Lossois (1972) refer to a transition betwen Attims's and Champagnat's model, but these models are not very similar and the interpretation seems wrong.

In a less tropical family, Temple (1975) recently recognized the same form of transition (between Scarrone's and Leeuwenberg's models) in two species of *Erica* (*E. scoparia* and *E. cinerea*). A tree member of this same family, *Arbutus unedo* L. (strawberry tree), shows a similar change determined by the environment. In an open, well-lighted habitat, the tree lacks a monopodial trunk and conforms to Leeuwenberg's model; in a closed, shaded habitat, a monopodial trunk is formed and Scarrone's model is represented.

Modules and units of extension

Both types of growth units are the result of rhythmic development of shoot systems, as has been established in the comparison of Leeuwenberg's and Rauh's models. In the former, the unit of growth (module) is established repeatedly by branching; these units are all equivalent in the architecture of the tree, with the activity of the apical meristem of each unit determinate and flowering terminal. In the latter, the unit of extension is determined by the indeterminate activity of a terminal meristem, with rhythmically produced increments and lateral flowering. However, axes are again all equivalent and orthotropic. Clearly, there are significant parallels between these two models, although it should not be implied that there is any necessary continuity from one to the other. Where both models occur within a single genus, however, this continuity is suggested.

Transitions and intermediate models

When one model combines characteristics of two others or is transitional in some features, we can refer to it as an "intermediate model," but we must recognize that this is an artificial concept derived from the parameters used in model recognition. In this sense, Scarrone's model, in which the trunk consists of units of extension, but the branches of modules, stands intermediate between Leeuwenberg's model (trunk and branches all modular) and Rauh's model (trunk and branches all units of extension). If there were any genetic continuity, one might expect to find similar numbers of representatives in a single family, but the figures in Table 9.1 show in the Apocynaceae a disproportionate representation of Leeuwenberg's (17 out of 50 species) and Scarrone's (6 out of 50) models compared with Rauh's (2 out of 50 species) model. Rauh's model in this family is known so far only for *Couma guianensis* Aubl. (Hallé & Oldeman, 1970) and *Alyxia clusiophylla* Guill. (Veillon, 1976).

Similarly in the Euphorbiaceae, there is only 1 recorded example (out of 40 species investigated) that conforms to Scarrone's model, although 6 conform to Leeuwenberg's and 15 to Rauh's (Table 9.1). The intermediate model is, therefore, exceptional in the Euphorbiaceae, even though this family is architecturally rich (Hallé, 1971).

Not surprisingly, it is even more difficult to find contrasts within one genus, but they exist in the Guttiferae and the Loganiaceae. In *Montrouzeria* (Guttiferae), endemic to New Caledonia, Veillon (1976) found *M. sphaeroidea* Planch. and *M. verticillata* Planch. to represent Leeuwenberg's model, but *M. cauliflora* Planch. to represent Rauh's model.

In the Loganiaceae, two genera illustrate the point. In *Anthocleista*, *A. procera* Leprieur and *A. nobilis* G. Don. are both African species representing Leeuwenberg's model, but *A. amplexicaulis* Baker, from Madagascar, represents Scarrone's model. The genus *Fagraea* is quite polymorphic and includes examples of modular models – *F. schlechteri* Gilg. & Ben. (Koriba), endemic to New Caledonia, and *F. crenulata* Clarke (Fagerlind), from Malaysia – as well as nonmodular models – *F. fragrans* Clarke (Aubréville), from Malaysia, and *F. racemosa* Jack. (Roux), from Australia and New Guinea. It should be mentioned that in Aubréville's model, although construction of the branches is sympodial, the units are not of determinate growth. Flowering is lateral, and growth of each unit continues indefinitely (albeit slowly) in the orthotropic position.

Experimental approaches

It is of interest to know if the natural distinction between modules and units of extension, which seems to depend on endogenous control, can be influenced either ecologically or experimentally so that one can be converted into another.

Natural transformation

Continuous activity of the terminal meristem of *Hevea*, otherwise naturally rhythmic, is observed in the so-called lampbrush form, described by Hallé & Martin (1968). Foxtailing of *Pinus* has been reported repeatedly (e.g., Kozlowski & Greathouse, 1970) and was noted as long ago as 1935 by Guinier in *Abies*.

The complete transformation of unit of extension into module has been recorded by Dupuy & Guédès (1969) in *Pinus pinaster* Ait. ssp. *maritima*, where the female cones are abnormally terminal.

Experimental transformation

The lampbrush form in *Hevea* was obtained experimentally by Hallé & Martin (1968) by cutting off the median leaflet of all leaves (i.e., by reducing the leaf area). In a similar experiment on cocoa, Vogel (1975) produced shoots with continued activity of the apical meristem by repeatedly pruning away all leaves before they reached a length of 4 cm. In *Impatiens balsamina* L., Nanda & Purohit (1966) claim to have produced determinate shoots with terminal flowers, instead of the normal indeterminate shoots with lateral flowers, by successively removing the axillary floral buds. In the reverse process, Médard (pers. comm.), working with cassava, has been able to induce nondeterminate (i.e., nonflowering) shoots, but growth can become rhythmic with successive units of extension.

These few examples are sufficient to show that there is a degree of plasticity in the functioning of meristems, either limited in modules or unlimited in units of extension, so that a degree of manipulation is possible.

Conclusions

This brief discussion shows the architectural possibilities inherent within trees constructed of modular units. These modules may be orthotropic or plagiotropic and are capable of producing tall rain forest trees. Some families are predominantly modular in their construction, as far as present observations indicate, notably Apocynaceae but also Loganiaceae and Boraginaceae. We may contrast them with amodular families, of which the Moraceae is a good example. In this family, of 14 studied species, examples from six architectural models have been observed – Attims's, Aubréville's, Cook's, Rauh's, Roux's, and Troll's – none of them with modular construction.

The other main growth unit of tropical trees has been termed a unit of extension, and the fundamental difference in the activity of the terminal meristem has been indicated. Normally the two types of units are distinct; one exception is *Schuurmansia heningsii* K. Schum., as observed by Hallé (1974) in New Guinea. It is organizationally typical of Leeuwenberg's model, except that each module is developed by a series of successive units of extension.

This serves to emphasize what should always be borne in mind in making use of architectural categorizations: There is considerable

230 *Marie-Françoise Prévost*

developmental plasticity of shoot organization in tropical trees. The experimental interruption of rhythms, or their imposition, emphasizes this. Nevertheless, in natural conditions quite clearly contrasted patterns of construction remain evident, and this is of obvious ecologic significance.

General discussion

Tomlinson: Have you any information about the cause of parenchymatization of apices of those modules that are determinate for this reason?
Prévost: We can imagine that the volume of the meristematic zone becomes too large.
Tomlinson: Do you have evidence that the shoot apex does increase during development of the individual module?
Prévost: No.
Oldeman: In several species of *Cordia* in French Guiana there are, at the apex of the trunk modules, all transitions between a functional inflorescence and a parenchymatization. It may be that for some physiologic reason the meristem is too small or too young to make a flower and ends up as parenchyma.
Philipson: In the development of inflorescences, where the lateral branches become great in size compared with the apex and are adjacent to it, the apex is, as it were, entirely used up in the production of laterals.
Hallé: In my opinion parenchymatization is due to the very rapid growth of the selected branches around the dying apex. In other words, branching precedes parenchymatization.
Nozeran: Parenchymatization of the apical growing point is probably due in large measure to the influence of the leaves proximal to the meristem. In *Pinus* the parenchymatization of the apices of brachyblasts is similar. These short shoots do normally not continue to grow. If leaves are removed at a very early stage, apical growth continues (see Chap. 18).

References

Aubert, B., & Lossois, P. (1972). Considérations sur la phénologie des espèces fruitières arbustives. *Fruits, 27*(4), 269–86.
Brunaud, A. (1970). Une mise à fleur incomplète? La transformation en épine du méristème terminal chez une Apocynacée: *Carissa macrocarpa* (Eckl.) A.DC. *Rev. Gén. Bot., 77*, 97–109.
– (1971). Recherches morphologiques sur les Apocynacées et Asclépiadacées, contribution à l'étude de phénomènes fondamentaux de la morphogénèse chez les végétaux supérieurs. Thesis, Université de Dijon. 155 pp.
Corner, E. J. H. (1952). *Wayside Trees of Malaya*, vol. I. Singapore: Government Printer.
Cremers, G. (1975). Sur la présence de dix modèles d'architecture végétative chez les Euphorbes malgaches. *C.R. Acad. Sci. Paris, 281*, 1575–8.
Dupuy, P., & Guédès, M. (1969). Homologies entre rameaux courts, rameaux longs et rameaux reproducteurs chez le pin maritime (*Pinus pinaster* Ait. ssp. *maritima* H. del Vill.): Notion de métamorphose chez les gymnospermes. *C. R. Sci. Soc. Biol., 163*(11), 2426–9.

Modular construction 231

Fagerlind, F. (1943). Die Sprossfolge in der Gattung *Randia* und ihre Bedeutung für die Revision der Gattung. *Ark. Bot.*, *30A*(7), 1–57.
Guinier, P. (1935). Une curieuse mutation chez le sapin (*Abies alba*): Les
sapins sans branches. *C. R. Sci. Soc. Biol.*, *119*(2), 863–4.
Hallé, F. (1969). Recherches sur l'architecture et la dynamique de croissance
des arbres tropicaux. *J. W. Afr. Sci. Assoc.*, *14*, 80–8.
– (1971). Architecture and growth of tropical trees exemplified by the Euphorbiaceae. *Biotropica*, *3*, 56–62.
– (1974). Architecture of trees in the rain forest of the Morobe District, New
Guinea. *Biotropica*, *6*, 43–50.
Hallé, F., & Martin, R. (1968). Etude de la croissance rythmique chez l'Hévéa
(*Hevea brasiliensis* Müll.-Arg. Euphorbiacées-Crotonoïdées). *Adansonia*, *8*,
475–503.
Hallé, F., & Oldeman, R. A. A. (1970). *Essai sur l'architecture et la dynamique
de croissance des arbres tropicaux.* Paris: Masson.
Kahn, S. (1975). Remarques sur l'architecture végétative dans ses rapports
avec la systématique et la biogéographie. D.E.A. Biol. Vég. Montpellier.
Kozlowski, T. T., & Greenhouse, T. E. (1970). Shoot growth and form of
pines in the tropics. *Unasylva*, *24*, 2–10.
Médard, R. (1973). Morphogenèse du manioc, *Manihot esculenta* Crantz (Euphorbiacées-Crotonoïdées): Etude descriptive. *Adansonia* (Ser. 2), *13*,
483–94.
Nanda, K. K., & Purohit, A. N. (1966). Experimental induction of apical
flowering in an indeterminate plant *Impatiens balsamina* L. *Naturwis-
senschaften*, *54*, 230.
Prévost, M. F. (1967). Architecture de quelques Apocynacées ligneuses.
Mém. Soc. Bot. Fr., *114*, 24–36.
– (1972). Ramification en phase végétative chez *Tabernaemontana crassa*
Benth. (Apocynacées). *Ann. Sci. Nat. (Bot.)*, Ser. 12, 119–28.
Temple, A. (1975). Ericaceae: Etude architecturale de quelques espèces.
D.E.A. Biol. Vég., Univ. Montpellier, 95 pp.
Veillon, J. M. (1976). Architecture végétative de quelques arbres de l'archi-
pel Neo-Calédonien. Thesis Université des Sciences et Techniques de
Languedoc. 300 pp. Montpellier.
Vogel, M. (1975). Recherche du détérminisme du rythme de croissance du
cacaoyer. *Café Cacao Thé*, *19*, 265–90.

10

Architecture of the
New Caledonian species of Araucaria

JEAN-MARIE VEILLON

O.R.S.T.O.M., Nouméa, New Caledonia

The genus *Araucaria* contains species that can be assigned to two of Hallé & Oldeman's (1970) architectural models. Massart's model is represented by *A. heterophylla* (native to Norfolk Island) and Rauh's model by *A. araucana* (native to Chile). Thirteen species of the genus (Table 10.1) are endemic to New Caledonia (Laubenfels, 1972), making the island the richest center for its study (Veillon, 1976).

Ecology and habit

The New Caledonian species grow on ultrabasic schistose and calcareous soils; 10 are restricted precisely to ultrabasic soils. One species, *A. montana*, grows in open vegetation on the ultrabasic and schistose soils of Grand Terre. Another species, *A. columnaris*, grows on calcareous soils in open vegetation. *A. schmidii* is a rain forest species on serito-schistose soils, restricted to Mount Panié in the northeast. This gives some indication of the range of habitat diversity.

When young, all species are typically cone-shaped (Guillaumin, 1950), but adult trees differ according to two main patterns: (1) the shape becomes round, as in *A. luxurians*, *A. muelleri*, *A. nemorosa*, and *A. rulei;* or (2) the shape becomes characteristically columnar, as in *A. bernieri*, *A. columnaris*, *A. schmidii*, and *A. subulata*. These differences can be partly explained by their architecture, which is largely determined by the short life span of self-pruning branch complexes.

Architecture

Vegetative parts

All species are alike in that they have an orthotropic trunk with indefinite, rhythmic growth and spiral phyllotaxy (Figs. 10.1–10.4). First-order branches are pseudoverticillate (Figs. 10.1F,

Fig. 10.1. *Araucaria bernieri* (Massart's model). *A.* Young tree, about 14 m high, illustrating conical form. *B.* Adult tree, about 30 m high, with three successive generations of branches, producing three nesting crowns. *C.* Detail showing position of second-generation, partially reiterated model, represented by plagiotropic branch, at base of first-generation branch axis. *D.* Detail of plagiotropic branch representing third-generation plagiotropic branch, at base of first-generation branch axis. *E.* Second-order axis end-on, showing acute dihedral arrangement of third-order axes. *F.* Basal pseudowhorl of seven axes, seen from above.

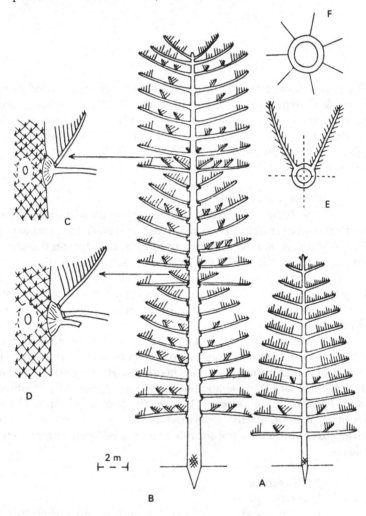

Fig. 10.2. *Araucaria columnaris* (Massart's model). *A*. Young sapling, 1 m high, establishing conical form. *B*. Young sterile tree, 15 m high, with some incomplete reiteration of older branches. *C*. Adult tree, 20 m high, with three successive nesting crowns. *D*. Detail showing position of second-generation axis (partial reiteration). *E*. Detail showing position of third-generation axis (further partial re-iteration). *F*. Pseudowhorl of seven axes at base of trunk, seen from above.

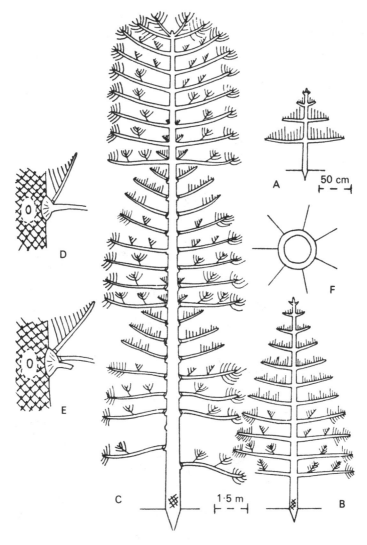

10.3*D*). Depending on branch orientation, two groups can be recognized, corresponding to the architectural models mentioned above. The adult trees of the first group have plagiotropic branches, corresponding to Massart's model (Table 10.1). Examples are illustrated in Fig. 10.1*A,B* (*A. bernieri*) and Fig. 10.2*A–C* (*A. columnaris.*). Adult trees of the second group have orthotropic branches, corresponding to Rauh's model. Examples are illustrated in Fig. 10.4*B* (*A. biramulata*). The distinctive feature of these species is the distal erection of the apex of the branch, revealing the essentially orthotropic nature of the branch meristem. This feature is also shown by the non-New Caledonian species *A. araucana*, *A. hunsteinii*, and *A. imbricata*. The

Fig. 10.3. *Araucaria rulei* (Rauh's model). A. Sapling, 60 cm high, with horizontal branches. B. Young tree, about 4 m high, with branches now more or less orthotropic. C. Adult tree, about 22 m high, with some partial reiteration along first generation of branches. D. Trunk axis viewed from above, showing pseudo-whorl of four branches.

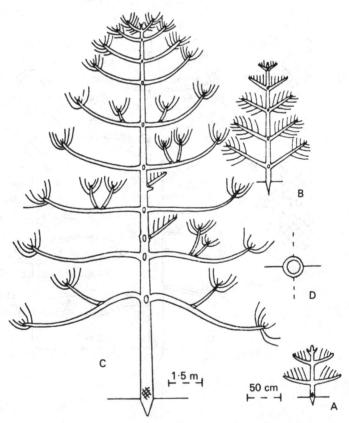

distal orthotropic orientation of the branch may appear only late in the development of the tree so that the sapling may closely resemble an example of Massart's model (e.g., Fig. 10.3A).

The ultimate (usually third-order) axes of the branch system, called "ramuli" (singular: ramulus), are characterized by determinate growth in length and diameter and are the predominant organs of photosynthesis. Some of them also bear reproductive organs. Depending on the species, ramuli on adult trees are arranged in two contrasting ways independent of the universal spiral phyllotaxis. In the first group, the ramuli have a dihedral arrangement (Fig. 10.1E), with the branch units in two lateral series, somewhat or markedly above the horizontal. Examples include *A. bernieri*, *A. humboldtensis*, *A. schmidii*, *A. scopulorum*, and *A. subulata* – all belonging to Mas-

Fig. 10.4. *Araucaria biramulata* (Rauh's model). *A*. Young plant, 1 m high. *B*. Adult tree, 22 m high, with successive nesting crowns. *C*. Detail of branched ramulus.

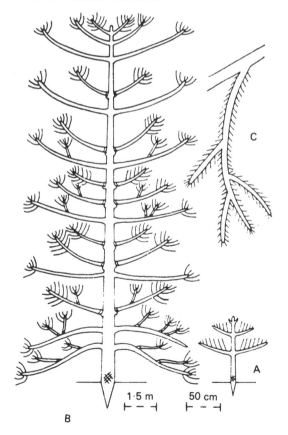

sart's model. The other group retains a spiral arrangement of ramuli in the adult phase, emphasizing the orthotropic condition (Figs. 10.3, 10.4), and includes all species listed in Table 10.1 as belonging to Rauh's model, both the exotic and New Caledonian species.

Sexuality

All New Caledonian species are monoecious; the cones are borne terminally on ramuli (Fig. 10.5). Female cones are at the ends of short ramuli, always in the top of the tree, corresponding to the apical zone of branches (Fig. 10.5A,B). Apparently the tree always produces female cones first; this has been confirmed on several specimens of *A. columnaris* about 18 years old. Male cones are produced by ramuli of varying ages, but always in the central and basal parts of the tree (Fig. 10.5A). The cones appear abruptly over a large part of the tree, with a distinct articulate attachment at which the cone is abscissed after the pollen is shed (Fig. 10.5C). The same ramulus continues as a photosynthetic organ by virtue of sympodial extension (Fig. 10.5D), but this growth is ultimately limited. The tree can be

Table 10.1. *Architecture of* Araucaria *species referred to in text*[a]

Species	Model
A. araucana (Molina) Koch	Rauh
A. bernieri Buchholz	Massart
A. biramulata Buchholz	Rauh
A. columnaris (Forster) Hooker	Massart
A. heterophylla (Salisb.) Franco (=A. excelsa [Lamb.] R. Br.)	Massart
A. hunsteinii K. Schum.	Rauh
A. humboldtensis Buchholz	Massart
A. imbricata Pav.	Rauh
A. laubenfelsii Corbesson	Rauh
A. luxurians (Brongn. & Gris) Laubenfels	Massart
A. montana Brongn. & Gris	Rauh
A. muelleri (Carr.) Brongn. & Gris	Rauh
A. nemorosa Laubenfels	Massart
A. rulei Müll.	Rauh
A. schmidii Laubenfels	Masart
A. scopulorum Laubenfels	Massart
A. subulata Vieill.	Massart

[a]All are endemic to New Caledonia, except *A. araucana* (South America), *A. heterophylla* (Norfolk Island), *A. hunsteinii* (New Guinea), and *A. imbricata* (South America).

divided into four distinct zones according to the distribution of male and female cones (Fig. 10.5A). The distal part (*a*) bears exclusively female cones; there is a transition zone (*t*) with branches bearing both kinds of cones; the central zone (*c*) bears exclusively male cones; and the basal zone (*b*) is sterile and consists of axes that either never were sexual or have lost this capability with age. There appears to be sexual specialization of branches, because male branches can never bear female cones, and vice versa.

Fig. 10.5. Sexuality in *Araucaria*. For detailed explanation, see text. A. Schematic representation of position of cones on tree. Areas *a*, *t*, *c*, and *b* correspond to following zones: apical (exclusively female); transitional (mixed male and female); central (exclusively male); basal (sterile). B. Position of female cones on lateral branch complex. C. Position of male cone on ramulus. D. Position of male cones on distal part of lateral branch complex, older branches showing sympodial growth.

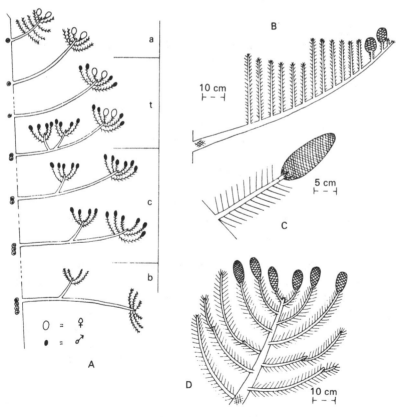

Root system
Limited observations on the architecture of the root system in *Araucaria* have been made on *A. rulei* in ultrabasic soils and *A. columnaris* in calcareous soils. Young trees show a pseudowhorl of four or five major horizontal roots, arising on the collar of the tap-root, about 30 cm below the soil surface. A second pseudowhorl becomes visible about 1 m below the first; this suggests rhythmic growth of the roots, although this has not been verified. The architecture of the root system remains the same in adult trees, but reiteration becomes evident in the form of additional sinkers or strongly developed vertical roots arising anywhere on the horizontal roots of the two pseudowhorled sets. They thus resemble parallel taproots.

Modification of the model
Two important modifications occur in architecture as the tree matures; both affect its shape. One is reiteration; the other is the production of successive crowns.

Reiteration
This process is essentially as defined by Oldeman (1974) and is always the result of damage to the tree (traumatization). Its features are illustrated in Fig. 10.6. Reiteration may be partial, when only part of the original model is repeated, or complete, when the whole model is repeated, except for the roots. Partial reiteration is illustrated in *A. rulei* in Fig. 10.6A. When the apex of a first-order branch is removed, new second-order branches are produced. These are orthotropic, as illustrated in Fig. 10.6B, which shows the branch end-on. This orthotropy reflects the nature of the branch system in Rauh's model. Complete reiteration may take place by the formation of new basal trunk axes, as is shown for *A. humboltensis* in Fig. 10.6C. An alternative method for reiteration of further models is by the epicormic stimulus for the development of previously dormant meristems on the upper surface of leaning trunks, as shown for *A. bernieri* in Fig. 10.6D. Distal, complete reiteration may also be induced if the leader is broken, as shown for *A. bernieri* in Fig. 10.6EF. The meristem involved originates from the base of one branch of the pseudowhorl closest to the break, but details are not known.

One exceptional species shows no reiteration. This is *A. muelleri*, which has very long branches (up to 4 m) that become flexuous with age. It always grows in strict conformity with the model.

Successive nesting crowns
Some species of *Araucaria* are distinguished by successive crowns that encase each other, like nests of tables; they have been described as having "nesting crowns" (Fig. 10.7). After the first gen-

Fig. 10.6. Various methods of reiteration in *Araucaria*. *A*. Partial
reiteration of axes on adult branch complex of *A. rulei*, cut off at *X*
(upper branch); normal appearance (lower branch). *B*. Pseudo-
whorled insertion of partially reiterated branches of *A. rulei*, seen
end on. *C*. Complete basal reiteration in *A. humboldtensis*. *D*. Complete reiteration on leaning trunk in *A. bernieri*. *E*. Complete reiteration after apical damage in *A. bernieri*. *F*. Detail showing site of
origin of renewal axis below broken leader in *A. bernieri*.

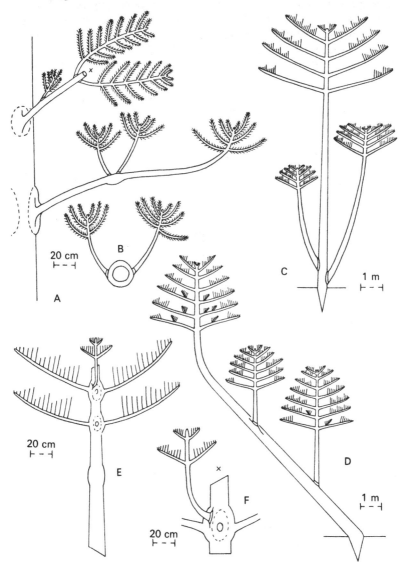

eration of branches is produced, they are lost by self-pruning but succeeded by several later branch generations arising from the original sites of branch insertion. As this involves production of new plagiotropic branches, partial reiteration of the model occurs. An example is provided in Fig. 10.1, for *A. bernieri*. Figure 10.1A shows a young tree, 14 m high, with the typical conical shape, representing the model precisely. Figure 10.1B shows an adult, 30 m high, with three generations of second-order axes, representing partial reiteration. The effect is that of a tree with three superposed crowns. Figure 10.1C shows a detail of the insertion of a reiterated plagiotropic axis, which originates at the base of a second-order axis of the first generation. Figure 10.1D shows the same process at a lower level, which

Fig. 10.7. Spatial distribution of male and female cones in *Araucaria*. For explanation, see text.

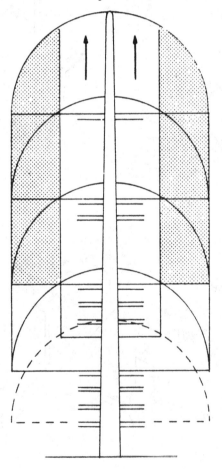

results in a third-generation axis, but again from the base of a second-order axis of the first generation and again a partial reiteration of the model.

The same features are shown for an adult *A. columnaris* (Fig. 10.2C) with an overall height of 21 m bearing three nesting crowns, representing three generations of second-order branches. All but three New Caledonian species listed in Table 10.1 develop nesting crowns; the exceptions are *A. humboldtensis, A. muelleri,* and *A. rulei.*

The development of nesting crowns has an influence on the distribution of cones in the developing tree, determined by the principles of sexuality established earlier. This is explained diagrammatically in Fig. 10.7. The central area delimited by the rectangle represents the spatial distribution of female cones on the tree; the shaded areas represent the region that bears male cones. Only the first-generation branches bear female cones, which are consequently restricted to the summit of the tree (the youngest crown). However, as the basal branches of successive crowns of all generations bear male cones, a much larger part of the tree is male. The oldest branches (in this diagram, the fourth generation) remain sterile. The shaded outline represents the former position of the most recently pruned crown.

Conclusion

The above brief description illustrates how tree form in *Araucaria* is determined by trunk-branch differentiation established by the architectural model the tree exhibits. It is conditioned by self-pruning. The addition of the reiterative process in the advent of further branching is readily recognized in this genus because of the very high level of model conformity and symmetry, the determinate nature of branches, and the precise process of abscission. The influence of these habit features on sexuality is precisely expressed. *Araucaria* thus supplies a model system for appreciating the preciseness of tree organization and the processes involved and lends itself to more detailed analysis and experimentation.

General discussion

Givnish: I was interested by your slides showing different species of *Araucaria* having differently shaped tree crowns: some cylindric, some rounded or cone-shaped, and others parasol or umbrella-shaped. Could you say more about the ecologic distribution or restriction of species with different crown shapes?

Veillon: As yet there is no clearly known relation between ecologic distribution and different crown shapes. Many species have a columnar shape and when they grow older become quite flat-topped. This is so regular a phenomenon that it is probably genetic rather than a reaction to ecologic conditions. The same applies to other forms of crowns. The conical or parasol-like forms are very regular. Of course, there is always an adjustment, but this would be a long-term evolutionary adjustment rather than an adjustment during the life of the tree.

Ashton: Do I understand that *Araucaria columnaris,* looking rather like the young foxtail pine trees that refuse to branch, eventually becomes flat-topped because of its genetic makeup?

Veillon: Yes, certainly, because the tabletop crown is a senescence characteristic.

Ashton: The reason I ask is because one got the impression in this species that the individual trees were spatially separate from one another, not growing in dense stands as one would expect from trees of such columnar shape. Do they maintain this isolation and columnar shape throughout the sexual stages of their life cycle and until they become senile, in contrast with most conifers? If male cones are all borne at the bottom of the column and the female cones at the top, pollination would be difficult in a dense stand. This is an unusual and interesting tree shape which may have some interest for Dr. Doyle. Could the spaces between the individuals of some such gymnosperm of the past have provided a cradle for angiosperm evolution?

Veillon: There are two species with exactly the same growth habit: *Araucaria columnaris,* which grows on isolated islands in monospecific populations, and *Araucaria bernieri,* which occurs in the rain forest.

Longman: This distribution of male and female cones, the female cones toward the top and the tips of the branches and the male in more proximal positions, is very common in temperate conifers of the Pinaceae, Cupressaceae, and other families. It is interesting to see that in the tropical Araucariaceae the monoecious habit is the same. I wonder whether this is a mechanism that tends to reduce selfing in these wind-pollinated trees.

Veillon: In large populations one may find two or three exceptions where the male sexuality covers the whole tree, but this does not seem to be correlated with the environment of New Caledonia.

Longman: It also sometimes occurs in *Pinus.*

Nozeran: The information given by Dr. Veillon can be complemented by information on the genus *Phyllanthus.* In different sections of this genus the distribution of male and female organs differs. For instance, in the group of which *Phyllanthus niruri* L. is an example, the male flowers appear first and subsequently at the end of the branches the female flowers, just as in *Araucaria.* The reverse is true for *Phyllanthus urinaria* L. and related species. In yet another species (*Phyllanthus amarus* Schum. & Thonn.) both kinds of flowers are mixed. The position of the flowers is linked to a whole complex of correlated phenomena in which the leaves probably play a particular role. The morphogenesis of *Araucaria* could be determined in the same way (i.e., by endogenous mechanisms).

Fisher: Is there a direct relation between the position of the nesting crown and the original periodicity of the leader apex? In other words, is there a correlation between the growth of the leader and the later development of nesting crowns?

Veillon: There is a variation among individuals of the same species in which the trunk growth pattern cannot be distinguished, some of them show nesting crowns and others a simple crown.

Whitmore: Are there other trees that have nesting crowns like these New Caledonian araucarias or is this something unique to New Caledonia and to *Araucaria?*

Veillon: Araucaria hunsteinii of New Guinea seems to make nesting crowns too, but not *A. cunninghamiana.*

Hartshorn: In the illustrations that accompany the key to the architectural models we see that the above-ground parts are much more complex and explicit than the below-ground parts, yet Dr. Veillon showed some slides of some interesting features of root systems in some *Araucaria* species and I would like to ask him if he knows about, or would speculate on, the relation between the above- and below-ground parts.

Veillon: The subterranean structures have been checked on two species and a total of less than 10 specimens.

Oldeman: It would be interesting to know if there is a correlation between aerial reiteration of the model and subterranean reiteration. If a new trunk is formed in the crown, does a new taproot grow below? The same model is always realized, so it would be nice if there was an underground part accompanying it.

Borchert: From Dr. Hallé's work (on rubber) and work on shoot growth in oak, it is clear that shoot growth rhythmicity is not necessarily accompanied by rhythms of root growth. In *Citrus* there are data showing that periods of root and shoot growth alternate (Marloth, 1949), so there cannot be the straightforward correlation implicit in this speculation of corresponding architectural models. Furthermore, root development is restricted by the physical nature of the soil (see Chap. 21).

Hallé: Yes, even in *Araucaria* there is a large number of tiers on the trunk, but there are just two tiers of lateral roots on the taproot. So the rhythm itself is quite different, and in some cases there is no rhythm at all in the taproot (e.g., in rubber).

References

Guillaumin, A. (1950). Formes de jeunesse des conifères en Nouvelle-Calédonie. *Not. Syst., 14,* 37–43.

Hallé, F., & Oldeman, R. A. A. (1970). *Essai sur l'architecture et la dynamique de croissance des arbres tropicaux.* Paris: Masson.

de Laubenfels, D. J. (1972). *Gymnospermes,* vol. 4 of *Flore de la Nouvelle-Calédonie et dépendances,* ed. A. Aubréville & J. F. Leroy. Paris: Muséum National d'Histoire Naturelle.

Marloth, R. H. (1949). Citrus growth studies. I. Periodicity of root growth and top growth in nursery seedlings and budlings. *J. Hort. Sci., 25,* 50–9.

Oldeman, R. A. A. (1974). L'architecture de la forêt guyanaise. *Mém. O.R.S.T.O.M.,* no. 73, 1–204.

Veillon, J. M. (1976). Architecture végétative de quelques arbres de l'archipel Néo-Calédonien. Thesis, Université des Sciences et Techniques du Languedoc, Montpellier.

11

Growth forms of rain forest palms

JOHN DRANSFIELD

The Herbarium, Royal Botanic Gardens, Kew, Richmond, Surrey, England

Throughout the lowland and lower montane tropical rain forests of Malesia and the Americas, palms are usually a conspicuous component of the forest. In Africa they are much less conspicuous, but the popular association of palms with tropical vegetation is generally valid. Of the more or less 2700 recorded palms (Moore, 1973), about 2000 can be regarded as rain forest species. Their importance in the structure of rain forests is generally appreciated, but studies of their ecologic role are few in number (e.g., Bannister, 1970). Palms are rarely absent from primary lowland vegetation in the humid tropics, some exceptions being various forms of peat swamp forest, heath forest, and limestone vegetation in Southeast Asia (Dransfield, 1969) and the coastal rain forests of East Africa. Not only are palms conspicuous in forest types, but they also occupy a wide range of niches within each type. In the following account, the growth forms of palms will be described and the relation between growth form and habitat discussed. Much of the information presented is based on personal observation in Southeast Asia and Africa.

Architecture
Basic models
Four basic architectural models can be distinguished among the palms (Hallé & Oldeman, 1970):

1. Unbranched monocarpic palms (Holttum's model), e.g., all species of *Corypha; Raphia regalis, Metroxylon salomonense,* and *Arenga pinnata* (Fig. 11.1)
2. Unbranched polycarpic palms (Corner's model), e.g., *Cocos nucifera, Areca catechu, Elaeis guineensis, Calamus manan,* and many others (Fig. 11.2)

Fig. 11.1. *Arenga pinnata*, monocarpic palm with basipetal sequence of inflorescence production. Aceh, northern Sumatra.

Fig. 11.2. *Pigafetta filaris*, unbranched polycarpic palm. In cultivation at Sibolangit, northern Sumatra.

Fig. 11.3. *Hyphaene compressa*, palm with trunk repeatedly forking. Witu, Kenya.

Fig. 11.4. *Eleiodoxa* (= *Salacca*) *conferta*, acaulescent, hapaxanthic swamp palm with terminal infructescence. Kuala Pilah, western Malaysia.

3. Branched palms (Tomlinson's model) that are either hapax-anthic, e.g. *Metroxylon sagu*, *Arenga brevipes*, all species of *Eugeissona* (Fig. 11.6), or pleonanthic, e.g., *Oncosperma tigillarium*, all species of *Ceratolobus*, *Ptychosperma macarthurii*, and many other

4. Dichotomously branched palms (Schoute's model), e.g., *Nypa fruticans* and *Hyphaene compressa* (Fig. 11.3)

These basic models are usually constant for each species, but some species may be represented by individuals showing different basic models. For example, most individuals of *Calamus manan* are soli-tary, unbranched palms, but clustered forms rarely occur; *Raphia farinifera*, though usually clustered and hapaxanthic, is occasionally solitary and monocarpic. From these basic architectural models has developed a series of basic growth forms.

Basic growth forms
There are four basic growth forms within the palms; these are not always easily distinguished from each other, but represent useful categories for describing palms in the field: (1) tree palms, those that reach the main forest canopy; (2) shrub palms, low palms found in the forest undergrowth; (3) acaulescent palms, those with subterranean or very short aerial stems; and (4) climbing palms.

It should be immediately apparent that, just as with dicotyle-donous plants there are sometimes difficulties in distinguishing among shrubs, trees, and climbers, so with the palms there are cer-tain species that behave as intermediates between categories. De-spite this, it is still thought valuable to use these terms.

Within tropical rain forest the most conspicuous palm growth form is the shrub palm; tree palm species are relatively few in number, though they may be locally gregarious and very conspicuous. In Southeast Asia, climbing palms may often be locally more numerous in species and individuals than any other of the categories. Acaules-cent species tend to be exceptional, sometimes occupying rather pre-cise and specialised habitats (e.g., swampy hollows and ridgetop podzols).

Usually each major forest type contains representatives of each of the basic growth forms. The palms apparently have successfully col-onised most of the larger niches of the forest, giving a range of forms with extremes such as tall trees over 50 m in height: e.g., *Ceroxylon alpinum*, reported by Humboldt & Bonpland (1805) to reach 60 m, and *Pigafetta filaris*, reaching 50 m (Dransfield, 1976b). Long climbers rep-resent some of the longest land plants: e.g., *Calamus manan*, recorded as 185 m long (Burkill, 1935). In contrast, small undergrowth pal-munculi scarcely reach 20 cm: e.g., *Iguanura palmuncula* (Beccari, 1904).

Types of flowering

Though pleonanthy is much more widespread among the palms, hapaxanthy produces such spectacular results that the phenomenon has attracted considerable attention and speculation. Hapaxanthic flowering occurs in only three major groups:

1. The Coryphoid group: all species of *Corypha* and *Nannorrhops*
2. The Lepidocaryoid group: all species of *Oncocalamus*, *Ancistrophyllum*, *Korthalsia*, *Plectocomia*, *Plectocomiopsis*, *Myrialepis*, *Eleiodoxa* (Fig. 11.4), *Eugeissona* (Fig. 11.5), and *Raphia*; all species of *Metroxylon* except for *M. amicarum*; and one or two species of *Daemonorops*
3. The Caryotoid group: all species of *Caryota* and *Wallichia* and most, but not all, species of *Arenga*

Where hapaxanthy is linked with monocauly, as in *Corypha*, monocarpy ensues. A special type of hapaxanthy occurs in the Caryotoid palms, where production of inflorescences is basipetal rather than acropetal. There is some evidence (pers. obs.) that the sequence of development of inflorescences in *Korthalsia* is also basipetal, but the inflorescences are confined to the uppermost nodes, as opposed to the Caryotoids, where successive inflorescences are produced in a basipetal direction from each of most of the nodes of the stem, even to the very base, resulting in inflorescences of subterranean origin.

The rarity of hapaxanthy and its striking biologic effect of presentation of flowers and fruit during one period only raise the question of which flowering method is primitive and which advanced. Dransfield (1976a) has shown the probability that in *Daemonorops calicarpa* hapaxanthy is a derived state (Fig. 11.6). The situation in other genera is not yet clear, and there are arguments in favour of both schools of thought. Tomlinson & Moore (1968) regard the whole argument about which flowering type is primitive and which advanced as redundant because of the morphologic equivalence of the lateral units of both hapaxanthic and pleonanthic flowering. Yet there is a major biologic difference between the gradual presentation of fruit throughout the mature life of the plant and massive fruit production during a short period, followed by death of the shoot. The period of fruit presentation in hapaxanthic *Arenga* spp. and *Caryota* may be much greater and more staggered than in other hapaxanthic species because of the sequential basipetal maturation of the inflorescences.

In secondary forest in Southeast Asia, the most abundant climbing palm (rattan) genera encountered are *Korthalsia*, *Plectocomiopsis*, and *Myrialepis*. In primary forest, *Korthalsia* is still widespread, but *Plectocomiopsis* and *Myrialepis* appear to be confined to gaps in the canopy, such as on landslips and areas where trees have fallen. Similarly, in the mountains, species of *Plectocomia* are most conspicuous

252 *John Dransfield*

Fig. 11.5. *Eugeissona utilis*, clustering hapaxanthic palm. Dead, leafless stems remain after fruit has fallen. Near Brunej Town, Brunei.

Fig. 11.6. *Daemonorops calicarpa*, acaulescent hapaxanthic palm. Langkat, northern Sumatra.

in areas where the forest has been opened. Seedlings of these four rattan genera are present in primary forest as rosettes, but it appears that for further development of *Plectocomia, Plectocomiopsis,* and *Myrialepis* direct sunlight is required. It is possible that hapaxanthy is an adaptation allowing greater possibilities of colonizing open habitats by presentation of a large quantity of fruit at one moment (see Chap. 27). In other hapaxanthic genera there is no obvious correlation between habitat and flowering method, but the rarity of hapaxanthy does suggest that generally pleonanthy is the more successful flowering method.

Modifications
Types of suckering
Basal branching is an extremely common feature in rain forest palms, and for a given species the character of solitary versus suckering is usually constant and often of value in field determination. In one species of palm, *Serenoa repens* (not a rain forest species: Fisher & Tomlinson, 1973), the branching system produced by the suckering is basically monopodial, with a horizontal stem producing from the nodes inflorescences, branches, or neither. In most other palms the branching system is basically sympodial, with axes repeatedly suckering. In species producing large clusters of stems, such as *Hyphaene* or *Oncosperma,* there is sometimes difficulty in interpreting whether shoots at the base of the clump are suckers or seedlings developing from fallen fruit. In the simplest type of sympodial suckering, a close, dense clump of suckers is built up that is apparently of indefinite growth. Good examples of this type are *Oncosperma tigillarium, Ceratolobus glaucescens, Caryota mitis, Pinanga kuhlii, Licuala spinosa, Phoenix reclinata,* and *Chrysalidocarpus lutescens.* There is some evidence from botanic gardens specimens that some closely clustering palms may go through a period of suckering activity early in their development, followed by a period in which suckers are not produced. Certain strains of *Cyrtostachys lakka* in cultivation in Kebun Raya, Bogor, Indonesia, appear to follow this pattern, though normally *C. lakka* continues to produce suckers throughout the life of the clump. Similarly, some strains of the date, *Phoenix dactylifera,* appear to sucker only when they are young, though this apparent periodicity in suckering behaviour may be due to disturbance by man in the removal of suckers and thus the depletion of potentially suckering nodes, there then being no further nodes capable of suckering. This phenomenon requires further investigation.

An elaboration of the close suckering habit is provided by the production of rhizomes, suckering resulting in open colonies. It is of interest that although *C. lakka* produces close clusters in cultivation, in the wild in peat swamp forest it tends to produce long rhizomes and open colonies. Other examples of open colonial palms are *Pinanga patula* and *Licuala paludosa* in Malesian peat swamp forest, *Bactris major* and *Bactris cruegeriana* in northern South America, and *Rhapis excelsa* in Indochina. In the examples cited, the rhizomes are of about the same diameter as the aerial stems. In *Arenga obtusifolia* (including *Arenga westerhoutii*) of the Sunda Shelf area, the lowermost nodes of the up to 40-cm diameter trunk produce leptocaul geotropic axes only 5–10 cm in diameter, which develop underground into rhizomes growing horizontally. New aerial stems are produced at distances up to 2 m from the parent trunk and involve a rosette stage with the increase in the diameter of the axis. A similar type·of suckering habit is found in an unnamed short-stemmed *Calamus* species from southern Sumatra.

In *Calamus*, a great range of clustering types can be found, from closely clustering species to species that spread by means of long rhizomes or stolons. *C. perakensis* of humic podzols on ridgetops in Malaya and Sumatra exemplifies the closely clustering habit. In *C. caesius* clustering is relatively close (not exceeding 10 cm lateral distance from the parent axis), but for every laterally growing axis that metamorphoses into an aerially growing axis, two axillary suckers are produced at its very base. An extreme is seen in *C. trachycoleus*, a closely related species from South Borneo, one of the few species of rattan cultivated on a commercial scale. In this species suckers develop as laterally growing stolons of greater diameter than the aerially growing stem, covered with spiny, bladeless, sheathing scale leaves; these stolons may grow for 3 m along the ground, rooting at the nodes and hence appearing almost as if perched on short stilt roots. When the stolon metamorphoses into an aerial stem, two alternate axillary shoots are produced just below the tapering region associated with the change to vertical growth. These two shoots develop into further laterally growing stolons. There is thus the potential for exponential increase in the number of aerial stems produced by this rapidly invasive river bank species, and this remarkable habit is probably the main reason for the great local success of this species as a plantation crop. It is interesting to compare the growth of *C. trachycoleus* with the related *C. caesius*; in the latter, because of the close-clustering habit, the axillary sucker shoots are effectively in competition with each other, although many of them remain dormant. Silviculture of this latter species involves much clearing of debris from the clump base in order to decrease competition effects among shoots and increase the amount of light reaching the sucker shoots. It is remarkable that the apparently highly successful river bank pioneer species *C. trachycoleus* is not more widespread in suitable habitats instead of being confined as it is to the Bornean river systems draining into the Java Sea. In contrast, *C. caesius* is widespread throughout Sumatra, Malaya, and Borneo. A further method of clustering is found in a few slender rattan species such as *C. javensis, C. heteroideus,* and *C. reinwardtii*: Short aerial stems in the forest undergrowth flop over and root, producing a section of short, thick internodes and large leaves in contact with soil and long, thin internodes and smaller leaves normally associated with the climbing stems, whose rooted portion also produces close suckers. When suitable supports like small trees are present, the aerial stems climb; the arching and rooting habit appears to be associated with lack of support. In Borneo an undescribed genus of rattans related to *Calamus* also has this habit, but is of massive construction, building up dense thickets on steep hillslopes by flopping and rooting.

Usually, colonies of clustering palms are built up after the production of a shoot of potentially maximum diameter for that species. In *C. caesius, C. trachycoleus,* and possibly also some species of *Eremospatha* (evidence from herbarium material in Kew), the colony is built up gradually, the first shoot being much thinner than the normal adult and not growing more than 2–3 m and then perhaps succumbing to predation or damage, but producing at the base suckers of greater diameter that grow into vigorous aerially growing shoots.

As emphasised by Fisher & Tomlinson (1973), clustering palms normally have two distinct phases in their growth: in early growth of the stem, proximal lateral buds develop into suckers; in later growth distal buds develop into inflorescences. *Serenoa repens* is an exception to this situation, as there is a continued production of suckers or inflorescences throughout the life of the horizontally growing axis. Another exception occasionally occurs in *Caryota mitis* (alluded to above), where lateral buds at the very base of the stem, which normally develop into suckers, may develop instead as inflorescences at the base of a basipetally flowering stem. Otherwise, when these lateral buds develop as suckers they prolong the life of the hapaxanthic flowering individual.

Very rarely, inflorescence buds in the distal areas of palm stems may develop as vegetative shoots instead, but such instances represent monstrosities. It is of interest to note that this last situation has been observed only in nonclustering palms; e.g., *Cocos nucifera* (Ridley, 1907), *Phoenix canariensis, Arenga pinnata,* and *Aiphanes erosa* (pers. obs.).

A curious kind of delayed suckering occurs in the rattan genus *Plectocomia; P. griffithii, P. sumatrana,* and *P. elongata* are all solitary species with hapaxanthic flowering. In large, mature individuals

small axillary shoots are produced at the base of the stem and at
nodes up to about 3 m above the base. These produce about three
leaves, and a series of roots that grow adpressed to the stem. These
bulbils remain more or less dormant until the ultimate hapaxanthic
flowering and collapse of the stem onto the forest floor, when they
are capable of developing into new solitary individuals. There is
scope for further investigation of this phenomenon. Similar rooting
bulbils have been observed in *Licuala gracilis* of Java and *Licuala
celebica* of Celebes and in *Pinanga spp.*, including *P. polymorpha* of
Malaya and *P. clemensii* of Borneo. It is not known whether these bulbils are of any significance in dispersal or vegetative reproduction. In
all these last four palms, flowering is pleonanthic and fruit production usually high. In *Plectocomia*, however, bulbil production may be
regarded as an adaptation offsetting the disadvantages in reproductive capability of hapaxanthy.

Aerial branching

Aerial branching, as opposed to suckering, has been recorded in a number of different palms. The most famous and spectacular example is *Hyphaene*; though not strictly a rain forest genus, one
species *H. compressa*, is found occasionally in marginal rain forest,
so the remarkable branching is considered here. There are about 12
species of *Hyphaene*, and all species regularly branch except *H.
ventricosa*; even this last may rarely branch. There appear to be
three methods of branching in the genus: aerial dichotomus branching, subterranean dichotomous branching, and apparent sympodial
suckering at the stem base. In one remarkable species from Somalia
and Arabia, *H. reptans*, the stem is prostrate but branches
dichotomously to produce a spreading network of stems (see illustration in Pichi-Sermolli, 1957). In the massive erect species, the
palm branches by successive forkings to produce a spreading
crown. In *H. compressa*, six seems the maximum possible number of
dichotomies, giving a crown made up of 64 apices. At each forking
and two shoots continuing the upward growth are of equal size and
suggest a true dichotomy. Schoute (1909), Tomlinson & Moore
(1966), Hallé & Oldeman (1970), and Tomlinson (1971) have all discussed in some detail the nature of the forking of *Hyphaene*. The
interpretation by Tomlinson seems the most reasonable – that the
branching does represent an equal dichotomy. Similar dichotomous
branching occurs in the mangrove palm, *Nypa fruticans* (Tomlinson,
1971); in *Chamaedorea cataractarum* (Fisher, 1974); and possibly in
Allagoptera arenaria (Tomlinson, 1967) and *Nannorrhops ritchiana*
(Tomlinson & Moore, 1966).

There are numerous reports of abnormal branching in commonly cultivated palms; several species, however, are usually found with branches. Examples are *Oncosperma fasciculatum* (pers. obs.); *Chrysalidocarpus lutescens*, *Chrysalidocarpus madagascariensis* var. *madagascariensis*, and *Chrysalidocarpus madagascariensis* var. *oleraceus* (Fisher, 1973; and pers. obs.); and a few species of *Pinanga* and *Licuala* (see under Types of Suckering). The morphology of branching in *Chrysalidocarpus* has been investigated by Fisher (1973) and has been shown to be nonaxillary and perhaps unique. In *Chrysalidocarpus* the only apparent selective advantage of the branching habit is in reproductive capacity, though it should be noted that in *C. lutescens* many of the branches are smaller in diameter than the axis subtending them and do not appear to produce inflorescences.

Korthalsia is the only rattan genus in which branching is a regular occurrence. All species examined in the field branch in the canopy. In undisturbed forest areas the stems of *Korthalsia* may develop into massive branching systems ramifying through the canopy. The branching is sympodial and axillary, the branch arising usually several metres below the stem apex. Branching does not necessarily occur in association with the flowering and consequent termination of growth of the axis; it often takes place long before flowering. There appears to be no well-defined relation between branch initiation and the distance from the apex – sometimes two or three branches may be produced on adjacent nodes – but this relation needs further investigation. Not only do species of *Korthalsia* branch aerially, but they invariably sucker at ground level; hence, the branching system eventually produced by one seed may contain as many as 30 apices spread over an area of about 2500 m^2 of forest canopy (*Korthalsia flagellaris* in peat swamp forest at Berbak in Sumatra).

The combination of hermaphrodite flowers and branching stems in *Korthalsia* may offset some of the limitations of hapaxanthy and give increased possibilities of dispersal, and the branching behaviour may have arisen through selection of this feature. Of the hapaxanthic rattans, *Korthalsia* is certainly the most abundant genus in the forests of Southeast Asia, with many widespread species (contrary to the opinion of Furtado, 1951).

Very rarely species of other genera of rattans branch in the canopy, but this seems always to be a result of injury.

Geotropic stems

Geotropy is a common feature of germinating palm seeds; in many species the cotyledonary sheath elongates, pushing the shoot apex into the ground, in extreme examples such as *Borassus* and *Hyphaene* up to 60 cm deep (Tomlinson, 1960). There is an obvious advantage to the seedling in this germination method, because the seedling can become firmly anchored in the ground and chances of desiccation are decreased. Even within humid rain forests many plants germinate in this way, though it is interesting to note that most of the rattans, very typical rain forest species, have adjacent ligular germination with no burying of the seedling shoot.

More unusual is geotropy in juvenile stems. Well-known examples are species of *Sabal, Acanthococos,* and *Orbignya cohune* (not strictly rain forest species but occasionally occurring within forest). In these three examples geotropic growth is followed by apogeotropic growth. *Raphia regalis,* of southeastern Nigeria, Cameroons, Gabon, and Angola, apparently has a habit similar to that of *O. cohune,* with the massive stem apex buried to depths of up to 1 m below soil level (Otedoh, 1972). In *Nypa fruticans* the juvenile stem grows downwards into mangrove mud to depths of about 1 m followed by horizontal growth and dichotomous branching (Tomlinson, 1971).

Furley (1975) has discussed the role of *O. cohune* in the development of soil profiles in Belize; he estimates 0.065 m³ soil disturbance per tree and from the density of trees per area and an estimated life span calculates the amount of soil turnover. Whatever the accuracy of these estimates, Furley's discussion points to the importance of *O. cohune* in the development of soil. It may be postulated that other geotropic-stemmed palms also play an important role in soil development. *Nypa fruticans* with its geotropic and then horizontally growing stems also probably plays a role in the stabilization of mangrove muds along the creeks and river mouths in which it grows.

Stilt roots

Stilt roots are found in several species of palms scattered through the major groups: e.g., many Iriarteoid palms; *Eugeissona minor* and several rattans in the Lepidocaryoid major group; and *Areca vestiaria, Nenga gajah,* and *Verschaffeltia splendida* in the Arecoid major group. *A. vestiaria* (Dransfield, 1974) and *E. minor* (Dransfield, 1970) are exceptional among stilt-rooted species in that the stems are clustered, and the roots support a whole branching system above ground (Fig. 11.7), though in the instance of *A. vestiaria* suckering occurs near ground level and stilt roots appear as a secondary phenomenon. All other stilt-rooted species are solitary. *E. minor* has very short stems, scarcely exceeding 25 cm in height before hapaxanthic flowering occurs, and hence appears as a stilt-rooted acaulescent species. In the Iriarteoid palms the seedling stem is very slender and successive nodes produce successively larger leaves and roots, the internodes increasing in diameter. Even from the seedling stage the plant is supported on stilt roots. In *V. splendida, A. vestiaria* (Fig. 11.8), and *N. gajah,* stilt roots are produced much later in the life of

Rain forest palms

Fig. 11.7. *Eugeissona minor.* View of lower part of plant. Bintulu, Sarawak.

Fig. 11.8. *Areca vestiaria,* with massive stilt-root production. Gunung Soputan, northern Celebes.

the individual, developing as a basal cone of roots surrounding the stem, which remains rooted in the ground. The significance of stilt roots to the plants is not always apparent. In *E. minor*, though commonest at the margins of swamps in *kerangas* (heath forest) in Sarawak and Brunei, the palms are found more rarely in a wide range of habitats, including hillslopes, ridgetops, and level ground in *kerangas* and lowland dipterocarp forest. This wide range of habitats seems to preclude any correlation between habitat and the stilt-rooted habit. Similarly in *A. vestiaria* stilt-root production is variable, some individuals being scarcely stilt-rooted, others having a cone of stilt roots up to 3 m long. At first it seemed that massive stilt-root production was correlated with occurrence of the palm on steep, unstable slopes, and in this way, stilt roots could be seen as an adaptation giving increased support and stability. However, on closer examination individuals without stilt roots and massively stilt-rooted individuals with a whole range of intermediates could be found in the same habitat and locality (Dransfield, 1974). *N. gajah* (Fig. 11.9) is confined to steeply sided small valleys in hill forest in south and central Sumatra; stilt roots are almost invariably produced and may have some function as supporting organs holding the short, massive trunk upright in possibly unstable soils; yet there are other non-stilt-rooted palms of similar habit occurring in the same habitat. In the Iriarteoid palms, though individual species may be confined to particular habitats (e.g., *Iriartea exorrhiza*, which is confined to swamps: Wallace, 1853), the stilt-rooted habit is found in a wide range of habitats from swamps to hillslopes. Without further examination of the plants growing in the field, it is difficult to discover the significance of the stilt roots. The stilt roots of *V. splendida* are regarded by Vesey-Fitzgerald (in Bailey, 1942), as an adaptation for self-support on steep, unstable hillslopes in the Seychelles. Two other Seychelles palms, *Roscheria* and *Nephrosperma*, are reported to produce some stilt roots, but not such massive roots as *Verschaffeltia*.

In summary, the significance of stilt roots, except perhaps in *V. splendida*, is not clear.

Climbing organs

Modifications of inflorescence and leaf are an integral part of the climbing habit of the rattans and other climbing palms. The simplest adaptation is that shown by *Chamaedorea elatior*, a climbing palm of Guatemala and Mexico; in this species the uppermost leaflets are reflexed and act as weak grapnels at the ends of the leaves. One of the most characteristic features of some of the rattans is an extension of the leaf rachis into a barbed whip or "cirrus" with reflexed grapnel spines. In the Americas one Cocosoid genus, *Desmoncus*, has the

Rain forest palms 261

Fig. 11.9. *Nenga gajah*. Lower part of trunk with stilt roots. Kepahiang, Bengkulu, southern Sumatra.

Fig. 11.10. *Geonoma sp*, palmlet with finely divided leaves growing alongside a stream liable to flooding. Bajo Calima, Buenaventura, Colombia.

climbing habit and bears leaves with cirri, each cirrus also with pairs of reflexed modified spinelike leaflets ("acanthophylls") grouped near the tip of the rachis. A similar arrangement of reflexed acantho-phylls on the cirrus is found in the African Lepidocaryoids *Ancis-trophyllum*, *Oncocalamus*, and *Eremospatha*. In the other Old World Lepidocaryoid climbing palms, acanthophylls are absent and cirri bear grapnel spines only; these spines are equivalent to epidermal emergences rather than leaflets. In some species of *Calamus* (sections *Macropodus*, *Afrocalamus*, *Rhombocalamus*, *Platyspathus*, and *Coleo-spathus*; see Furtado, 1956), cirri are absent and a second climbing organ of different origin is present. This is the "flagellum" and con-sists of a very slender whiplike axis clothed in bracts bearing reflexed grapnel spines, the base of the flagellum inserted in an oblique posi-tion near the mouth of the leaf sheath. The flagellum is a sterile inflorescence, the homology being borne out by the position of the flagellum and the fact that for any leaf sheath flagella and inflores-cences are mutually exclusive (Beccari, 1908). It has long been noted (Beccari, 1908; Furtado, 1956) that in *Calamus* flagellate species do not bear cirri, and cirrate species are without flagella. There are, how-ever, a few Bornean and Sumatran species related to section *Pla-tyspathus* that regularly bear both cirrus and flagellum (e.g., *Calamus ulur*, *C. pogonacanthus*, and *C. eriocanthus*), and one species in sec-tion *Coleospathus* (*Calamus spathulatus*) in the juvenile state may have a short cirrus at the leaf tip as well as a well-defined flagellum.

Rheophytes

Two species of *Pinanga*, *P. calamifrons* and *P. rivularis* (which are possibly conspecific), are highly characteristic of the vegetation of stable rocks by large rapids on the larger rivers of Borneo. Here they grow with several fine-leaved or fine-leafleted plants, such as *Homonoia riparia*, *Hygrophila saxatilis*, and *Dipteris lobbiana*, and themselves have leaves finely divided into single-fold leaflets of a texture unusually tough for the genus *Pinanga*. Unlike most species of *Pinanga*, which are forest undergrowth species, they appear to be tolerant of exposure to full sunlight. The coincidence of these two palms with rheophytes also with fine leaves suggests that *P. calami-frons* and *P. rivularis* are obligate rheophytes and that the fine dissec-tion of the lamina is an adaptation to this habit. However, several other species of *Pinanga* (e.g., *P. beccariana* and *P. canina*) have simi-larly fine-leafleted leaves – the latter in the juvenile phase only – yet are confined to podzolic soils of ridgetops. Several other species of palms are characteristic of river banks (e.g., *Arenga undulatifolia* in Borneo), but they do not behave as rheophytes capable of withstand-ing complete submergence (as evinced by occurrence of the two *Pinanga* spp. below the flood debris line).

An apparently undescribed species of *Geonoma* (Fig. 11.10) in the Pacific lowlands of Colombia has a habit very similar to that of *Pinanga rivularis* and grows in a similar habitat (H. E. Moore, Jr., pers. comm.).

Distichous palms
Three palms have crowns with leaves arranged in two ranks; they are *Oenocarpus distichus, Orania disticha* (Burret, 1936), and the widespread cultivated palm *Wallichia disticha*. The parallel in habit between these palms and *Ravenala madagascariensis* is most striking. The significance of the distichous habit is not understood.

Vegetative reproduction from inflorescences
In three Lepidocaryoid palms, a remarkable method of vegetative reproduction occurs where inflorescences root at the tip and produce new shoots. These are *Salacca flabellata, Salacca wallichiana* (Furtado, 1949), and *Calamus pygmaeus* (Ridley, 1907).
Ridley gives an illustration and description of the process in *C. pygmaeus*, a local species of northwestern Borneo. In this species stems are very short and closely clustering. The inflorescences are adnate to the leaf sheaths as in most *Calamus* spp. and are flagelliform, up to 2 m in length, arching out with the ultimate branches lying on the surface of the soil. The ultimate area of the main inflorescence axis in contact with leaf litter grows into a shoot, which roots and when the rest of the inflorescence rots, becomes established as a separate plant. This method of vegetative reproduction in *C. pygmaeus* is not vivipary, as shoots develop from axes rather than flowers and both staminate and pistillate inflorescences can produce new plantlets. In *S. flabellata*, inflorescence rooting is similar to that in *C. pygmaeus*, but has only been observed in the staminate plant. This situation shows a remarkable metamorphosis of the determinate inflorescence axis into an indeterminate axis. The development of the inflorescence in these three species requires further study. In *C. pygmaeus* fruit is frequently produced, there being an abundance of staminate and pistillate plants in populations so far examined. In *S. flabellata* pistillate plants have only once been found (Whitmore, pers. comm.), and rooting from inflorescences might be an adaptative compensation for poor sexual reproduction.

Role of Hyphaene
An interesting marked interaction between the growth form of a palm and forest ecology is shown by *Hyphaene compressa* in East Africa. What follows is only speculation based on field observations; the role of the doum palm in the ecology of the lowlands of East Africa presents an important field for further study.

H. *compressa* is an extremely conspicuous and abundant palm in the lowlands of East Africa. Its natural habitat is not known for certain, but it is thought to be the margins of forest along watercourses in seasonally dry to semiarid areas of East Africa. It appears to have undergone great expansion and have colonised areas cleared for farmland; in some instances, these areas would normally support a climatic climax vegetation of dry evergreen tropical forest. In such areas H. *compressa* produces a sort of savanna with a diffuse, interrupted canopy of palm crowns. The maintenance of much of this doum palm savanna appears to depend on frequent burning of the grass understorey. In a few areas of north coastal Kenya (e.g., near Witu, where tracts of doum palm savanna were included within the boundaries of newly demarcated forest reserves), the highly characteristic form of H. *compressa* can be seen to be an important factor in the reestablishment of forest in the absence of fire (Fig. 11.3). In possessing a branching system sometimes with as many as 64 heads, H. *compressa* apparently produces an effective shading and amelioration of climate under its canopy and also a possible perch for potential seed dispensers. This allows the reestablishment of secondary forest, and eventually the palms are engulfed by fast-growing trees such as *Trichilia emetica* and *Terminalia hildebrandtii*. Rawlins (1958) suggests that *Hyphaene*, with its branched crown, acts as an effective nucleus for the establishment of banyan figs, in whose shade seedlings of other tree species then develop. Personal observations suggest that the role of strangling figs is not essential to the reestablishment of forest under doum palms and that the protection produced by the branching crown is sufficient for reestablishment in this area, the rainfall of which is only marginally suitable for forest plant cover.

Role of densely colonial palms

Another palm occurring in sufficient abundance to have a marked effect on forest composition is the bertam palm of Malaya, *Eugeissona tristis*. Bertam produces a horizontally growing, sympodially branched, massive, woody, subterranean rhizome; and from the short apogeotropic branches grow acaulescent tufts of quite massive, densely spiny, pinnate leaves and erect hapaxanthic inflorescences. The leaves do not dehisce neatly from the stems, but tend to flop down and rot while still attached at the base to the stem. A tuft of bertam is hence surrounded by a mass of slowly decaying spiny leaves, and this, in conjunction with the dense crown of leaves, apparently produces a shading and mechanical barrier to regeneration of tree species (see Wong, 1959; Dransfield, 1970). In ridgetop forest in the western part of Malaya, where *E. tristis* is moderately abundant, opening of the canopy by selective logging appears to encourage the growth of bertam, which is resistant to damage and not killed by full sunlight; an almost worthless forest develops, with regeneration of timber trees being almost excluded by the dense colonies of bertram.

Eleiodoxa conferta, Nypa fruticans, and *Metroxylon sagu* are other examples of colonial palms that produce dense stands to the exclusion of most other plants.

Conclusion

The purpose of this discussion has been to indicate the great range of form in tropical rain forest palms. In some instances, it is possible to correlate growth forms with habitat and to indicate possible selective advantages of the respective growth forms. In most instances, however, so little is known about the interactions of palms and their environment that no obvious correlations can be indicated. There is great scope for further study of this extremely important group.

General discussion

Whitmore: One of the habitats in the eastern tropics in which palms are extremely rare is where there is recolonization of an abandoned area of shifting cultivation into which simple-leaved trees characterized by *Macaranga* come in large numbers. Could you comment on whether the absence of palms from this particular niche is the case in South America and Africa as well as Asia? In other words, are there more secondary palms in these large, open, hot clearings in these other two parts of the tropics?

Dransfield: I think that in the recolonization of an old area of shifting cultivation, palms come in after about 4 or 5 years. It is interesting that the palms that do come in are the hapaxanthic rattans, like *Plectocomiopsis, Myrialepis,* and *Korthalsia.* They get there, but I suspect they do require some amelioration of the climate before they can become established. I cannot answer for South America. I think there are relatively few, if any, strictly secondary forest palms. *Pigafetta filaris* is an example in the Celebes of a palm that is apparently a strict heliophile when it is young. Seedlings deposited in bird dung in primary forests do not develop beyond the two- or three-leaf stage, whereas seeds deposited in dung in the open develop rapidly.

Whitmore: Apart from these three climbing genera you mention for Asia, the only example I can think of is *Caryota mitis.*

Dransfield: C. mitis is certainly a rather weedy palm characteristic of secondary areas.

Whitmore: So, there really are no other palms, certainly not the little tree Arecoids.

Dransfield: Definitely not. These are palms that are very characteristic of forest undergrowth. Opening up the canopy leads to scorching and death.

Borchert: In the disturbed deciduous tropical forests of Colombia a certain palm is becoming very common as a nuisance in the pastures. Do you know the species?

Dransfield: This is *Acrocomia antioquensis*, a spiny palm characteristic of these arid grasslands.

Ashton: You mentioned *Eugeissona minor* and the problem of stilt roots; you also mentioned *Calamus pygmaeus*, which sends out a long inflorescence and roots apically. These examples might serve as a springboard for suggesting that one needs to understand fully the rather large number of constraints imposed on the range of variation among palms compared with many angiosperm families. Perhaps, for instance, one cannot understand these two species very well without knowing more about their root systems. Are the root systems of palms restricted in the extent to which they can spread in the soil, especially in rain forest species? If they are, then peaty soils, like those in which both the above-mentioned species grow, may present the palms with a problem in that they've got a very restricted volume of soil they can utilize. If palms have some way of growing away from the central stem and putting out a new "nodum," from which roots can spread, this may be greatly advantageous to them. In *E. minor* the original stem sequentially dies and rots away from the base, so that finally the stem apex is supported by a series of tall root "pilings."

Dransfield: *E. minor*, of course, is hapaxanthic, so the development of the shoot system on top of the stilt roots is sympodial. The area of exploited ground is as restricted as it would be if it were without stilt roots because the column of stilt roots is really quite confined. The rooting system of *C. pygmaeus* is extremely extensive; in fact we have a very nice collection at Kew by Ridley of which half the specimens are roots. Indeed, I wish we knew more about palm roots, and my impression is that there is a most extraordinary range of root types, sometimes in the same species. In the American palm *Cryosophila* something like six different root types can be distinguished (pers. obs.). There are thorns on the stems, some branching; stilt roots; tiny pneumatophores that are lateral roots on the stilt roots; subterranean roots that are plagiotropic and some that are geotropic. There is an enormous field for further detailed investigation.

Tomlinson: Holttum (1955) has discussed and possibly answered the point raised by Dr. Ashton. Palms obviously are limited in the extent to which their roots can grow laterally in the soil because they have no mechanisms for secondary thickening. Therefore, no matter how much they branch distally, there eventually is a proximal bottleneck. This could account for their limited architecture.

A possible explanation of why suckering (basal branching) in palms is so common, when it seems to produce closely juxtaposed and competing trunks, has been offered by Henry S. Horn (pers. comm.). He calculated that for a given area of soil exploited by a given root biomass several more trunks could be supported close together than if the same root biomass were made available to dispersed but not competing trunks. This brings the neglected aspect of root morphology, and especially the biologic significance of root dimensions, further into prominence.

Hartshorn: Stilt roots in Iriarteoid palms may be related to the degree of shade tolerance in the adaptive strategies of these species and in establishing their position in the canopy or subcanopy. For example, the Iriarteoid palms in Central America and part of South America are not nearly as shade-tolerant as some of the Geonomoid palms, such as *Welfia georgii*, which initially makes its stem below ground and later develops the aboveground trunk. In contrast, *Iriartea* and *Socratea* grow much faster than *W. georgii*, and as they are subcanopy palms they have a structural support problem. The evolution of stilt roots aids these Iriarteoid palms in growing rapidly to their canopy position.

Dransfield: I don't quite see the necessity for having stilt roots to reach into the canopy. I admit that the Iriarteoid palms are remarkable. Instead of establishment growth building up a maximum diameter stem in the rosette stage, there is a gradual buildup in stem diameter, each successive node being perched on the successively longer stilt roots. In stilt-rooted palms like *Areca vestiaria*, an undergrowth species that is relatively intolerant of disturbance and apparently has no requirement to extend into the canopy, I can see no adaptive significance for the roots except perhaps as support on disturbed soil on unstable slopes.

Oldeman: I think that stilt roots of *Iriartea* can be explained by the need to get air and mineral solutions into the tree without the need to accentuate mechanical theories to establish their adaptation.

Givnish: If stilt-rooted palms are generally shade-tolerant, it should not come as much of a surprise: The growing, above-ground apices of the stilt roots are not likely to survive dry air or sunny conditions. So the question is: Why is it advantageous to get the crown away from the soil? It is my impression that stilt-rooted palms are more common in soils that are occasionally waterlogged, where, if the apex of a palm stayed at ground level for a long time, it might be damaged. There are thus two strategies: One is sit and wait while developing diameter, and the other is to grow vertically a little bit, become supported, grow a little farther, become supported, grow, and so on.

Dransfield: This explanation would be very elegant, but with Iriarteoid palms, the same species may range from swamps to hilltops. As far as I can see after looking, albeit briefly, at Iriarteoids, no correlation can be made between stilt roots and habitat.

References

Bailey, L. H. (1942). Palms of Seychelles islands. *Gentes Herbarum, 6,* 1–48.

Bannister, B. A. (1970). Ecological life cycle of *Euterpe globosa* Gaertn. In *A Tropical Rain Forest: A Study of Irradiation and Ecology at El Verde, Puerto Rico,* ed. H. T. Odum & R. F. Pigeon, pp. B-299–314. Oak Ridge, Tenn.: U.S. Atomic Energy Commission.

Beccari, O. (1904). *Wanderings in the Great Forests of Borneo.* Tr. E. Gigliolo; rev. and ed. F. H. H. Guillemard. London: Constable.

– (1908). Asiatic palms, Lepidocaryeae. I. The species of *Calamus. Ann. Roy. Bot. Gard. Calcutta, 11.*

Burkill, I. H. (1935, 1966). *A Dictionary of the Economic Products of the Malay Peninsula,* vols. 1, 2. Governments of Malaysia and Singapore.

Burret, M. (1936). Neue Palmen aus Neuguinea. IV. *Notizbl. Bot. Gart. Mus. Berl., 13,* 317–32.

Dransfield, J. (1969). Palms in the Malayan forest. *Malay Nat. J., 22,* 144–51.

– (1970). Studies in the Malayan palms *Eugeissona* and *Johannesteijsmannia.* Ph.D. thesis, Cambridge University.

– (1974). New Light on *Areca langloisiana. Principes, 18,* 51–7.

– (1975). A remarkable new *Nenga* from Sumatra. *Principes, 19,* 27–35.

– (1976a). Terminal flowering in *Daemonorops. Principes, 20,* 29–32.

– (1976b). A note on the habitat of *Pigafetta filaris* in North Celebes. *Principes, 20,* 48.

268   *John Dransfield*

Fisher, J. B. (1973). Unusual branch development in the palm *Chrysalidocarpus. Bot. J. Linn. Soc.*, *66*, 83–95.
- (1974). Axillary and dichotomous branching in the palm *Chamaedorea. Amer. J. Bot.*, *61*, 1046–56.
Fisher, J. B., & Tomlinson, P. B. (1973). Branch and inflorescence production in saw palmetto (*Serenoa repens*). *Principes*, *17*, 10–19.
Furley, P. A. (1975). The significance of the Cohune Palm *Orbignya cohune* (Mart.) Dahlgren on the nature and in the development of the soil profile. *Biotropica*, *7*, 32–6.
Furtado, C. X. (1949). Palmae Malesicae. 10. The Malaysan species of *Salacca. Gard. Bull. Singapore*, *12*, 378–403.
- (1951). Palmae Malesicae. 11. The Malayan species of *Korthalsia. Gard. Bull. Singapore*, *13*, 300–24.
- (1956). Palmae Malesicae. 19. The genus *Calamus* in the Malayan Peninsula. *Gard. Bull. Singapore*, *15*, 32–265.
Hallé, F., & Oldeman, R. A. A. (1970). *Essai sur l'architecture et la dynamique de croissance des arbres tropicaux.* Paris: Masson.
Holttum, R. E. (1955). Growth habits of monocotyledons: Variations on a theme. *Phytomorphology*, *5*, 399–413.
von Humboldt, A., & Bonpland, E. (1805). *Plantes équinoxiales*, vol. 1. Paris: Levrault.
Moore, H. E. (1973). The major groups of palms and their distribution. *Gentes Herbarum*, *11*, 27–141.
Otedoh, M. O. (1972). The rediscovery of *Raphia regalis* in Nigeria. I. Cyclostyled paper presented at eighth annual conference of the Agricultural Society of Nigeria, Calabar, July 1972.
Pichi-Sermolli, R. E. G. (1957). Una carta geobotanica dell'Africa orientale (Eritrea, Etiopia, Somalia). *Webbia*, *13*, 15–132.
Rawlins, S. P. (1958). Some notes on the exotic and indigenous palms occurring along the coastal region of North Kenya from the Tana River to the Somal Border. Unpublished, Kew Herbarium Library.
Ridley, H. N. (1907). Branching in palms. *Ann. Bot.*, *21*, 417–22.
Schoute, J. C. (1909). Ueber die Verästelung bei monokotylen Bäumen. II. Die Verästelung von *Hyphaene. Rec. Trav. Bot. Neerland.*, *6*, 211–32.
Tomlinson, P. B. (1960). Essays on the morphology of palms. 1. Germination and the seedling. *Principes*, *4*, 56–61.
- (1967). Dichotomous branching in *Allagoptera? Principes*, *11*, 72.
- (1971). The shoot apex and its dichotomous branching in the *Nypa* palm. *Ann. Bot. Lond.*, *35*, 865–79.
Tomlinson, P. B., & Moore, H. E. (1966). Dichotomous branching in palms? *Principes*, *10*, 21–9.
- (1968). Inflorescence in *Nannorrhops ritchiana* (Palmae). *J. Arnold Arbor.*, *49*, 16–34.
Wallace, A. R. (1853). *Palm Trees of the Amazon and Their Uses.* London: van Voorst.
Wong, Y. K. (1959). Autecology of the bertam palm *Eugeissona triste* Griff. *Malayan Forester*, *22*, 301–13.

12

Araliaceae: growth forms and shoot morphology

W. R. PHILIPSON

Botany Department, University of Canterbury,
Christchurch, New Zealand

The Araliaceae are a well-defined, homogeneous family of moderate size, occurring in all tropical and many temperate regions. The family is best represented in Asia and Polynesia, where its members often form a noticeable part of the woody vegetation at all altitudes. Its contribution to the tropical rain forests of Africa and the Americas is negligible, though at higher altitudes in those continents there is some increase in the number of genera and species. The family may well represent a very ancient lineage, but it includes several genera, especially *Schefflera*, *Polyscias*, and *Osmoxylon*, which are vigorously diversifying at the present time. Corner (1964) cites this family as typical of his concept of a small-statured, large-leaved, pachycaulous tree; yet, in spite of this prevalent aspect, the family exhibits a surprising range of form. Even the thin twigs and willowlike leaves of typical leptocaul trees are represented by *Pseudopanax simplex* (Forst.f.) Philipson. The deciduous habit occurs even in the tropics (*Sciadodendron*) in this overwhelmingly evergreen family, and the distichous phyllotaxis of *Hedera* is most exceptional. Diversity extends to floral morphology, for such aberrant features as superior ovaries and tubular corollas occur, and in *Osmoxylon* whole flowers are transformed into pseudofruits that play a special role in pollination by birds (Beccari, 1877–90). However, the present account will be confined to the organography of vegetative shoots.

Lateral appendages
Foliage leaves
The foliage leaves are extremely variable in form, but compound leaves of both digitate and pinnate construction may be considered typical. Not infrequently the leaves, whether digitate or pinnate, are compound to the second, third, or even fourth degree. The

269

lateral rhachides of doubly pinnate leaves may be "subtended" by pinnae on the main rhachis (e.g., *Aralia, Arthrophyllum*); more rarely, similar pinnae may also occur above the insertion. However, the lateral parts of the leaf do not arise in the axils of these pinnae, as all leaflets lie in the same plane and must be interpreted morphologically as discrete parts of a single blade. Complex digitate leaves, such as those of *Schefflera heterophylla* (Seem.) Harms, bear well over 100 leaflets and simulate the delicate branch systems of leptocaul trees. Highly subdivided leaves provide transitions between pinnate and digitate types of compounding. It is meaningless to categorize leaves like those of *Mackinlaya celebica* (Harms) Philipson, in which the central leaflet may be repeatedly subdivided, as digitate or pinnate.

Leaves with an articulation between the petiole and a single blade are common, and these can be regarded as unifoliolate. However, several genera have leaves with a single blade without any demarcation at the top of the petiole (e.g., *Hedera, Dendropanax*), but even in such genera lobing may be so pronounced that species with a compound leaf occur (e.g., *Trevesia, Osmoxylon*).

The leaf usually clasps the stem by means of a basal sheath supplied with many vascular traces. The sheath may be prolonged as two stipules or as a single ligule, and this may be extended around and up the petiole as crests, which may in turn be vascularized (*Osmoxylon*).

The diversity of leaf form may be increased by heteroblastic development. Some remarkable examples occur in New Zealand, the best known and most striking being the lancewood (*Pseudopanax crassifolium* (Sol.) C. Koch), in which the long-persistent, unbranched juvenile form bears leaves that are narrowly linear, reflexed, rigid, and up to 1 m long, whereas the foliage of the adult crown is speading and lax, with the blades lanceolate and much shorter (about 15 cm long). In *Pseudopanax simplex* the juvenile leaves are more complex than those of the adult, being digitately compound with the leaflets finely lobed, in marked contrast to the simple lanceolate blades of the mature tree. Heteroblastism is better known in these temperate trees, but occurs also in the tropics. For example, photographs of a species of *Gastonia* in Mauritius sent to me by M. J. E. Coode show a striking change in leaf form between juvenile and adult plants. The size and complexity of leaf shape may also vary along a shoot produced by one flush of growth, and simple and compound leaves occur apparently in random sequence in *Pseudopanax gilliesii* Kirk.

Cataphylls

Bud scales may occur in the main axes, which thus show intermittent growth (*Oreopanax*: Borchert, 1969), or may be confined to lateral shoots. Cataphylls are not necessarily present on shoots that are intermittently active, as their dormant buds may be enclosed by stipular sheaths rather than by cataphylls (*Schefflera:* Koriba, 1958). Shoots that in their vegetative phase never develop cataphylls frequently terminate in inflorescence buds heavily invested with scales or bracts. Axillary buds when furnished with basal cataphylls probably undergo a period of dormancy: i.e., the branches produced from them are proleptic (Tomlinson & Gill, 1973). This is confirmed for several temperate species (Philipson, 1972), but has rarely been determined for tropical examples (*Oreopanax:* Borchert 1969; *Arthrophyllum:* pers. obs.). Nor is it known whether the branching of species in which cataphylls do not occur even on lateral shoots is sylleptic, though this certainly appears so in *Mackinlaya*, where leafy laterals may soon overtop the terminal inflorescence. Two types of cataphylls occur in the Araliaceae (Philipson, 1972). Many are clearly equivalent to stipular sheaths, of which the lamina is rudimentary (e.g., *Meryta*); in others it is the lamina that forms the greater part of the scale (e.g., *Pseudopanax*).

The occurrence of resting buds protected by cataphylls in trees of tropical rain forest has often occasioned comment. Such coverings are more common on lateral buds and are then to be related to growth form rather than to climate. If the normal habit of a species is for lateral branches to grow out close to the main apex, cataphylls will be lacking at the base of lateral shoots (according to the principle of syllepsis: Chap. 7). On the other hand, if the habit of growth is such that lateral branching is delayed, it is likely that these proleptic shoots will be provided with cataphylls, as can be seen in the lateral innovation shoots of *Arthrophyllum*. In a group of woody plants that has spread from the tropics into a more severe temperate climate, these covered lateral buds may assume a protective role. It is a curious fact that many north temperate trees continue their growth each year by lateral branches. This is necessarily the case when flowers or inflorescences are terminal, but also occurs in many species in which the terminal buds abort (Büsgen & Münch, 1929; Romberger, 1963). It is suggested that the occurrence in the tropics of covered lateral buds in trees whose main branches lack such protection may explain the prevalence of perennation in north temperate trees by means of their lateral buds. Trees already possessing covered lateral buds might prove to be preadapted to a life in more seasonal climates, as they might more readily develop hardiness.

Branch systems

The branch systems that bear the lateral appendages are usually of great simplicity. Shoot elongation stops with the development of terminal inflorescences, further extension growth being from one or more lateral buds. Of even simpler form is the single-trunked monocarpic *Harmsiopanax ingens* Philipson, of the New Guinea high-lands. This plant is equivalent in habit and habitat to the dendroid *Senecios* of the African mountains or the *Espeletias* of the Andes. Its growth form is very similar to that of many biennial umbellifers (e.g., *Daucus, Pastinaca*), except that it develops an aerial stem and vegetates for several years before flowering. It may be no coincidence that so many floral and fruit characters of *Harmsiopanax* are closer to the Umbelliferae (where it was originally placed) than to the Araliaceae (Philipson, 1970, 1973). Since *Harmsiopanax* is intermediate between the two families, it may represent a survival of the stock from which both arose. The evolution of herbs from woody ancestors has been described as "miniaturization" by Hallé & Oldeman (1970). One origin of the Umbelliferae may have been by miniaturization of a dendroid plant such as *Harmsiopanax*. Several umbellifers are megaherbs that grow in subalpine habitats corresponding to that of the tropical montane dendroid plants, but in higher latitudes. ("Megaherb," or "megaphorb," is a term applied by European ecologists to giant herbs that characterize fertile habitats in the subalpine zone.) Examples are *Eryngium* of the European Alps and *Aciphylla* of the Southern Alps of New Zealand. Another striking example is the subantarctic *Stilbocarpa*, but at these latitudes the megaherb habit occurs at sea level. This genus also lies in the uncertain zone between the two families (Philipson, 1970), and this supports the hypothesis that the Umbelliferae arose as giant subalpine herbs by reduction from the woody Araliaceae. A similar origin of tropical African herbaceous *Senecios* by reduction from tropical subalpine dendroid species has been postulated by Mabberley (1974).

Other single-stemmed araliads are not monocarpic. In *Anakasia simplicifolia* Philipson, the inflorescences are axillary, and consequently the main axis can continue to produce new foliage leaves indefinitely. The stems of these species, like those of *Harmsiopanax ingens*, are monopodial, but sympodial single-trunked trees also occur among tropical araliads. *Brassaiopsis glomerulata* (Bl.) Regel, species of *Aralia*, and *Mackinlaya* form erect apparently unbranched treelets with inflorescences overtopped by leafy shoots. The single trunk is a sympodium composed of a succession of lateral shoots that emerge from buds below the terminal inflorescences. Older individuals of all these species may eventually branch because more than one bud may become active under an inflorescence.

The clustering of several neighboring trunks produced from underground rhizomes is a habit that occurs in *Tetrapanax* and some shrubby species of *Schefflera*. On a smaller scale it is found in *Stilbocarpa*, a subantarctic genus of three large herbaceous species, and a similar habit occurs in subherbaceous species of other genera (*Aralia, Panax*).

The indefinite growth of the main shoots is rare in the family. Most araliads with axillary inflorescences also have their principal axes ending in inflorescences. However, a more strict relegation of reproduction to a lateral position, either axillary or on short shoots, occurs in *Plerandra stahliana*, some species of *Schefflera, Trevesia,* and *Stilbocarpa*, as well as in *Anakasia*, as already mentioned. Well-differentiated vegetative short shoots are comparatively rare in the family.

In considering the total range of growth forms found within the family, notice should be taken of the prevalence of the epiphytic habit and the occurrence of climbers and scramblers. For example, the adhesive adventitious roots of *Hedera* are familiar (though unmatched in any other araliad genus), and many species of *Schefflera* cling by specialized roots. *Acanthophora scandens* Merr. scrambles by means of grappling hooks, raising itself up by a succession of arching branches in the manner of a briar rose. *Pentapanax elegans* Koord., of volcanic peaks in east Java, and the widespread *Acanthopanax tripartitus* (L.) Merr. also sprawl in this way, the former almost achieving tree stature. Sucker shoots are produced prolifically from the roots of the rice-paper tree (*Tetrapanax*), and *Osmoxylon borneense* and related species spread to form extensive thickets on stream banks by the rooting of their drooping branches.

The growth forms discussed above together with two others to be described in more detail below, can be related to the models of tree architecture proposed by Hallé & Oldeman (1970). Of the 23 models described by these authors, 7 are here recorded for the Araliaceae, and at least one araliad (which is not a tropical tree) displays an additional architectural type. These are summarized diagrammatically in Fig. 12.1 and listed in Table 12.1. Little is known of the range of growth forms found in single dicotyledonous families. Hallé (1974) records 7 models in the small family Icacinaceae, and Hallé & Oldeman (1970) consider the 6 models they record for the Lecythidaceae as a rich variation. The Araliaceae is a family of about the same size, so that with at least 8 models its range of habit can be accepted as considerable. It should be noted, however, that with the exception of the two forms now to be described, the architecture shown is always relatively simple. The following two growth forms are more unusual.

The genus *Arthrophyllum* occurs abundantly from the Andamans and Indo-China to the Philippines and New Guinea. The growth forms of some species are not unusual, but most species conform to the following specialized arrangement. The primary axis bears a spiral of large pinnate leaves and does not branch until flowering occurs. At that time sylleptic lateral branches emerge from the upper leaf axils. These branches bear terminal and often also lateral inflorescences, and their leaves are distinctly different from those of the main stem. They are smaller, often being reduced to a single leaflet,

Fig. 12.1. Growth forms found in Araliaceae. Each is attributed to one of models of Hallé & Oldeman (1970), as follows: *A*, Leeuwenberg; *B*, Holttum; *C*, Corner; *D*, Chamberlain; *E*, Tomlinson; *F*, Rauh; *G*, *H*, Not attributable to any model; *I*, Champagnat.

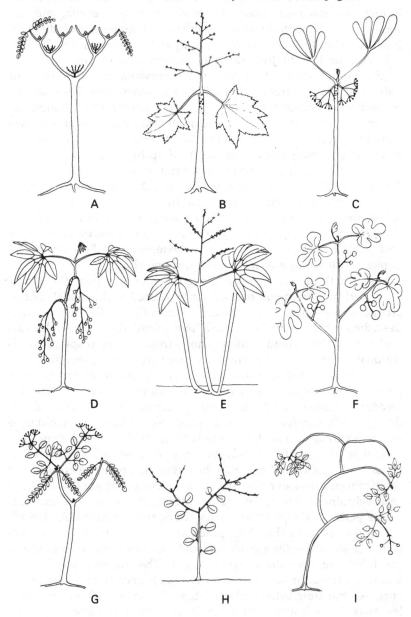

and most arise in decussate pairs but come to lie in a single plane owing to the twisting of their petioles (Fig. 12.2). These specialized floriferous branches form a tuft around the main apex, which aborts, and after fruiting the branches are shed. Extension growth of the tree is by one (or more) proleptic lateral bud(s) below the flowering branches. From these branches large pinnate leaves in spiral sequence are produced. They therefore resemble the original trunk, which they extend sympodially. The lateral branches have been described as leaflike, mainly because they hold their leaves in one plane, have scattered vascular bundles with little or no secondary thickening, have an arc-shaped sheathing base, and are soon abscised.

In relating the growth form of *Arthrophyllum* to the models of Hallé & Oldeman (1970), if the lateral branches are interpreted as a terminal inflorescence, the tree conforms readily to Leewenberg's model. However, the amount of foliage borne by them and their axillary position make this interpretation doubtful. Consequently, the tree form in this genus may call for some extension of the scheme of these authors.

The second unusual growth form also occurs in a genus whose

Table 12.1. *Architectural models of Hallé & Oldeman (1970) known to occur in the Araliaceae*

Model	Example
Leeuwenberg	*Gastonia spectabilis* (Fig. 12.1A)
	Many other araliads
Holttum	*Harmsiopanax ingens* (Fig. 12.1B)
Corner	*Plerandra stahliana* (Fig. 12.1C)
	Anakasia simplicifolia
Chamberlain	*Brassaiopsis glomerulata* (Fig. 12.1D)
	Other species of *Brassaiopsis*
	Aralia dasyphylla
	Mackinlaya schlechteri
Tomlinson	*Schefflera* sp. (Fig. 12.1E)
	Stilbocarpa spp. (herbaceous)
Rauh	*Trevesia sundaica* (Fig. 12.1F)
	Plerandra spp.
Champagnat	*Pentapanax elegans* (Fig. 12.1I)
	Acanthopanax trifoliatus
	Aralia scandens
Not conforming to any named model	*Arthrophyllum diversifolium* (Fig. 12.1G)
	Pseudopanax anomalum (Fig. 12.1H)

276   *W. R. Philipson*

other species have simpler habits. *Pseudopanax anomalum* (Hook.) Philipson is quite unlike its relatives in its branch morphology, and though it is a shrub of the temperate regions it is included here to complete the range of growth forms known among the Araliaceae (Fig. 12.1*H*, 12.3). The main axes of the mature shrub are determinate in their growth: Their apices abort after one season's elongation (Fig. 12.1*H*). Their lateral appendages are either exclusively small scales or some toward the base may have expanded photosynthetic blades. All, or at least the greater number of the foliage leaves, are borne on very condensed lateral shoots that arise in the axils of the scale leaves in the second season of growth. As these short shoots usually bear a solitary leaf, the second-season axes appear to bear leaves in a normal phyllotactic pattern (Fig. 12.3). The umbels are borne on the short shoots, usually terminating them. Extension growth takes place

Fig. 12.2. Habit of *Arthrophyllum diversifolium*.

from one to few lateral buds behind the aborted apex, from which fresh cataphyll-bearing long shoots emerge. The main axes are, therefore, sympodial, composed of a succession of determinate sections; the lateral leaf- and flower-bearing short shoots are very distinct from them. It is perhaps unrealistic to expect a temperate shrub to fit into a scheme based on tropical trees, and it is, therefore, not surprising that *P. anomalum* does not conform to any of the models of Hallé & Oldeman (1970). It will be interesting to see whether other extensions of the scheme put forward by these authors are required to fit other temperate shrubs and, in particular, other divaricating shrubs of New Zealand.

Fig. 12.3. Twig of *Pseudopanax anomalum*. Shoots of current season (unstippled) bear scale leaves and a few small foliage leaves; 1-year-old shoots (stippled) bear long shoots above and short shoots below (latter bear foliage leaves); 2-year-old shoots (cross-hatched) are similar, but short shoots more frequently bear two (or more) foliage leaves; 3-year-old shoots (black) are similar. All long shoots are determinate.

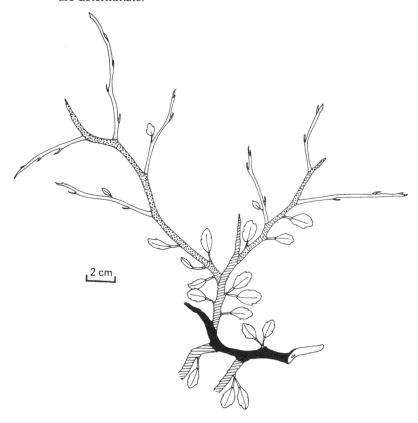

2 cm

Heteroblastic development may affect the habit of growth of a tree in much the same way as we have seen it affects foliage morphology (Philipson, 1963). In its early life *Pseudopanax crassifolium* is a single-stemmed tree, but later it develops a crown. The simultaneous change of growth form and of leaf morphology is extremely striking. The change is not unlike the onset of branching caused in, for example, *Gastonia spectabilis* by the termination of the erect unbranched sapling because of flowering. However, the loss of apical dominance in *P. crassifolium* is rarely due to the formation of an inflorescence, the main apex usually being overwhelmed by pseudowhorls of vigorous lateral branches that form the crown. The process is similar in some respects to that described for *Hevea* (Hallé & Martin, 1968), with the addition of a heteroblastic leaf change. There may also be some similarity between this erect juvenile form and the orthotropic juvenile phases discussed by Roux (1968), the adult crown of *Pseudopanax* being composed mainly of branches arising from lateral (but orthotropic) buds.

Terminology

The Araliaceae have been presented as an example of plasticity in growth form within a family of woody plants. Some general morphologic topics arising from this account will now be considered.

Characteristics of buds

In any rapidly developing subject, with new ideas coming forward, it becomes necessary to agree on the use of terms. There has not always been time for this in the expanding field of tree form. For example, in two of the most influential contributions, the term "articulated" is applied to shoots with quite different meanings. In Hallé & Oldeman (1970) articulated shoots are those built from successive units, whereas Tomlinson & Gill (1973) use the term to refer to shoots with intermittent or rhythmic growth. However, confusion has arisen only by a misinterpretation of the French word *"article"*; substitution of the word "module" clarifies the situation (Chap. 9). The characteristics of buds will be used to illustrate in outline the advantages and dangers of a rigid terminology.

To define the nature of a bud, its position, morphology, and behavior must each be considered, and none of these are independent of one another. As regards position, buds may be related to lateral appendages (whether axillary or not) or they may be quite unrelated to them. Sattler (1974) discussed the varied usages of the term "adventitious bud," and Büsgen & Münch (1929) recorded conflicting definitions of the term "epicormic bud." An adventitious bud may either be exogenous or endogenous (Thompson, 1952), and it may be solitary or a member of a series. No system for describing these positional features of buds has general acceptance, and the same may be said of their morphologic features. The young primordia of a bud may be covered by specialized leaves, parts of foliage leaves, or not at all. The nodes between some or all of the cataphylls may elongate eventually, with the result that the foliage leaves stand in pseudowhorls (as in many species of *Rhododendron*). Buds may be differentiated into floral and foliage buds, and these may be quite separate; or the foliage buds may be lateral and included within the cataphylls of the floral buds (i.e., so-called compound buds occur). Large sections of the genus *Rhododendron* are characterized by these bud arrangements and others by the more general arrangement with flowers and leaves borne on the same shoots and, therefore, arising from the same buds. In the area of bud behavior, the terms, "sylleptic" and "proleptic" were revived with an expanded meaning by Tomlinson & Gill (1973) to fill a gap in the terminology that became obvious once it had been pointed out. As already remarked, there is some uncertainty about the terms to be used in describing the continuous or intermittent growth of shoots. Neither do there appear to be any generally accepted terms for another aspect of bud behavior. It is, of course, well known that in some trees shoot primordia representing the whole of the next season's flush of growth are held within the resting bud, whereas in others primordia of only some of the leaves are present, later leaves being initiated after bud break. Terms such as "shoot primordia complete" and "shoot primordia incomplete" suggest themselves. But such clear-cut contrasting terms may not be helpful in a situation without clear discontinuities. All intermediate states may occur. In coining terms, a balance may have to be struck between the rigidity of formally defined terms and the flexibility needed when describing many biologic situations. As the subject develops, I hope it will be possible, on the one hand, to avoid the glossaries of terms characteristic of wood anatomy, as these may have stultified the natural development of that subject. On the other hand, progress will be hampered without some general agreement on usage (see Chap. 7).

The concept of the shoot
    As the Araliaceae are a family with considerable plasticity of growth form, it is of interest to consider how far they conform to the classic concept of shoot morphology, especially as the validity of this concept is currently questioned. In proposing a new concept of the shoot, Sattler (1974) makes two principal criticisms of the classic outlook: (1) that it does not provide for changes in the relative position

of organs, and (2) that it excludes the concept of organs intermediate between the generally accepted categories.

The first of these criticisms is cogent only if a narrow definition of the classic concept is adopted. If the shoot is considered to consist of an axis upon which lateral appendages arise and to branch by buds situated in the axils of these appendages, then difficulties will be encountered. But such a narrow definition is clearly insufficient. Quite commonly buds do arise in positions other than axillary, but with very few exceptions (to be considered later), these extraaxillary buds arise on a plant whose organography is already in conformity with the classic plan. Such nonaxillary buds may be related in some other way to the lateral appendages; more usually they may be quite unrelated to them. These adventitious buds may be endogenous (as in *Couroupita*: Thompson, 1952); they may arise on the hypocotyl (*Linaria*: Champagnat, 1961; *Rhizophora*: Gill & Tomlinson, 1969) or on the lateral appendages (Dickinson & Sattler, 1974, 1975; Sattler, 1975). In all these plants the classic organography precedes the more or less extreme variations resulting from the activity of adventitious buds. Such examples of the plasticity of angiosperm morphogenesis are of great interest, and careful studies of them such as those of Dickinson and Sattler are essential. But their occurrence calls not for the abandonment of the classic concept, but rather for its slight extension in ways that must always have been in the minds of the classic morphologists themselves.

The second proposition of Sattler, however, may be of more importance. Do organs intermediate between shoots and leaves exist? Is homology semiquantitative? (Sattler, 1966, 1974). As flower-bearing branches of *Arthrophyllum* have been cited as a branch system with many leaflike characters (Harms, 1894), the case for organs intermediate between shoots and leaves should be examined here.

It is commonplace that branch systems may resemble fronds and that leaves may resemble branch systems (see Chap. 18). They may share functions, such as the display of photosynthetic surfaces, and their common problems may be solved in similar ways. In *Arthrophyllum* the branches bear leaves in one plane; the axis resembles the leaf base, petiole, and rhachis of a compound leaf both morphologically and anatomically, and is soon abscised. Other even more convergent examples are well known (Roux, 1968; Hallé & Oldeman, 1970). But this convergence does not lead to intermediate organs. The form of an organ may be related to its function, but the distinction between a shoot and its lateral appendages is concerned with neither form nor function, but with their developmental relation. The essence of the classic concept of the shoot lies in the apical meristem. Here the axis prolongs itself and leaf primordia arise. I know of no instance in a woody angiosperm where an apex and its appendages need be confused (the situation may well be different in Pteriodophyta: Bierhorst, 1973, 1974). It is true that branches may arise in regular phyllotactic sequences without axillant appendages, but I believe this to be confined to inflorescences such as some capitula and the racemes of crucifers. Also, experimental procedures can modify the subsequent development of a primordium, so that it departs from the classic pattern. By demonstrating this latent capacity, these results emphasize that similar departures do not occur naturally. The most remarkable leaf known to me, *Chisocheton* (Sattler, 1974), combines epiphyllous inflorescences with the persistence of apical extension growth. Nevertheless, its relation to the parent shoot is entirely that of a leaf. It is remarkable only in combining two unusual features. This unusual combination occurs in a few species of an otherwise normal genus.

It is of interest and importance to record such departures from the general run of organography, but we should recognize their rarity. They would be more significant if they succeeded in establishing themselves as a new departure from which diversification evolved. Such dynasties do not seem to have been set up, at least among woody plants, for all the models of tree architecture described by Hallé & Oldeman (1970) conform to the classic concept, and I would confirm this for the Araliaceae. For these reasons, the classic concept is valuable and is in no need of replacement.

Some deviations exhibited by herbaceous plants are of general interest because they demonstrate that angiosperm organography has potentialities for profound departure from the classic model. These are few in number and of limited scope, but Jong & Burtt (1975) are right in emphasizing their significance. The plants concerned belong to the genera *Streptocarpus* and *Utricularia* and the families Lemnaceae, Rafflesiaceae, and Podostemaceae. Much remains to be learned about the morphology of these plants, but the detailed study of *Streptocarpus* reveals strikingly anomalous organography.

Angiosperms are extraordinarily versatile, but their morphologic adaptations are circumscribed by their basic organography. We must be alert to recognize departures when they occur, but they are very rare, especially among woody plants.

General discussion

Tomlinson: Does *Harmsiopanax ingens* have axillary buds that never develop?
Philipson: I don't know, but it is probable, because of the two other species of the genus: One is also monocaulous, and the other may grow after flowering, especially if damaged.

Tomlinson: But you have never seen *Harmsiopanax ingens* branching even when damaged?

Philipson: I have never seen it branching. I shouldn't be surprised if it did occasionally, though I think its general habit is not to do this.

Oldeman: Concerning the eventual transitions between axis and leaves, I think you are quite right in saying that we can use these terms in the classic concept because such transitions are rare. I think, however, that because there are big compound leaves with secondary thickening, a terminal meristem, and an indeterminate number of leaflets, from which the only thing lacking to call them branches are their axillary buds, you cannot negate the existence of transitions. Intermediates do exist, though they are rare.

Philipson: This part of my talk was condensed, but I acknowledge that there are structures that have intermediate features and could become very shoot-like. I would insist, however, that their origin in the apex is entirely like that of a leaf. I know that by experimentation one can influence the fate of a primordium, but this simply emphasizes the fact that in nature this does not happen. This, therefore, makes it more extraordinary that the plant has this capacity but never uses it.

Whitmore: You showed us four rather intriguing genera of the Araliaceae, including a herbaceous one, that you said were intermediate between Araliaceae and Umbelliferae. You really did not tell us in what respects you thought these two were different and why you consider that these families are so distinct?

Philipson: The two families together form a distinct group. The taxonomic differences between them are principally in fruit characters. In the Umbelliferae the fruit is a dry seed that divides into two mericarps. In the Araliaceae the fruit is a berry. These link up with other characters or nobody would consider them to be families, but no other character is strictly correlated to these fruit characters. In *Harmsiopanax*, one of the four I mentioned, the fruit is dry and divides up into two, although in a slightly different way from the method in the Umbelliferae. I cannot help feeling that *Harmsiopanax* is really as much an umbellifer as it is an araliad. This is one of the reasons that encouraged me to suggest that it might represent stock from which both families had arisen. *Stilbocarpa* looks even more like an umbellifer than it does an araliad, but it has berries, and for that reason it is put in the Araliaceae. There are anatomic reasons why you could put it in the other family, and I personally would hesitate to decide which it is.

Whitmore: What about the herbaceous one in the subantarctic islands?

Philipson: That has berries and, therefore, is technically an araliad, but it looks very like a member of the Umbelliferae.

Whitmore: So, is it correct to say that the two families really differ only in a single character with the corresponding groups, one woody and one herbaceous?

Philipson: Yes. It is quite convenient to consider the Umbelliferae as a mainly herbaceous, temperate family distinguished from a mainly woody tropical family, but whether it is taxonomically justifiable is a matter of opinion.

Givnish: Dr. Philipson, could you comment on the relative frequency of araliad trees that have leaves of different complexity at different stages of their life cycle? Is there a greater tendency for such species to have a simple-leaved juvenile form and a compound-leaved adult form, or vice versa?

Philipson: No, I cannot comment. In New Zealand the situation is clear because there is a division between those species that become more complex in the adult phase and those species that become less complex. In the tropics there is no information about relative frequencies.

References

Beccari, O. (1877–90). *Malesia: Raccolta di Oservazioni Botaniche Intorno alle Piante dell' Archipelago Indo-Malese e Papuano*. Genoa: Tip. del R. Instituto sordo-muti. 3 vols.

Bierhorst, D. W. (1973). Non-appendicular fronds in the Filicales. In *The Phylogeny and Classification of the Ferns*, ed. A. C. Jermy, J. A. Crabbe, & B. A. Thomas, pp. 45–57. London: Academic Press.

– (1974). Variable expression of the appendicular status of the megaphyll in extant ferns with particular reference to the Hymenophyllaceae. *Ann. Missouri Bot. Gard.*, *61*, 408–26.

Borchert, R. (1969). Unusual shoot growth pattern in a tropical tree, *Oreopanax* (Araliaceae). *Amer. J. Bot.*, *56*, 1033–41.

Büsgen, M., & Münch, E. (1929). *The Structure and Life of Forest Trees*. tr. T. Thomson. London: Chapman & Hall.

Champagnat, M. (1961). Recherches de morphologie descriptive et expérimentale sur le genre *Linaria*. *Ann. Sci. Nat. (Bot.)*, Ser. 12, 2, 1–170.

Corner, E. J. H. (1964). *The Life of Plants*. London: Weidenfeld & Nicolson.

Dickinson, T. A., and Sattler, R. (1974). Development of the epiphyllous inflorescence of *Phyllonoma integerrima* (Turcz.) Loes.: Implications for comparative morphology. *Bot. J. Linn. Soc. Lond.*, *69*, 1–13.

– (1975). Development of the epiphyllous inflorescence of *Helwingia japonica* (Helwingiaceae). *Amer. J. Bot.*, *62*, 962–73.

Gill, A. M., & Tomlinson, P. B. (1969). Studies on the growth of red mangrove (*Rhizophora mangle* L.). Habit and general morphology. *Biotropica*, *1*, 1–9.

Hallé, F. (1974). Architecture of trees in the rain forest of Morobe District, New Guinea. *Biotropica*, *6*, 43–50.

Hallé, F., & Martin, R. (1968). Etude de la croissance rythmique chez l'Hévéa (*Hevea brasiliensis* Müll.-Arg. Euphorbiacées-Crotonoïdées). *Adansonia* (Ser. 2), *8*, 475–503.

Hallé, F., & Oldeman, R. A. A. (1970). *Essai sur l'architecture et la dynamique de croissance des arbres tropicaux*. Paris: Masson.

Harms, H. (1894). Araliaceae. In *Die Natürlichen-Pflanzenfamilien*, ed. A. Engler & K. Prantl, vol. 3, pp. 1–62. Leipzig: Engelmann.

Jong, K., & Burtt, B. L. (1975). The evolution of morphological novelty exemplified in growth patterns of some Gesneriaceae. *New Phytol.*, *75*, 297–311.

Koriba, K. (1958). On the periodicity of tree-growth in the tropics, with reference to the mode of branching, the leaf-fall, and the formation of the resting bud. *Gard. Bull. Singapore*, *17*, 11–81.

Mabberley, D. J. (1974). Branching in pachycaul Senecios: The Durian theory and the evolution of angiospermous trees and herbs. *New Phytol.*, *73*, 967–75.

Philipson, W. R. (1963). Habit in relation to age in New Zealand trees. *J. Indian Bot. Soc.*, *42A*, 167–79.

– (1970). Constant and variable features of the Araliaceae. In *New Research in Plant Anatomy*, ed. N. K. B. Robson, D. F. Cutler, & M. Gregory, pp. 87–100. London: Academic Press.

284   W. R. Philipson

- (1972). Shoot differentiation in the Araliaceae of New Zealand. *J. Indian Bot. Soc.*, *50A*, 188–95.
- (1973). A revision of *Harmsiopanax* (Araliaceae). *Blumea*, *21*, 81–6.
Romberger, J. A. (1963). *Meristems, Growth, and Development in Woody Plants.* USDA Forest Services Tech. Bull., no. 1293, pp. 1–214.
Roux, J. (1968). Sur le comportement des axes aériens chez quelques plantes a rameaux végétatifs polymorphes: Le concept de rameau plagiotrope. *Ann. Sci. Nat. (Bot. Biol. Veg.)*, *Ser. 12, 9*, 109–255.
Sattler, R. (1966). Towards a more adequate approach to comparative morphology. *Phytomorphology*, *16*, 417–29.
- (1974). A new concept of the shoot of higher plants. *J. Theor. Biol.*, *47*, 367–82.
- (1975). Organverschiebungen und Heterotopien bei Blütenpflanzen. *Bot. Jb.*, *95*, 256–66.
Thompson, J. McLean (1952). A further contribution to our knowledge of cauliflorous plants with special reference to the cannon-ball tree (*Couroupita guianensis* Aubl.). *Proc. Linn. Soc. Lond.*, *163*, 233–50.
Tomlinson, P. B., & Gill, A. M. (1973). Growth habits of tropical trees: Some guiding principles. In *Tropical Forest Ecosystems in Africa and South America: A Comparative Review*, ed. B. J. Meggers, E. S. Ayensu, & W. D. Duckworth, pp. 129–43. Washington D.C.: Smithsonian Institution Press.

13
A quantitative study
of Terminalia branching

JACK B. FISHER

Fairchild Tropical Garden, Miami, Florida

Terminalia branching is distinctive of tropical trees and is represented by trees with horizontal, usually well-separated tiers of
branches borne on a single trunk. Each tier consists of series of leafy
rosettes at the end of erect short shoots, the rosettes well separated
on the horizontal branch complex, which branches sympodially at its
margin. In its most precise expression it has been categorized by
Hallé & Oldeman (1970) as Aubréville's model, which they recognized as occurring in at least 13 dicotyledonous families. A distal tendency to the same kind of sympodiality of lateral branch complexes
may be found in other trees, but belonging to other models (e.g., Attims's, Rauh's). Aubréville's model is, however, distinguished by
the immediate inception of sympodial branching of the *Terminalia*
type on the lateral axes produced by a rhythmically growing trunk.
The name comes from the pantropical genus *Terminalia* (Combretaceae), which has over 200 species of trees and shrubs, many of
which exhibit this form of branching. The most widely planted example is *T. catappa*, a native of the coastal regions of Southeast Asia.
Although this species is the familiar model for *Terminalia* branching,
there is little quantitative information available about it. The present
study was undertaken to give a better understanding of correlative
growth and form in a tropical tree and to serve as a descriptive foundation for experimental studies now in progress that will shed light
on the control mechanisms that maintain such a very regular and
repeating growth pattern.

Materials and methods

Seeds of *Terminalia catappa* L. were germinated in a greenhouse and grown in either full sun or partial shade for the first year.
Two-year-old seedlings were planted out along a north-south row in

a well-drained, sandy field with full sun exposure. Measurements of growth and morphology were made primarily on these 30 plants in their third year, after 1 year of establishment growth in the field. At the end of the third year's growth, the trees were 4–5 m tall with three to five branch whorls. Growth observations and measurements were done in the field on tagged branches and leaves numbered with ink. Lengths were taken to the nearest 0.5 cm once per week during the summer and fall seasons of active growth (July to November). Angular measurements of leaf and branch orientation were taken in the morning with a circular protractor weighted with a plumb. Many seedlings and established trees of unknown ages were also examined. Material for anatomic study was fixed in FAA, and sections were stained with safranin and chlorazol black E as modified from Robards & Purvis (1964). All data were taken from plants growing in Miami, Florida (lat. 25° 47′ N). In this region, there are a well-defined hot, rainy season (June to October) and a marked dry season (November to May), with January and February being the coolest months. Supplemental irrigation was used during the first dry season after seedlings were planted in the field.

General morphology and architecture
Seedling

Germination is epigeal, with early and rapid development of a taproot system. The first foliage leaves after the cotyledons are alternate. Phyllotaxy is a 3/8 right- or left-handed spiral. Every leaf, including both cotyledons, subtends a single bud, as is true of all subsequent leaves on the adult. The seedling axis may produce branches in the first year when in full sun. Seedlings in deep shade may grow for many years with little or no branching. The basic patterns of shoot branching and growth can be well established in the second year of seedling development.

Adult tree

The following descriptions are based on plants 3 years old or older in which the basic architecture is apparent. This distinctive growth habit was described in a general, but accurate, way by Raciborski (1901), Corner (1952), Hallé & Oldeman (1970), and Tomlinson & Gill (1973). They all recognized the presence of an erect (orthotropic) leader axis and distinct groups or tiers of horizontal (plagiotropic) lateral branches. These branches are composed of a series of discrete sympodial units.

Leader

This central axis is a continuation of, and similar to, the seedling axis. It is vertical, with alternate foliage leaves in a 5/13 spiral that is either left- or right-handed with equal frequency. There are long and short internode regions. In short internode regions, groups of lateral branches arise from the axils of three to five adjacent leaves. The details of internode length and growth of the leader, presented in Figs. 13.10 and 13.11, will be described in more detail later. The widely separated tiers of branches produce the distinctive "pagoda" branch pattern that has long been recognized (Troll, 1937; Corner, 1952; Chap. 25).

Branches

The horizontal branches are composed of a series of repeating sympodial units, including the first unit arising directly from the axillary bud of a foliage leaf on the leader axis. Stages of the development of these units are illustrated in Fig. 13.1.

The sympodial unit consists of one long internode, a pair of small, subopposite to alternate foliage leaves, and then a short segment of short internodes bearing a series of larger alternate foliage leaves. There are no scale leaves. The long first internode and the first few short internodes are essentially horizontal, but the distal internodes become relatively vertical. Therefore, the sympodial unit may be conveniently divided into two parts: the horizontal unit and the vertical unit (Fig. 13.1*A*).

The horizontal unit, the proximal part of the sympodial unit, consists of the long first internode and leaves 1–5. The axis begins to curve at leaf 3, and the genetic sequence of leaves is often obscured at maturity with an enlarged leaf 3 appearing distal to a reduced leaf 4 (Fig. 13.1*B*,*C*).

The vertical unit, the distal part of the sympodial unit, consists of leaf 6 and all succeeding leaves. All internodes are short and approximately three to four leaves are produced in the vertical unit during the initial growth flush of the sympodial unit. Therefore, the sympodial unit as a whole has an average of 8.3 mature foliage leaves (Table 13.3, old resting stage). The vertical unit produces more leaves after a rest period in second and later growth flushes. This axis is vertical, and 7–9 mature leaves are produced (average $= 9.0$ in the second flush of small trees, 7.4 in later flushes of old trees; $n = 25$). The phyllotaxy is 5/13 in old vertical units, and the foliar spiral can be followed proximally into the leaf scars of the first flush. However, in the original sympodial unit and in later flushes, no leaf is superimposed over another (the same orthostichy) as there are never 14 leaves present at any one time.

The average lengths of internodes and leaves of a sympodial unit are given in Fig. 13.12*A*. Because the vertical unit initially is not exactly vertical, the orientation of leaves on a sympodial unit with respect to gravity was determined. The angle of displacement of the

288   Jack B. Fisher

leaf axil from the upright was taken by viewing the axis from the apical bud. Therefore, this angle represents the orientation of the leaf insertion (and its axillary bud) on the sympodial unit axis with respect to gravity and does not correspond to the degree of bending of the axis. The distribution of orientations for leaves 1–7 is shown in Fig. 13.2. It is important to note that leaves 1 and 2 are almost horizontal, leaf 4 is always on the upper side, and leaves 3 and 5 are almost always on the lower side of the sympodial unit. This general arrangement can be seen in Fig. 13.1C.

New sympodial units mainly arise from the axillary buds of leaves

Fig. 13.1. Sympodial units at different stages of development, viewed from side. A. Sympodial unit bud emerging from axil of leaf 3 of preceding sympodial unit, mature leaves cut off. B. Sympodial unit near end of growth of long first internode, leaves 1 and 2 almost mature; a sympodial unit at earlier stage is arising from axil of leaf 5, mature leaves of preceding unit cut off. C. Same as B, but viewed from top. D. Two sympodial units at same stage near end of growth of long first internode, arising from leaves 3 and 5 of preceding unit in which inflorescences arise from axils of leaves 6 and 7. All leaves numbered in sequence; i, long first internode; infl, inflorescence; sb, sympodial unit bud.

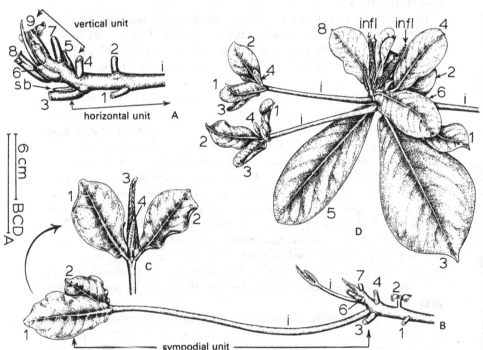

Fig. 13.2. Frequency of leaf orientation on sympodial unit with respect to leaf insertion. 180°, directly down; 90°, horizontal. Data for leaves 1–7 given. See text for procedure in determining leaf orientation.

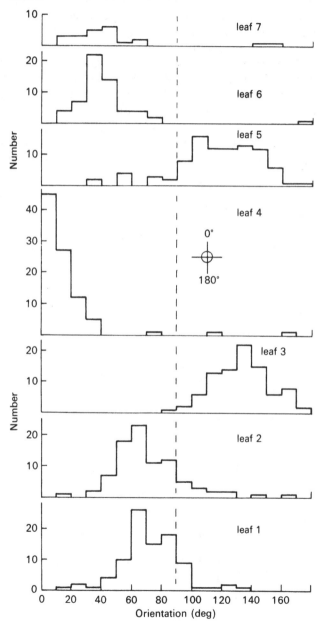

3 and 5 (Fig. 13.3), which means that most new units arise from the lower side of the preceding unit axis (Fig. 13.4). The newly initiated unit is densely covered by trichomes, as are all developing leaves and internodes. The long first internode elongates mostly from the distal region of the internode as indicated by displacement of ink marks on the growing internode.

The length of successive horizontal units, mainly consisting of the long first internode, tends to decrease irregularly along a branch (e.g., sympodial units 2–7 in Table 13.1*B*). In addition, sympodial unit 3 is usually long and dominant over unit 5 (Table 13.1*D*); further details are given under Branching Pattern.

During periods of flowering, newly developed sympodial and older vertical units produce racemose inflorescences from axillary buds (Fig. 13.1*D*). In old vertical units, inflorescences are produced

Fig. 13.3. Frequency of new sympodial unit and inflorescence positions with respect to sequence of leaf subtending them. Data based on 238 inflorescences and 238 sympodial units.

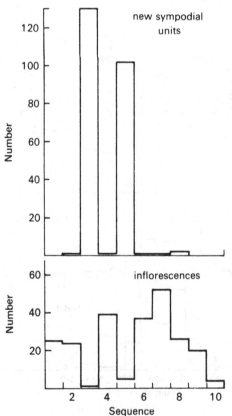

from many buds, all of which are borne on the vertical unit axis. In younger sympodia, inflorescences mainly arise from the axils of leaves 1, 2, 4, and 6–10 (Fig. 13.3). These represent buds on the upper side of the sympodial unit axis (Fig. 13.4). Inflorescences were never observed on the leader axis. The flowering period is not well defined and varies from tree to tree. Inflorescences may be present on all sympodial units of a tree while there are none present on other trees at the same time. Flowering has been observed sporadically from March (beginning of active growth season) through November (end of growth season) on different branches and individuals.

Branch orientation
     The branch complexes of a tree become progressively more horizontal. The first sympodial unit of a newly initiated branch forms an acute angle with the vertical leader axis. The relation of branch orientation, taken as the displacement from the vertical of the long internode of the first sympodial unit vs. the vertical distance of its insertion on the trunk from the leader apex, is presented in Fig. 13.5. The direction of gravity and not the leader axis determined the vertical in all data, and the angle was taken in the midregion of the long first internode. The data in Fig. 13.5 demonstrate that branches

Fig. 13.4. Frequency of new sympodial unit and inflorescence orientations with respect to angle of insertion of subtending leaf, all leaf positions grouped together for 126 inflorescences and 126 sympodial units.

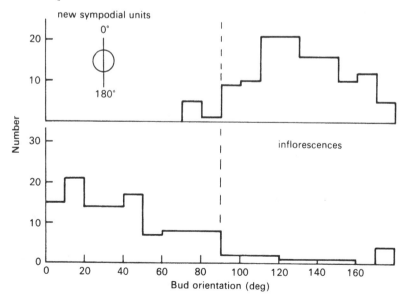

become approximately horizontal with increasing distance from the leader apex. Increasing distance also corresponds to increased age of branches. In Fig. 13.7, the changes in branch orientation after 4–5 months are plotted. The line indicates no change in branch angle. The smaller initial angles show the greatest change with time. A more complete description of changing branch orientation is presented in Fig. 13.6. The long internodes of successive sympodial units on one branch are plotted at bimonthly intervals. Internodes 1–8 constitute the main branch axis, and 1–3 make up a secondary branch. All parts of the branch become more horizontal, with the greatest change in the youngest distal parts.

The changing orientation may be due to three causes: increased branch weight related to branch growth, the active bending caused by reaction (tension) wood, or a combination of these two factors. Young horizontal branches were sectioned transversely in the mid-region of the long first internode. Chlorazol black E staining of the

Fig. 13.5. Branch orientation plotted against vertical distance of branch insertion from leader shoot apex. Angle of mid region of long first internode adjacent to leader axis with respect to gravity was measured for 56 branches.

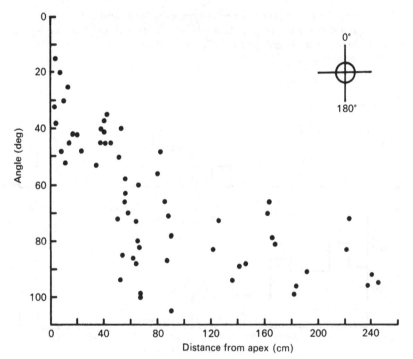

inner gelatinous walls of reaction fibers clearly located areas of reaction wood. The fibers tend to occur in small arcs on the lower side of the axis near the pith and in larger arcs on the upper side further from the pith. Therefore, reaction wood seems to be produced on the lower side of young branches and on the upper side of old branches. These initial observations indicate that early reaction wood forms on the lower side of nonhorizontal branches and on the upper side of horizontal branches. The observations tentatively support, but do not in themselves prove, the active role of reaction wood in pulling the branch down to the horizontal and then supporting the growing horizontal branch by tension on the upper side. Increased secondary growth may (Fig. 13.9F) or may not (Fig. 13.9E) be associated with tension wood on the upper side of old internodes (horizontal units).

The orientation of old branches shows some variation from tree to tree. In the relatively uniform group of young trees examined (Figs. 13.5–13.7), the branch orientation tended toward 90°. However, the orientation of all the old branches on some large trees was found to be about 60°. This indicates that there is a degree of genetic variation in the final orientation of branches among trees in Miami.

Fig. 13.6. Changes with time in orientation of same lateral branch with eight sympodial units and a secondary branch from it with three sympodial units. Angles of midregion of long first internodes with respect to gravity taken in July, September, and November. Internodes become more horizontal with time.

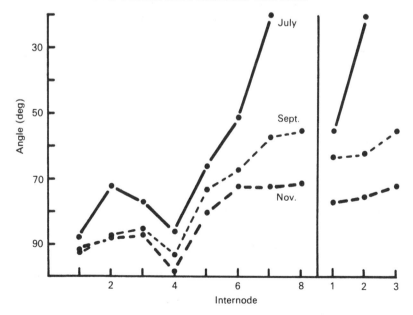

Branching pattern
Lateral branches are normally initiated from the axillary buds of three to five adjacent leaves of the leader axis. Further buds may grow out from distally adjacent nodes, but these typically abort. As the leader has a 5/13 phyllotaxy, the angle between successive branches is approximately 138.5°. This branch divergence angle applies only to the long first internode adjacent to the leader axis.

Each lateral branch subsequently subdivides or forks as a result of the outgrowth of new sympodial units from leaves 3 and 5 (= unit 3 and unit 5, respectively). The first fork is typically equal, but in later forks, unit 3 predominates and alternates in its direction within the plane of branching (i.e., left-right-left, etc.). This zigzag pattern is shown in Fig. 13.8B,E, where the diagrams indicate the sequence of units, and seems to be related to the phyllotactic spiral of the unit. In 100 forks starting with old unit 3, 88% of the new units 3 had the direction of the spiral opposite that of the old unit (12% were the same as the old unit). This nonrandom reversal of the phyllotactic spiral is presumably directly related to the pattern of units 3, but not the pattern of units 5. Only 43% of new units 5 had the reverse direction of the spiral of the old unit 3. Each sympodial unit may produce two new units (Fig. 13.8A) or it may only produce one unit 3 (Fig.

Fig. 13.7. Change with time in angle of midregion of long first internode adjacent to leader axis for 80 lateral branches. Initial angles are plotted against angles of same branches after 4 or 5 months. Line indicates equal initial and final angles.

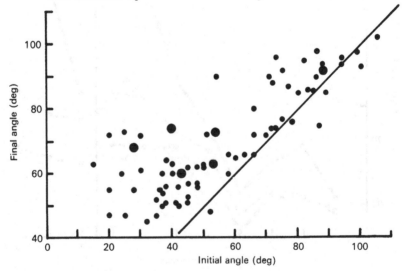

Fig. 13.8. Actual and model lateral branches. Fork angles and sympodial unit sequence indicated. *A*. Actual branch with most fork angles greater than the 62° average. *B*. Model of same branch (*A*) using equal fork angle of 62°; sympodial unit sequence of original shown; 3, new unit from leaf 3; 5, new unit from leaf 5. *C*. Model of same branch (*A*) using unequal fork angle as indicated in diagram. *D*. Actual branch with various fork angles. *E*. Model of right side of same branch (*D*) using unequal fork angles as indicated in diagrams; sympodial unit sequence of actual shown; *n*, sympodial unit from an old vertical unit, leaf number uncertain. *F*. Actual branch with most fork angles less than the 62° average. *A*, *D*, and *F* drawn from photographs of branches taken from below.

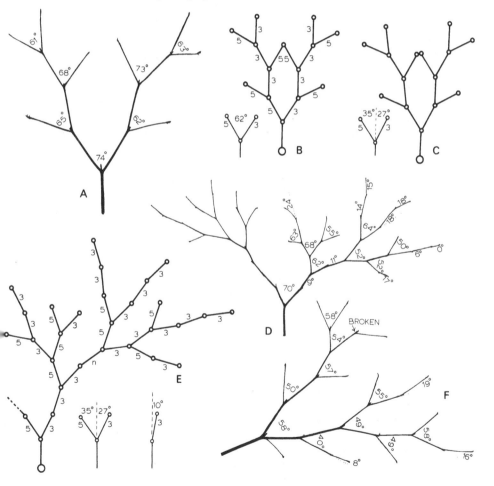

13.8E,F). A single new sympodial unit is especially common at the periphery of older branches.

The average fork angle in 100 randomly selected forks is 61.4° taken in the plane of the branch. In some branches, most fork angles tend to be larger than the average (Fig. 13.8A), and in others most angles tend to be smaller (Fig. 13.8F). The fork is not symmetric; in 50 forks, the average divergence angle of unit 3 from the axis of its old unit is 26.8° (see Fig. 13.8C,E). The average angle of divergence of 50 single new sympodial units from the axis of their old units is 9.8° (see Fig. 13.8E). Such single new sympodial units are always units 3 in actively growing branches.

Preliminary models were constructed using actual sympodial unit sequences and average angles of divergence in an effort to simulate the original branches. In these initial attempts, equal horizontal unit lengths were used, although the early formed sympodial units on a branch are longer than later formed ones, and unit 3 is usually longer than 5 (Table 13.1). The simulation could be greatly refined if the pattern of progressive sympodial unit shortening (Table 13.1) was understood. Equal fork angles of 62° produce the idealized model in Fig. 13.8B. Unequal fork angles, which are related to differences caused by unit sequence, produce the model in Fig. 13.8C, which is closer to the original in form. Similar unequal fork angles and the 10° average angle for single units produce the model in Fig. 13.8E, which closely approximates the form of the right side of the actual branch in Fig. 13.8D. These simple models can closely duplicate the actual

Table 13.1. *Variations in length of sympodial units from proximal to distal ends of single branch complex: averaged data for main axis of 30 lateral branches*

| | Sympodial unit sequence | | | | | | |
	1	2	3	4	5	6	7
A. Length[a] of unit 3[b] forming main axis of complex (cm)	40.8	35.2	35.0	31.9	30.2	27.5	25.4
B. Difference in lengths of old and new units 3 (cm)		5.7	0.2	3.1	1.7	2.7	2.1
C. Difference (B) as % old unit		13.9	0.5	9.0	5.2	9.1	7.7
D. Difference in lengths of units 3 and 5 at each fork (cm)		0.7	3.7	2.6	1.6	2.9	0.03
E. Difference (D) as % unit 3		2.0	10.7	8.0	5.2	10.4	0.1

[a]Length of horizontal unit only.
[b]Unit 3, sympodial unit subtended by leaf 3; unit 5, sympodial unit subtended by leaf 5.

branch, but they cannot yet predict branch structure. At the present stage of observation and description, there does not seem to be a predictable pattern in the occurrence of single new sympodial units within a branch system (vs. its frequent occurrence at the periphery of a branch system). Neither is there an obvious pattern in the site of change from equal forking (in the proximal region of a branch) to the dominance of unit 3 over unit 5 (Table 13.1E). The features of number and degree of dominance of new units at a fork may be related to environmental factors such as shading, to vigor of branch growth, to distance from the leader; or they may be simply random events.

The simplicity and repetitive sympodial pattern of branches would indicate that further modeling could well predict branching patterns in this species. Four main features of branch pattern must be included in any such model: (1) the angles of divergence of new sympodial units are related to the sequence and number of new units; (2) units 3 and 5 alternate in positions along the branch; (3) unit 3 dominates in an unequal fork and forms the main axis; (4) sympodial units (more precisely the horizontal units) decrease in length with distance from the leader, but in an uneven way, as indicated in Table 13.1C,E.

Old tree

The distinctive form of the adult tree is usually lost or obscured with increasing age. The single terminal leader is usually lost, with one or more new leaders arising from dormant buds on the older vertical leader axis. New leaders are also initiated from older vertical units of branches. Such an old vertical unit apex becomes active and behaves like the original leader. In trees with many leaders, the leaders tend to fan out away from the center, presumably owing to a shading effect, and lose their distinctly vertical orientation. Thus, the pagoda pattern of branch tiers is lost, and the crown becomes ever wider and more complex, with many leader axes.

Sympodial units still remain the basic growth unit in newly initiated branches on the leader axes and on the old branches. Old branches tend to initiate only single new units (from leaf 3) and thus become unforked at their growing ends (as in Figs. 13.8D and 13.9A). New units rarely arise at the tips of the oldest and lowest branches, which end in old but growing vertical units. After many years of growth, an old vertical unit may produce a new sympodial unit (Fig. 13.9B), complicating the repeating sympodial pattern of a branch. Most leaves in the crown arise from older vertical units. These continue to have periodic growth flushes. Secondary growth of the branches occurs predominantly in the horizontal units. The vertical units continue to elongate by the production of new internodes, but have little secondary growth compared with the horizontal units (cf.

298   *Jack B. Fisher*

Fig. 13.9. Old sympodial units in side view and transverse sections of their axes. *A*. Distal end of lower branch showing terminal unit in its second year of growth; leaves in both vertical units were produced in second year. *B*. Detail of old vertical unit that produced new sympodial unit in its initial growth from axil of leaf 3 and another new sympodial unit several years later; drawn in winter after all mature leaves abscised. *C*. Detail of old branch after many years of secondary growth; horizontal units (*hu*) have increased in diameter, and vertical units (*vu*) have increased in length with relatively little thickening. *D*. Transverse section of old vertical unit with relatively thin wood and large, irregular pith region due to leaf gaps. *E*. Transverse section of old horizontal unit with small, circular pith region and relatively thick wood showing slight indication of eccentricity; section oriented with upper side toward top of page. *F*. Transverse section of another old horizontal unit in which there is clearly increased amount of wood on upper side (toward top of page), which is associated with arcs of reaction wood fibers. Numbers, leaf sequence; *b*, bark; *i*, long first internode; *lg*, leaf gap; *p*, pith; *w*, wood.

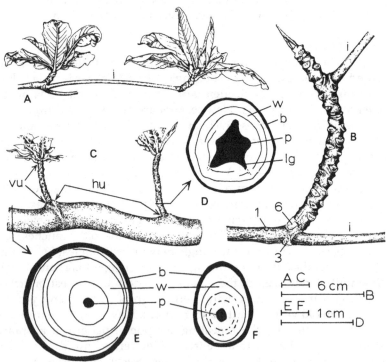

Fig. 13.9*D*,*E*). In old branches, the different thickness of horizontal and vertical units is striking (Fig. 13.9*C*). The secondary xylem of the horizontal units is often eccentric, with varying degrees of increased growth on the upper side of the axis (Fig. 13.9*F*).

Growth in the adult tree
Leader
The growth of the terminal bud is periodic, with alternating flushes of active growth and inactive or rest periods. The internodes of the leader axis vary greatly in length, and its leaves exhibit some variation in length, but less than internodes on a total length basis (Fig. 13.10*A*,*B*). There are usually one or two growth flushes (visible as cycles of internode length) per year, as in Fig. 13.10*B*. However, a few trees did not show a flush period during an entire year (Fig. 13.10*A*, previous year). The inactive period is usually one of slow growth, with few to no new leaves formed, or a period with no visible growth. During this stage, the leader resembles an old vertical unit. The longest leaves seem to be produced either before (Fig. 13.10*A*) or after (Fig. 13.10*B*) the longest internodes. The growth rate of one leader was taken every 2 weeks for 22 weeks (Figs. 13.10*D*, 13.11*A*) that included the end of one flush (approximately leaves 1–15), a short rest period (leaves 16–18), and a second flush (leaves 19–40) before a prolonged winter period of general inactivity, as in all trees. Most mature leaves remain attached for the entire growing season and then fall sporadically over many weeks in the winter. Leaves become bright red just before abscission, starting first with the older leaves. The biweekly growth rate is adjusted in Fig. 13.10*D* so that the growth rate at the time of leaf maturation is plotted against the leaf sequence. Therefore, Fig. 13.10*D* presents the growth rate of the leader at the end of leaf growth for that particular leaf. The most rapid period of shoot growth, leaves 18–29, occurs before the period of most rapid leaf expansion, leaves 29–33 (Fig. 13.10*C*). These leaves have a growth period of 4 weeks and precede branch positions. Growth rate on a time basis is presented in Fig. 13.11*A*. The production of lateral branches (the first sympodial units) on the leader occurs at the end of a flush, although some rare exceptions have been noted. The leaves at the end of a flush, in the region of shortening internodes, subtend the lateral branch buds. The length of these leaves may be average (Fig. 13.10*A*) or the longest of a flush (Fig. 13.10*B*). Thus, there are usually one or two whorls of lateral branches produced every year, although sometimes no lateral branches are initiated (Fig. 13.10*A*, previous year).

The fundamental relation between lateral branching and leader growth is complex. Branches grow out after their subtending leaves

Fig. 13.10. Length of successive leaves and internodes on leader axis of two trees, together with growth rate for one. A. Leader in which branches develop at end of present season's growth but none in previous season (leaf lengths unavailable). B. Leader in which branches develop twice within present season's growth at end of two growth flushes. C. Average period of leaf growth (period to nearest week from 0.5 cm to final length) of B. D. Growth rate of B plotted against those leaves that completed growth in 2-week period; incomplete rest is present when leaves 16 and 17 mature. Circle, developed branch; dotted circle, aborted branch bud.

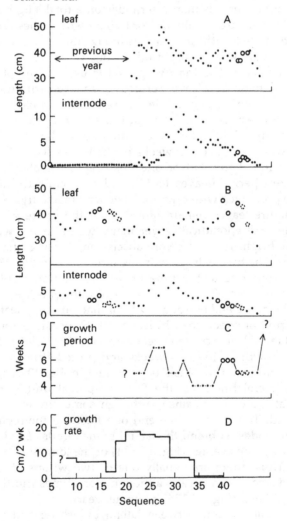

are half grown, so that branches are first obvious at the very end of the flush (cf. Fig. 13.10*B,D*). The buds subtended by leaves 35–37 were first noticeably enlarged at the end of week 17 (Fig. 13.11*A*). The first long internode of these lateral branches matured during weeks 20–22 (Fig. 13.11*A*) or 1–2 weeks after the leaves matured. However, these same leaves are first visible (≥1–2 cm) during the period of most active shoot growth (Fig. 13.11*A*). The leaves that subtend branches are first visible during the active period of leaf expansion as determined by the number of newly visible leaves (Fig. 13.11*B*) and the number of growing leaves (≥1 cm) in the terminal leader bud (Fig. 13.11*C*). As the leaf subtending the first branch is leaf 35 and there are about 19 growing leaves in an active leader (Table 13.2), leaf 35 was initiated at the shoot apex at about the time leaf 26 completed growth. This means that leaves initiated during the growth flush will subtend branches that develop at the end of the flush. The growth period of a foliage leaf (the time from 1 cm to maturity) in a flush

Fig. 13.11. Growth of same leader shown in Fig. 13.10*B* on bi-weekly basis. *A.* Growth rate of axis. *B.* Number of newly visible leaves (>1 cm). *C.* Total number of growing leaves present (>1 cm). Shaded region indicates appearance (in *A* and *B*) or growth (in *C*) of leaves 35–40, which subtend branches.

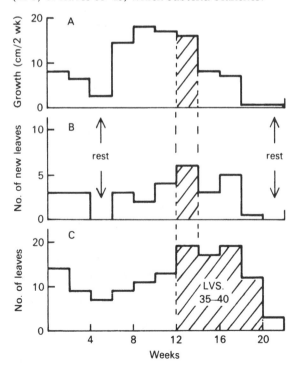

ranges from 4–7 weeks (Fig. 13.10C). Other trees had leaf growth periods averaging about 7–9 weeks during inactive growth and 3–4 weeks during rapid growth. In prolonged rests after branching (as with leaves after 42 in Fig. 13.10C), the growth period of a leaf may be 14–20 weeks or more during the summer and fall growing season. The period is greater when the terminal bud enters a rest period at the beginning of winter. During the winter rest period (January–February) when all or most mature foliage leaves have fallen, resting terminal buds are found in what appear to be various stages of flushing. Some buds have long adjacent proximal internodes; others have a series of short ones. Therefore, in winter, some leader buds are adjacent to the youngest whorl of lateral branches and others are far removed.

During prolonged and winter rests, young dormant leaf primordia may abscise, but most leaf primordia remain arrested in development. They are protected by a dense indumentum and include numerous tanniniferous cells. When growth resumes, the long-dormant leaf primordia resume growth and develop into foliage leaves. The first few leaves to expand in the spring may be distorted or injured and abscise early.

Potted plants grown in a warm greenhouse continue active leader growth throughout the winter months. Therefore, if day length plays a role in winter rest, it can be overcome by favorable growth conditions.

The number of leaf primordia in resting and active (flushing) leader buds are presented in Table 13.2. Although there are about 5 more growing leaves in the active leader, there are only 2 more young primordia (<2 cm) than in the resting bud. As approximately 25–40 leaves are produced in a flush, the apex must produce in one flush two to three times the number of growing leaves found at any one time. Therefore, it appears that the number of leaf primordia varies only slightly in the different phases of growth and rest. This

Table 13.2. *Number of developing leaves in terminal leader bud during periods of rest and active growth: averages of 10 apical buds*

| Type of leader | No. of immature leaf primordia | | | |
	Total	<2 cm	≥2 cm	>5 cm
Active	19.4	15.4	4.0	2.4
Resting	14.2	13.1	1.1	0.0

indicates a relatively close balance between leaf production at the apex and expansion of new leaves.

Branches

The first sympodial unit of a lateral branch appears to grow in the same way as later units. However, quantitative growth data were collected from distal units only.

The new sympodial unit grows actively for a limited time and produces about eight foliage leaves (Table 13.3) before becoming temporarily inactive. The growth period for each leaf (the time from 1 cm to maturity) ranges from 3 weeks (leaves 1 and 2) to 5 weeks (last leaves), as illustrated in Fig. 13.12*B*. The youngest leaf primordia remain immature for many weeks or months in a prolonged rest period similar to the inactive leader bud. The resting bud of the vertical unit may produce more leaves in a second distinct flush during the same growing season, remain at rest until the following year, or continue to produce several more leaves over an extended period (semirest) within the present season. The degree of bud activity determines the growth period of those leaves in Fig. 13.12*B* that have a question mark. Growth periods in older vertical units range from 4 weeks to 9 or 10 weeks within one flush. The bud becomes inactive after seven to nine foliage leaves are produced (not counting the occasional few initial deformed primordia). The old vertical units normally have only one flush per year. Most leaves remain on the tree for the entire growing season and fall unevenly in the winter season (January–February). There is considerable variation from tree to tree in the degree and timing of leaf abscission. When growth begins in

Table 13.3. *Number of mature and developing leaves per sympodial unit at different growth stages in late summer (October) and early spring (March): average of 20 units*

Stage of growth	No. of leaves		
	Mature	Developing (total)	Developing (<2 cm)
Late summer			
Young growing unit	1.25	13.7	10.6
Older growing unit	6.8	14.4	12.0
Old resting unit	8.3	11.4	11.2
Early spring			
Old unit, just starting to grow	0.0	16.2	13.0
Old unit, actively growing	3.5	20.3	13.9

Fig. 13.12. Length of successive leaves and internodes on new and old (second season) sympodial units, together with growth period of each leaf. *A*. Average leaf and internode length in 20 new and 15 old sympodial units. Broken lines, average of 10 units. *B*. Average period of leaf growth (period to nearest week from 0.5 cm to final length) in 13–26 new and 2–6 old units. Broken lines, averages of two units; arrow, rest period of uncertain duration.

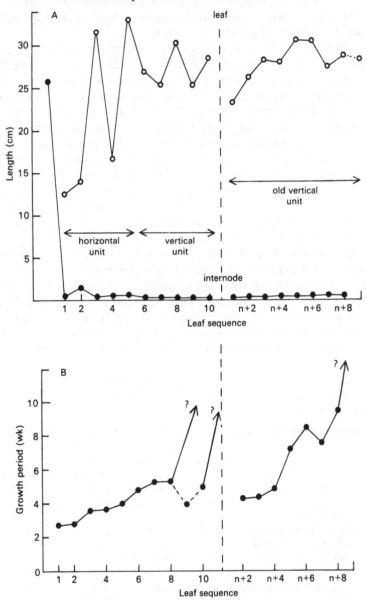

early March, the first two to four leaf primordia usually abort or are deformed. They are not true scale leaves, but rather show all the characteristics of developing foliage leaves. Such leaf abortion does not occur on those few vertical units in which the last-formed leaves of the previous year remain attached. New sympodial units are not initiated from the axils of the first few new leaves produced in the spring on the distal old vertical units.

The number of developing leaves in the summer on an inactive vertical unit bud is only about two less than the number in a very young sympodial unit bud (Table 13.3), although the former has seven more mature leaves than the latter. This indicates that the number of leaf primordia remains relatively constant. In the first growth flush of spring (early March), the number of leaf primordia <2 cm long within the buds of old units are similar both at the start of and during the flush (Table 13.3). Therefore, leaf initiation at the apex appears to be in approximate balance with leaf expansion during differing phases of sympodial growth.

A detailed study was made of the correlative growth of the long first internode, the leaves below and above its insertion, and the first three leaves of the new sympodial unit (Fig. 13.13). Complete growth data for two adjacent units were plotted on the same time axis as in Fig. 13.13 showing the lower (subtending) and upper (next older) leaves of the old unit, the growth of the long internode, and leaves 1–3 of the new unit. As data were taken once a week, the exact point of maturation (final length) was approximate. In an effort to have a more accurate reference point for growth comparisons, the point of 75% final length was determined on the curve. The 75% point is roughly within the rapid phase of the growth curve, and so the time of the 75% point determined by the straight-line approximation is more accurate than the timing of the point of growth cessation (100% final length). The differences in timing of the point of 75% final length for different leaves were taken relative to that point for the internode. In the example given in Fig. 13.13, the lower and upper leaves (*d* and *e*) mature 1.4 and 1.2 weeks before the long internode. The first two leaves of the new sympodium (*a* and *b*) mature at about the same time (−0.1 and 0.0) as the internode. Leaf 3 (*c*) matures about 1 week (0.9) after the internode. The average values for 16 such relative growth differences are given in Table 13.4. These data show clearly that the growth of the long internode is closely related to the pair of subopposite leaves at its distal end (leaves 1 and 2). The subtending leaf (lower) develops almost 2 weeks before the internode, and leaf 3 develops 1 week after it. The implications of these correlations in the control of growth are cited under Discussion.

The old vertical unit continues to grow for an indefinite period, with seasonal production of new leaves and short internodes. In rare cases, usually in old trees, the old vertical bud becomes active and new internodes elongate. The axis then assumes the growth and morphologic characteristics of the leader. Old vertical units may occasionally produce new sympodial units (Fig. 13.9B). However, the relation between the induction of new sympodial units on old vertical units and the growth of these vertical units was not examined.

Discussion

The present study fully supports the description of *Terminalia* branching given by Hallé & Oldemann (1970). Their architectural model was defined by the following features: periodic (rhythmic)

Fig. 13.13. Growth of long first internode of sympodial unit, its subtending leaf (lower), next leaf above it (upper), and first three leaves of new unit (1–3). Points of 75% final length are indicated. Relative differences in timing of these 75% points with respect to internode are given for *a–e* in this one sympodial unit. Thus, leaf 2 matures at almost same time as internode ($b = 0.0$), and leaf 3 matures about 1 week afterward ($c = 0.9$). Average values of 16 such units are given in Table 13.4.

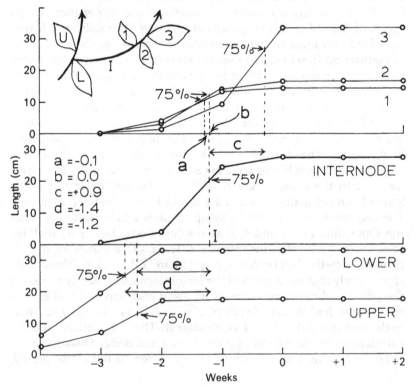

growth of the leader axis (trunk), a monopodial leader, plagiotropic branches composed of distinct sympodial units (*articles* or modules), and lateral inflorescences. These features of growth and form can now be related to each other quantitatively to define this architectural model more accurately and in a way that can be examined experimentally in the future.

Periodicity of growth and leaf production
Both the leader and branch sympodial units undergo repeated periods of flushing and inactive growth. There is usually one flush per growing season in each sympodial unit and often more than one in the leader. Although the initiation of a branch is related to leader growth, the initiation of new sympodial units on branches is independent of leader growth. Most branch growth occurs during the rest of the leader apex, thus overtopping the leader for a time, and seems unaffected by the leader flush. The growth of the leader and the older individual units may be termed "intermittent" (Koriba, 1958). In West Africa, where new leaves grow out before the old ones fall, *Terminalia catappa* was included in the category "periodic growth, leaf-exchanging" by Longman & Jeník (1974). However, when there is an interval between leaf fall and leaf growth, as in Miami, this same species should be categorized under "periodic growth, deciduous" (Longman & Jeník, 1974) similar to those species with leafless periods of many months (e.g., *Terminalia ivorensis*). The growth of young branches by continuous initiation of new units during the growing season may be termed "manifold" (Koriba, 1958). The crown as a whole grows throughout the season, but the individual sympodial units do not. The overall winter rest and leaf abscis-

Table 13.4. *Differences in growth periods with respect to long first internode of sympodial unit: averages of 16 sympodial units, all relative to 75% final length of internode and leaves*

	Relative difference[a]	
Leaf	Weeks	Days
a (leaf 1)	−0.08	− 0.5
b (leaf 2)	+0.13	+ 0.9
c (leaf 3)	+1.08	+ 7.5
d (lower leaf)	−1.74	−12.1
e (upper leaf)	−1.49	−10.4

[a]Procedure for determining a − e for one sympodial unit is shown in Fig. 13.13.

sion are superimposed upon the pattern of intermittent and asynchronous growth of the leader and unit apices as evidenced by the various stages of flushing and resting present in different meristems at the onset of winter rest. Therefore, although leaf fall is seasonal, shoot growth is best considered as only secondarily seasonal. Flowering occurs throughout the growing season. The winter rest period in Florida (January-February) occurs at a time of short day length and cool temperatures, and during part of the dry season. The last environmental factor seems least important, as the new foliage expands in early March although the dry season does not end until June. A similar but longer and more gradual winter deciduous period (May to October) was reported for this species in Rio de Janeiro, Brazil (Koriba, 1958). However, in Singapore and southern Malaya, *T. catappa* has two periods of leaf fall (January-February and July-August), which was associated with two dry periods (Corner, 1952; Koriba, 1958). In Ceylon, the same species is also deciduous twice a year (February-March and August-September) with a winter dry season and a summer *rainy* season (Wright, 1905). It would be useful to have phenologic data for this species in other geographic areas to improve understanding of such differences in periodicity. There are many examples in which the seasonality of a species differs in differing latitudes and climates (Alvim, 1964). Other species of *Terminalia* are deciduous during the dry season in West Africa (e.g., *T. superba* and *T. ivorensis:* Taylor, 1960), whereas *T. lucida* is deciduous during "heavy rains" (October-November) in Costa Rica (Allen, 1956). Seedlings tend to lack the periodicity of leaf fall of the adults (Coster, 1923; Longman & Jeník, 1974).

The rhythmic growth of leaders appears to be basically independent of season, although there is a tendency for synchrony of leaf production at the start of the growing season. On the other hand, initiation of lateral branches is directly related to the leader flush. The exact relation of branching to flushing is not obvious because branch buds and their subtending leaves are initiated during the flush and elongate at the end of the flush. Only future experimental work can clarify the causal relation between the leader flush and timing of branch inception and outgrowth. Presently, observations seem to support the theoretic interpretation of sylleptic branching offered by Tomlinson & Gill (1973). In *Terminalia*, the threshold value for bud release is reached in the leader during flushes, and the lateral buds develop precociously into sylleptic branches. Thereafter, every new lateral sympodial unit attains the threshold value during its early growth and initiates a new unit in a continuing sequence during the growing season. It should be noted that branch buds are initiated and expand after the maximum growth rates occur in the flush of the leader (Figs. 13.10D, 13.11A). In *Betula* and *Alnus*, Champagnat

(1954) showed a closer correlation between maximum shoot elongation and development of sylleptic branches (*rameaux anticipés*). However, compared with *Terminalia*, these temperate species have delayed branch growth with respect to the leader, and exact homologies may be inappropriate.

Leaf initiation and expansion are in close balance so that approximately the same number of small leaf primordia (those < 2 cm) are present in flushing and inactive buds of leaders and sympodial units (Tables 13.2, 13.3). Naturally, the total number of growing leaves is greater during the flush. No leaves are initiated during prolonged bud inactivity; the bud apex is truly inactive. The apex is most active in initiating leaves during the flush. A similar balance was observed in *Rhizophora* (Gill & Tomlinson, 1971), although this species lacks distinct flushes. This is in contrast to *Camellia* (Bond, 1942), whose rate of leaf production remains rather constant, resulting in a maximum number of primordia during, and a minimum number immediately after, a flush. In *Oreopanax* (Borchert, 1969), *Theobroma* (Greathouse, Laetsch, & Phinney, 1971), and *Hevea* (Hallé & Martin, 1968) leaf initiation does not occur between flushes, but it is unclear exactly when leaves are initiated during the flushing period and whether the number of primordia varies in the apical bud during the flush. In temperate species like *Fraxinus* (Gill, 1971) and in the short shoots of *Ginkgo* and *Acer* (Critchfield, 1970, 1971), leaf initiation occurs after expansion, and presumably there is a minimum number of primordia during the flush.

Although the resting leader bud and the vertical unit buds may look like short shoots because of their shortened internodes, they do not develop like short shoots of temperate species. In those species studied, leaves on short shoots always develop from primordia that have overwintered in the bud (Critchfield, 1970, 1971). Leaves on long shoots develop from the last-formed, overwintering primordia together with the precocious primordia formed during the same season. In *Terminalia*, the leaf primordia in a resting leader bud as well as newly initiated leaves contribute to the elongated portion of the stem. On the other hand, not all the primordia in the resting bud of a vertical unit expand in the flushing of that bud. The youngest primordia are, therefore, present in two successive resting buds before they finally expand.

Sympodial branching

The articulated nature of the lateral branches (composed of sympodial units) has long been recognized in *Terminalia* (references cited in Hallé & Oldeman, 1970). The sympodial unit with a long first internode and without basal bud scales is a sylleptic branch (Tomlinson & Gill, 1973). All other buds on both the leader and the older

310   Jack B. Fisher

units grow out as either inflorescences or proleptic branches that
have bud scales and a short first internode. As noted above, there is a
close relation between sylleptic branching and flushing, both in the
leader and in lateral branches. Sylleptic branching is continuous at
the periphery of a branch complex by the repeated outgrowth (flushing) of new sympodial units during the growing season. This induction of new units is interrupted by the winter rest and is not immediately resumed in the distal units of the previous season until well
into the new growing season. In addition, the degree of sylleptic
branching (number of new units) declines at the periphery of old
branches. These features indicate that a minimum vigor, the threshold value for bud release (Tomlinson & Gill, 1973), in the old unit is
necessary for the initiation of new units. This would explain why old
vertical units only rarely produce new units and why vigorous seedlings may produce branches without distinct flushes of the leader. In
the growing sympodial unit the first leaves of the flush (leaves 3 and
5) subtend the new sylleptic branch, and in the leader the last leaves
of the flush subtend the new sylleptic branches. However, both the
leader and the sympodial unit are similar in that only four or five
leaves expand within one flush after the most distal branch (Fig.
13.10, 13.12).

The sympodial unit predictably changes in behavior and morphology during its development, and two parts are recognized. The horizontal unit is plagiotropic, dorsiventral (anisophyllous leaves 1–5:
Figs. 13.1D, 13.12A), and exhibits gravimorphism (Fig. 13.4). The
vertical unit becomes orthotropic, radially symmetric (isophyllous
leaves 6 and greater: Fig. 13.12A), and lacks gravimorphism. The
horizontal unit thickens greatly, establishing the branch axis, while
the vertical unit functions as the leaf-bearing part of the branch.

The old vertical unit is similar to the inactive leader in having relatively short internodes. It can similarly (but rarely) produce new
sympodial units, and can itself become a new leader. However, only
buds of the vertical unit and upper side of the horizontal unit produce inflorescences, which never occur on the leader.

Experimental data have shown that the leader apex affects the vertical units, as the vertical unit becomes the leader when isolated by
bark girdles or removal of the leader apex (Hallé & Oldeman, 1970,
pp. 67–8). Conversely, the lateral branches also affect the growth of
the leader, as older branches reduce the growth of the leader in seedlings of T. ivorensis (Damptey & Longman, 1965; Longman & Jeník,
1974). The lack of response in Ocotea, which shows Terminalia
branching, to leaf pruning and decapitation was inconclusive (Howard, 1969).

The closely correlated growth of the long first internode with its supradjacent leaves (Table 13.4) indicates that the growth of this internode may be regulated by these leaves, as was shown for supradjacent leaves in *Helianthus* (Wetmore & Garrison, 1966). It may also indicate that growth of the long internode is influenced by the growing distal regions of the new unit and not by the proximal old unit. Obviously, an experimental approach is needed to verify such relations.

Branch orientation

The entire branch becomes nearly horizontal with age after its initial oblique orientation. This change occurs most obviously by the bending of the first proximal long internode. However, each horizontal unit only becomes horizontal with time, owing to sequential bending of the long internode (Fig. 13.6). In small trees the whorls of horizontal, dorsiventral branches give the tree its characteristic pagoda shape (Troll, 1937; Corner, 1952). The distribution of reaction fibers in the proximal internode indicates that reaction wood plays a role in establishing the mature angle, as it does in other woody species (Scurfield, 1973).

Branching patterns

The regular pattern of forking is determined by a number of factors: the number of new sympodial units at each fork; the usual dominance of unit 3; the alternating direction of the phyllotactic spiral in successive units, which results in the zigzag pattern of the main axis; the determination of fork angle by the positions of the new units; the greater length of unit 3 over 5; and the decreasing unit length and increased frequency of single units with distance from the leader axis. Unfortunately, these factors are not yet quantitatively predictable, although trends are clear. The simple models presented (Fig. 13.8) show that the regular pattern of repeating units in *Terminalia* can be reproduced with accuracy once the growth of sympodial units is better understood. The approach to computerized tree modeling given by Honda (1971) appears promising. His model takes into account most of the factors listed above, except for the alternating dominance of one branch unit (unit 3) and the increased frequency of single branch units at the periphery of the lateral branch. The major obstacle to a more accurate quantitative description of branch pattern is the ever-present environmental and genetic diversity in the population of branches studied. Branches sampled from a grafted clone of similar age might overcome this problem.

The branch system is highly efficient in presentation of leaves to vertical solar radiation. Branches are tiered and do not overlap much within a tier because of the 138.5° branch divergence angle. Within a branch, the majority of leaves are borne horizontally on the old vertical units. Because of the phyllotaxis (5/13) and the number of leaves per unit, no leaf is directly above another (i.e., in the same orthostichy). In addition, the lengths of the horizontal unit and the leaf are about the same. The large, downward-directed leaves 3 and 5 on new horizontal units occur at the growing ends of an old branch. They are basically at the periphery of the tree crown and unshaded for much of their functional lives.

Meristems and tree form

The young adult tree that fully expresses its architectural form was referred to as the "microclimatic" stage of the individual by Hallé & Oldeman (1970). At this stage, *Terminalia* has five types of shoot meristems: (1) leader apex, either flushing or inactive; (2) inhibited lateral buds on the leader, all vegetative; (3) growing apices of new sympodial units; (4) old vertical unit apices, either flushing or inactive; and (5) inhibited lateral buds on sympodial units, either vegetative or reproductive. Inhibited lateral buds may become active leaders (new proleptic branches) or inflorescences, but never new sympodial units (sylleptic branches). The vertical unit may become a leader. Only flushing leaders and growing horizontal units (new sympodial units) normally initiate sympodial units. In all cases, the behavior of these meristems is closely interrelated.

Considerable repetition of the basic architectural model (termed "reiteration" by Oldeman, 1974) in old trees in their "macroclimatic" stage (Hallé & Oldeman, 1970) is possible because of the diversity in meristem behavior. The shape of the old tree changes greatly and may superficially seem not to follow the architectural model of the young adult because of increasing crown size and the effects of environmental injury. Other forest species of *Terminalia* (e.g., *T. ivorensis, T. superba, T. amazonia, T. lucida,* and *T. pyrifolia*) have distinctive crown shapes and branching characteristics in aged individuals as described by Taylor (1960), Allen (1956), and Corner (1952). In these species and in *T. catappa,* the tiered nature of lateral branches is lost and the crown becomes enlarged and rounded or flat-topped by the production of many new leader axes from old lateral branches and the original leader axis. In effect, the trunk supports a crown that might be thought of as representing many young adult trees of various ages, thus reiterating the basic model for the species. Lower branches fall away, leaving a long, often fluted and buttressed trunk. The resulting tree form is altered, but the architectural principles of *Terminalia* branching remain the basis for tree growth.

Acknowledgments

I am grateful to Karen Fisher, James French, Stanley Harvey, and Janice Wassmer for help in collecting field data; to Dennis Stevenson for assisting with photography; and to Priscilla Fawcett for drawing Figs. 13.1 and 13.9.

General discussion

Givnish: The basic zigzag pattern of *Terminalia* is important because otherwise the plant would have its branches overlapping quite rapidly. However, if the plant shows absolute zigzagging (i.e., dichotomy with either axis aborting alternately), it should rapidly become like an octopus, with isolated strands running radially in various sectors but forming a very incomplete peripheral canopy. This suggests either that there should be some deviation from the exact pattern or that, if there is really very little deviation, *Terminalia* is not adapted very well to competition in certain environments. This latter may be part of the reason why *Terminalia* is restricted to open habitats.

Fisher: What you said about the importance of the zigzag pattern is true, but this is the complication in trying to predict branching patterns. In *Terminalia* you have an extreme regularity of distinct units, and it is a dorsiventral system. The entire system is in one plane, making it easy to devise models. The first few bifurcations are equal and there is no dominance of one branch over another. Toward the periphery of the canopy, a radial gradient of increased dominance becomes expressed, resulting in a zigzag effect. What you say is true in an adaptive sense, but the tree is not responding directly in a competitive sense. The tree is growing under internal control mechanisms, and I think it would be more useful to think of the underlying control mechanisms. Of course, those mechanisms are under selective pressures.

Whitmore: I was intrigued by the statement you made that the phyllotaxis alternates on successive clusters of leaves. Is it a common occurrence that adjacent pseudowhorls of leaves on a stem have reverse phyllotaxis?

Fisher: You are looking at the phyllotaxis of successive branches. It is simplest to think of the sympodial units as being made by an erect axis and consider the relation as that of a parent axis with successive daughter axes. In *Terminalia* there is a discordancy (i.e., a tendency for the branch to have a phyllotaxis opposite to that of the parent axis). This has been studied in quite a few herbaceous species (Gómez-Campo, 1970), but I do not know of work on woody species. Some herbaceous species are predominantly concordant or discordant (i.e., where the branches have the same phyllotactic spiral as the main axis or they are all opposite). In other plants there is a random effect with an equal distribution of phyllotactic directions.

Janzen: Are you aware of the work done by a statistician in India on the productivity of right- and left-handed palms?

Fisher: You are referring to some of the work done by Davis (1963) in which he looked at large samples of coconut palm trees on a plantation and, in fact, found a statistical correlation between right-handed spirals and left-handed spirals and productivity. The direction of the spiral was random (nongenetic), but the yield was not. In suckering palms and *Strelitzia*, which are basically sympodial with lateral branches growing from a main axis, the distribution is approximately random, with a 50-50 chance of change in phyllotactic spiral, so that phyllotaxis is not influenced by a parent axis (Fisher, 1976). In *Terminalia*, where there is approximately 80% change to an opposite direction of phyllotaxis, the new sympodial unit is being affected by the old sympodial unit.

Ashton: It is interesting to compare the crowns of very old trees in Miami and also the efficiency of the arrangement of leaves of the trees you are looking at with those of the same species growing in a more constantly humid climate. I have a strong impression that, in the younger trees of comparable size to the ones you are mainly working with, there is much more overlap of the leaves, the leaves are larger, and the overlap of the leaves between each shoot is much greater than you describe and illustrated.

Fisher: It would be very interesting to see if there is a correlation with climate.

Ashton: But, this poses a problem in old trees because, as we saw very well in the photographs, their horizontal branches, though more ascending than those of the earlier stages, are, nevertheless, much closer together. Consequently, it is difficult to visualize how so many leaves could be stacked together unless, of course, the leaves toward the trunk are dropped. The lower branches come to have no leaves on them at all and die. What I have observed is that when the trees develop their rounded old crown, the leaves on average get smaller, but the lengths of the leafless internodes do not become commensurately shorter, so that the clumps of leaves are more isolated and allow a more structured arrangement of leaves through the crown than you see in the young, tiered stages. In view of this change within one species, a condition rather typical of secondary forest – a flat arrangement of large leaves – to a more diffuse leaf arrangement in a hemispheric crown typical of mature phase forest canopy, is there a change in wood structure with age?

Fisher: I have not looked for changes in wood structure with age. The wood structure is diffuse, porous, and rather uniform, so it would be difficult to see any obvious structural changes without quantitative analysis. The vertical units of old branches within the crown also tend to abort quickly; the bigger ones slowly decline, producing fewer as well as smaller leaves per whorl. Finally they abort, leaving a long lateral branch with only the distal portion leafy. It might be simply a light-intensity effect. If one looks at seedlings or saplings growing under dense shade, no branches are produced. They have a typically unbranched appearance, and branching seems to occur only when light intensity increases. The same phenomenon occurs in the crown of the tree, as I illustrated, with many vertical shoots extending without branches in the shady interior part of the crown, but as they reached higher light intensity regions there was an induction of the lateral branches. As this is a coastal species naturally and grows in full sun, it does not normally compete for light.

Tomlinson: What variation occurs in the development of sylleptic shoots in the sympodial system? Do they always complete a particular cycle of growth or can they rest at any stage of development?

Fisher: During the growing season, unless they are injured by insects, they will always complete their growth. If a whorl of new branches is initiated at the end of the growing season and, therefore, an active shoot has winter inhibition and leaf drop imposed upon it, sympodial units are often only partially developed and normally abort during the winter. They appear to dry up and die back. This again indicates that the winter rest period is superimposed upon the natural rhythm of the plant. There is considerable abortion of actively growing sympodial units, but only at the beginning of the winter rest period.

Tomlinson: In relation to the way the plagiotropic complex itself branches, what are the factors that determine whether one rather than two branches grow out? Is this simply a matter of competition or is the process more positive?

Fisher: The process seems to be related to the general vigor of the plant. Normally at the early stages of sympodial initiation there are always two expanding buds in the axil of leaf 3 and leaf 5. So, initially there are two enlarging buds. At this early stage there is either dominance of the first-formed bud (i.e., that in the axil of leaf 3) and abortion of the second one (leaf 5), or they both grow out. Whether one or two buds grow out is determined relatively late, and well after bud induction. Thus competition seems to be important.

Tomlinson: If I understood you correctly, there appears to be a correlation between the vigor (i.e., the growth rate of the orthotropic axes) and the ability to produce a plagiotropic branch. Do you think the phenomenon is the same in the two kinds of axes, orthotropic and plagiotropic?

Fisher: I believe you are asking if there is a correlation between the vigor of growth and the production of either the sylleptic branches on orthotropic shoots or the sympodial units with long internodes on plagiotropic complexes. There seems to be good support in *Terminalia* for the suggestion that you and Dr. Gill (Tomlinson & Gill, 1973) made that sylleptic or precocious branching is related to a threshold value of the bud, which has to be exceeded. Although the new sympodial units seem to appear at the end of the flush, as I pointed out, I think they are actually initiated during the maximum growth period of the flush. Therefore, they would seem to appear at the time of the maximum vigor of the flush (i.e., that is when the branches are truly determined), but we see them growing out only toward the end of the flush. The same relation would apply to the lateral branches, where there is the occasional occurrence of an erect short shoot, the vertical unit, producing a new sympodial unit at a later stage in its existence. Here this seems to be correlated with an increased vigor of the vertical unit. Further support is provided by observations on seedlings, as sylleptic branching occurs in very rapidly growing seedlings, but without a correlation with growth periodicity of the parent axis. This indicates that the seedling exceeds a threshold value before rhythmic growth of the terminal bud is established.

Tomlinson: Do you know the circumstance under which the erect, vertical branch units can be transformed into a vigorous new model or new orthotropic axis?

Fisher: In other words showing reiteration, the vertical unit becoming the leader axis? All the existing observations by myself and those reported in the literature (reviewed in my chapter) indicate that it usually is related to loss of leader apical dominance. The same effect is produced if the main orthotropic leader is decapitated or if there is a bark girdle blocking external phloem transport. One vertical unit is transformed directly into a new leader.

Longman: We have shown similar responses in young seedlings of *Terminalia ivorensis*, which has a very similar growth habit to *Terminalia catappa*. The leading shoot continues to grow while it produces the branches, and Dr. Damptey and I found we could make it grow faster if we took the branches off. Thus there was definitely some "lateral dominance," some effect of the laterals inhibiting the growth of the leading shoot, the opposite of classic apical dominance. On the other hand, when we tried taking off leaves 1 and 2 to try to influence this extremely long and very rapidly growing first internode, the plagiotropic sympodial unit of *Terminalia*, we couldn't influence it at all (Damptey, 1964). I should also like to make two comparisons with temperate plants in relation to this question of sylleptic branches produced at the same time as the leader is growing most rapidly. Do you agree that in *Betula, Larix, Salix*, etc., and in temperate pome fruits, this is true of what are sometimes called "feather shoots"? The most extensive production of these feather shoots is in the middle of the growing season when the leader is growing most rapidly, and, therefore, we have to think not only of the usual idea of competition, but also of correlation between leader growth and branching.

Borchert: Dr. Champagnat has done extensive work on this phenomenon, which is documented in the *Encyclopedia of Plant Physiology* under the title of *"rameaux anticipés"* (Champagnat, 1965).

Hallé: Is it possible to unify three phenomena that have been described today? First, the slow activity of the apical meristem owing to the presence of phyllomorphic branches shown by Dr. Nozeran; second, the periodic rest of the leader in *Terminalia;* and third, apical parenchymatization in Apocynaceae as described by Mlle. Prévost. It seems to me that the three cases are the results of the same phenomenon, which is sylleptic development of the lateral branches. Can they be interpreted as three aspects of the same phenomenon?

Fisher: I think you could well say that. There are three different methods of inducing lateral buds to develop precociously: by abortion of the apex, by production of parenchyma instead of meristematic tissue, or by a reduction of meristematic activity as in *Terminalia*. It is also interesting to note that if lateral buds undergo a prolonged rest in *Terminalia* they can never be induced to grow out as plagiotropic branches, but always grow out as a new leader and later that new leader develops sylleptic branches. Once the bud has undergone a period of inhibition, it cannot be induced to grow out as a new sympodial unit; it has to pass into the orthotropic phase first.

Stein: Is this inhibition due to the fact that there has been further growth at the other end? What kind of inhibition are we talking about?

Fisher: Presumably the lateral bud inhibition on both the orthotropic leader and the orthotropic unit of the branch system is a form of apical dominance, because those lateral buds are immediately released when their terminal bud is decapitated, but they develop by prolepsis and they reproduce a similar (orthotropic) axis.

Janzen: Terminalia catappa is grown as an exotic in many parts of the tropics. It commonly establishes itself just behind the beach but does not penetrate feral communities. On islands (e.g., Puerto Rico) there are a number of places where it goes up the rivers and ravines, into the forest and acts like a *Cecropia* or balsa tree. In those circumstances, which I suspect resemble closely the habitat it occupies in southern India, where it probably is a native tree, big old adults do not have the life form you illustrated. Rather, they form a large-trunked, tall tree that maintains the pagoda life form well up into its crown. It looks more like a "normal" forest tree. The examples shown of old trees growing on a boulevard or in a fairly open situation differ from old trees growing in the forest. Consequently, what one accepts as the shape of an old tree depends on the environment.

Fisher: As Dr. Longman is more familiar with *Terminalia* as forest trees in

Africa he might answer that question. Do you see reiteration in large specimens of *T. superba* or *T. ivorensis?*

Longman: Yes, you certainly see reiteration with more than one reiterated model in big forest trees. You also find that *T. ivorensis* as saplings in plantations are very nonsynchronous in their development. One is showing leader growth while another is just stopping or completely dormant. Subsequently more synchrony can be found, with all individuals leafless together, for example. In natural or disturbed forest only one big set of branches is often to be seen on large specimens, with all the lower ones gone. There may sometimes be one-sided development, causing the tree to lean, so that it doesn't resemble *Terminalia* without close examination.

Oldeman: Terminalia amazonica is the same. In French Guiana it is a big forest tree with a large crown reaching 50–60 m, with a lot of reiterated models, each resembling the "shrub" Dr. Fisher illustrated, but in the crown of a big tree and originating from previously dormant meristems.

Fisher: So in the reiteration of something like *T. amazonica* you would still be able to distinguish separate leaders with tiers of branches in the crown?

Oldeman: Yes, but not in the biggest trees because reiteration is so prolific that each of the many models is miniaturized and most often imperfect.

Alvim: Dr. Fisher, do you know of any treatment that would convert an orthotropic into a plagiotropic branch, as you have shown in *Terminalia* and as is common in cacao? In cacao it seems to be related to the diameter of the plagiotropic branch because it occurs only when the branch is relatively thick. It also seems to be related to water supply because if you peg a branch to the ground many of the lateral buds coming from this plagiotropic branch become orthotropic. The origin of the orthotropic branch seems to be an axillary bud. Is there any causal study of this?

Nozeran: In every case of cuttings of *Theobroma cacao* the buds on plagiotropic branches gave rise to branches that remained plagiotropic. The only way vegetative orthotropic axes developed was by regeneration near the base of the tree. This differentiation is further accentuated because plagiotropic cuttings not only give plagiotropic axes but also exclusively plagiotropic roots. The orthotropic taproot appears only very much later.

Alvim: Very often in a plant about 20 years old, orthotropic shoots can develop from axillary buds of a plagiotropic axis laid horizontally on the ground. I have seen one example of two branches of the same age, from the same site, one plagiotropic and the other orthotropic.

Tomlinson: Does the terminal meristem of that plagiotropic branch ever become orthotropic?

Alvim: No.

Nozeran: In the leaf axils of the orthotropic relay axis there is a swollen basal cushion with generally two buds. The development of these buds has been studied. When they are isolated from the neighboring correlation complexes by pruning, they will produce orthotropic or plagiotropic or intermediate branches. The same thing can be shown at the base of the primary plagiotropic branches, very near to their point of attachment to the parent orthotropic axis.

Oldeman: In a closely related species such as *Theobroma speciosum*, which I studied in French Guiana, orthotropic shoots are more readily produced than in *T. cacao*, because plagiotropic differentiation is not as strict.

Nozeran: Dr. Alvim, is it an existing bud that produces orthotropic axes in *Theobroma?*

Alvim: I presume so; I have never seen an adventitious bud in *T. cacao.*
Nozeran: Dedifferentiation is often induced in callus. It is then necessary to
have extensive growth of the plagiotropic extremity of the branch so that the
callus or bud is removed from the orthotropic meristem, before dedifferentiation will occur to produce an orthotropic axis. This is not an experiment I
have verified personally.
Longman: We should consider here Dr. Jeník's reminder that the tropical
forest is big and diverse enough to give us many examples rather than just
one. We may describe the change in cocoa when it produces fan branches as
phase change, implying that the older parts are now mature and it is rather
difficult and rare to get "rejuvenation." But with many other trees if cuttings
(especially short ones) are taken from different parts of the stem (e.g., from
the main stem and from branches), their "rejuvenated" growth can be very
similar. For example, single node cuttings of *Terminalia ivorensis* and *Terminalia superba* grow vertically and appear to branch similarly whether they
are taken from the sylleptic or proleptic segment of the plant. The difference
in branch habit seen within the intact tree is, therefore, presumably imposed by correlation by the various growing parts, and not by phase change.
On the other hand, in *Triplochiton* (Sterculiaceae) you can get from the same
plant cuttings that will grow either erect or at an oblique angle, and this may
turn out to involve phase change.
Fisher: We have to establish whether plagiotropy is determined simply by
correlative growth or whether it is habitual. In *Araucaria,* as provided by M.
Veillon's description of New Caledonia species, lateral branches provide the
classic example of habitual plagiotropism in which there is a prolonged
"memory" of the meristem that maintains plagiotropic behavior (topophy-
sis), no matter what happens to the branch. The branch may be rooted but
will always maintain dorsiventral symmetry. With some of the herbaceous
species that show phase change (e.g., *Hedera helix,* in which juvenile and
adult forms have different geotropic responses), experiments have shown
that roots established close to the meristem can change phase (e.g., from orthotropy to plagiotropy). This response is apparently related to gibberellin
production by roots (Frydman & Wareing, 1974). Proximity of the source of
gibberellin to the apex can change the reaction of the meristem. I think one
has to be very cautious when dealing with isolated cuttings from trees,
because of the newly imposed effect of root proximity.
Longman: Yes, I agree, though I think this is what one is trying to do when
isolating a small portion of the tree. One is trying to ask if the changes that
have occurred are just temporary, due to internal competition or other circumstances within the shoot system, or are something rather permanent, as
in the example of *Hedera helix,* where it is difficult to reverse a response. It
can be done, as you say with gibberellin, and also with high temperature
and various other techniques, but it is not a common occurrence. On the
other hand, the change from juvenile to mature is more easily effected.
Janzen: In this discussion of descriptive morphology of trees, it is relevant to
consider functional aspects. One cannot describe teeth without considering
the question of what is being eaten.
Tomlinson: This will be discussed later in this symposium.

References

Allen, P. H. (1956). *The Rain Forests of Golfo Dulce.* Gainesville: University of Florida Press.

Alvim, P. de T. (1964). Tree growth periodicity in tropical climates. In *The Formation of Wood in Forest Trees,* ed. M. H. Zimmermann, pp. 479–95. New York: Academic Press.

Bond, T. E. T. (1942). Studies in the vegetative growth and anatomy of the tea plant (*Camellia thea* Link.) with special reference to the phloem. I. The flush shoot. *Ann. Bot. Lond. N.S., 6,* 607–30.

Borchert, R. (1969). Unusual shoot growth pattern in a tropical tree, *Oreopanax* (Araliaceae). *Amer. J. Bot., 56,* 1033–41.

Champagnat, P. (1954). Recherches sur les "rameaux anticipés" des végétaux ligneux. *Rev. Cytol. Biol. Végt., 15,* 1–53.

– (1965). Physiologie de la croissance et de l'inhibition des bourgeons: dominance apicale et phénomènes analogues. In *Encyclopedia of Plant Physiology,* ed. W. Ruhland, vol. XV/1, pp. 1106–64. Berlin: Springer.

Corner, E. J. H. (1952). *Wayside Trees of Malaya,* 2nd. ed. Singapore: Government Printer. 2 vols.

Coster, C. (1923). Lauberneuerung und andere periodische Lebensprozesse in dem trockenen Monsun-Gebiet Ost-Java's. *Ann. Jard. Bot. Buitenzorg, 33,* 117–89.

Critchfield, W. B. (1970). Shoot growth and heterophylly in *Ginkgo biloba. Bot. Gaz., 131,* 150–62.

– (1971). Shoot growth and heterophylly in *Acer. J. Arnold Arbor., 52,* 240–66.

Damptey, H. B. (1964). Studies of apical dominance in woody plants. M.Sc. thesis, University of Ghana.

Damptey, H. B., & Longman, K. A. (1965). Main stem and branch growth in *Terminalia ivorensis* A. Chev. *J. W. Afr. Sci. Assoc., 10*(1), 69.

Davis, T. A. (1963). The dependence of yield on asymmetry in coconut palms. *J. Genet., 58,* 186–215.

Fisher, J. B. (1976). Development of dichotomous branching and axillary buds in *Strelitzia* (Monocotyledonaea). *Can. J. Bot., 54,* 578–92.

Frydman, V. M., & Wareing, P. F. (1974). Phase change in *Hedera helix* L. III. The effects of gibberellins, abscisic acid and growth retardants on juvenile and adult ivy. *J. Exp. Bot., 25,* 420–9.

Gill, A. M. (1971). The formation, growth, and fate of buds of *Fraxinus americana* L. in central Massachusetts. *Harvard Forest Paper,* no. 20, pp. 1–16.

Gill, A. M., & Tomlinson, P. B. (1971). Studies on the growth of red mangrove (*Rhizophora mangle* L.). 3. Phenology of the shoot. *Biotropica, 3,* 109–24.

Gómez-Campo, C. (1970). The direction of the phyllotactic helix in axillary shoots of six plant species. *Bot. Gaz., 131,* 110–15.

Greathouse, D. C., Laetsch, W. M., & Phinney, B. O. (1971). The shoot-growth rhythm of a tropical tree, *Theobroma cacao. Amer. J. Bot., 58,* 281–6.

Hallé, F., & Martin, R. (1968). Etude de la croissance rythmique chez l'Hévéa (*Hevea brasiliensis* Müll.-Arg. Euphorbiacées-Crotonoïdées). *Adansonia* (Ser. 2), *8,* 475–503.

Hallé, F., & Oldeman, R. (1970). *Essai sur l'architecture et dynamique de croissance des arbres tropicaux.* Paris: Masson.

320   *Jack B. Fisher*

Honda, H. (1971). Description of the form of trees by the parameters of the tree-like body: Effects of the branching angle and the branch length on the shape of the tree-like body. *J. Theor. Biol.*, *31*, 331–8.

Howard, R. A. (1969). The ecology of an elfin forest in Puerto Rico. 8. Studies of stem growth and form and of leaf structure. *J. Arnold Arbor.*, *50*, 225–62.

Koriba, K. (1958). On the periodicity of tree-growth in the tropics. *Gard. Bull. Singapore*, *17*, 11–81.

Longman, K. A., & Jeník, J. (1974). *Tropical Forest and Its Environment*. London: Longmans.

Oldeman, R. A. A. (1974). L'architecture de la forêt guyanaise. *Mém. O.R.S.T.O.M.*, no. 73, 1–67.

Raciborski, M. (1901). Ueber die Verzweigung. *Ann. Jard. Bot. Buitenzorg*, *17*(2), 1–67.

Robards, A. W., & Purvis, M. J. (1964). Chlorazol black E as a stain for tension wood. *Stain Tech.*, *39*, 309–15.

Scurfield, G. (1973). Reaction wood: Its structure and function. *Science*, *179*, 645–55.

Taylor, C. J. (1960). *Synecology and Silviculture in Ghana*. Edinburgh: Nelson.

Tomlinson, P. B., & Gill, A. M. (1973). Growth habits of tropical trees: Some guiding principles. In *Tropical Forest Ecosystems in Africa and South America: A Comparative Review*, ed. B. J. Meggers, E. S. Ayensu, & W. D. Duckworth, pp. 129–43. Washington, D.C.: Smithsonian Institution Press.

Troll, W. (1937). *Vergleichende Morphologie der höheren Pflanzen*, Band 1:1. Berlin: Borntraeger.

Wetmore, R. H., & Garrison, R. (1966). The morphological ontogeny of the leafy shoot. In *Trends in Plant Morphogenesis*, ed. E. G. Cutter, pp. 187–99. London: Longmans.

Wright, H. (1905). Foliar periodicity of endemic and indigenous trees in Ceylon. *Ann. Roy. Bot. Gard. Peradeniya*, *2*, 415–517.

Part IV: Roots, leaves, and abscission

14

Roots and root systems in tropical trees: morphologic and ecologic aspects

JAN JENÍK

Institute of Botany, Czechoslovak Academy of Sciences,
Pruhonice, Czechoslovakia

In the period of Theophrastus (about 300 B.C.) several writings are known to deal with "rhizotomy," a branch of herbalism describing excavated underground organs of plants. Most of this activity referred to roots, rhizomes, and bulbs of medicinal herbs. Theophrastus himself, while describing the Persian Gulf, mentioned trees that were "all washed by the sea up to their middle and held by their roots like a polyp." This may be the first reference to tropical tree roots, presumably to those of *Rhizophora*.

After the general stagnancy of botany in the Middle Ages, new books on roots and atlases of roots were published in the eighteenth and nineteenth centuries. More detailed works on various aspects of root morphology and physiology (mainly on aerial roots in mangroves) appeared toward the end of the nineteenth century (e.g., Goebel, 1886; Jost, 1887; Janse, 1897). Examination of other aerial roots and so-called abnormal root forms continued, and the first excavations and experimental work were undertaken in Java by Coster (1932, 1933). Since then reliable observations in tropical Africa, Southeast Asia, and South America have gradually accumulated, particularly after 1950.

The objective of this chapter is to outline a comprehensive picture of forms and environmental relations in tropical tree roots. Earlier reviews of this subject (e.g., Richards, 1952; Kerfoot, 1963; Schnell, 1970–71) showed problems associated with this limitless task.

Although polymorphism of tropical roots has attracted many botanists in the past, available evidence and description are both inadequate and uneven. In certain species, root forms were examined very thoroughly, but for the majority of tropical trees the root system is entirely unknown. Mangroves were frequently investigated, and much has been published about the diversity of rain forest species that show stilt roots and buttresses. On the other hand, little is known about the roots in freshwater swamps and savanna woodlands. Little of the past work has treated the root system of the tree in its entirety, and the ambiguous term "root system" has been applied to limited parts and temporary stages of the complex and dynamic structure.

General concepts and terms

Much confusion derives from different interpretation of two basic terms: root and root system. In the following analysis we shall distinguish (1) a "root," represented by a single axis that was generated by one apical meristem; (2) a "root system," represented by the totality of roots within a single tree, regardless of their morphogenetic origin; and (3) a "root subsystem," represented by any defined part of the root system (e.g., a parent root with all laterals, a set of stilt roots, a stilted peg root, the totality of crown roots).

Most of the terms referring to various kinds and parts of the roots were initially defined narrowly on the basis of particular examples, but later became more generally applied. Ultimately, their original meaning was shifted and obscured. A good example is the term "mycorrhiza," which was originally coined by Frank (1885) to express the composite double organ of the tree and fungus. The same term was later used by physiologists in the sense of mycotrophy, and morphologists were forced to use the tautologic expression "mycorrhizal root" or invent new words like "mycoclena."

Obviously, complex biologic phenomena require adequate complex terms, but whenever possible overlapping and homonymous usage of analytic and synthetic words should be avoided.

Morphogenetic aspects

In describing the progressively developing feature of tropical tree roots, two terms appear useful: distal versus proximal root. The "distal roots," including the thin ultimate branches (i.e., terminal roots), are situated at the periphery of the branched system, display little or no secondary thickening, and perform fundamental physiologic functions: absorption of nutrients and synthesis of organic compounds. The "proximal roots" are thickened root axes that are situated near the base of the trunk and, in gymnospermous and dicotyledonous trees, show marked secondary thickening.

Any explanation of the branching process requires a distinction between parent and lateral root. The system of branching roots can be described in terms of "root of the first, second, etc. order." The terms "primary roots," "secondary roots," etc., often used in the same sense, should be avoided, as they are frequently applied to the site of origin of the root primordia. "Primary roots" are roots that developed at the root pole of the embryo and all those that subsequently arise from root primordia in an acropetal sequence. "Secondary roots" develop from primordia originating in various secondary tissues of both root and shoot and from the callus of injured organs.

"Adventitious roots" is another term for secondary roots with the last-mentioned meaning (see Esau, 1953). However, many foresters and botanists use the term adventitious root in a narrow sense for root organs developed on the above-ground tree trunk and limbs (see Köstler, Brückner & Bibelriether, 1968). If this definition is accepted, another term is needed for roots that develop in progressively differentiated roots in the soil. In tropical trees, terrestrial and aerial adventitious roots of both root and shoot origin may be similar, and their distinction is next to impossible.

Morphologic aspects

As a first approach to tropical tree roots one may distinguish between thick (coarse) and thin (fine capillary) roots. Coarse roots in a progressive stage of secondary thickening are called "skeletal roots," and their subsystem near the base of the tree is called the "root skeleton." The subsystem of capillary roots also is specially named in some languages (e.g., *"kornevaya motchka"* in Russian).

A tree root is normally a cylindric, more or less tapering body. Radial growth of horizontal roots is often enhanced on the upper side, creating flattened or tabular roots. The enlarged junction between surface root and the tree base is termed a "root spur" or a "buttress" if the transition region is tall and prominent. A well-developed vertical root in the center of the system is called a "taproot." Other major vertical roots are called "sinker roots" or "sinkers."

Adventitious roots arising on the lower portion of the bole and penetrating the ground at certain distance from the bole are termed "stilt roots" (also "prop roots" or "strut roots"). Flattened stilt roots form flying buttresses. Roots possessing a sharp-pointed apex are named "root spines." Roots developing a pronounced vertical loop represent "knee roots." Another distinct root form is the "peg root," which grows vertically upward from a horizontal root. Two kinds of peg roots can be distinguished: one with determinate longitudinal and radial growth (e.g., the simple, pencil-like roots of *Avicennia* spp.) and the other with continuous growth (e.g., the spindle-shaped or conical peg roots of *Sonneratia* spp.). A polyaxial peg root that generates positively geotropic laterals has been termed "stilted peg root" (Jeník, 1970a).

Epiphytic shrubs and trees such as *Ficus* spp. develop "strangling roots" that attach closely to the surface of the host tree, gradually anastomose, and create a continuous sheath. Figs are also known for their free-hanging roots, which in some species, develop into firm and anchored "column roots."

The polymorphism of tropical tree roots is also observed in the anatomic structure and growth capacity of distal rootlets: Thick polyarch roots possessing the potential for longitudinal and radial growth are termed "macrorrhizae"; thin diarch roots of determined longitudinal and radial growth are called "brachyrrhizae" (Jeník & Sen, 1964). In some tropical trees these two kinds of distal roots may not be clearly distinguished; species such as *Rhizophora racemosa* (see Attims & Cremers, 1967) and *Aeschynomene elaphroxylon* (Jeník & Kubíková, 1969) display marked differentiation.

Physiologic aspects

Roots of tropical trees, like those of temperate region trees, are often given names indicating various physiologic attributes. For example, tree roots have been described as breathing (or respiratory), feeding (or absorption), excretory, parasitic (or haustorial), symbiotic, mycotrophic, and strangling.

The terminology of breathing roots in the tropics is particularly vague. Most frequently they have been called "pneumatophores," which is a rather confusing term referring to a wide array of biologic organs in algae, vascular plants, insects, and even fish. For this reason, we have suggested a more specific equivalent: the "pneumorrhiza" (Jeník, 1970b; Longman & Jeník, 1974). Incidentally, de Granville (1974) used a rather similar term – pneumatorhiza – for a quite different structure in palms that had earlier been called a pneumathode by Yampolsky (1924).

From the point of view of their growth physiology, roots are often assumed to be negatively or positively geotropic – and sometimes aerotropic, phototropic, hydrotropic, or chemotropic – though seldom with any experimental backing. In addition to the above-mentioned distinction between macrorrhizae and microrrhizae, terminal roots of tropical trees can be divided as follows: (1) roots with determinate growth, (2) roots with indeterminate or continuous growth, (3) roots with rhythmic growth, and (4) roots with pseudoarticulate growth. The last occurs frequently when the root apex dies off and a lateral root assumes the role of the leading axis.

Ecologic aspects

More than any other organs of tropical trees roots occur in very diverse environments. Terrestrial, aerial, aquatic, and internal roots can sometimes be distinguished within a single tree. The presence of crown humus, small water tanks between opposite leaf bases, or disintegrating tissues of the tree can create special microhabitats. An individual root may pass through more than one kind of environment; e.g., stilt roots or pneumorrhizae that have successively either aerial and soil portions or vice versa. (An aerial root cannot be regarded automatically as a breathing root.)

The surface of the ground provides the boundary for making the obvious distinction between underground (or subterranean) and above-ground roots. Above-ground roots are not necessarily aerial roots, as some may grow inside soil pockets or decaying bark of the trunk. Roots that grow at the surface of the ground are called superficial roots, though this may be rather uncertain in the case of roots interlacing litter layers. Underground roots may be described as shallow or deep. Referring to the orientation in space, roots of tropical trees are vertical, oblique, or horizontal organs.

Examples of diversity

Attributes of tree roots referred to in the preceding paragraphs combine in various ways and change with diurnal, seasonal, and ontogenetic periods. The complex nature of roots is best expressed by enumeration of their "analytic" morphologic, morphogenetic, and physiologic features. Where supporting evidence is lacking, however, physiologic terms should be avoided and morphologic ones preferred. In making analyses emphasis should be given to roots and root subsystems that reflect the major phases of ontogeny and environmental relations of the tree species.

Stilt roots

Stilt roots develop on the bole as adventitious roots that pass through the air and root usually in the ground. Their aerial portion is either cylindric or compressed to form a flying buttress. In the ground the stilt roots branch freely, thus anchoring a tall tree in soft soil. In various species the morphology of stilt roots differs a great deal (Jeník, 1973). Their characteristics are affected by the height of their emergence on the trunk, by anatomic characters of their primary body, and (in gymnospermous and angiospermous trees) by the rate of secondary thickening. Unless injured by external factors, these roots seldom branch in the aerial zone, but in mature and senescent trees they tend to anastomose and even form root grafts, as in *Rhizophora*.

Though in a particular tree or even species the adaptive value of stilt roots may not be straightforwardly explained, their significance for the mechanical support of a tall and unstable tree in soft ground is evident. In addition, the "snowshoe effect" and "short circuit" in the nutrition of the shoot whose underground roots are frequently damaged by suffocation in poorly aerated flooded soil can be expected (Jeník, 1973). However, Kunkel's (1965) assumption that stilt roots lift the tree upward and prevent suppression by surrounding trees is unlikely.

Tabular roots and buttresses
Large, shallow, horizontal roots of trees such as *Piptadeniastrum africanum* or *Xylocarpus granatum* display irregular radial growth resulting in flattened skeletal roots that grade into the trunk as buttresses. (A few temperate species show similar, but much less spectacular, organs: e.g., *Quercus, Ulmus, Populus* spp.) Many attempts have been made to explain the occurrence of these organs (Senn, 1923; Richards, 1952; Chalk & Akpalu, 1963). It is generally agreed that buttresses provide mechanical stabilization, but their rarity outside the tropics is less easily explained. Smith (1972) suggested that in temperate regions the high surface/volume ratios of buttressed trunk would be selected against because of damaging temperature fluctuations.

Spine roots and root spines
Both monocotyledonous and dicotyledonous tropical trees develop sharp-pointed above-ground organs whose identity as roots was first recognized in palms and later in African dicotyledons (van Tieghem, 1905; Jeník & Harris, 1969). The latter authors studied spine roots in eight species, mainly in *Bridelia micrantha*. It appeared that the spine roots arise in close proximity with stilt roots and that some of them can resume their longitudinal growth and become stilt roots. A recent study of root spines in a palm is provided by McArthur & Steeves (1969).

Peg roots and stilted peg roots
Peg roots in mangrove and freshwater swamp species develop as laterals of shallow, horizontal roots. They are common both in dicotyledonous and in monocotyledonous trees (mainly palms). The peg roots are negatively geotropic and grow above the ground surface. Their morphology was first described by many authors in *Avicennia* spp., which develop pencil-like slender organs, and in *Sonneratia* spp., which can create conical organs up to 2 m in height (*S. caseolaris* in New Guinea: Percival & Womersley, 1975). In freshwater swamps peg roots were described by Ogura (1940a,b) and by Jeník (1970b, 1971c). The latter author newly described stilted peg roots in *Xylopia staudtii*, which appear to be a more advanced form of a pneumorrhiza (Jeník, 1970a). Simple peg roots have also been observed in

many species of palms, but only recently has de Granville (1974) published a detailed account of these organs. The anatomic structure of peg roots has been found rather varied, but in most cases ventilating tissues and various pneumathodes are markedly developed. Physiologic function has been experimentally examined by Ernould (1921), Chapman (1944), and Scholander, van Dam, & Scholander (1955), who confirmed the role of these organs in gaseous exchange.

Knee roots and root knees

In both mangrove and freshwater swamps some tree species develop aerial loops of roots protruding above the surface of the waterlogged soil. Several papers attempt to explain the growth physiology and function of these pneumorrhizae (Troll, 1930; MacCarthy, 1962; Oldeman, 1971) but further excavations and experiments are still needed. Generally, there are two ways in which knee roots form (see Longman & Jeník, 1974): (1) the apex of a shallow horizontal root alternately emerges above and returns below the surface of the ground, thus creating a series of loops ("serial knee roots"); (2) a shallow horizontal root produces a lateral that forms a single aerial loop and then grows downward in the soil ("lateral knee root"). Earlier descriptions automatically assumed the presence of the serial type, which can be safely confirmed only in *Bruguiera* spp. (see Gill & Tomlinson, 1975). Our excavations in African freshwater swamps suggest that *Ceriops tagal, Lumnitzera racemosa, Mitragyne ciliata,* and *Symphonia globulifera* generate only lateral knee roots; the same root form is identified by Oldeman (1971) in a species of *Eschweilera* in South America.

At the top of the aerial loop formed by the knee root, enhanced secondary growth frequently results in the formation of a thickened protuberance. This should be clearly distinguished from the root knee well known in the subtropical *Taxodium distichum* and developed also in the African *Bequaertiodendron magalismontanum*, in which the root apex never comes above the ground surface but the aerial organ develops by enhanced local activity of the cambium on the upper side of a horizontal skeletal root.

Free-hanging roots and columnar roots

Free-hanging roots are abundant among American Araceae, but in dicotyledonous trees this root form occurs mostly as a juvenile stage of the columnar roots. Though column roots were observed for many years, the functional, anatomic, and morphologic details of their morphogenesis were recognized only recently by Zimmermann, Wardrop, & Tomlinson (1968) in *Ficus benjamina*. Initially, slender free-hanging roots of this species become anchored and form secondary tissues in which tension wood is developed. This results in contraction. Ultimately, the aerial roots grow thick and make sizable props for the limbs of the crown. Vegetative extension of some *Ficus* spp., notably of *F. benghalensis*, can result in the apparent "groves" whose trunks are actually column roots.

Occasionally, free-hanging roots occur over tree trunks and branches in trees of the floodplains and misty mountains. Gill (1969) recorded many species of this kind in the elfin forest of Puerto Rico. Jeník (1973) observed short hanging roots in *Spondianthus preussii*, *Myrianthus serratus*, and a *Pachystela* sp. in West Africa. A transition between columnar roots and stilt roots can be seen in *Protomegabaria stapfiana*, where in swampy soils adventitious roots gradually develop on the leaning trunk and create efficient props.

Strangling roots

These organs develop in epiphytic shrubs and trees (*Ficus, Schefflera, Clusia, Posoqueria, Metrosideros* spp.) that start their life in the crown of a host tree, sending roots over the surface of the host trunk toward the soil. After the roots have reached the ground, the aerial parts thicken, branch and anastomose, and create a compact casing that may eventually kill the host tree. As suggested by Gill & Tomlinson (1975), competition for light and nutrients must play a role in the ultimate death of the supporting tree. Observation in Aburi Garden, West Africa, showed that *Ficus leprieuri* could kill a large host tree within 30 years. As seen in the Fairchild Tropical Garden, Florida, figs display a wide range of transitions among free-hanging, strangling, and columnar roots.

Symbiotic roots

There are many kinds of symbiotic roots and root tissues that develop as a result of interaction between the tree root and bacteria, algae, fungi, other vascular plants, and many groups of animals. As anticipated in ecologic descriptions, endomycorrhizae are very common in trees of both humid and dry tropics. But ectomycorrhizae are relatively rare organs in tropical rain forest; hitherto only members of the Caesalpiniaceae, Dipterocarpaceae, Myrtaceae, and Fagaceae have been reported as ectotrophs (see Meyer, 1973). The anatomy of tropical ectomycorrhizae shows many unusual structures, like the double mantle composed of different fungal tissues (Jeník & Mensah, 1967), but little physiologic research has been done. One recent discovery refers to tropical mycorrhizae: The fungus can pass nutrients from decaying litter straight to the tree without preceding mineralization by bacteria. Progress has been also achieved in the study of nodules in leguminous tropical trees (see Harris, Allen, & Allen, 1949) and of coralloid roots in cycads that harbor nitrogen-fixing blue-green algae (Nathanielsz & Staff, 1975).

Xylopodia
Many tropical trees growing in seasonally dry savanna woodlands develop large woody organs, the "xylopodia" (singular, xylopodium) or "lignotubers," situated in the soil. These arise by swelling of the root collar, by enhanced growth of the upper part of the taproot, and/or by action of contractile roots that pull down the basal part of the stem bearing adventitious roots and buds. Xylopodia have been described from Africa by Lebrun (1947) and from Brazil by Rawitscher (1948). In the savannas of West Africa prominent lignotubers of *Cochlospermum planchonii, Piliostigma thonningii, Terminalia avicennioides,* and *Burkea africana* have been excavated by Lawson, Jeník, & Armstrong-Mensah (1968).

Development of root systems
The outstanding root forms mentioned in the preceding section should not obscure the fact that the majority of roots in tropical trees develop as normal terrestrial organs comparable to those encountered in most of the temperate forest trees. The internal organization of root tissues and the architecture of terrestrial skeletal and capillary roots are much the same as in common European and North American hardwoods (Jeník, 1957; Lyford & Wilson, 1964). Many basic forms of aerial roots observed in the tropics, on the other hand, do in fact occur in the temperate zone. For example, in swampy woodlands of Central Europe tabular roots, stilt roots, and even peg roots (in *Salix pentandra*) can occasionally be observed. It is the greater frequency, number, and variability of these root types that makes tropical woody species remarkable. However, the spectacular aerial and superficial roots seem to have diverted botanists from studying the terrestrial organs of tropical trees so that data on entire root systems are seldom available.

Genotypic adaptations
Observations on widely distributed species provide enough evidence that there is an inherent tendency to create a particular root form, subsystem, and overall root system. This genetic potential is, however, not necessarily expressed in every instance. For example, certain aerial roots (e.g., peg roots, knee roots, or stilt roots) represent facultative organs that may not develop in all habitats. This is shown notably by species such as *Anthocleista nobilis* and *Xylopia staudtii,* which can grow both in well-drained and in waterlogged soils; the subsystem of their pneumorrhizae never occurs in mesic

conditions, but only on hydromorphic soils. In the mangrove *Laguncularia racemosa* peg roots develop only in sites with a particular regimen of fluctuating tidal water.

Genotypic adaptations are observed also in terrestrial roots (e.g., taproots and sinkers) that are developed by certain species only if the soil layering, moisture, and competition permit. The inherited features are always better reflected in young trees; at a greater age environmental stresses may prevail as decisive factors.

Phenotypic modifications

Extreme environmental factors often cause irregular growth of roots and distorted development of root systems. Hurricanes, fire, flooding, fungal diseases, insects, and browsing animals can all create phenotypic modifications. For example, injury in mangrove species produces irregular branching of aerial roots or even development of unexpected root forms; e.g., the knee roots in *Avicennia marina* (McCusker, 1970) and *Euterpe oleracea* (de Granville, 1974). If palatable, stilt roots of many rain forest trees may be heavily damaged and subsequently transformed into unusual organs. In the African emergent species *Tarrietia utilis*, branching of the flying buttresses is frequently caused by unknown browsing animals. Large horizontal roots running for great distances above the surface of the ground, which have been observed frequently in *Entandrophragma angolense*, also could be explained as a phenotypic modification in response to strong wind action (Jeník, 1971b).

Changes with age

During the life span of a particular tree, changes take place within individual roots as they undergo differentiation. In addition, partial subsystems develop in mutual relations, partly independently. Thus the whole root system undergoes profound changes and must be considered in terms of a series of developmental stages. However, most of the studies of terrestrial roots of tropical trees refer to the juvenile stage. Coster's pioneer work (1932, 1933) was carried out in young plantations, and excavations by many tropical foresters and agriculturists also examined young trees and seedlings. Many observations have been made on roots of tree crops, such as *Theobroma cacao* (van Himme, 1959; Dyanat-Nejad, 1970).

Two emergent African rain forest trees, *Chlorophora excelsa* and *Aucoumea klaineana*, have been examined with respect to the development of their root system in various stages of their growth (Mensah & Jeník, 1968; Leroy-Deval, 1974). In both species the young root system was dominated by a prominent taproot, but gradually, large, horizontal skeletal roots with deeply reaching sinkers were developed. Similar characters were found in two tree species of the South American rain forest by Förster (1970). Another detailed examination of development of the root system was conducted by Gill & Tomlinson (1971, 1977) in *Rhizophora mangle*. Subsystems of pneumorrhizae in *Xylopia staudtii*, *Laguncularia racemosa*, and *Anthocleista nobilis* were studied by Jeník (1970a,c, 1971c).

Layering of roots

The layering of the root biomass and distribution of roots in space are far more complicated in tropical forests than elsewhere (Longman & Jeník, 1974). Above-ground roots occur in various types of aerial subsystems in the tree layer. Crown roots develop mainly in the wet humus caught in clusters of leaves, in persisting leaf bases, and in branch forks. They can be particularly abundant in crowns of small pigmy trees (e.g., *Pycnocoma* and *Ouratea* spp.) in the understory of the rain forest. Many palms (e.g., *Raphia*, *Mauritia*, *Metroxylon*, and *Cryosophila* spp.) also produce crown roots within the persistent leaf bases. Free-hanging roots, columnar roots, and clasping roots occupy other layers of the above-ground space.

The ground surface in the rain forest can be regarded as the site of a layer of freely growing roots (see Gill, 1969; Leroy-Deval, 1974). Indeed, it is this superficial layer that led ecologists to generalize about the shallow-rootedness of tropical trees (Chevalier, 1916). Quantitative assessment of the distribution of roots within the soil appeared much later and showed that there is great variation according to species, soil, and type of forest. Taproots and sinker roots of many species reach a depth of 3–5 m in latosol and ochrosol; their layering can be compared to that of the temperate forest trees (see the excavations described by Mensah & Jeník, 1968; Förster, 1970; Lawson, Armstrong-Mensah, & Hall, 1970; Klinge, 1973a,b; Huttel, 1975). On the other hand, in hydromorphic soils with permanently waterlogged subsoil trees develop a large plate of superficial terrestrial roots.

In savanna woodlands, the layering of tree roots is strongly affected by competition from grasses and by the source of water. Van Donselaar-Ten Bokkel Huinink (1966) showed that in habitats receiving only rainwater, roots always spread in the surface soil layer, whereas in habitats with groundwater available the roots may grow down to 10 m (see also Maxwell, 1972). Roots of savanna grasses occupy the surface layer, and tree roots spread underneath those of grasses, at a depth of about 20–30 cm (Lawson, Jeník, & Armstrong-Mensah, 1968).

Accretion of mineral and organic soil in floodplains seems to stim-

334   Jan Jeník

ulate root modifications that can successfully grow toward the newly
created surface. *Mitragyne ciliata* was found to build up several layers
of superimposed knee roots (Jeník, 1967).

Classification

In the preceding sections we have used terms like "type" or
"kind," suggesting that at the individual root and root subsystem
level we could not avoid a simple classification. So far we have not attempted to classify the entire root system of tropical trees.

In Coster's papers (1932, 1933) the classification units are only
vaguely described. Wilkinson (1939) was strongly influenced by the
classification of root skeletons used by German foresters (Büsgen &
Münch, 1927) and employed categories like (1) taproot system, (2)
heartroot system, and (3) plate root system, which were later recognized also by Bibelriether (1966) and Köstler et al. (1968). Much earlier Büsgen (1905) recommended a division of the capillary root subsystem according to the size and density of the distal branches. In
this division (see also Kubíková, 1967) two categories were coined:
(1) extensive and (2) intensive root systems. Obviously, these categories refer more properly to the subsystem.

In a new approach to the classification of tropical tree root systems,
Krasilnikov (1968, 1970) developed an earlier classification of Cannon
(1949) based entirely on temperate plants. Cannon defined three
main types of plant root system: (1) a primary root system (derived
exclusively from branching of the seminal root), (2) an exclusively
adventitious root system, and (3) a primary-adventitious root system
(of mixed origin).

In tropical trees, the first category most probably does not exist at
all. Among dicotyledons, most species sooner or later develop terrestrial and aerial adventitious roots. Monocotyledons retain their seminal root only for a short time in the juvenile stage. Krasilnikov obviously used a narrower concept of adventitious roots than that
employed in this review, which allowed him, for example, to establish a type of "primary root system." However, both *Avicennia officinalis* and *Sonneratia acida*, included by him in this type, generate
their pneumorrhizae from adventitious primordia on old horizontal
roots, as well as from primary roots.

Tentative organization models

The preceding description of the diversity of roots and root
subsystems suggests that any attempt at classification of the entire
root system of tropical tree species is bound to be incomplete and
premature. Much broader knowledge is needed to cover the abundance of tropical tree roots, to incorporate the details of anatomic

structure, to recognize the main stages of their morphogenesis, and to illustrate the effect of the environment.

We find it impossible to classify the root system of tropical trees according to a few diagnostic features and to establish a hierarchic system of types. At the present stage it seems more adequate to describe a set of organization models that can serve as units for further comparison and observations. Though inspired by the models of tree architecture (Hallé & Oldeman, 1970), these organization models merely distinguish instances of root systems frequently observed in tropical forests.

The following criteria are used in the description of organization models: (1) capability of the root axes for secondary thickening, (2) general structure of the root skeleton in adult specimens, (3) relations of terrestrial and aerial root subsystems, (4) changes of roots during the life span of a tree, (5) occurrence of "abnormal" root forms assumed to represent genotypic adaptation, (6) occurrence of certain phenotypic modifications caused by the environment.

In the following survey, the limited space available has been used only for concise description and mainly diagrammatic illustration (Figs. 14.1, 14.2).

The models are named according to the species that are either common in the tropics or referred to in the literature. All described models are known to develop in a number of tropical species, but only one example is given.

Root systems with axes capable of secondary thickening

1. Model *Chlorophora excelsa* (Fig. 14.1A). Terrestrial root system, in the juvenile phase dominated by a taproot; in adult trees, shallow horizontal roots develop deeply rooted sinkers; terminal roots are transformed into endomycorrhizae (Mensah & Jeník, 1968).

2. Model *Cariniana pyriformis* (Fig. 14.1B). Underground root system dominated by long, thick, surface horizontal roots; oblique and vertical laterals display limited longitudinal and radial growth; small root spurs develop at the trunk base (Förster, 1970).

3. Model *Piptadeniastrum africanum* (Fig. 14.1C). Prevailingly terrestrial roots; flattened surface roots protrude above the ground surface (tabular roots); prominent buttresses join the obconical trunk with tabular roots (Jeník, in prep.).

4. Model *Xylocarpus mekongensis* (Fig. 14.1D). Large horizontal roots generate protuberances of localized secondary thickening to form root knees, similar to those in subtropical *Taxodium distichum* (Gill & Tomlinson, 1975).

5. Model *Uapaca guineensis* (Fig. 14.1E). Juvenile terrestrial root

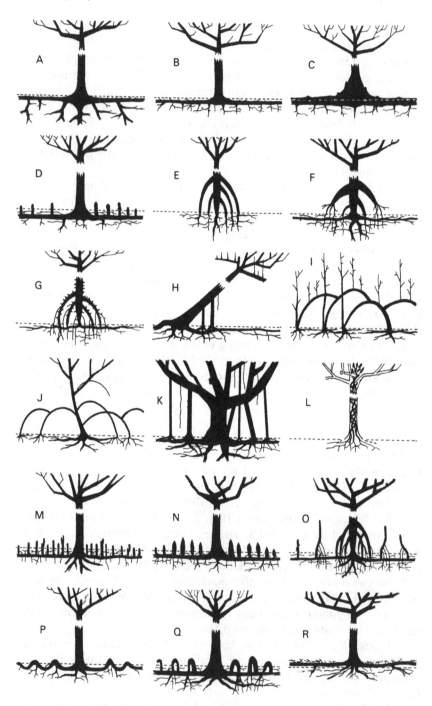

system is completed and successively substituted by prominent stilt roots arising on the trunk and anchoring in the ground; skeletal stilt roots are cylindric or taper toward the ground surface (Jeník, 1973).

6. Model *Tarrietia utilis* (Fig. 14.1*F*). Adult trees have underground root system completed by flattened flying buttresses with broomlike branching on the distal end; frequently injured by browsing animals (Jeník, 1973).

7. Model *Bridelia micrantha* (Fig. 14.1*G*). Subsystem of terrestrial roots is successively substituted by subsystems of adventitious roots developed from root spines; spine roots are richly represented on the surface of the trunk (Jeník & Harris, 1969).

8. Model *Protomegabaria stapfiana* (Fig. 14.1*H*). Mainly terrestrial root system develops in topsoil of swamps; leaning stems generate adventitious free-hanging roots, which become columnar roots propping the tree (Jeník, 1973).

9. Model *Scaphopetalum amoenum* (Fig. 14.1*I*). Thin trunks are curved down and terminal twigs develop adventitious roots; epicormic shoots arise on the top of the arch and later repeat rooting in the soil; large polycormic phanerophyte is formed (krummholz) (Jeník, 1969).

10. Model *Spondianthus preussii* (not illustrated). Usual development of underground root subsystems; later, subsystem of short, free-hanging aerial roots develops on twigs and the bole; fog and flooding enhance the formation of adventitious roots (Gill, 1969; Jeník, 1973).

11. Model *Rhizophora mangle* (Fig. 14.1*J*). Subsystem of short-lived primary terrestrial roots is present only in the early juvenile stage; later, adventitious roots emerge on the shoot and branches to create arching loops that root in the soil; ultimate capillary rootlets have strictly determinate radial growth (Attims & Cremers, 1967; Gill & Tomlinson, 1971, 1975, 1977).

12. Model *Ficus benjamina* (Fig. 14.1*K*). Branched roots of seedling develop deep terrestrial roots; free-hanging roots arise on limbs of crown and, upon reaching the ground, develop into columnar roots; each columnar root displays a deeply rooted subsystem of terrestrial roots (Zimmermann et. al., 1968).

Fig. 14.1. (*facing page*) Major organization models of root systems in dicotyledonous tropical trees. *A, Chlorophora excelsa; B, Cariniana pyriformis; C, Piptadeniastrum africanum; D, Xylocarpus mekongensis; E, Uapaca guineensis; F, Tarrietia utilis; G, Bridelia micrantha; H, Protomegabaria stapfiana; I, Scaphopetalum amoenum; J, Rhizophora mangle; K, Ficus benjamina; L, Ficus leprieuri; M, Avicennia germinanas; N, Sonneratia alba; O, Xylopia staudtii; P, Bruguiera gymnorrhiza; Q, Mitragyna stipulosa; R, Alstonia boonei.*

13. Model *Ficus leprieuri* (Fig. 14.1L). Epiphytic seedlings develop superficial roots on the host's trunk; strangling roots successively penetrate the soil distally and create proximally an anastomosed mantle over the host's bole; in addition, a subsystem of free-hanging and columnar roots may develop (Richards, 1952).

14. Model *Avicennia germinans* (Fig. 14.1M). Shallow underground skeleton has thin sinker roots; pencil-shaped peg roots arise on the upper part of the horizontal roots; radial growth of peg roots is limited (maximum, 1 cm thickness), not branching above the ground unless injured (Chapman, 1944).

15. Model *Sonneratia alba* (Fig. 14.1N). Polymorphic underground roots develop negatively geotropic peg roots; above the ground, peg roots slowly elongate and with secondary thickening result in formation of conical organs (Troll & Dragendorff, 1931).

16. Model *Laguncularia racemosa* (Fig. 14.3). Shallow terrestrial skeleton develops club-shaped underground and above-ground peg roots; terminal head of peg roots generates ventilating tissues and dwarfed ventilating rootlets, mainly at level of the ground surface (Jeník, 1970c).

17. Model *Xylopia staudtii* (Fig. 14.1O). Mature root system consists of terrestrial roots and richly branched stilt roots arising on the stem; in swampy habitats shallow horizontal roots generate stilted peg roots represented by aerotropic peg roots with positively geotropic laterals entering the soil (Jeník, 1970a).

Fig. 14.2. Major organization models of root systems of monocotyledonous tropical trees, lacking secondary thickening. *A. Elaeis guineensis; B, Mauritia flexuosa; C, Socratea exorrhiza; D, Cryosophila aculeata; E, Pandanus candelabrum.*

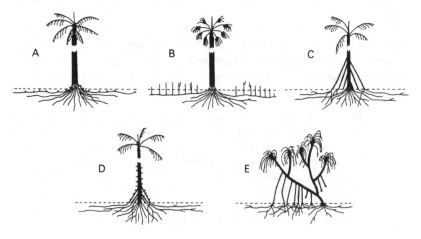

Fig. 14.3. *Laguncularia racemosa*, Miami, Florida. Root system exposed at low tide. *Top*. Root system largely exposed by tidal runoff at side of mosquito drainage channel. *Bottom*. Detail showing erect, club-shaped pneumatophores originating as branch roots from larger horizontal roots. (Photograph by J. B. Fisher)

18. Model *Bruguiera gymnorrhiza* (Fig. 14.1*P*). Terrestrial subsystem persists by efficient regeneration of adventitious branches on old root axes; surface horizontal roots develop serial knee roots with thickened aerial loops (Troll, 1930).

19. Model *Mitragyna stipulosa* (Fig. 14.1*Q*). Underground roots branch and spread down to the depth of waterlogged soil; shallow roots generate single lateral knee roots that protrude above the ground level and spread distally in the soil depth (MacCarthy, 1962).

20. Model *Alstonia boonei* (Fig. 14.1*R*). Very shallow root skeleton develops short sinkers penetrating into the subsoil; substantial part of both coarse and fine roots develops as superficial mat deposited above the ground surface (Jeník, in prep.).

Root systems with axes lacking secondary thickening

21. Model *Elaeis guineensis* (Fig. 14.2*A*). Adventitious roots radiate from the stem base, partly spreading underground, partly creeping in the topsoil; in waterlogged soil shallow roots generate pneumathodes as short laterals with aerating tissues (Yampolsky, 1924).

22. Model *Mauritia flexuosa* (Fig. 14.2*B*). Vertical and oblique terrestrial roots penetrate into the subsoil; shallow horizontal roots develop pencil-shaped peg roots with ventilating tissues (de Granville, 1974).

23. Model *Socratea exorrhiza* (Fig. 14.2*C*). Subsystem of terrestrial roots is completed, both in mesic and swampy habitats, by aerial roots; these roots develop as a cone of stilt roots surrounding the stem base (Jeník, 1973).

24. Model *Cryosophila aculeata* (Fig. 14.2*D*). Adult root system is represented by terrestrial roots radiating into both subsoil and surface horizon; stem bears numerous root spines and spine roots; near the ground aerial roots can develop into small stilt roots (Chandler, 1912; McArthur & Steeves, 1969).

25. Model *Pandanus candelabrum* (Fig. 14.2*E*). In adult trees the root system is dominated by aerial adventitious roots that arise at progressively higher levels; aerial roots develop into firm stilt roots and branch richly at the ground surface level (Gill & Tomlinson, 1975).

General discussion

Whitmore: There is an old story that circulates around the Asian tropics that most big trees which have taproots do not have buttresses and vice versa. I don't think there is very much evidence, but it is a statement well entrenched in the literature. Would you comment?

Jeník: It is true, at least in African emergent trees. Trees that develop prominent taproots and maintain them in the adult stage do not develop prominent buttresses. On the other hand, buttressed emergent trees have mostly
very shallow, horizontal roots with quite short sinkers.

Whitmore: So, it could be correct. One does occasionally see roots excavated
on the sides of logging roads. In the course of 10 years or so, I have seen
quite a lot of exceptions to this generalization, which made me question it.
This led me to think that maybe the generalization was made without much
evidence. For instance, I have a photograph (Fig. 14.4) of an *Intsia palembanica* on the side of the road up to Kedah Peak in Malaya. The tree had a
massive taproot and at the same time buttresses. I have also seen *Calophyllum* without buttresses and without a taproot.

Jeník: I think the East Asian knowledge derives mostly from work by Wilkinson (1939), where he used the classic classification of German foresters,
which recognizes three types: (1) the so-called flat or plate root system, (2)
the taproot system, and (3) the so-called heartroot system. Wilkinson added
two types, the one with buttresses that somehow excluded the possibility of
taproots, and there was a fifth.

Ashton: I wanted to question whether root systems can be classified into
some kind of upside-down architectural models. They would seem to me
from observation in the field to be too plastic, but I recognize the possibility
for the existence of very distinct models. A particular species seems to be
able to change its root pattern quite pronouncedly in different habitats. In
Southeast Asia, along ridgetops in submontane and mossy forests,
frequently one sees in the genus *Syzygium,* and indeed in genera of other
families, that the superficial slender roots grew upward, up the trunk, in

Fig. 14.4. *Intsia palembanica.* Root system exposed by road excava-
tion. (Photograph by T. C. Whitmore)

a net or weft, and *behind* these roots are the colonizing mosses. The mosses are not there first and then the roots; it is the other way around. I have always seen these roots to be very wet. I wonder if this has been seen elsewhere in mossy forests.

In the Dipterocarpaceae there is a basic pattern of plank buttresses that divide and end in a superficial root system. It is interesting to compare two soils, both of them deep so that the root system can develop well, one of which is a very porous infertile soil and the other a fertile but also freely draining soil. In the fertile soil one would find that the superficial root system is very sparse, but, arising vertically below the buttresses, there is a very well-developed system of "carrots" descending into the soil, perhaps for 3 m or more. On infertile soil it is exactly the other way around: The "carrots" don't exist, and there may be only one or two relatively fine descending roots, but a very dense superficial root system.

Jeník: I have not seen roots ascending living neighboring trees, but I have seen many places where roots grew over the stump of a decaying tree in the vicinity, especially under the decaying bark. As far as root plasticity is concerned, I agree it is probably not only waterlogging and drainage, but mostly also accessibility of nutrients, that determine the pattern of a root system.

Philipson: I wish to add a few observations to the question of the interrelation between buttresses and taproots. It was my experience in the Amazon region that buttressed trees have no significant root system immediately below the trunk. If you cut through the buttresses the tree was completely unattached to the ground. I can visualize a *Parkia* that had enormous buttresses but was sitting at ground level. No doubt it was connected from the buttresses into the soil, but there was no downward continuation of the trunk.

Jeník: I think the right explanation would be that we have to expect many possibilities, as we are dealing with an enormous variety of species.

Oldeman: If we look at the models of the aerial system in trees, we can say that the model develops a trunk with branches organized in a certain stereotyped way. When the tree grows up, one may get the phenomenon of reiteration, when other organ complexes, with the same structure are formed. In your slides of root systems, I think I saw in big trees what you call sinkers. I have often asked myself if there is, in a young tree, a more or less stereotyped root organization and subsequently subterranean reiteration. What are the criteria you use to describe different systems and do you think that these different systems can be reiterated?

Jeník: There is a pronounced tendency to create a substitute root system in very old trees, but not in the way you suggested: always in such a way that new adventitious roots arise from the old parts and substitute for them so that the old skeletal roots decay into a sort of rubbish. On the contrary, there is substitution by another generation of big roots that come from the base of the tree; so there really is a rejuvenation, but not in the sense you suggested.

As far as the criteria for establishing the models are concerned, the situation in roots is different. Persistence of primary root systems and occurrence of adventitious root systems are the most important (i.e., whether the original seminal root branches and persists or whether it is partially or completely substituted for by a secondary root system). Second in importance is the occurrence of some other particular root subsystem (e.g., an aerial system or an underground subsystem). In general, the main features are the ratio or representation of primary and secondary root systems to the occurrence of all specialized and genetically determined root subsystems. This follows the system established by Cannon in 1949, an American ecologist who wrote about the classification of roots, and later work partly by Krasilnikov (1968, 1970). These schemes were not particularly suited to the tropics, but I have adapted them to apply to the actual diversity of tropical roots.

Janzen: As roots and rotting tree trunks were mentioned, I would like to propose the following hypothesis (Janzen, 1976) and ask you if anyone has tested it. The hypothesis is that hollow cores in adult trees are adaptive to those adult trees in that the hollows are sites for nesting animals, microbial metabolism, fungal growth, and a variety of other kinds of activities that will generate nitrogen and other minerals as a form of fertilizer inside the base of the tree. Some of the roots from the same tree could get into this fertilizer, and thus the hollow core is a device for nutrient trapping by the adult tree. Therefore, one would then look at a hollow core in a tree not as a pathologic event, but rather as something the tree might be programmed to produce at an age when it has a large robust trunk.

Jeník: I do not know if the symbiotic phenomena you are referring to are obligate, but I should mention again the confusion that exists because mycorrhizae and nodulation of leguminous trees and proliferation with galls of insects are not regarded as separate phenomena. I am pretty sure that the symbiotic processes within the root system of tropical trees are extremely complex and possibly favor an improvement of the habitat. There are some distinctive mechanisms. There is a recently developed South American theory that the mycorrhizal fungi of tropical roots more or less directly pass nutrients from decaying material to the tree. This represents a new approach to symbiosis in tropical trees and mycorrhizae in general, but I don't know anything about the general idea that you have in mind.

Addicott: A related phenomenon has been known for some time. Developing roots will sometimes follow the channels left by the decay of roots of previous plants. C. H. Muller (1946) in his extensive studies of root development in guayule gave several examples.

Kira: Felling a number of trees in the Southeast Asian rain forest, I became aware that there were two types of buttresses or stilt roots. In most buttressed trees, the main trunk tended to taper downward until it disappeared entirely at the ground level; there was no round central core in the ground level cross section of buttresses (Fig. 14.5A). The same thing occurred in stilt roots; the main trunk almost vanished at ground level, support being provided only by stilt roots. In some cases, on the other hand, the main trunk remained even at the ground level, where a distinct disk was seen at the center of the cross section (Fig. 14.5B). In the former case, the buttresses or stilt roots began to develop apparently in a very early stage of growth when the tree was still very small, whereas in the latter case the abnormal growth began after a certain period of normal growth. Both cases could be found in the same species growing in the same forest stand. Do you have any idea about how these differences take place?

Jeník: Yes, that reminds me of a very interesting theory forwarded by Smith (1972), who published the first comprehensive evolutionary explanation for the significance of buttresses. The buttresses provide support and offer a rational solution to mechanical stabilization. One must ask why the same solution does not occur in the temperate forest. His explanation relates to the question of surface/volume ratio in connection with heat exchange. He

344   *Jan Jeník*

thinks that in the typical case the original cylinder is substituted for by an obconical axis with buttresses. However, this has quite a large surface as compared with the volume, and this is, of course, exceptionally sensitive to either frost or excessively high temperature. So, he says, there has been selection in the temperate region against this kind of otherwise rational solution for tree establishment. The same consideration applies to the existence of stilt roots in the tropics but not in temperate regions. So the question is related to ratios compared with the fluctuation of temperature. I think it's a good idea.

Givnish: There remains the classic explanation: Buttresses occur in areas that prevent trees from growing taproots deep enough to stabilize themselves on shallow soils. They are of particular mechanical advantage, with tall trees (Henwood, 1973). There *are* trees in the temperate zone with buttresses, and a few of these occur in areas where buttresses are expected because of a high water table. The most notable example is American elm, which develops small buttresses and is largely restricted, particularly in the Midwest, to low, swampy areas and ancient lake bottoms. Clearly birch twigs can withstand freezing quite well and yet they are considerably thinner than most buttresses. I don't think the surface/volume theory is tenable.

Fisher: In most temperate areas where root buttressing occurs there is danger neither of fire nor of severe freezing. Another of Smith's points is limitations imposed by the characteristics of the bark, which are indirectly related to surface/volume ratios; he felt that an important aspect is the anatomic nature (thin, green vs. thick, nongreen) of the bark in buttressed trees of both tropical and temperate climates. Up to now people have paid little attention to this topic. There are certain physiologic and physical properties of thin barks that allow root buttressing, whereas the same properties in temperate regions may have a negative selective value. I believe he would counter your examples by saying that the danger of damage by fire or freezing, which is related to surface/volume, is eliminated in certain temperate regions.

Jeník: The existence of buttresses in temperate regions is perfectly true. In Europe, poplars, *Ulmus,* and certain *Quercus* species do develop slight buttresses. Clearly the situation is not directly comparable in contrasted latitudes; in the tropics the evolutionary situation is much more complicated.

Del Tredici: Dr. Jeník, you made no mention of the ability of some roots to

Fig. 14.5. Cross section of two types of root buttresses in tropical trees. See text for explanation.

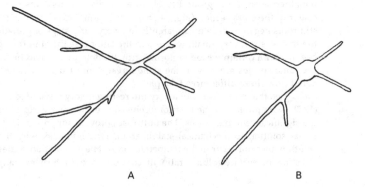

A   B

sucker, in particular adventitiously. In terms of classification of roots this would appear to be an important criterion to consider, especially as a suckering root system would have to remain close to the surface.

Jeník: This process definitely comes into my models. I should note that there aren't many suckering rain forest trees, but savanna trees frequently show it, especially in relation to this phenomenon of xylopodia.

Hartshorn: Dr. Jeník, do you have some ideas about the selective pressures that lead to the adaptation of these very interesting and complex stilted peg roots that you showed us?

Jeník: Yes, I think they represent the most advanced type of peg root. They more or less fulfill in the most economical way the purpose of breathing roots, if I may speak teleologically. In *Xylopia staudtii* the peg root itself is no more than 6 mm thick and can be as much as 1.5–2 m high. So it is an extremely slender structure penetrating above the thick surface litter, and it can be maintained in an erect position only by means of these stilts. It is a very well-developed kind of pneumorrhiza.

Whitmore: Dr. Jeník, in relation to the question of correlation between model and function discussed earlier, there is the interesting observation by Nye & Greenland (1960) and by several workers in northern Thailand (Land Development Department, 1970) that there are changes in the amounts of minerals at different soil depths at different stages of the shifting cultivation cycle. It appears that those few roots which penetrate to depth bring nutrients up. Nye & Greenland pointed out that even a small transfer will have a considerable effect over long periods and that this is an important counteractant to leaching. Is this ability part of the syndrome of characters of pioneer species which man is using in the bush fallow phase of shifting cultivation? Does Dr. Jeník know whether pioneer species as a class do have deep roots, as I have no evidence from Southeast Asia, apart from the Thailand studies?

Jeník: I don't know. Consider that the species of disturbed sites would be perhaps *Harungana madagascariensis, Bridelia* spp., *Macaranga* spp. I cannot answer. *Musanga* I know has a very characteristic stilt root system that is fairly shallow.

Ashton: Just a little addition to Dr. Whitmore's point. Many secondary forest trees of the group of models that could be called "pagoda trees" have persistent and rather prominent taproots. I have noticed in particular in Southeast Asia *Alstonia angustiloba, Endospermum malaccense,* and some *Elaeocarpus* spp. (Aubréville's model).

Hallé: On the other hand, I saw recently a species of *Elaeocarpus* with stilt roots.

Janzen: Why do root crowns remain more or less centered on the trunk itself ? For what reason should the roots stay in the region of the adult tree?

Tomlinson: Perhaps Walter Lyford can answer that question.

Lyford: There are many trees in New England whose roots go out 25 m, but I can't really answer the question.

Borchert: Normally roots do extend beyond the diameter of the crown, but there is a limit to root extension, because roots have to be supplied with assimilates by the crown, and a certain functional balance has to be maintained.

Janzen: By the same token there shouldn't be any vines 350 m long, and there are such examples.

Oldeman: The root growing out from the tree has to maintain transport capacity by producing wood because only the youngest is functional, the older wood dies. If you extended the root enormously in a distal direction, it

would result in a wide root at the foot of the tree. As younger wood is
needed to replace the older wood that dies off, this cannot go on indefinitely.
Janzen: Why doesn't it keep the old wood in the root interior alive?
Oldeman: Because every tissue is mortal.

References

Attims, Y., & Cremers, G. (1967). Les racines capillaires des palétuviers dans
  une mangrove de Côte d'Ivoire. *Adansonia (Ser. 2), 7,* 547–51.
Bibelriether, H. (1966). Die Bewurzelung einiger Baumarten in Abhängig-
  keit von Bodeneigenschaften. *Allg. Forst., 47,* 3–7.
Büsgen, M. (1905). Studien über die Wurzelsysteme einiger dikotyler Holz-
  pflanzen. *Flora, Regensburg, 95,* 58–94.
Büsgen, M., & Münch, E. (1927). *Bau und Leben unserer Waldbäume.* Jena:
  Fischer.
Cannon, W. A. (1949). A tentative classification of root systems, *Ecology, 30,*
  542–8.
Chalk, L., & Akpalu, J. D. (1963). Possible relation between the anatomy of
  the wood and buttressing. *Commonwealth For. Rev. Lond., 42,* 53–8.
Chandler, B. (1912). Aerial roots of *Acanthorhiza aculeata. Trans. Proc. Bot.
  Soc. Edinb., 24,* 20–4.
Chapman, V. J. (1944). 1939 Cambridge University Expedition to Jamaica. 3.
  The morphology of *Avicennia nitida* Jacq. and the function of its pneuma-
  tophores. *J. Linn. Soc. Bot. Lond., 52,* 487–533.
Chevalier, A. (1916). *La Forêt et les bois du Gabon.* Paris: Challamel.
Coster, C. (1932). Wortelstudiën in de Tropen. I. De jeugdontwikkeling van
  het wortelstelsel van een zeventigtal boomen en groenbemesters. *Tectona,
  Buitenzorg, 25,* 828–72.
– (1933). Wortelstudiën in de Tropen. III. De zuurstofbehoefte van het wortelstelsel. *Tectona, Buitenzorg, 26,* 450–97.
van Donselaar-Ten Bokkel Huinink, W. A. E. (1966). Structure, root systems
  and periodicity of savanna plants and vegetations in Northern Surinam.
  Wentia, Amsterdam, 17, 1–162.
Dyanat-Nejad, H. (1970). Controle de la plagiotropie des racines latérales
  chez *Theobroma cacao* L. *Bull. Soc. Bot. Fr. Mém. Paris, 117,* 183–92.
Ernould, M. (1921). Recherches anatomiques et physiologiques sur les
  racines respiratoires. *Mém. Acad. Roy. Belg. Cl. Sci. (Ser. 2), 6,* 1–52.
Esau, K. (1953). *Plant Anatomy.* New York: Wiley.
Förster, M. (1970). Einige Beobachtungen zur Ausbildung des Wurzelsys-
  tems tropischer Waldbäume. *Allg. Forst. Jagat. Ztg., 141,* 185–8.
Frank, A. B. (1885). Ueber die auf Wurzelsymbiose beruhende Ernährung
  gewisser Bäume unterirdische Pilze. *Ber. Dtsch. Bot. Ges., 3,* 128–45.
Gill, A. M. (1969). The ecology of an elfin forest in Puerto Rico. 6. Aerial
  roots. *J. Arnold Arbor., 50,* 197–209.
Gill, A. M., & Tomlinson, P. B. (1971). Studies on the growth of red mangrove *(Rhizophora mangle* L.). 2. Growth and differentiation of aerial roots.
  Biotropica, 3, 63–77.
– (1975). Aerial roots: An array of forms and functions. In *The Development
  and Function of Roots,* ed. J. G. Torrey & D. T. Clarkson, pp. 237–60. New
  York: Academic Press.

– (1977). Studies on the growth of red mangrove (*Rhizophora mangle* L.). 4. The adult root system. *Biotropica, 9,* 145–55.

Goebel, K. (1886). Ueber die Luftwurzeln von *Sonneratia. Ber. Dtsch. Bot. Ges., 4,* 249–55.

de Granville, J. J. (1974). Aperçu sur la structure des pneumatophores de deux espèces des sols hydromorphes en Guyane. *Cah. O.R.S.T.O.M. (Ser. Biol.), 23,* 3–22.

Hallé, F., & Oldeman, R. A. A. (1970). *Essai sur l'architecture et la dynamique de croissance des arbres tropicaux.* Paris: Masson.

Harris, J. O., Allen, E. K., & Allen, O. N. (1949). Morphological development of nodules on *Sesbania grandiflora* Poir., with reference to the origin of nodule rootlets. *Amer. J. Bot., 36,* 651–61.

Henwood, K. (1973). A structural model of forms in buttressed tropical rain forest trees. *Biotropica, 5,* 83–93.

van Himme, M. (1959). Etude du système radiculaire du Cacaoyer. *Bull. Agric. Congo Belge, 50,* 1541–600.

Huttel, C. (1975). Root distribution and biomass in three Ivory Coast rain forest plots. In *Tropical Ecological Systems* (Ecological Studies 11), ed. F. B. Golley & E. Medina, pp. 123–30. New York: Springer-Verlag.

Janse, J. M. (1897). Les endophytes radicaux de quelques plantes Javanaises. *Ann. Jard. Bot. Buitenzorg, 14,* 53–201.

Janzen, D. H. (1976). Why tropical trees have rotten cores. *Biotropica, 8,* 110.

Jeník, J. (1957). Kořenový systém dubu letního a zimního (*Quercus robur* L. et *Q. petraea* Liebl.). (Zusammenfassung: Das Wurzelsystem der Stiel- und Traubeneiche.) *Rozpravy Čs. Akad. Věd., Praha,* Řada MPV, *67*(14), 1–85.

– (1967). Root adaptations in West Africa trees. *J. Linn. Soc. Bot. Lond., 60*(381), 25–9.

– (1969). The life-form of *Scaphopetalum amoenum* A. Chev. *Preslia Prague, 41,* 109–12.

– (1970a). Root system of tropical trees. 4. The stilted peg-roots of *Xylopia staudtii* Engl. et Diels. *Preslia Prague, 42,* 25–32.

– (1970b). The pneumatophores of *Voacanga thouarsii* Roem. et Schult. (Apocynaceae). *Bull. I.F.A.N. Dakar (Ser. A.), 32,* 986–94.

– (1970c). Root system of tropical trees. 5. The peg-roots and the pneumathodes of *Laguncularia racemosa* Gaertn. *Preslia Prague, 42,* 105–13.

– (1971a). Root structure and underground biomass in equatorial forests. In *Productivity of Forest Ecosystems* (Proceedings, Brussels Symposium, 1969), pp. 323–31. Paris: UNESCO.

– (1971b). Root system of tropical trees. 6. The aerial roots of *Entandophragma angolense* (Welw.) C.DC. *Preslia Prague, 43,* 1–4.

– (1971c). Root system of tropical trees. 7. The facultative peg-roots of *Anthocleista nobilis* G. Don. *Preslia Prague, 43,* 97–104.

– (1973). Root system of tropical trees. 8. Stilt-roots and allied adaptations. *Preslia Prague, 45,* 250–64.

– (1974). Adventivní kořeny u nahosemenných a dvouděložných dřevin. (Summary: Adventitious roots in gymnospermous and dicotyledonous woody plants.) *Acta musei silesiae, Opava (Ser. Dendrologia), 23,* 153–64.

Jeník, J., & Harris, B. J. (1969). Root-spines and spine-roots in dicotyledonous trees of tropical Africa. *Oesterr. Bot. Z., 117,* 128–38.

Jeník, J., & Kubíková, J. (1969). Root system of tropical trees. 3. The heterorhizis of *Aeschynomene elaphroxylon* (Guill. et Perr.) Taub. *Preslia Prague, 41,* 220–6.

Jeník, J., & Mensah, K. O. A. (1967). Root system of tropical trees. 1. Ectotrophic mycorrhiza of *Afzelia africana* Sm. *Preslia Prague, 39,* 59–65.

Jeník, J., & Sen, D. N. (1964). Morphology of root system in trees: A proposal for terminology. *Abstracts, Tenth International Botanical Congress, Edinburgh,* pp. 393–4.

Jost, L. (1887). Ein Beitrag zur Kenntnis der Athmungsorgane der Pflanzen. *Bot. Ztg., 45,* 601–6, 617–27, 633–42.

Kerfoot, C. (1963). The root systems of tropical forest trees. *Commonwealth For. Rev. Lond., 42,* 19–26.

Klinge, H. (1973a). Root mass estimation in lowland tropical rain forest of central Amazonia, Brazil. I. Fine root masses of a pale yellow latosol and giant humus podzol. *Trop. Ecol., 14,* 29–38.

– (1973b). Root mass estimation in lowland tropical rain forest of central Amazonia, Brazil. II. "Coarse root mass" of trees and palms in different height classes. *An. Acad. Brasil Cienc. Rio de Janeiro, 45,* 595–609.

Köstler, J. N., Brückner, E., & Bibelriether, H. (1968). *Die Wurzeln der Waldbäume.* Berlin: Parey.

Krasilnikov, P. K. (1968). On the classification of the root system of trees and shrubs. In *Methods of Productivity Studies in Root Systems and Rhizosphere Organisms,* ed. M. S. Ghilarov et al., pp. 106–14. Leningrad: Nauka.

– (1970). Klassifikaciya kornevych system derevev i kustarnikov. *Lesovedenie Moscow, 3,* 35–44.

Kubíková, J. (1967). Contribution to the classification of root systems of woody plants. *Preslia Prague, 39,* 236–43.

Kunkel, G. (1965). Der Standort: Kompetenzfaktor in der Stelzwurzelbildung. *Biol. Zentralb. Leipzig, 84,* 641–51.

Land Development Department (1970). *International Seminar on Shifting Cultivation and Economic Development in Northern Thailand.* Bangkok: Land Development Department.

Lawson, G. W., Armstrong-Mensah, K. O., & Hall, J. B. (1970). A catena in tropical moist semi-deciduous forest near Kade, Ghana. *J. Ecol., 58,* 371–98.

Lawson, G. W., Jeník, J., & Armstrong-Mensah, K. O. (1968). A study of a vegetation catena in Guinea savanna at Mole Game Reserve (Ghana). *J. Ecol., 56,* 505–22.

Lebrun, J. (1947). *La végétation de la plaine alluviale au sud du Lac Edouard.* Brussels: Inst. Parcs Nat. Congo Belge. 2 vols.

Leroy-Deval, J. (1973). Les anastomoses racinaires chez l'Okoumé (*Aucoumea klaineana*). *C. R. Acad. Sci. Paris, 276 (Sér. D.),* 3425–8.

– (1974). *Structure dynamique de la rhizosphère de l'Oukoumé dans ses rapports avec la sylviculture.* Nogent sur Marne: Centre Technique forestier Tropical.

Longman, K. A., & Jeník, J. (1974). *Tropical Forest and Its Environment.* London: Longmans.

Lyford, W. H., & Wilson, B. F. (1964). Development of the root system of *Acer rubrum* L. *Harvard Forest Paper,* no. 10, pp. 1–17.

MacCarthy, J. (1962). The form and development of knee roots in *Mitragyna stipulosa. Phytomorphology, 12,* 20–30.

Maxwell, D. (1972). Luano catchments research project. 3. Root range investigations. *Water Resources Research Lusaka, WR 16,* 1–10.

McArthur, I. C. S., & Steeves, T. A. (1969). On the occurrence of root thorns on a Central American palm. *Can. J. Bot., 47,* 1377–82.

McCusker, A. (1970). Knee roots in *Avicennia marina* (Forsk.) Vierh. *Ann. Bot. Lond., 35*, 707–12.

Mensah, K. O. A., & Jeník, J. (1968). Root system of tropical trees. 2. Features of root system of Iroko (*Chlorophora excelsa* Benth. et Hook.). *Preslia Prague, 40*, 21–7.

Meyer, F. H. (1973). Distribution of ectomycorrhizae in native and manmade forests. In *Ectomycorrhizae*, ed. G. C. Marks & T. T. Kozlowski, pp. 79–105. New York: Academic Press.

Muller, C. H. (1946). *Root development and Ecological Relations of Guayule.* USDA Tech. Bull., No. 923. 114 pp.

Nathanielsz, C. P., & Staff, I. A. (1975). On the occurrence of intracellular blue-green algae in cortical cells of the apogeotropic roots of *Macrozamia communis* L. Johnson. *Ann. Bot. Lond., 39*, 363–8.

Nye, P. H., & Greenland, D. J. (1960). The soil under shifting cultivation. *Tech. Commun. Commonw. Bio. Soil Sci., 51*, 1–156.

Ogura, Y. (1940a). New examples of aerial roots in tropical swamp plants. *Bot. Mag. Tokyo, 54*, 327–37.

– (1940b). On the types of abnormal roots in mangrove and swamp plants. *Bot. Mag. Tokyo, 54*, 389–404.

Oldeman, R. A. A. (1971). Un *Eschweilera* (Lecythidaceae) à pneumatophores en Guyane française. *Cah. O.R.S.T.O.M. (Ser. Biol.), 15*, 21–7.

Percival, M., & Wormersley, J. S. (1975). Floristics and ecology of the mangrove vegetation of Papua New Guinea. *Bot. Bull. Lae, Papua New Guinea, 8*, 1–96.

Peyronel, B., & Fassi, B. (1957). Micorrize ectotrofiche in una cesalpiniacea del Conga Belga. *Attix Accad. Sci. Torino, 91*, 569–76.

Rawitscher, F. (1948). The water economy of the vegetation of the campos cerrados in Southern Brazil. *J. Ecol., 36*, 237–68.

Richards, P. W. (1952). *The Tropical Rain Forest.* Cambridge: Cambridge University Press.

Schnell, R. (1970–71). *Introduction à la phytogéographie des pays tropicaux.* Paris: Gauthier-Villars. 2 vols.

Scholander, P. F., van Dam, L., & Scholander, S. I. (1955). Gas exchange in the roots of mangroves. *Amer. J. Bot., 42*, 92–8.

Senn, G. (1923). Ueber die Ursachen der Brettwurzel-Bildung bei der Pyramiden-Pappel. *Verhandl. Natur. Ges. Basel, 35*, 405–35.

Smith, A. P. (1972). Buttressing of tropical trees: A descriptive model and new hypotheses. *Amer. Nat., 106*, 32–46.

van Tieghem, P. (1905). Sur les Irvingiacées. *Ann. Sci. Nat. (Bot.) Paris (Ser. 9), 1*, 247–320.

Troll, W. (1930). Ueber die sogenannten Atemwurzeln der Mangroven. *Ber. Dtsch. Bot. Ges., 48*, 81–99.

Troll, W., & Dragendorff, O. (1931). Ueber die Luftwurzeln von *Sonneratia* Linn. f. und ihre biologische Bedeutung. *Planta Berlin, 13*, 311–473.

Wilkinson, G. (1939). Root competition and sylviculture. *Malayan Forester, 8*, 11–15.

Yampolsky, C. (1924). The pneumathodes on the roots of the oil palm *Elaeis guineensis* Jacq. *Amer. J. Bot., 11*, 502–15.

Zimmermann, M. H., Wardrop, A. B., & Tomlinson, P. B. (1968). Tension wood in aerial roots of *Ficus benjamina* L. *Wood Sci. Tech., 2*, 95–104.

15

On the adaptive significance of compound leaves, with particular reference to tropical trees

THOMAS J. GIVNISH
Department of Biology, Harvard University, Cambridge, Massachusetts

For the purposes of this discussion, I arbitrarily define a "compound leaf" as a collection of leaflets arranged about a deciduous twig (Fig. 15.1). Here "leaflet" is construed, in a deliberately vague fashion, as being a leaflike organ that has no bud in its axil. It should be noted that my concept of leaf complexity differs somewhat from that of a few morphologists. For example, some authors (e.g., see Fries, 1909) interpret the seemingly simple blades of *Cercis* or *Bauhinia* as compound and representing several fused leaflets. Others view the seemingly compound palm leaf as simple because the leaflets are not articulated. These differences of opinion are, however, irrelevant to this discussion. The conclusions I shall draw will depend not on strict morphologic typology, but on functional grounds. Thus, in most cases these conclusions will apply to leaves that "look" compound, without distinguishing between unifoliolate and simple leaves or between articulate and nonarticulate leaflets.

Distribution of compound-leaved trees and previous interpretations

Woody species bearing compound leaves are common in certain arid communities such as deserts, parklands, and seasonal rain forests. They are less common in moist or shady areas, where broader, simple leaves predominate. This pattern, and the fact that leaf size generally tends to decrease toward drier areas, has led to the development of a variety of explanations for the adaptive value of compound leaves. All these have had some central notion that the leaflets into which a compound leaf is divided behave like so many

Written originally as part of a doctoral thesis for the Department of Biology, Princeton University.

small simple leaves and, thus, should advantageously allow greater light penetration into diffuse canopies (Horn, 1971), decrease the heat load on photosynthetic tissue (Gates, Alderfer, & Taylor, 1968), or increase the efficiency with which water is used in photosynthesis (Parkhurst & Loucks, 1972). Whatever may be the adaptive significance of effectively small leaves, if compound leaves are optically and convectively equivalent to sets of smaller simple leaves, the question that immediately arises is: Why should plants in arid areas evolve compound leaves with small leaflets rather than small simple leaves?

One may argue that different groups with different genetic proclivities may simply arrive at the same solution along different routes – some via small simple leaves and others via finely divided compound leaves. However, this argument is convincing only if species with effectively small simple or compound leaves show roughly the same pattern of distribution. This is not the case. For example, in moving from lowland rain forest toward either savanna or montane rain forest, we find a tendency in both cases toward effectively smaller leaves. However, whereas in the savanna these leaves tend to be compound, they are almost entirely simple in montane rain forest (see below). Compound leaves also appear to be virtually absent

Fig. 15.1. Compound (*left*) and simple (*right*) leaves of same effective size differ mainly in that "twig" (rachis) bearing leaflets of compound leaf is promptly deciduous, whereas shoot bearing simple leaves is usually persistent. In addition, rachis is usually nonwoody and of determinate growth, whereas comparable shoot is usually woody and of indeterminate growth.

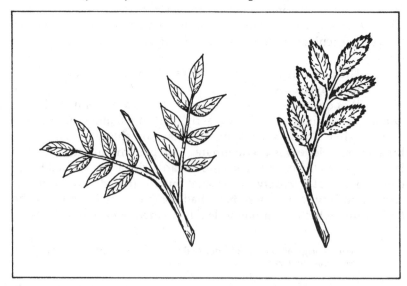

from many habitats, such as chaparral, elfin forest, and tundra, where small simple leaves abound. On the basis of distributional data alone, compound and simple leaves do not appear to be adaptively equivalent.

However, it must be remembered that the occurrence of compound leaves is restricted taxonomically to a small group of families (Table 15.1). One possible cause for differences in the distribution of simple and compound leaves might be that families that characteristically bear compound leaves are generally not found, for reasons other than leaf form, in areas otherwise favoring effectively small

Table 15.1. *Major families of angiosperms with compound leaves*

Order[a]	Family	Order	Family
Palmales	Palmae[b]	Rhamnales	Vitaceae[b]
Ranunculales	Ranunculaceae[c] Berberidaceae[b] Lardizabalaceae	Sapindales	Staphyleaceae Melianthaceae Connaraceae[b] Akaniaceae
Papaverales	Papaveraceae[c] Fumariaceae[c]		Sapindaceae[b] Hippocastanaceae Burseraceae[b]
Juglandales	Juglandaceae[b]		Anacardiaceae[b] Simaroubaceae[b]
Dilleniales	Paeoniaceae		Rutaceae[b] Meliaceae[b]
Theales	Caryocaraceae		Zygophyllaceae[b]
Malvales	Bombacaceae[b]	Geraniales	Oxalidaceae[c] Geraniaceae[c]
Capparales	Capparidaceae[b] Cruciferae[c] Moringaceae	Umbellales	Umbelliferae[e] Araliaceae[b]
Rosales	Cunoniaceae[b] Rosaceae[b] Leguminosae[b]	Scrophulariales	Oleaceae Bignoniaceae[b]
		Dipsacales	Caprifoliaceae Valerianaceae[c]
Proteales	Proteaceae[b]		

[a]According to Cronquist (1968).
[b]Major woody families with appreciable number of species with compound leaves.
[c]Major herbaceous families with appreciable number of species with compound leaves.

leaves. For example, some families might lack tolerance to cold temperatures and be able to survive only in hot, dry climates. If compound-leaved families were so restricted (and that would have to be shown), any correlation between climate and leaf complexity could be argued to be spurious. Two principal arguments can be developed to counter this possibility: the "convergence" argument and the "functional" argument.

The convergence argument

If we can show that many different, distantly related families share the same distribution, it becomes difficult to argue that traits restricted to particular families can help create a general pattern whose unifying theme is the compound leaf. The adaptive value of compound leaves can be demonstrated most convincingly if (1) variation in leaf complexity exists within families and (2) simple- and compound-leaved species within such families also show the same distribution. Particularly important are examples that demonstrate that a family is not physiologically limited to areas where compound leaves are common and show that species occurring outside those areas tend to have simple leaves. The more variation between and within families, or even between and within taxa of lower rank, that we can correlate with particular environmental factors, the less likely is the possibility that differences in leaf complexity depend mainly on historical or genetic accidents rather than on environmentally conditioned selective pressures.

The functional argument

Distributional data alone can never actually demonstrate that compound leaves are adaptive; leaf complexity might merely be a mechanical or physiologic response to particular sets of conditions. It would be hazardous, for example, to argue that bleeding is adaptive merely because most animals bleed when cut. To suggest that a trait is adaptive, we must show that (1) the trait significantly affects the function of an organism and (2) the effect of the trait is to increase the organism's competitive ability and potential to leave offspring. Thus, if we can show that the functional or physiologic differences between simple and compound leaves favor the occurrence of compound-leaved species in just those habitats in which they are found, the inference that compound leaves do have adaptive value and do affect the distribution of species bearing them becomes hard to avoid.

In this chapter, I shall attempt to develop various aspects of these arguments, particularly the latter and more important one. I shall rely principally on testing the ecologic predictions that follow from the basic physiologic and structural differences between simple and compound leaves. The question on which I shall focus is: In what ways do compound and simple leaves of the same effective size differ?

Adaptive value of compound leaves

According to the definition in use here, compound leaves are simply collections of leaflets arranged on deciduous twigs. The leaflets thus differ from simple leaves of the same size and packing only in that the twig on which they are borne is deciduous. An important question is: Under what circumstances is it adaptive to throw away branches? Although it may at first appear maladaptive for a plant methodically to throw away branchwork it has only recently built, there are a few compensating factors that might make such twig shedding advantageous in certain situations. For example, a shed twig does not transpire during a dry season. A branch that will be thrown away early can be built inexpensively and thus confer an advantage wherever extensive branching is not the best strategy. These two points suggest that compound leaves can be an adaptation to seasonal drought or an adaptation to rapid vertical growth.

Adaptation to seasonal drought

A tree with compound leaves can regularly shed its highest-order branches: the deciduous twigs or leaflet-bearing axes. These, because of their high surface/volume ratio and low suberization, are a large potential source of residual water loss after leaf abscission. Walter (1971) showed that transpiration takes place from even the highly corticated stem and branches of woody plants in arid areas and suggested that such transpiration may limit the occurrence of woody plants with permanent above-ground parts in areas where the soil dries out completely between rainy seasons. This conclusion is supported by the data of Orshan (1953), which indicate that plants in the Middle East tend to shed more of their above-ground parts in areas with longer or more severe dry seasons.

Because of their high surface/volume ratio and low suberization, the highest-order axes should have the greatest potential rate of water loss per unit weight of any branch. Regularly shedding these axes, perhaps as compound leaves, should be the most economical method of reducing residual stem transpiration. Compound leaves should thus be advantageous in areas with a pronounced dry season.

The greater the length and intensity of a dry season, the greater should be the adaptive advantage of compound leaves. This should be true particularly in warm regions where the potential evaporation from woody surfaces is large. At the same time, we should not expect to find any unusually large numbers of trees or shrubs with compound leaves in areas that, though droughty, favor plants with evergreen leaves. Compound leaves should be less common in areas where the building cost of twigs is high, as might occur in areas of low productivity or nutrient-poor soils (Monk, 1966; Janzen, 1974). We thus expect to find large numbers of woody species with compound leaves in such seasonally arid environments as savannas, thorn forests, warm deserts, and subtropical forests.

Savannas, thorn forests, warm deserts
Trees that are deciduous, compound-leaved, and often thorny are characteristic of these three warm, seasonally arid habitats. In the Sonoran Desert, for example, almost all the dominant nonsucculent plants bear ephemeral compound leaves with minute leaflets. Among these are the many green-stemmed leguminous trees found along sandy washes: the mesquites (*Prosopis*), palo verdes (*Cercidium*), smoke tree (*Olneya*), and Jerusalem thorn (*Parkinsonia*); the last species sheds its leaflets but retains the petiole as part of its drought-adapted photosynthetic system. Leguminous trees with compound leaves also form an integral part of many tropical parklands: The flat-crowned acacias (many, but not all, of which are deciduous) and fiery-blossomed erythrinas almost symbolize the African savannas of Rhodesia, Tanzania, and the Sudan. Another striking element of African savanna and thorn scrub, the swollen-trunked baobab (*Adansonia digitata*), also puts forth an ephemeral foliage of compound leaves.

In South Africa, Bews (1925, 1927) found a uniform increase in the percentage of tree species with compound leaves in moving from evergreen forests in the well-watered regions to drought-deciduous tree velds and scrubs in the more arid areas (Table 15.2). Bews noted that deciduous and compound leaves become even more common in the drier regions of northern Rhodesia. Significantly, he also found compound leaves to be quite uncommon in southwestern South Africa where, although little rain falls during the summer months, the vegetation is nevertheless evergreen. I have not yet tested whether the eastward increase in the percentage of species with compound leaves is due, over the entire gradient, to an increase in the percentage of species with deciduous compound leaves, but I have analyzed the flora of one area near the dry end of Bews's gradient, the Witwatersrand. There, only 37 out of 85 simple-leaved species are deciduous, whereas 24 out of 30 compound-leaved species shed their leaves (Tree Society of Southern Africa, 1964). The percentage of compound-leaved species that shed their leaves becomes even greater if one does not include species with nearly simple, bifoliate or trifoliate leaves; in that case, 16 out of 17 species are deciduous.

Subtropical forests
In subtropical forests experiencing a marked yearly dry spell, trees with compound, often deciduous leaves are frequently dominant, particularly in the exposed forest canopy (Beard, 1944, 1955). According to Beard, compound leaves predominate in the upper stories of such forests, making up one-third to two-thirds of the total; simple leaves predominate in the lower storys. Compound leaves tend to be more common in the more seasonal of Beard's vegetation types. A probably significant example is provided by *Tabebuia heterophylla*, a species of the seasonal forests of the West Indies that is unique there in having different races that are either simple- or compound-leaved. The compound-leaved race is found on Puerto Rico and is deciduous; the simple-leaved races are found on Puerto Rico and on Dominica, Guadeloupe, and Martinique and are apparently evergreen (Little & Wadsworth, 1964; Little, Woodbury, & Wadsworth, 1974; R. Howard, pers. comm.). Thus the intraspecific variation in *T. heterophylla* parallels the expected interspecific variation in leaf complexity with environment.

Leaf fall in tropical or subtropical forests can be either facultative or obligate, responding to either an actual drought or the prospect of one. In the tropical and subtropical forests, compound-leaved trees frequently tend to be numerically concentrated in a few large families, such as the Anacardiaceae, Bignoniaceae, Burseraceae, Melicaeae, and Leguminosae. In these families, there is often also a

Table 15.2. *Percentage of various leaf types found in some forest habitats of South Africa*

Habitat	No. of species	% of species with simple leaves			% of species with compound leaves	% of species with succulent forms
		>3 in. long	1–3 in. long	<1 in. long		
Moist subtropical forests	60	75	19	0	6	0
Mesophytic forests	150	32	47	3	18	0
Dry tree veld and scrub	500	11	34	13	42	5

Source: Data from Bews (1925, 1927).

marked tendency toward deciduousness. However, in other families in which compound leaves are common, such as the Cunoniaceae, Rutaceae, Sapindaceae, and Simaroubaceae, this tendency is not clear or is reversed. More research may be needed to determine whether trees in these families are evergreen or facultatively deciduous. However, even if a large percentage of species in these families should prove to be evergreen, it would not necessarily mean that compound leaves are familial quirks bereft of adaptive significance, as there are other advantages provided by compound leaves. For example, members of evergreen families or evergreen members of largely deciduous families might be rapidly growing, early successional forms that use large compound leaves as cheap, throwaway branches. Among the native leguminous trees of Puerto Rico, for example, all species bear compound leaves: Of the minority that are evergreen, most are either very rapidly growing species (*Inga vera, Machievium lunatum, Ormosia krugii, Pithecellobium arboreum,* and *Pterocarpus officinalis*) or have leaf axes so reduced that they are as short as, or shorter than, ordinary leaf petioles (*Cynometra portoricensis, Hymenaea courbaril, Inga laurina,* and *Pithecellobium unguis-cati*). Only 2 out of 31 species, *Lonchocarpus glaucifolius* and *Stahlia monosperma,* are neither evergreen and fast growing nor deciduous and thus do not fit the above two categories. Indeed, in the Leguminosae one might trace an evolutionary trend (Fig. 15.2) from extensive compound leaves in drought-deciduous or rapidly growing evergreen genera like *Acacia* or *Albizia* to highly reduced compound leaves in slow-growing evergreen genera such as *Cynometra* and *Hymenaea.* This trend would culminate in the partially fused leaflets of the functionally simple leaves of *Bauhinia,* an evergreen genus mainly of the rain forest understory, and in the entirely simple leaf of *Cercis,* a genus of slow-growing shrubs of the sheltered understory of north temperate forests.

Familial uniformity in leaf type, far from indicating the nonadaptive and conservative nature of a trait, may in fact reflect and result from a major ecologic radiation of the family along one or more lines of adaptive innovation posed by the leaf type. The large and evergreen palm family may represent just such an adaptive radiation based on the importance of compound leaves for certain branching and successional strategies (see below). The tree fern and araliad families may represent similar radiations based on the use of compound leaves as throwaway branches; more will be said about this later.

Fig. 15.2. Evolution of leaf complexity in Leguminosae. Most woody legumes have extensive compound leaves (*A*) and are either drought-deciduous or rapidly growing, sparsely branched pioneers; typical genera include *Acacia* and *Albizia*. In species that are slow-growing and evergreen, however, there is a trend toward reduction of compound leaf to nearly simple, bi- or trifoliate condition (*B*) in genera like *Hymenaea* and *Cynometra*. This trend culminates in apparently fused leaflets of *Bauhinia* (*C*), a genus made up largely of slow-growing, evergreen trees of rain forest understory, and in entirely simple leaves of *Cercis* (*D*), a genus of deciduous, but sheltered, slow-growing shrubs found in understory of north temperate forests.

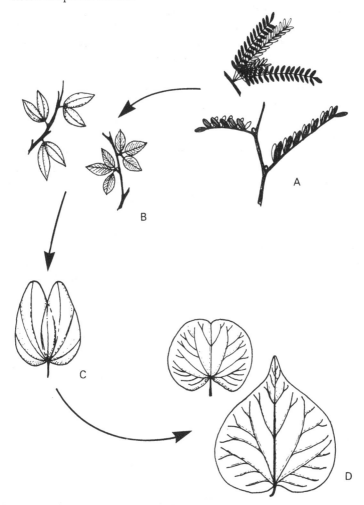

360   Thomas J. Givnish

Tropical rain forest

Compound leaves are more common, at least in the neotropics, in lowland rain forest than in montane rain forest, and are more common in the canopy of lowland forest than in the understory. In Brazilian lowland rain forest, for example, Cain, Castro, Pires, & da Silva (1956) showed that the proportion of tree species bearing compound leaves increases with the height and exposure of each stratum, even though the leaflets of such leaves are of about the same size as simple leaves found at the same height (Table 15.3).

Most rain forests experience some short periods of dryness during the course of a year; at these times, the trees in the upper strata are more exposed to desiccation from dry air and intense sunlight than are those in the lower stories. Hence, the percentage of deciduous leaves should be greater in the canopy than in the understory; we would also thus expect an increase in the percentage of compound leaves in moving from the understory to the canopy. Such an increase should be at the expense of species with simple evergreen leaves and to the advantage of species with deciduous compound leaves.

Data that confirm this prediction and corroborate the findings of Cain et al. (1956) are given by Loveless & Asprey (1957) and Asprey & Loveless (1958) for dry forests in Jamaica and by Beard (1946) for dry seasonal forest in Trinidad (Table 15.4). In every habitat except Ja-

Table 15.3. *Stratal distribution of tree species with simple and compound leaves in neotropical rain forest, Mucambo, Brazil*

Height of crown above ground (m)	Simple leaves			Compound leaves		
	No. of species	% of species	Average leaf area (cm^2)	No. of species	% of species	Average leaflet area (cm^2)
>30	31	63	56.7 ± 34.8	18	37	47.7 ± 81.8[a]
8–30	61	73	77.2 ± 60.0	22	27	93.5 ± 141.5
2–8[b]						
0.25 – 2	19	86	85.8 ± 76.9	3	14	101.7 ± 43.9

[a] Average leaf area without *Didymopanax morototoni* is 28.8 ± 26.5 cm^2.
[b] Only 3 species in this height range; data too scanty to warrant inclusion.
Source: Data from Cain, Castro, Pires, & da Silva (1956).

Table 15.4. *Distribution (percent) of compound- and deciduous-leaved species, by habitat and strata*

Stratum	Dry evergreen thicket		Evergreen bushland		Cactus scrub		Dry seasonal forest	
	Compound	Deciduous	Compound	Deciduous	Compound	Deciduous	Compound	Deciduous
Jamaica								
Mesophanero-phyte (8–30 m)	30[a] *34*	50 *49*						
Microphanero-phyte (2–8 m)	9 *5*	21 *14*	26 *21*	40 *33*	40 *59*	30 *27*		
Nanophanero-phyte (0.25–2 m)	0 *0*	20 *13*	8 *24*	25 *15*	14 *3*	57 *85*		
Trinidad								
Emergent							58 *68*	42 *64*
Canopy							27 *27*	11 *26*

[a] Roman numbers, percentage of species with leaves of given kind; italic numbers, percentage of individuals with leaves of given kind.

Source: Data on Jamaica from Loveless & Asprey (1957) and Asprey & Loveless (1958). Data on Trinidad from Beard (1946).

maican cactus scrub, the percentages of species bearing compound leaves and deciduous leaves decrease in parallel in going from the most exposed story to the one below it. Asprey & Loveless (1958) tabulated the leaf-shedding behavior of each species they studied; these data allow us to test whether the increase in the percentage of species bearing compound leaves in the upper storys is due to an increase in the number of deciduous compound-leaved species (Table 15.5).

In Table 15.5, the cross tabulation for each stratum in a given habitat represents the percentage of species with, respectively, evergreen compound, deciduous compound, evergreen simple, or deciduous simple leaves. The cross tabulations marked Δ were obtained by subtracting the table corresponding to the less exposed layer from that of the more exposed layer; they thus reflect the relative changes in the representation of various leaf types that occur in moving from the lower story to the upper. As expected, the vertical increase in compound leaves comes mainly at the expense of species with evergreen simple leaves (large negative ES entry) and at the gain of species with deciduous compound leaves (large positive CD entry), with little change in the percentage of species with either deciduous simple or evergreen compound leaves, at least in dry evergreen thicket and evergreen bushland. The number of species with compound leaves in cactus scrub is too small to permit a similar analysis: Δ' shows how the Δ table could change if only one of the supposedly evergreen compound-leaved species were instead facultatively deciduous. It is remarkable, however, that the understory is much more deciduous than the overstory. In the open cactus scrub, this might be because shorter trees and shrubs are almost as exposed as the taller trees and yet have less capacity to develop deep root systems to enable themselves to endure droughts.

Grubb, Lloyd, Pennington, & Whitmore (1963) found the proportion of tree species bearing compound leaves to decrease from 36% to 9% in moving from lowland to montane rain forest in Ecuador. Beard (1944, 1955) recognized a similar tendency throughout Central America and the West Indies. Brown (1919) documented this same trend on Mount Maquiling in the Philippines (Table 15.6). These altitudinal trends are to be expected as a result of the decreased temperature, light intensity, and soil fertility, and the increased humidity and rainfall that prevail in montane forests, factors that favor the evergreen habit. Data summarized for Puerto Rico (source: Little & Wadsworth, 1964; Little et al., 1974) indicate that (1) 19% of species occurring below 1200 f., and 15% of those occurring above 1200 f. have compound leaves; and (2) over 80% of the increase in the percentage of species bearing compound leaves toward the lowlands is

Table 15.5. *Cross tabulation of Jamaican dry forest species, by leaf complexity and deciduousness (see text for discussion)*[a]

Dry evergreen forest

Msph (20)

	E	D
C	.00	.30
S	.50	.20

Δ

	E	D
C	−.09	.30
S	−.20	−.01

Mcph (33)

	E	D
C	.09	.00
S	.70	.21

Evergreen bushland

Mcph (47)

	E	D
C	.11	.15
S	.49	.25

Δ

	E	D
C	.03	.15
S	−.18	.00

Nph (12)

	E	D
C	.08	.00
S	.67	.25

Cactus scrub

Mcph (10)

	E	D
C	.30	.10
S	.40	.20

Δ

	E	D
C	.16	.10
S	.11	−.37

Nph (7)

	E	D
C	.14	.00
S	.29	.57

Δ'

	E	D
C	.06	.20
S	.11	−.37

[a] Each entry represents fraction of species in a given stratum with leaf form and behavior cited. Abbreviations: E, evergreen; D, deciduous; C, compound; S, simple; Msph, mesophanerophyte (8–30 m); Mcph, microphanerophyte (2–8 m); Nph, nanophanerophyte (0.25–2 m). Number of species in parentheses.

due to an increase in the percentage of deciduous compound-leaved species. The altitudinal trend on tropical mountains is particularly significant when viewed against that found in the lowlands. Whereas effective leaf size decreases in moving from lowland rain forest toward either savanna or montane rain forest, the tendency is for these effectively small leaves to be compound in savanna and entire in montane forest. The adaptive advantage of compound leaves thus does not appear to result from their effectively small size alone.

In summary, compound leaves are more common in the upper, exposed storys of rain forests than in the more sheltered lower stories, and are more common in lowland rain forest than in montane or elfin forest. The increase toward the drier habitats in the percentage of species bearing compound leaves appears largely to be a result of the increase in number of species with deciduous compound leaves.

North temperate deciduous forests
Winter in much of the northeastern United States is effectively also a dry season, as low soil temperatures make water largely unavailable (Walter, 1973). Although water may be difficult to extract under cold winter conditions, the potential evaporative loss of water from twigs would not be nearly so great under such conditions as it would be during a dry season in the warm tropics and subtropics. As a result, although several tree species in the Northeast have compound leaves, the percentage they make up in the flora is not nearly so great as, for example, in Puerto Rico. Of the trees that do bear compound leaves, many occur in early succession and seem to employ their leaves as cheap throwaway branches (see below).

Compound leaves tend to become rare toward the poles (Table 15.7): This is probably because (1) evaporation is even further retarded by cooler winters (although this might be more than offset by winter's greater length and harsher winds); (2) snow can shelter

Table 15.6. *Percentage of species bearing compound leaves, as a function of altitude and level in canopy: Mount Maquiling*

Stratum	Virgin lowland dipterocarp forest	Midmountain forest	Mossy forest
Upper	32	17	0
Middle	17	15	
Lower	13		

Source: Data from Brown (1919).

small plants from desiccation (this would not necessarily hold for tall trees, the most obvious exceptions to the rule proposed being the mountain ashes, *Sorbus*); and (3) discarding twigs may be profligate if primary production is low or mineral nutrients rare. The trend away from compound leaves in more poleward climes is not confined to the angiosperms; toward the north one also finds that deciduous conifers with simple needles, like tamarack and Siberian larch (*Larix*), replace the more southerly deciduous "compound"-needled conifers, like bald-cypress (*Taxodium*) and possibly dawn-redwood (*Metasequoia*).

In temperate forests of the southern hemisphere, which experience mild oceanic winters and are largely evergreen, compound leaves are rare. In New Zealand, for example, only one tree species (*Sophora*) is deciduous and it has compound leaves. In southern California, where the winter rainfall and topography favor evergreen trees and shrubs in the coastal chaparral, very few evergreen plants have compound leaves; those that do include a few species of *Berberis* and one of *Chamaebatia*. Many chaparral species bear evergreen simple leaves even though they belong to families that typically have compound leaves; these include *Rhus integrifolia*, *Rhus laurina*, and *Rhus ovata* in the Anacardiaceae (*Rhus trilobata* has compound leaves and is deciduous); *Cneoridium* in the Rutaceae; and *Adenostoma*, *Cercocarpus*, and *Heteromeles* in the Rosaceae (Munz, 1974). Indeed, the only important compound-leaved tree in the chaparral region, the California walnut (*Juglans californica*), grows along stream bottoms and is de-

Table 15.7. *Latitudinal trends in percentage of woody angiosperm species bearing various leaf types*

Latitude	Leaf type			
	Entire	Toothed	Lobed	Compound
Northern Alaska[a]	63.0	28	5.0	4
Alaska[a]	55.0	31	10.0	4
Northeastern Minnesota[b]	24.0	48	14.0	14
Carolinas[c]	29.0	44	11.0	16
Puerto Rico, Virgin Islands[d]	74.5	7	0.5	18

[a]Data from Hultén (1968).
[b]Data from Rosendahl & Butters (1927).
[c]Data from Radford, Ahlee, & Bell (1968)
[d]Data from Little & Wadsworth (1964); Little, Woodbury, & Wadsworth (1974).

ciduous. Compound leaves are uncommon in most areas having a Mediterranean-type climate with winter rainfall: Evergreen sclerophyllous shrubs with simple leaves dominate much of the Californian chaparral, Chilean matoral, Mediterranean macchia, South African fynbos, and Australian mallee. Some exceptions, such as *Grevillea* and *Hakea*, may be using their compound leaves as throwaway branches in areas regularly swept by fire.

The deciduous compound leaf is only one kind of plant strategy that can reduce drought damage. The least extreme strategy is simple stomatal closure; this may escalate to the shedding of leaves and/or the shedding of leaflets and leaf axes of compound leaves. Cladoptosis (the shedding of twigs and branches) is a more extreme strategy carried out by many plants in arid regions; it can be either facultative or obligate. For example, the drought-resistant creosote bush (*Larrea divaricata*) self-prunes its own branches in the face of extreme heat or drought, starting with the highest and most exposed twigs and working downward. Other desert plants, such as *Encelia farinosa*, are partially or entirely suffrutescent, their entire above-ground mass dying back regularly during each dry season. Orshan (1953) proposed a life-form scheme for desert plants based on the extent of above-ground tissue shed during the dry season. Based on his experience in the Middle East, Orshan's scheme is designed to reflect a gradation of adaptations to enduring drought. This gradation begins with the shedding of leaves and branches, progresses to suffrutescence, and leads ultimately to the annual habit in which the entire plant is shed. Orshan found that plants tended to shed more of their above-ground mass as the dry season to which they were exposed increased in length and severity (Table 15.8). The deciduous compound leaf seems a natural addition to the less extremely adapted end of Orshan's spectrum of hydroeconomic forms.

Adaptation for rapid vertical growth

Certain tree species appear to specialize in reproducing within and competing for gaps that open in forest canopies. For such plants, a high rate of vertical elongation is essential if they are to reach the canopy. This, however, raises a conflict in growth strategy between branching and not branching. Branching too extensively would divert energy from leader growth into short-lived side branches that quickly become useless to a rapidly elongating plant (Fig. 15.3). Not branching at all would concentrate all growth on the leader, but would at the same time reduce the potential for growth by reducing the total leaf area deployed.

Large compound leaves arranged in a rosette about a single central axis seem an ideal solution to this problem. This arrangement offers the following advantages:

1. Compound leaves are cheap, throwaway branches. The rachis of a compound leaf derives no small portion of its strength from turgor pressure and fibrous material, and hence should not be as costly to build as a woody branch bearing the same load. If a plant were to branch extensively, then the initially more costly woody twigs and branches could be economical in the long run, as they are a potential core for more massive mechanical tissue that could support future foliage, spread their initial cost over several sets of leaves, and reduce the cost of future support. However, if extensive branching is not adaptive, woody twigs are an unnecessarily durable and expensive investment. Extensive branching is surely not adaptive for a plant that must elongate rapidly to reach a gap in the canopy or to keep up with rapidly growing competitors, as branch after branch must be discarded to minimize self-shading (Fig. 15.3).

2. Compound or deeply dissected leaves can cover a large area rather inexpensively, particularly when conditions call for effectively small leaves. A simple linear leaf with the same area and effective size as a bipinnate or tripinnate one would have roughly the same total length of support tissue arranged in effectively much longer, and hence more expensive, lever arms. Because the strength of a branch or vein varies with the cube of its diameter, whereas its cost varies as the square (Givnish & Vermeij, 1976), it should be more

Table 15.8. *Hydroeconomic spectra of dominant species in arid regions of Middle East*

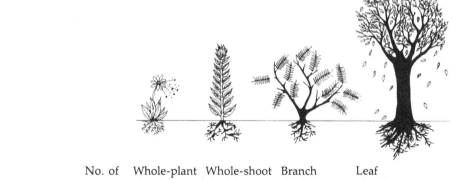

Region	No. of species	Whole-plant shedders[a]	Whole-shoot shedders	Branch shedders	Leaf shedders
Mediterranean	90	30	37	17	17
Irano-Turanian	70	47	30	19	4
Saharo-Sindian	71	60	13	27	0

[a]Each entry represents percentage of species in region having given growth form.
Source: Data from Orshan (1953).

economical to coalesce several long, subparallel linear segments into a pinnate arrangement with a single midrib than to support the segments individually (Fig. 15.4).

3. The nonwoody green rachis of a compound leaf can help pay for and further reduce its overhead costs by photosynthesizing.

4. Most importantly, a rosette plant with one or few vertical axes and compound leaves should have a high rate of vertical growth, both absolutely and relative to other plants with similar crown diameter and density, as it has ultimate apical dominance and large total leaf area.

Fig. 15.3. In early successional vegetation, a premium is put on rapid vertical growth and strong apical control. Plants that invest heavily in long, expensive side branches (A) can fall behind in height competition because they divert energy away from leader and trunk growth and into branches that quickly become shaded and of no further use (bands on either side of diagram represent crowns of neighboring plants at successive time intervals, t, t_2, and t_3). Plants with strong apical dominance (B) do not tie up energy in expensive side branches with short effective lifetimes, but concentrate resources on permanently useful growth of stem and leader. As side branches are, in any case, short-lived in early successional and light gap vegetation, they need not be made durable enough to serve as core for future growth or to resist attacks of insects or other agents. Compound leaves with nonwoody axes and extensive total area seem ideal throwaway branches.

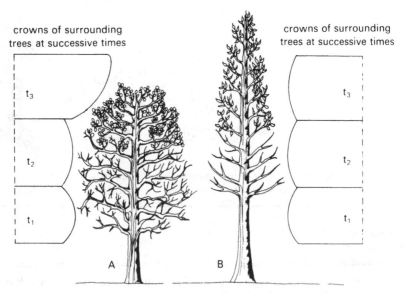

crowns of surrounding
trees at successive times

t_3

t_2

t_1

crowns of surrounding
trees at successive times

t_3

t_2

t_1

A   B

Throw-away branches in tropical forests

"Parasol trees," with rosettes of leaves arranged on a solitary axis, are common in many tropical and subtropical forests. Many of the smaller ones that remain in the understory, the "cabbage heads" (German: *Schopfbäumchen*) have broad entire leaves. However, many others have large compound leaves and seem particularly adept at invading gaps that open in the forest canopy; they also tend to occur particularly in areas where the canopy is chronically disturbed and open. One entire family, the palms, seems to have specialized along these lines (Chap. 11). Most palms live in the understory, lower levels, and canopy of tropical forest. In relatively stable and well-drained rain forests, pinnately leaved palms frequently spread their often enormous fronds in lighted areas caused by holes in the canopy. Most of these have trunks and grow toward the canopy; others are trunkless, and remain shrubs of the understory. A number of species, the rattans, have become scrambling vines.

Arborescent palms are notably common in areas where winds, un-

Fig. 15.4. Long, linear simple leaves (*right*) can be mechanically less efficient in covering given radial sector near plant than a compound or deeply dissected leaf (*left*) of same effective size and total area. Simple leaves are supported on separate and relatively long lever arms and thus require relatively large amount of support tissue. In compound leaf, lever arms are shortened somewhat and, more importantly, subparallel axes of support are coalesced, leading to greater economy, because strength of vein increases more rapidly than its diameter or cost (see Givnish & Vermeij, 1976).

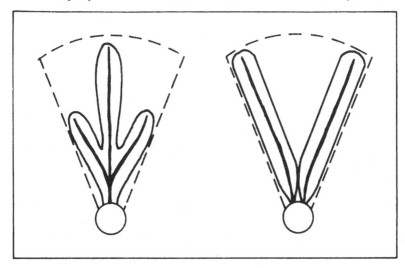

stable soils, or flooding frequently open holes in forest canopies. Thus, for example, palm brakes of *Euterpe* (Puerto Rico, 650–70 m), *"Areca"* (probably *Euterpe;* Guadeloupe, 1000 m), and *Ceroxylon* (Venezuela, 2000 m) occupy steep hillsides in tropical America where winds batter the canopy and ground slippage undermines the trees (Beard, 1944, 1955). In areas of the neotropics that are flooded seasonally, palm swamps develop. The canopy of these swamps is open and, with the understory, is commonly occupied by palms such as *Mauritia, Attalea, Roystonea,* and *Oenocarpus.* Similar formations develop in Australia (Webb, 1959, 1968).

In neotropical areas where seasonal waterlogging alternates with soil desiccation, two other palm-dominated formations, marsh forest and palm forest, develop. In the former, 50% of all trees and 75% of those in the lower story are pinnately leaved palms such as *Manicaria, Jessenia,* and *Euterpe.* In palm marsh, fan palms stand out above a thicket of bushes (Beard, 1944, 1955). There seems to be some tendency, in fact, for fan palms to grow as emergents or in open scrub. For example, the emergent oasis palms of our Southwest deserts, *Washingtonia filifera* and *Washingtonia robusta,* are fan palms, as are the cabbage and saw palmetto of Florida scrub. The Caribbean thatch palm, *Thrinax,* tends to grow exposed in relatively open coastal forests; *Acoelorrhaphe* is another Floridian fan palm that grows in exposed positions or in open hammocks. Fan palms, with their compact crowns, appear well adapted for the mechanical stress and desiccation of winds in exposed situations and seem poorly equipped for growing up into light gaps. The compact crown of a fan palm would be susceptible to shading by a single branch of a competitor. If, however, fan palms are restricted to these open conditions (and that remains to be seen), they hardly have a monopoly of them: The pinnate fronds of *Euterpe* and *Cocos* are able to weather the winds of exposed hillsides and shores, and the date palm *(Phoenix)* of African oases seems to endure the dry desert winds quite well.

Tree ferns share many of the architectural characteristics of palms and tend to replace them in ecologically analogous positions at higher elevations in montane rain forest and in certain flooded habitats in the lowlands. The Araliaceae, a family largely made up of sparsely branched, rapidly growing trees with large, branchlike compound leaves (Chap. 12), is another group of gap-invasive species at higher elevations in the tropics.

In the lowlands, Hartshorn & Orians (pers. comm.) found that many of the tree species that invade forest gaps and openings at La Selva, Costa Rica, develop strong leaders with encircling spirals of compound leaves. Of the simple-leaved species that reproduce in gaps, many employ a subtle strategy that appears to go one step beyond the throwaway branch and essentially amounts to building a branch for nothing. As saplings, these trees develop a drooping leader that acts as a broad branch; this straightens slowly at the base as it lengthens at the tip, so that it acts as a self-renewing leaf surface which continuously metamorphoses into a vertical support, thus yielding both for the price of one. Hallé & Oldeman (1970) describe such axes generally as "mixed."

Throwaway branches in temperate forests
Single-stemmed parasol trees are not found in temperate regions, perhaps because it is too risky to wager the life of a plant on the overwintering success of a single terminal bud (Corner, 1968). However, there are multistemmed but sparsely branched north temperate counterparts; these appear to be found typically in early successional vegetation. Sumacs (*Rhus* and *Toxicodendron*), Kentucky coffee tree (*Gymnocladus dioica*), devil's walking stick (*Aralia spinosa*), Hercules' club (*Zanthoxylum clava-herculis*), certain ashes (*Fraxinus*), and the introduced tree-of-heaven (*Ailanthus*) are all rapidly growing trees and shrubs of early successional situations, each bearing few main branches and numerous pithy compound leaves as throwaway "branches." Walnuts (*Juglans*) are interesting "early successional" trees that create and maintain their own gaps in surrounding vegetation by poisoning their neighbors. Other compound-leaved species, like elderberry (*Sambucus*), box elder (*Acer negundo*), and ash (*Fraxinus*), are common in floodplain forests, where the frequently disturbed and open canopy favors opportunistic growth. Other rapidly growing, sparsely branched trees of early succession, such as the catalpas (*Catalpa speciosa, Catalpa bignonioides*) and introduced empress tree (*Paulownia tomentosa*), do not have compound leaves, but may reap the same kinds of benefits by placing large simple leaves on branchlike petioles. Similar advantages are probably also gathered by the various sparsely branched trees of tropical succession, such as *Cecropia, Musanga, Macaranga*, and *Ochroma*.

Compound leaves in herbaceous plants
Weedy herbs and perennials must face many of the same evolutionary problems as early successional trees, in that they too must compromise between extensive branching and rapid vertical growth. Particularly when such plants are large and growing vigorously (and thus possess the resources to generate leaves of large area) and when the optimal effective leaf size is small, compound leaves should be favored. Although a compound leaf can confer no advantage as a throwaway branch if all a plant's branches are herbaceous, a pinnate or palmate arrangement can allow a consolidation of support tissue

and a shortening of supportive lever arms (see above). This can permit greater mechanical efficiency in covering space near a plant's stem than would many radial, subparallel leaves, especially when such leaves are long and narrow (Fig. 15.4). It is interesting to note that many herbaceous families that display compound leaves, such as the Umbelliferae, Papaveraceae, Fumariaceae, and Ranunculaceae, typically have leaves that are divided into fine, often almost linear segments. The one disadvantage a compound leaf may pose to such plants is that it may commit too much energy to each leaf unit, and thus may reduce the flexibility and increase the vulnerability of the plant's growth strategy.

Summary and outlook for future research

Compound leaves appear to be adaptive in at least two sorts of environmental contexts: in warm, seasonally arid situations that favor the deciduous habit, and in light gap and early successional vegetation where rapid upward growth and competition for light favor the cheap throwaway branch. Compound leaves are to be expected on mechanical grounds in either situation, particularly when the optimal effective leaf size is small and the rate of growth large. To test further the bases and predications of the above theory, the following areas of research appear important:

1. To evaluate quantitatively the advantage of compound leaves in seasonally arid situations, we need to know more about the comparative building costs of rachides and woody twigs, about the potential for water loss from twigs, and about the energetic cost of the roots needed to supply that water (Givnish & Vermeij, 1976). With these data, we can begin to calculate whether deciduous compound leaves have an energetic advantage in such areas.

2. We need to know more about the leaf-shedding behavior of tropical trees and the moisture stress they experience if we are to investigate the significance of compound leaves in seasonal forests and the upper story of rain forests. We also need to know more about the branching strategy and demography of leaves and branches of plants of early succession and light gaps, particularly in the tropics, if we are to evaluate the importance to such plants of compound leaves as throwaway branches. Studies of forest dynamics, such as those of Hartshorn, Oldeman, and Whitmore (Chaps. 23, 26, 27), may cast light on the vertical pattern in leaf complexity within forests insofar as they yield data on the early growth strategies of plants found in different storys in the mature phase; if most emergents begin growth as gap colonizers, this may help explain the high incidence of compound leaves among such species.

3. One interesting prediction of the models presented here, which can be tested easily in the field, is whether compound leaves become less common as their building cost increases, as might occur in areas with nutrient-poor soils. It should be relatively simple to determine whether compound leaves are less common for example, on leached podzols in the tropics than on the richer latosols: A specific approach to this problem might be to look at the relative incidence of compound leaves in typical lowland, heath, and peat swamp forests of the Old World tropics.

4. Perhaps the most exciting area of research suggested by this chapter is the possibility of analyzing that part of the adaptive radiation within and among plant families which depends on their characteristic leaf complexity. The advantages cited for compound leaves and the characteristic ecologic habit found in families like the Palmae and the Araliaceae suggest that these families may have originally radiated on the basis of the adaptive innovation of compound leaves as throwaway branches. The possibility that certain leaf forms may open up new adaptive zones and encourage the radiation of various groups is of great relevance to studies of the evolution of the angiosperms (Hickey & Doyle, 1977; see also Chap. 1). As shown above, the diversity of leaf complexity within families like the Leguminosae may also allow us to trace how the modification of leaf form can allow entry to, and supremacy in, different ecologic roles.

A priori, we might expect that the taxonomically conservative character of leaf complexity should be a less easily modified trait than, for example, leaf width. To change leaf width would require only the prolonging or shortening of meristem activity, but to change leaf complexity would require a major reorganization of the pattern of growth. Even though leaf complexity is a rather conservative trait, its taxonomic restriction need not obscure the patterns predicted, and its conservatism may actually illuminate other patterns. For example, the compound, slightly divided leaf of the creosote bush (*Larrea divaricata*) may at first appear paradoxic in a slow-growing evergreen shrub. However, in the context of its compound-leaved family, the Zygophyllaceae, *Larrea*'s highly reduced, functionally simple leaves contrast appropriately with the more extensive compound leaves found in *Zygophyllum*, a largely deciduous and often somewhat suffrutescent genus of the arid Middle East that makes up almost half the family (van Huyssteen, 1937; Cronquist, 1968; Hunziker, Palacios, De Valisi, & Poggio, 1972; see also Orshan, 1954, for *Zygophyllum dumosum*'s use of compound leaves in a winter-summer heterophyllous strategy). Much interesting work remains to be done on the evolutionary role of leaf complexity in the radiation of tropical families like the Rutaceae, Sapindaceae, and Anacardiaceae; I firmly believe that an integrated study of familial strategies of leaf complexity, branching pattern, and deciduousness will prove a rewarding approach to this research.

Acknowledgments

I gratefully acknowledge the comments and constructive criticism of my adviser, John Terborgh; Henry Horn, Richard Howard, and Jane Menge also read and commented on an earlier draft of this chapter. I also wish to thank Gary Hartshorn and Gordon Orians for allowing me allude to their Costa Rican research.

General discussion

Whitmore: Undoubtedly there is a decrease in pinnate leaves as one goes from lowland to montane tropical rain forest. Most people would probably agree that there are more pinnate leaves in trees at the top than at the bottom of the canopy of a wet lowland tropical rain forest, but I am not terribly happy in your explanation of the adaptive significance of this. What do you say, for example, to the observation that three of the commonest big trees in the wet lowland rain forest of New Guinea, *Pometia pinnata, Dracontomelum puberulum,* and *Intsia bijuga,* are all pinnate and that none of them is deciduous?

Givnish: I do not claim that all species in the upper story ought to have compound leaves or be deciduous. Compound leaves have other advantages (e.g., as throwaway branches), but as yet I have no reason to believe that these advantages apply more in the upper than in the lower story. This is why I have analyzed the characteristics of the *additional* percentage of species with compound leaves in the upper story. In answer to your direct question regarding specific tree species, I am not familiar enough with their ecology to say whether they are using compound leaves as throwaway branches or whether they derive any advantage from it if they do.

Whitmore: You told us that pinnate leaves are advantageous in the top of a rain forest canopy because they can be dropped in the dry season. However, there are these three very common evergreen tree species in New Guinea. The same thing can be seen in Malaya, not quite so strikingly. You need to think of another functional explanation. Have you another suggestion?

Givnish: Of course, not all upper story species with compound leaves in *any* rain forest are deciduous – there are advantages to compound leaves other than minimizing water loss during drought. It is possible that, because of the course of forest succession and the life history strategy of emergents, upper story species may use compound leaves as throwaway branches more frequently than species in the understory. We do not yet know enough about how emergents reach the upper story, how they invade in light gaps (see Chap. 26), or whether they develop as sparsely branched, rapidly growing saplings, to say whether the preponderance of pinnate-leaved species in the canopy is an artifact of their earlier growth strategy.

Tomlinson: I'd like to raise two points. First, why is a throwaway strategy of any particular size of unit efficient? What is the difference between throwing something away in small bits rather than as one big bit? Surely the significant parameter is ontogeny. A simple leaf represents a small unit that can be produced in large numbers, whereas in the development of a compound leaf, the unit is much bigger and the ontogenetic organization much more involved. Dealing in units of smaller currency, there is much greater plasticity. The second question relates to stratification. If there is a significance in the incidence of compound leaves in different strata within the forest, why is there not more heteroblastic diversity in the individual ontogeny of the tree? A small tree has to grow through a succession of strata: We know that marked heteroblastic change involving the morphology of leaves is possible (Chap. 12).

Givnish: First, clearly a compound leaf is a much bigger investment and a greater risk than a large number of simple leaves of the same total area, but put out one at a time, and this contrast is important. This may be partly the reason why compound leaves do not occur in low-productivity environments like tundra or elfin forest. Where they do occur in dry areas, there seems to be a fair amount of productivity. The question I have tried to answer is: If compound and simple leaves can have the same characteristics with respect to convection, water loss, and optics, what are the adaptive differences between them? With compound leaves, a plant regularly tosses away its highest-order branches; with simple leaves, the plant holds on to them and they become cores for future growth. This is part of a compromise; the question is: How much support should a plant put into the leaves themselves rather than into primary support tissues like stems and branches? Second, the araliads described by Dr. Philipson do indeed give many examples of seedling or adult heterophylly, with an ontogenetic trend from simple leaves to compound and occasionally the other way round. The Sterculiaceae provide another example. This phenomenon may not be very common because the transition from a compound leaf to a simple leaf is probably a difficult one simply because it involves a major reorganization of meristem activity.

Jeník: A distinctive aspect of compound leaves is their potential for movement. This is evident in compound leaves in a seasonal or dry climate. The work of Holdsworth (1959), who worked especially with *Bauhinia* leaves, is relevant here. The whole petiole may move, or the two leaflets may move independently. In a more divided leaf there are still further possibilities for movement, either as a protection against insolation or to shed water and so enhance transpiration. The really remarkable advantage seems to be to some extent the possibility for movement of the leaflets, governed by pulvini. The advantages in relation to climate may also extend to the upper canopy of the rain forest.

Givnish: The reason I did not bring pulvini into the argument is that I did not see any a priori reason why simple leaves could not also have pulvini that would allow leaf movement and collapse under dry conditions. If someone can find a physiologic difference between compound and simple leaves that puts simple leaves at a disadvantage with regard to this function, it would certainly suggest another potential advantage of compound leaves. However, many chaparral species (e.g., *Arctostaphylos*) do have simple leaves with marked heliotropic movement, which keeps them oriented edge-on to the sun.

Alvim: One advantage of compound leaves, particularly pinnate leaves, is related to heat dissipation, as indicated by Dr. Jeník, because there is greater convection and a better water supply to their leaflets. There are data (Gates

& Papian, 1971, pp. 80, 83) indicating a higher transpiration rate per unit leaf area from leaflets in certain environments. Single large leaves offer more resistance to water loss and heat dissipation and may reach very high temperatures in direct sunlight, sometimes leading to burning.

Givnish: This is a classic argument for the adaptive value of compound leaves and depends on a comparison between a large simple leaf and a compound leaf of similar area. In this comparison, the compound leaf will allow air to move around it much more easily and will not have to lose as much water to stay cool. Again, what is the difference between a compound leaf and a collection of simple leaves with the same area and the same effective size and resistance to convection? They have the same total area, they have the same convective properties, they lose heat in the same way, they let light filter through the plant crown in the same way, and they lose water in the same way. The big difference to me is that in one case the branch bearing the photosynthetic tissue persists, perhaps for inclusion as a core for future support tissue, and in the other case it is shed. I do not think the argument that compound leaves are adaptive in certain situations because they are effectively small can be used directly.

Sarukhán: I would like to consider your ideas on the adaptive significance of compound leaves within the context of secondary succession. You clearly imply in your proposition that the more xeric an environment becomes the greater the proportion of species possessing compound leaves in the community. Therefore, in secondary succession there would be an initial maximum of compound-leaved species that should decrease as succession advances. However, thinking of most of the components of the first 5 years of secondary succession in the Mexican tropics, all the names that come to my mind are species with simple leaves (e.g., *Waltheria brevipes, Bixa orellana, Heliocarpus donnell-smithii, Helicteres guazumaefolia, Aegiphila deppeana, Tabernaemontana alba, Hamelia erecta*). Later pioneer stages also contain many simple-leaved species among the dominant species (e.g., *Luehea speciosa, Apeiba tibourbou, Ochroma lagopus, Cordia* spp., *Croton* spp.). Only 20% of 170 species observed to occur in a 5-year secondary succession possess compound leaves (Sarukhán, 1964). It seems to me that this situation does not fit well in your arguments.

Givnish: First, the connection I drew between successional status depended on structural, not hydroeconomic, arguments. Increased dryness in early successional habitats should further favor compound leaves only to the extent it favors deciduousness. Second, in streamside or levee successions in the Amazon basin there are very large-leaved, simple or deeply lobed species like *Cecropia* or *Ochroma*. Certainly this appears to be a xeric environment, but most of the areas I have seen have a shallow water table because they are so close to the river or stream. In Peru, John Terborgh and I made some casual observations on the branching strategies used by early successional plants that relate to this. We examined a disturbed hillside in montane rain forest, with a dry crest and a stream at the bottom – the site had been opened by a landslip and appeared about 5–10 years old. At the lower, wet end of the landslip we found plants like *Cecropia*, araliads, and various other megaphyllous monolayer-type trees. Further up the slope toward the fairly xeric crest, the megaphyll monolayers are replaced by mesophyll or even microphyll multilayer species like *Trema*, which have a large number of overlapping branches, layers, and small leaves. To us this indicated that, in the tropics, when a plant has sufficient moisture it can afford to be a large-leaved monolayer, but where conditions are drier the advantages of decreased heat and evaporative loads tend to favor smaller-leaved multilayers, but not compound leaves. I am not saying that all early successional plants ought to have compound leaves; I am saying that they ought to be found frequently.

Stevens: In the definition of a compound leaf you have almost got proof in your premises. Why can you not consider as one group those species that are so numerous in the lower canopy or subcanopy in the tropics that have plagiotropic branch systems, just like compound leaves? This will disrupt statistics with regard to a stratification analysis, but it may lead to a better understanding of the advantage of a compound leaf, per se. With regard to your more xeric seasonal habits, you are probably more or less right, but in the lowland tropics I think the situation is a bit different.

Givnish: If the plagiotropic branches are nonwoody and determinate or nearly so, I would consider them functionally as compound leaves.

Hallé: Compound leaves occur mainly on large twigs and on vertical growing twigs; small simple leaves mainly occur on horizontal slender twigs. Consequently, if you find so many compound leaves in dry open vegetation, in my opinion, it is because in such vegetation most of the axes are orthotropic. On the other hand, if you find so many simple small leaves in the understory of the rain forest, it is because plagiotropy is so common in this particular kind of environment.

Givnish: That means you still have to determine why stems should be orthotropic in one environment and plagiotropic in the other.

Tomlinson: This leads very naturally into a discussion of the point raised earlier (Chap. 12) that there are really two concepts of branch organization: a biologic one and a morphologic one. There are morphologic branches that biologically function like leaves and are also throwaway. Examples in the Hallé-Oldeman system are seen in Cook's model. The morphologic definition of a leaf and a branch should not enter into an ecologic consideration, and I agree with Professor Philipson that the morphologic rules of position are never broken in higher woody plants. One of the illuminating things in Dr. Givnish's presentation is that this distinction is made very clear.

Ng: Surely the main difference between a compound leaf and a branch behaving like a compound leaf is that a compound leaf is an organ of determinate growth and, therefore, limited in its capability. It cannot grow so big as to fill any gap available to it.

Givnish: There are a few exceptions; for example, in some Meliaceae, the compound leaf has a terminal meristematic zone that remains active and continues to produce extensions of the rachis and additional leaflets.

Oldeman: In Cook's model, which occurs in precise ecologic situations, several kinds of organs behave in an intermediate way. Branches exist that are not quite determinate and yet not quite of limited growth. They behave quite like leaves, but have sufficient plasticity to be opportunistic. Many of these phyllomorphic twigs are so specialized that they have definite growth, as Professor Nozeran's work (Chap. 18) with *Phyllanthus urinaria* and other small plants has indicated.

Givnish: One should not think that flexible and more complex strategies are necessarily better. There are some conditions under which stereotyped behavior has a definite advantage, and others under which flexible behavior is favored. One should not wonder at how a "lowly" or "primitive" organism manages to survive in the face of competitors with apparently more sophisticated strategies, but should ask what allows the simpler organisms to succeed in the environments in which they are found.

Baker: I am surprised that nobody today has mentioned C_3 and C_4 plants. I believe most C_4 plants have simple leaves, and maybe these are the ones that endure and are adapted to living through the dry seasons, whereas the ones with compound leaves are C_3 and shed their leaves during the dry season.

Givnish: I would agree that C_4 plants ought to be less deciduous than C_3 plants. However, there are some deciduous C_4 treelets in Hawaii (Pearcy & Troughton, 1974), although these do have simple leaves.

Janzen: Just a comment not to neglect the discussion of leaf movements. How flexible are true branches?

Tomlinson: In certain *Phyllanthus* species the simple leaves of specialized branches do exhibit sleep movements, a further step toward the biologic character of compound leaves.

Janzen: That is certainly a minority among most true branches.

Ashton: I refer to some measurements of leaf movements we have made in *Macaranga hypoleuca* (Chap. 25). Quite a lot of other secondary forest trees with large leaves can serve as examples (e.g., some of the Rubiaceae of the *Sarcocephalus-Nauclea* alliance and a species of the genus *Saurauja* that grows on river banks and moist gaps in the rain forest). However, these have no pulvinus on the leaf at all; quite regularly, when exposed to full sun, the leaves hang down; in effect, they wilt. Biologically the result must be the same. This is not with reference to a drought, just a normal sunny day.

Borchert: I want to continue where Dr. Tomlinson left off and raise a basic question concerning the biologic implication of the whole hypothesis you propose. You try to ascribe the whole phenomena of the distribution of compound leaves essentially to one basic adaptive strategy: namely, that of the throwaway branch. Does this not ignore the fact that growing trees are fairly complex systems, with interplay of a great number of different factors: morphologic ones, as Dr. Hallé has indicated, and physiologic ones, as have been suggested with reference to heat loads, leaf movements, and energetic considerations. Is it permissible even to seek for one single cause to explain such phenomena? The same question should be raised more often in coming discussions because it has a basic implication in the property of trees as dynamic systems.

Givnish: Obviously there is no simple solution. The discussion indicates that, aside from compound leaves being advantageous as throwaway branches, they may have other advantages, such as their power of movement. The bulk of a large compound leaf represents a large energetic risk for a plant, and this may limit the distribution of such leaves. However, I refuse to retreat to a position of claiming that, because there are so many factors and we cannot hope to understand them all, there is no point in drawing conclusions from what little we already do know. A multiplicity of phenomena need not imply a multiplicity of effects. For example, simply because small leaves occur in both dry and nutrient-poor habitats, one cannot claim that two basically different causal phenomena are involved without violating Occam's razor. Indeed, both patterns appear to have a common basis (Givnish & Vermeij, 1976).

References

Asprey, G. F., & Loveless, A. R. (1958). The dry evergreen formations of Jamaica. II. The raised coral beaches of the north coast. *J. Ecol.*, 46, 547–70.

Beard, J. S. (1944). Climax vegetation in tropical America. *Ecology*, 25, 127–58.

– (1946). The natural vegetation of Trinidad. *Oxford For. Mem.*, *20*, 1–152.
– (1955). The classification of tropical American vegetation-types. *Ecology*, *36*, 89–100.
Bews, J. W. (1925). *Plant Forms and Their Evolution in South Africa.* London: Longmans.
– (1927). Studies in the ecological evolution of the Angiosperms. I. *New Phytol.*, *26*, 1–21.
Brown, W. H. (1919). *Vegetation of the Philippine Mountains: The Relation between the Environment and Physical Types at Different Altitudes.* Manila: Department of Agriculture and Natural Resources, Bureau of Science Publication, no. 13, 1–434.
Cain, S. A., Castro, G. M. de O., Pires, J. M., & da Silva, N. T. (1956). Application of some phytosociological techniques to Brazilian rain forest. *Amer. J. Bot.*, *43*, 911–42.
Corner, E. J. H. (1968). *The Life of Plants.* New York: Mentor Press.
Cronquist, A. (1968). *The Evolution and Classification of Flowering Plants.* Boston: Houghton Mifflin.
Fries, R. E. (1909). Zur Kenntnis der Blattmorphologie der *Bauhinien* und verwandter Gattungen. *Arch. Bot.*, *8*(10), 1–16.
Gates, D. M., Alderfer, R., & Taylor, S. E. (1968). Leaf temperatures of desert plants. *Science*, *159*, 954–5.
Gates, D. M., & Papian, L. E. (1971). *Atlas of Energy Budgets of Plant Leaves.* New York: Academic Press.
Givnish, T. J., & Vermeij, G. J. (1976). Sizes and shapes of liane leaves. *Amer. Nat.*, *110*, 743–78.
Grubb, P. J., Lloyd, J. R., Pennington, T. D., & Whitmore, T. C. (1963). A comparison of montane and lowland rain forest in Ecuador. I. The forest structure, physiognomy, and floristics. *J. Ecol.*, *51*, 567–601.
Hallé, F., & Oldeman, R. A. A. (1970). *Essai sur l'architecture et la dynamique de croissance des arbres tropicaux.* Paris: Masson.
Hickey, L. J., & Doyle, J. (1977). Early Cretaceous evidence for angiosperm evolution. *Bot. Rev.* (in press).
Holdsworth, M. (1959). The effects of day length on the movements of pulvinate leaves. *New Phytol.*, *58*, 29–45.
Horn, H. S. (1971). *The Adaptive Geometry of Trees.* Princeton: Princeton University Press.
Hultén, E. (1968). *Flora of Alaska and Neighboring Territories.* Stanford: Stanford University Press.
Hunziker, J. H., Palacios, R. A., De Valisi, A. G., & Poggio, L. (1972). Evolucion en el genero *Larrea*. In *Sobretiro de las Memorias de Symposia del I Congreso Latinoamericano y V mexicano de Botanico*, pp. 265–78. Santa Cruz: Sociedad Botánica de México.
van Huyssteen, D. C. (1937). *Morphologisch-systematische Studien über die Gattung Zygophyllum mit besonderer Berücksichtigung der afrikanischen Arten.* Ph.D. thesis, Friedrich-Wilhelms-Universität, Berlin.
Janzen, D. H. (1974). Tropical blackwater swamps, animals, and mast fruiting in Dipterocarpaceae. *Biotropica*, *6*, 69–103.
Little, E. L., & Wadsworth, F. H. (1964). *Trees of Puerto Rico and the Virgin Islands*, vol. 1. Washington, D.C.: GPO.
Little, E. L., Woodbury, R. O., & Wadsworth, F. H. (1974). *Trees of Puerto Rico and the Virgin Islands*, vol. 2. Washington, D.C.: GPO.
Loveless, A. R., & Asprey, G. F. (1957). The dry evergreen formations of Jamaica. I. The limestone hills of the south coast. *J. Ecol.*, *45*, 799–822.

380   *Thomas J. Givnish*

Monk, C. D. (1966). An ecological significance of evergreenness. *Ecology, 47,* 504–5.

Munz, P. A. (1974). *A Flora of Southern California.* Los Angeles: University of California Press.

Orshan, G. (1953). Notes on the application of Raunkiaer's system of life forms in arid regions. *Palestine J. Bot., 6,* 120–2.

– (1954). Surface reduction as a hydroecological factor. *J. Ecol., 42,* 442–4.

Parkhurst, D. F., & Loucks, O. L. (1972). Optimal leaf size in relation to environment. *J. Ecol., 60,* 505–37.

Pearcy, R. W., & Troughton, J. J. (1974). C_4 photosynthesis in tree-form Euphorbias in wet tropical sites in Hawaii. *Year Book Carnegie Inst., 73,* 809–12.

Radford, A. E., Ahlee, H. E., & Bell, C. R. (1968). *Manual of the Vascular Flora of the Carolinas.* Chapel Hill: University of North Carolina.

Rosendahl, C. O., & Butters, F. K. (1927). *Trees and Shrubs of Minnesota.* Minneapolis: University of Minnesota Press.

Sarukhán, J. (1964). Estudio sucesional de un área talada en Tuxtepec, Oax. In *Contribuciones al estudio ecológico de las zonas cálidohúmedas de México.* Inst. Nac. Inv. For. México, Pub. Esp. nó. 3, pp. 107–72.

Tree Society of Southern Africa. (1964). *Trees and Shrubs of the Witwatersrand.* Johannesburg: Witwatersrand University Press.

Walter, H. (1971). *Ecology of Tropical and Subtropical Vegetation.* Edinburgh: Oliver & Boyd.

– (1973). *Vegetation of the Earth.* New York: Springer-Verlag.

Webb, L. J. (1959). A physiognomic classification of Australian rain forests. *J. Ecol., 47,* 551–70.

– (1968). Environmental relationships of the structural types of Australian rain forest vegetation. *Ecology, 49,* 296–311.

16

Abscission strategies
in the behavior of tropical trees

FREDRICK T. ADDICOTT
Department of Botany, University of California, Davis, California

Abscission is the process whereby plants shed various structures. Any discrete structure can be found to be abscised by one or another species. Leaves, leaflets, stipules, stems, buds, bark, flowers, flower parts, fruits, and seeds are some of the structures commonly abscised. Abscission is under physiologic control and is influenced by a number of internal factors (e.g., water, nutrition, and hormones). Abscission is also sensitive to environmental factors and, indeed, is often triggered by changes in the environment. The involvement of many structures and many physiologic factors in abscission makes the process a complex one. The result of abscission can almost invariably be considered as having some advantage to the plant. Consequently, the various kinds of abscission behavior can be viewed as strategies of value in the survival of the individual or of the species. This chapter summarizes the salient factors in abscission and gives particular attention to the ways in which abscission is of benefit to tropical trees.

Morphology
The processes of abscission are usually restricted to definite regions called "abscission zones." Typically, these occur at the proximal end of an organ, such as a leaflet, pulvinus, petiole, or pedicel. In stems that are abscised the abscission zone is immediately distal to a node. Within the abscission zone the cells of all tissues remain relatively undifferentiated. The cells are smaller and richer in cytoplasm; cell walls show less secondary thickening than in adjacent tissues (Addicott, 1954). For example, resin canals and lactifers may not develop in the abscission zone, and heavily thickened fiber cells are commonly absent (Addicott, 1945). The dense cytoplasm is rich in organelles and is able to function in abscission (i.e., secrete the

necessary hydrolytic enzymes) at almost any stage in the development of the subtending organ. Separation is brought about by the hydrolysis of cell wall constituents, primarily pectins and celluloses, an observation made by the early plant anatomists (see Eames & MacDaniels, 1947). The hydrolytic enzymes are secreted by the cells of a separation layer that becomes active within the abscission zone. Some time after the commencement of the separation process, cell divisions commonly take place across the abscission zone, typically immediately proximal to the separation layer. These cell divisions lead to the development of one or more protective layers that eventually become continuous with the periderm in adjacent portions of the plant. Such observations led von Mohl (1860) to point out that there are two major aspects of the process of abscission: separation and protection. Each of these has its benefits to the plant.

Physiology
The physiology of abscission can be viewed in several lights. In overview, it is a correlation phenomenon. Events or changes in an organ often have a decisive influence on the abscission of the organ. In general, a healthy leaf, fruit, etc., inhibits its own abscission. An injured, diseased, or aborted organ is usually abscised. Further, abscission is a correlation phenomenon in another sense: It is often a reflection of competition among the organs of a plant. Those leaves and fruits that are less well able to compete with more vigorous leaves and fruits are abscised. When the amounts of nutrients become limited, they will be transported to the most vigorous sinks, often with the result that the weaker sinks are abscised. Similarly, under the stress of water or mineral nutrient deficiency, older leaves and sometimes a portion of the flowers or fruits will be abscised. The result of such abscission is to maintain a kind of homeostasis within the plant that tends to keep the remaining organs of the individual in balance with each other.

There are several important physiologic factors that affect abscission, including carbohydrate and nitrogen metabolism, respiration, and plant hormone synthesis. High carbohydrate levels within the plant tend to delay abscission and make it more difficult. Among other effects, high levels of carbohydrates lead to thicker cell walls in the abscission zone. In contrast, low levels of carbohydrates lead to weaker cell walls and easier abscission. High levels of nitrogen within the plant tend to retard and inhibit abscission. Low levels of nitrogen favor abscission. High-nitrogen plants are high-auxin plants and are, in general, quite vigorous and less inclined to abscise young fruits and leaves. The greater leaf retention by young trees can be ascribed, at least in part, to the more favorable nitrogen supplies usually available to young trees (see Addicott & Lyon, 1973).

Abscission is an oxidative process requiring the energy of respiration. Any condition that prevents respiration in the abscission zone also prevents abscission. It is now apparent that energy is required for the synthesis of hydrolytic enzymes – synthesis that is initiated when stimuli for abscission promotion reach the abscission zone. These functions explain the advantage to the plant of maintaining dense cytoplasm in the cells in the abscission zone. Recent electron microscope studies have disclosed the presence of a large number of mitochondria. As abscission approaches, the dictyosomes increase in activity; their vesicles are believed to be the structures by which the hydrolytic enzymes are carried from the cytoplasm to the cell walls (Gilliland, Bornman, & Addicott, 1976).

More than 40 years ago it was shown that the plant hormone auxin was a strong inhibitor of abscission (Laibach, 1933). It now appears to be the principal messenger from subtending organ to the abscission zone, conveying the information that all is well in the organ. When the flow of auxin to the abscission zone becomes reduced through senescence or injury, or is lost (e.g., by removal of a leaf blade from the petiole), the process of abscission is initiated in the abscission zone. A second hormone that affects abscission is abscisic acid. This carries a direct message: Increasing amounts of abscisic acid reaching the abscission zone promote and speed the process. Production of abscisic acid tends to increase as leaves age and is greatly increased in response to water stress and stress of other kinds (see Alvim, Alvim, Lorenzi, & Saunders, 1974; Boussiba, Rikin, & Richmond, 1975). Thus, the two principal hormonal factors tending to promote abscission are a decline in auxin and an increase in abscisic acid. Conversely, high auxin levels and low abscisic acid levels are associated with delay or inhibition of abscission. Another hormone that influences abscission is ethylene. Increased amounts of ethylene are released by many species following injury or other stimuli that promote abscission. Ethylene is strongly promotive of respiration, apparently by increasing the permeability of tissues to oxygen. In those species in which it functions, its role in abscission appears to be that of a secondary messenger, released by the same stimuli that initiate abscission and acting to increase respiration, thereby speeding the synthesis of the hydrolytic enzymes. Two other plant hormones can also influence abscission. These are gibberellin and cytokinin. They act primarily to strengthen sinks; young fruits that are high in gibberellin or cytokinin are unlikely to be abscised. Application of gibberellin or cytokinin to a young fruit usually promotes fruit development and prevents abscission. On the other hand, organs that are relatively low in gibberellins and cytokinins are weak sinks and are likely to be abscised. Collectively, these five hormones are the principal chemical messengers that control the pattern of abscission within a plant (see Addicott, 1970).

In addition to the internal physiologic factors, numerous environmental factors influence abscission in one or more ways. These include light intensity, quality, and photoperiod; temperature; water relations; wind; soil factors, including deficiencies and toxicities; atmospheric factors; and biotic factors such as insects, fungi, and bacteria. The environmental factors influence the internal physiology of the plant in many ways, affecting carbohydrate levels, mineral nutrition, respiration, and hormone levels. The effects of the environmental factors on abscission appear to be largely indirect, through their effects on the physiology of the plant. The plant "integrates" messages conveyed by environmental stimuli and responds with homeostatic behavior that often includes abscission.

Leaf abscission

With respect to the abscission of a leaf, the factors discussed in the preceding sections have direct application. As long as the leaf is vigorous and well supplied with light, minerals, and water, its abscission is inhibited. However, after a leaf becomes senescent or is damaged or diseased, the process of abscission is initiated. It may be of interest to note that the blade typically has rather strong abscission-inhibiting properties. For example, it was necessary to remove more than 75% of the blade of a *Citrus* leaf in order to accelerate abscission (Livingston, 1950). The abscission of individual leaves is an important aspect of the homeostasis of a tree. Leaves are retained much longer when conditions are favorable, as they usually are for a young tree. As the tree matures and reaches the limits of its mineral resources and available sunlight, leaves are retained for shorter periods.

Many species of trees that grow in climates having an alternation of seasons have developed foliage abscission patterns that appear to be beneficial adaptations. In temperate climates, the autumnal deciduous habit is a conspicuous feature of trees growing over vast areas. The current flora appears to have originated during the Cretaceous some 100 million years ago, when the present climatic pattern of alternate hot summers and cold winters became established (Axelrod, 1966). Earlier, a deciduous *Glossopteris* flora had existed in a similar climate of the Carboniferous in Gondwanaland (Plumstead, 1958). The autumnal abscission of these deciduous trees appears to be induced by the advent of cold temperature and/or decreasing day length. The net effect is to defoliate and place the tree in a dormant condition during the period of unfavorable weather. If all goes well, the senescent leaves are abscised and the scars protected by the development of a periderm before pathogenic organisms can gain entrance.

In the tropics, many species are deciduous in regions of alternate wet and dry seasons. For these species the advent of the dry season triggers leaf fall, which commonly results in complete defoliation. The period for which the trees remain leafless varies with species and with the length of the dry period. With these species the triggering factor is clearly moisture stress, probably acting through a marked increase in abscisic acid. In many parts of the tropics, the dry season is of variable length and in some years may not occur. A number of species respond accordingly by retaining their leaves in wet years and abscising them only in dry years. Such species have been termed "facultatively deciduous" by Merrill (1945). Bews (1925) described such species as showing an "irregular deciduous tendency." Observation of such species in Africa and in Southeast Asia indicates that they are very adaptable to the local climatic conditions and to year-to-year climatic variations.

In South Africa, most of the thorn veld species are leafless for short periods in most years, and some are regularly deciduous (e.g., *Erythrina* spp., *Commiphora* spp., *Bersama* spp., *Brachystegia* spp.). The deciduous tendency is much less common among the species of the coast belt (Bews, 1925). In northeastern Brazil, the deciduous trees of the Caatinga abscise their leaves during the dry season and refoliate after the rains resume (Alvim, 1964). Perhaps the most extreme example of this kind of abscission behavior is shown by *Fouquieria* spp. in the deserts of southwestern North America. These remain leafless until the advent of rains, after which they rapidly leaf out to remain in foliage for a few weeks until soil moisture is exhausted. The leaves are abscised rapidly after the soil becomes dry, and the plants remain leafless until the next rain. Thus, in most years, the plants are in leaf for only a few weeks; but should there be more than one period of rain, the plants will leaf out after each period (Darrow, 1943). *Fouquieria* is also of interest because of its unusual spines that develop from the lower part of the petiole and midvein of the primary leaves. When the primary leaves become senescent, a separation layer develops along the entire length of the petiole and a portion of the midvein, with the result that the blade and the upper portions of the petiole are abscised (Henrickson, 1972).

In Trinidad, most deciduous species are leafless for relatively long periods; however, a few species, including *Hymenaea courbaril*, *Platymiscium trinitatis*, *Copaifera officinalis*, and *Albizia caribea*, are leafless for only a few days or weeks in the beginning of the dry season and then leaf out again well before the end of the dry season. Other species abscise only a portion of their foliage during the dry season. Defoliation was not observed during the wet season in Trinidad (Beard, 1946).

An extensive study of leaf abscission habits of trees in Singapore and other locations in Southeast Asia was made by Koriba (1958). He found several patterns of deciduous behavior including (1) trees that are completely deciduous (holodeciduous); (2) trees that shed only a portion of their leaves from some branches or sections of the crown (semideciduous); (3) trees in which all the leaves abscise over a period of a few days but begin leafing out before even this brief period of abscission is complete (vice-deciduous); and (4) trees that annually abscise all the previous years' leaves, but not until the next years' leaves have appeared. This habit appears identical with vernal leaf abscission of many subtropical evergreens, discussed below. Koriba pointed out that these patterns of leaf fall are far from constant and vary with the geographic locations and with the annual variations in weather. For example, species that are deciduous in northern Malaya are often semideciduous in Singapore. In the relatively uniform wet tropical climate of Singapore, deciduous species show various patterns of defoliation, the most common being once a year; but several, such as *Couroupita guianensis*, *Elateriospermum tapos*, *Peltophorum pterocarpum*, *Pentaspadon officinale*, and *Terminalia catappa*, defoliate twice a year. In contrast, *Ficus caulocarpa* defoliates three times a year, and *Heritiera* spp. tend to defoliate once every 2 years. A comparison of leaf fall habits of several trees at two widely separated locations in Southeast Asia, Ceylon and eastern Java, shows that the period of leaf fall comes in December-January in Ceylon and in August-September in eastern Java. This behavior is correlated with the advent of the dry season in the two locations.

The foregoing discussion draws attention to drought as a stimulus for leaf abscission and defoliation. This is a stimulus to which many tropical trees (although far from all) are responsive. It seems apparent that the defoliation response is not necessarily an indication of a need for a period of dormancy on the part of the trees. Although possibly such a need may exist in some species, it is clear that in many species new growth emerges soon after the loss of the old leaves, and in some cases even simultaneously with the loss of the old leaves. The most probable explanation of this behavior is that in all the species concerned, the abscission mechanism is sensitive to moisture stress, and abscission follows the advent of such stress. It would appear that with some species the amount of stress sufficient to induce leaf abscission is not enough to prevent the development of buds and new leaves once the factors of competition and inhibition from old leaves have been eliminated.

There are a number of evergreen species, such as *Citrus* spp., *Theobroma cacao*, and *Eucalyptus* spp., in the tropics and subtropics that commonly retain each leaf for several years. With these species, leaf fall is sporadic and tends to come in waves a short time after a period of hot, dry weather (see Alvim, Machado, & Vello, 1974). Between these waves, the rate of leaf fall is low. In these species, it is the older, disadvantaged leaves that are abscised following the periods of water stress. In southern California *Citrus* also shows a strong peak of leaf abscission in correlation with flowering in the spring (Erickson & Brannaman, 1960).

Among the tropical deciduous trees a few species shed their leaves in the rainy season. For example, in Java, *Spondias mangifera* sheds leaves in the wet season of January to April, and *Tetrameles nudiflora* sheds in December. In Malaya, a species of *Melanorrhoea* sheds its leaves in October-November at the beginning of the rainy season. It is difficult to understand this behavior except by assuming that such species are unusually sensitive to the low light intensities and high humidities that accompany the rainy season (Koriba, 1958).

In the temperate regions, there is substantial evidence that photoperiod is one of the important environmental stimuli promoting leaf abscission of deciduous trees (Addicott & Lyon, 1973). Alvim (1964) observed that in tropical latitudes having seasonal differences in day length, leaf abscission of a number of species occurs during short days. For example, *Hevea brasiliensis* and *Erythrina* spp. lose their leaves from January to March in Costa Rica (lat. $10°$ N) but from July to September in Peru (lat. $10°$ S).

An interesting experiment on this point was performed by Murashige (1966). He observed that *Plumeria* growing in Hawaii (lat. $21°$ N) began losing leaves as the days shortened in October and were completely defoliated by the end of January. When he interrupted the nights with artificial light, simulating a long-day condition, the plants retained their leaves and grew vigorously during the "winter" months.

A number of tropical and subtropical tree species that may properly be classed as evergreen may still have a complete change of foliage each year. Abscission of the previous year's leaves occurs with the advent of the most favorable growing period as the new growth is developing on the tree. Species showing this vernal leaf abscission include the live oaks (*Quercus* spp.), *Cinnamomum camphora*, *Magnolia grandiflora*, *Grevillea* spp., and *Persea americana*. Vernal abscission appears to be triggered by the development of many new sinks on the tree in the form of actively growing vegetative or flowering buds. These sinks become rich sources of auxin and other hormones, and the older leaves weaken as hormone sources. Such a shift in the pattern of hormones appears to account for the leaf abscission.

Flower and young fruit abscission

A number of trees have adopted the strategy of producing, or at least initiating, far more flowers than the tree is capable of maturing into fruit. Such species may show considerable abscission of immature flower buds; or, if the flowers are retained through anthesis, there may be considerable abscission of aborting young fruit a few days after anthesis (e.g., Sweet, 1973). The number of flowers may be vastly greater than the final number of fruits; for example, the avocado (*Persea americana*) commonly develops no more than two fruits for each cluster of a thousand or more flowers. After anthesis, the aborting flowers and their stalks abscise rapidly, in a kind of disarticulation, until the only stalks remaining are those of the one or two developing fruits. Many flowers abscise calyx, petals, anthers, stigma, and styles rather promptly after anthesis. The abscission of these organs appears correlated with fertilization and with the rapid increase in secretion of plant hormones (especially auxin, gibberellin, and cytokinin) by the endosperm and embryo. Many species have adopted the strategy of abscising these organs promptly after their principal function has been served. Other species appear able to tolerate the senescent petals, etc., and do not abscise them. Some species retain chlorophyllaceous floral bracts and calyx and obtain considerable photosynthate from them for the development of the fruit (e.g., *Gossypium* spp.).

With some species, any young fruit that is retained through the early period of possible abscission is not abscised and is kept to maturity. With such species, if stress occurs while the fruit is developing, the fruit may suffer but will not be abscised. With other species, developing fruit may be abscised at almost any time if stress develops from either internal conditions or adverse environmental factors. Thus, various abscission strategies have evolved in relation to flowering and fruit development.

Mature fruit and seed abscission

In various ways abscission is involved in the dissemination of seeds. Fleshy fruits that are attractive to various animals are often abscised as the fruit becomes mature and edible. Abscission may permit the fruit to fall to the ground or merely loosen it to the extent that it can be easily dislodged by the interested herbivore. Dry fruits are usually retained on the tree, but dehisce in a process that appears to be anatomically and cytologically identical to abscission. Dehiscence of legumes and even of some fleshy fruits can be explosive and project seeds for some distance. Abscission of the fleshy fruits facilitates seed dispersal by making the fruit more easily eaten. Birds are among the better-known agents to which trees have adapted their fruit abscission and seed dispersal strategies (Stanley & Kirby, 1973). However, it would be difficult to find a seed dispersal strategy more certain to succeed than that of the species of *Acacia* in Africa whose abscised fruits are very attractive to elephants. Once the seeds are ingested, few factors are able to stop the transport of the seed to wherever the elephant may care to go. During the period of transport the seed coats are softened, facilitating germination, and when the seeds reach the ground they are provided with a substantial, localized application of fertilizer (Douglas-Hamilton & Douglas-Hamilton, 1975).

The mangroves have an interesting strategy in connection with their seed dispersal. The embryo is held in the fruit until well after germination, and is in various stages of development in the different genera at the time of separation. In *Rhizophora* the seedling may reach a length of 50 cm before it is released from the parent plant (Warming, 1909; Egler, 1948; van der Pijl, 1972).

Stem and branch abscission

The phenomena of stem abscission almost exclusively involve the abscission of branch stems (shoots) of various sizes and degrees of development. However, it may be noted in passing that a few species (e.g., the tumbleweed, *Psoralea argophylla*: Becker, 1968) abscise the main stem at the ground level, but this pattern of abscission is quite rare. More common in both tropical and temperate zone trees is abscission of shoot tips that abort after a period of growth. Also quite common is abscission of small twigs with or without the leaves in place, a phenomenon called "cladoptosis." Somewhat larger branches, up to 1 m in length and several centimeters in diameter, may be abscised by some species. This kind of abscission results from physiologic activity in a separation layer within a conspicuous abscission zone. In other species branch abscission is more a matter of attrition. The dead, lower branches eventually rot and fall away. The efficiency with which the wounds left by such branch abscission can be sealed varies greatly from species to species. For many species the wounds are a port of entry of serious infections. The pattern of branch abscission is one of the major factors determining the architecture of a tree.

Shoot tip abscission

For most trees shoot growth is an intermittent process, periods of active growth alternating with periods of rest or dormancy. In many instances this alternation is annual, correlated with alternations in temperature and in rainfall. In some species growth involves two or more flushes of activity a year. For example, *Xanthophyllum curtisii* of Malaya has two flushes of growth per year; *Citrus* spp. may have three or four. There is little information on the factors responsible for multiple annual growth flushes (but see Romberger, 1963); however, present evidence suggests that a flush of shoot growth terminates because of a temporary inability of the roots to supply sufficient water and minerals to maintain a high shoot growth rate. Growth is delayed until the roots are once again able to supply the materials for a flush of growth (see Chap. 21).

As development slows toward the end of a period of growth, any of several changes may take place at the shoot tip, according to the species. The most familiar change is the development of bud scales that enclose the shoot tip during the period of dormancy. Another kind of change is directly pertinent to the present discussion because it involves the abortion of the shoot tip followed shortly by its abscission. In *Citrus*, which is presumably fairly typical, as the end of a period of growth approaches, growth of the terminal few millimeters of the shoot ceases, but the adjacent leaves continue to expand. Obviously the shoot tip has ceased to be an effective sink, whereas the nearby leaves remain strong sinks. An abscission layer develops in the stem tissues immediately distal to the node of the last vigorous leaf, and the aborted shoot tip is abscised. The bud that is now terminal on the shoot is morphologically the axillary bud of the last leaf. In *Citrus* it is a very small bud hardly discernible without magnification. When the next flush of growth occurs, it is this bud that develops into the new shoot. The net result of this tip abortion and abscission is a strategy in which the most terminal bud of the shoot spends the dormant period in a small, relatively undeveloped stage and in a protected location. It would seem a somewhat more efficient strategy than that involving the development of a set of protective scales (see Chap. 12). Another consequence of shoot tip abscission is that morphologically the shoots develop in a sympodial manner, in contrast to the monopodial growth pattern of shoots that enclose the apical bud in scales for the dormant period. The sympodial growth habit can result in trees of striking form, such as the seemingly dichotomously branched crown of *Sapium discolor* and the "pagoda" form of *Alstonia scholaris* (Koriba, 1958).

Branch abscission

Several patterns of branch shoot abscission are recognizable. One involves the annual abscission of all the small leafy branchlets on a tree and can be considered a morphologic variation of the deciduous leaf abscission habit. Trees shedding branchlets in this manner include *Taxodium distichum* and *Metasequoia glyptostroboides.* In a somewhat similar manner the deciduous genus *Larix* abscises its leaves in clusters attached to short shoots. The deciduous leaf abscission of *Glossopteris* of the Carboniferous period involved clusters of leaves on short shoots. Some evergreen conifers follow such patterns when leaves are abscised; *Sequoia* and *Sequoiadendron* abscise branchlets with the leaves attached, and *Pinus* abscises its leaves in the characteristic clusters of usually two to five needles.

Many trees abscise small branchlets one to a few years old, in a kind of self-pruning habit. Such branchlets are often swollen at the base with a conspicuous abscission zone. The abscission layer may function only weakly, and the assistance of a mechanical force such as wind may be required for abscission of the branchlets. Abscission of the individual leaves on a branchlet may precede branchlet abscission in varying degrees. There is considerable variation within and between species in this habit; the abscising branchlet may carry with it all its leaves in fairly green condition. More commonly, many of the leaves fall before the time of branchlet abscission. In other species the branchlets do not break away until after the leaves have fallen and the branchlets have been dead for some time (van der Pijl, 1952; Millington & Chaney, 1973).

The architecture of many trees is influenced by the abscission of primary branches along the main trunk. The "candelabra" trees of the Euphorbiaceae are well-known examples. The distinctive architecture of members of the genus *Araucaria* results from abscission of primary branches followed by different patterns of secondary growth (Chap. 10). In forest trees the abscission of primary branches occurs from the base upward as a result of the crowding and shading of adjacent vegetation (e.g., Ashton, 1964). An excellent example of this is shown by *Agathis australis,* which in the crowded conditions of the New Zealand forest abscises all its primary branches up to a height of 15–20 m. The result is a massive bole of highly prized timber. Growing in the open of a city park, *A. australis* retains its branches down to the ground (Licitis-Lindbergs, 1956).

In some leafless species the mature size is maintained by abscission (disarticulation) at the nodes. Examples of this habit are found in *Ephedra,* and are quite common in the Euphorbiaceae.

A few plants abscise the entire shoot. One such example is a tumbleweed, *Psoralea argophylla*. It is a woody perennial legume that sends up shoots annually; at the end of the growing season the shoots are abscised by the development of abscission layers at the ground level (Becker, 1968). A somewhat similar habit is shown by some members of the Vitaceae: *Cyphostemma juttae* of Africa annually abscises all its shoots back to the short, very fleshy, main stem.

The foregoing discussion has served to direct attention to the role of stem and branch abscission in the life of trees. Branch abscission can have much the same adaptive benefits as leaf abscission by helping to maintain the homeostasis of the individual. Further it is one of the important factors influencing the morphology and architecture of trees.

Bark abscission

Among the many patterns of bark development exhibited by trees, several are characterized by eventual abscission of the outer layers (Chattaway, 1953; Borger, 1973). Perhaps the most impressive examples of bark abscission are found in *Eucalyptus* spp., which annually exfoliate rather thick sheets that fall away in long rolls. In contrast are species of *Combretum, Commiphora,* and *Bursera,* whose abscised bark is thin, papery, and almost transparent. Many species abscise bark in scales or patches in varying sizes and shapes. Conspicuous among these are species of *Cedrus, Picea, Pinus* (e.g., *P. ponderosa*), *Carya,* and *Platanus.* With few exceptions bark is abscised only from the main trunk and larger branches. One of the exceptions is *Arbutus menziesii*, which annually abscises a brittle, paper-thin layer from all but the youngest parts of the tree.

Commonly, bark abscission occurs in hot, dry weather following the period of active growth (Chattaway, 1953). Presumably, such abscission results from a combination of pressure from the recently expanded inner layers of bark and drying and shrinkage of the outer, dead layers. Separation results from the rupture of thin-walled, dead cells. The many studies of bark abscission have disclosed no evidence of physiologic or metabolic activity associated with bark separation. Consequently, bark abscission is of little interest to the plant physiologist, but remains a fascinatingly complex subject to the plant anatomist.

Botanists have had difficulty determining what benefits, if any, bark abscission could have for a tree; many species survive successfully without it. The suggestion that bark abscission could be a deterrent to the establishment of epiphytes has not been universally accepted, as it is largely a phenomenon of the older portions of a tree, leaving the younger portions open to infection, and some trees that show little bark abscission may also have few epiphytes. Undoubtedly bark chemistry as well as other factors in addition to bark abscission could serve to deter epiphytes. It appears to the writer that bark abscission can hardly fail to be of some benefit by helping, at least, to cleanse the trunk and branches of noxious organisms of many kinds. It may be of interest to record in passing a correlation observed in California. *Arbutus menziesii*, which annually abscises bark from all its branches that are 1 year old or more, is not known to be parasitized by mistletoe, although numerous other tree species in the same habitat are so parasitized.

Summary

The attributes of abscission can be summarized as follows:

1. Abscission is a primitive character, known since the Paleozoic era.
2. Abscission occurs widely throughout the higher plants and may affect any organ.
3. Typically abscission results from the enzymatic hydrolysis of cell wall constituents; some abscission may be simply mechanical breakage.
4. Abscission is responsive to a variety of physiologic factors, and is often delicately poised and very responsive to environmental stimuli.
5. General benefits of abscission include: helping to maintain homeostasis (i.e., a balance among vegetative organs and between vegetative and reproductive organs); helping to keep a balance between the plant and the environment; facilitating the recycling of mineral nutrients and the maintenance of desirable soil qualities; protection from infection, etc., through sealing of scars, etc.

The benefits of leaf abscission are:

1. Removal of senescent leaves
2. Removal of injured or infected leaves
3. Removal of excess foliage from stressed plants
4. Defoliation of deciduous trees
5. Recycling of mineral nutrients
6. Protection, by the development of leaf scars and spinescent petioles

The benefits of flower, fruit, and seed abscission are:

1. Removal of excess flowers
2. Removal of flower parts after function
3. Removal of aborted, diseased, or excess fruits
4. Separation of mature fruit
5. Dehiscence of mature fruit
6. Dispersal of seeds

The benefits of branch abscission are:

1. Shoot tip abortion, which facilitates termination of growth
2. Removal of excess crown structure (deramification), whose benefits are the same as for leaf abscission.

The benefits of bark abscission are:

1. Hinderance or prevention of epiphyte establishment
2. Contribution to recycling of nutrients

General discussion

Borchert: You showed a slide of *Quercus suber* having all brown leaves in the spring and in the next slide the new flush was being initiated and yet the old leaves were still green. Was the second one a young plant?

Addicott: Yes, the second slide was from a tree about 12 years old. The first slide was of a somewhat older tree and growing in a more exposed situation.

Alvim: One of the economically most significant types of abscission is that of fruits of intermediate size from young trees. It is impossible to control this loss in spite of a lot of experimental work. Do you have any idea about the mechanism involved, and is there any hope of controlling this? It occurs in crops like oranges, cacao, mango, coconut.

Addicott: What you describe sounds identical with the "June drop" of *Citrus* and deciduous fruits in the northern hemisphere. Hormones and regulators have been completely ineffective in controlling this drop. Horticulturists think it results from an insufficient number of leaves for the fruit load, but I doubt this is the entire answer. My personal speculation is that if we could find a way to stimulate the processes of assimilate translocation, we would go a long way toward solving the problem.

Oldeman: In Surinam, young but fully productive avocado trees were treated with some synthetic hormone-like product because there was a large percentage of abscission. This kept nearly the whole crop on the tree.

Addicott: Also, with deciduous fruit trees, auxin regulators can improve fruit set, but such chemicals have been effective only to counteract some brief difficulty as might result from a period of cold weather. Major fruit set problems have not yielded to hormone applications.

Ng: Holttum (1930) had a theory that the life span of a leaf is relatively fixed according to its intrinsic senescence. He tried to test this in an environment like Singapore, which is relatively constant throughout the year. But one might be able to prove this better in a growth chamber. Is there any experimental evidence in support of this theory?

Addicott: By removing competing structures – upper leaves, lower leaves, fruits – a leaf can be kept functional indefinitely. Environmental factors offset leaf retention: Almost invariably young trees, rich in nitrogen, will hold their leaves longer; a mature tree or a plant that is starved of nitrogen will abscise its leaves sooner.

Tomlinson: There is no such thing as a fixed leaf age; abscission is a phenomenon of growth. We tended in our discussion to relate it to stress. In an ever-growing tree, like *Rhizophora*, phenologic studies have shown that abscission is a very important process in maintaining a constant number of leaves in each shoot. During periods of rapid growth there is rapid leaf loss; during periods of slow growth there is slow leaf loss. So the age of a leaf could be quite short or quite long, depending upon factors that are environmentally modified (Gill & Tomlinson, 1971).

Ng: My next question is: "What is the effect of secondary thickening on the life of the leaf?" The leaf does not undergo secondary thickening, as far as I know; yet it is attached to a stem that does undergo secondary thickening. What is the effect of growth of the stem on life of the leaf in a plant like *Rhizophora?*

Tomlinson: Araucaria would be a better example. The base of persistent leaves is often stretched enormously. There are a lot of leaves that have some secondary thickening at the base; botanists of temperate regions often fail to recognize this.

Addicott: In many species the current crop of leaves is connected with the current crop of xylem and phloem and the current crop of roots. If there is a massive development that would involve degeneration of younger phloem and xylem with which that leaf was connected, then the leaf might degenerate. Dr. Zimmermann may have further comments.

Zimmermann: This is certainly a good textbook generalization, but like all textbook generalizations it has to be modified to accommodate the many exceptions. For example, in many long tree roots, the proximal part may have no current growth ring (e.g., Wilson, 1964). Therefore, old phloem must conduct sugars out to the root apex. In stems of deciduous trees with missing or partial growth rings there must be older functioning phloem. *Tilia* is reported to have phloem that conducts for more than one season, although all leaves drop in autumn. There must be a path from the youngest phloem to older layers. We have demonstrated this anatomically by ciné analysis.

Tomlinson: Dr. Addicott, in explaining something of the physiologic mechanism of abscission, you mentioned the anatomic changes that result in hydrolysis of the middle lamella. Does this same mechanism apply to the abscission of woody parts, as must be common in tropical trees?

Addicott: Yes. All the indications are that the same mechanism of abscission functions in woody plants as in herbaceous ones. Our study of abscission in the *Hibiscus* flower pedicel showed identical ultrastructural changes with what others have found in leaf abscission of *Phaseolus* and *Coleus.* The weakening of tissues that occurs when woody branches are abscised almost certainly involves secretion of enzymes such as pectinases and cellulases.

Tomlinson: Does this mechanism work in secondary xylem?

Addicott: Abscission involving secondary xylem has not yet been examined by electron microscopy, but there is every reason to expect that wherever there are living cells in the abscission zone, they will secrete hydrolytic enzymes at the time of abscission. Evidence is already in the literature of the softening and plasticizing of lignified walls of primary tissues.

396   *Fredrick T. Addicott*

Longman: Dr. Addicott mentioned abscission as being triggered by a large number of factors and referred to periodicity of leaf shedding in the tropics. We have considered the latter in relation to the periodicity of flushing, and one or two points may be helpful in clarification. If you distinguish only between the evergreen and the deciduous condition, physiologic difficulties present themselves. Dr. Jeník and I recognized another category, which we called "leaf exchanging" (Longman & Jeník, 1974). Dr. Addicott showed an example of this in *Quercus suber*, where the old leaves were senescing or falling and the new buds flushed at approximately the same time (Longman & Coutts, 1974). This situation holds in quite a number of tropical trees (e.g., *Terminalia catappa* in West Africa) and applies to those that exchange their leaves once a year and those that do it more often (usually twice). From the physiologic point of view, it is easy to appreciate that the two processes are linked, senescence of the mature leaves probably releasing the terminal buds from predormancy. In the evergreen case, as Dr. Ng mentioned, there is often a tendency for leaf fall to occur after flushing, which may be related to competition between new and old leaves. Conversely, in deciduous trees leaf fall happens long before flushing, and the two processes may not be closely linked. In all these cases, there is the rather surprising general phenomenon that a peak period for bud break tends to occur *before* the rainy season starts. Seasonal peaks of leaf fall may also be found at about the same time, and again in the first half of the dry season.

References

Addicott, F. T. (1945). The anatomy of leaf abscission and experimental defoliation in guayule. *Amer. J. Bot.*, *32*, 250–6.
– (1954). Abscission and plant regulators. In *Plant Regulators in Agriculture*, ed. H. B. Tukey, pp. 99–116. New York: Wiley.
– (1970). Plant hormones in the control of abscission. *Biol. Rev.*, *45*, 485–524.
Addicott, F. T., & Lyon, J. L. (1973). Physiological ecology of abscission. In *Shedding of Plant Parts*, ed. T. T. Kozlowski, pp. 85–124. New York: Academic Press.
Alvim, P. de T. (1964). Tree growth and periodicity in tropical climates. In *The Formation of Wood in Forest Trees*, ed. M. H. Zimmermann, pp. 479–95. New York: Academic Press.
Alvim, P. de T., Machado, A. D., & Vello, F. (1974). Physiological responses of cacao to environmental factors. *Rev. Theobroma (Brasil)*, *4*(4), 3–25.
Alvim, R., Alvim, P. de T., Lorenzi, R., & Saunders, P. F. (1974). The possible role of abscisic acid and cytokinins in growth rhythms of *Theobroma cacao* L. *Rev. Theobroma (Brasil)*, *4*(3), 3–12.
Ashton, P. S. (1964). Ecological studies in the mixed dipterocarp forests of Brunei State. *Oxford For. Mem.*, no. 25, 1–75.
Axelrod, D. I. (1966). Origin of deciduous and evergreen habits in temperate forests. *Evolution*, *20*, 1–15.
Beard, J. S. (1946). The natural vegetation of Trinidad. *Oxford For. Mem.*, no. 20, 1–152.
Becker, D. A. (1968). Stem abscission in the tumbleweed, *Psoralea*. *Amer. J. Bot.*, *55*, 753–6.
Bews, J. W. (1925). *Plant Forms and Their Evolution in South Africa*. London: Longmans. 199 pp.

Borger, G. A. (1973). Development and shedding of bark. In *Shedding of Plant Parts*, ed. T. T. Kozlowski, pp. 205–36. New York: Academic Press.

Boussiba, S., Rikin, A., & Richmond, A. E. (1975). The role of abscisic acid in cross-adaptation of tobacco plants. *Plant Physiol.*, *56*, 337–9.

Chattaway, M. M. (1953). The anatomy of bark. I. The genus *Eucalyptus*. *Aust. J. Bot.*, *1*, 402–33.

Darrow, R. A. (1943). Vegetative and floral growth of *Fouquieria splendens*. *Ecology*, *24*, 397–414.

Douglas-Hamilton, I., & Douglas-Hamilton, O. (1975). *Among the Elephants*. London: Collins & Harvill. 285 pp.

Eames, A. J., & MacDaniels, L. J. (1947). *An Introduction to Plant Anatomy*, 2nd ed. New York: McGraw-Hill. 427 pp.

Egler, F. E. (1948). The dispersal and establishment of red mangrove, *Rhizophora*, in Florida. *Caribbean Forester*, *9*, 299–320.

Erickson, L. C., & Brannaman, B. L. (1960). Abscission of reproductive structures and leaves from orange trees. *Amer. Soc. Hort. Sci. Proc.*, *75*, 222–9.

Gill, A. M., & Tomlinson, P. B. (1971). Studies on the growth of red mangrove (*Rhizophora mangle* L.). 3. Phenology of the shoot. *Biotropica*, *3*, 109–24.

Gilliland, M. R., Bornman, C. H., & Addicott, F. T. (1976). Ultrastructure and acid phosphatase in pedicel abscission of *Hibiscus*. *Amer. J. Bot.*, *63*, 925–35.

Henrickson, J. (1972). A taxonomic revision of the Fouquieriaceae. *Aliso*, *7*, 439–537.

Holttum, R. E. (1930). On periodic leaf-change and flowering of trees in Singapore. *Gard. Bull. Straits Settlements*, *5*, 173–206.

Koriba, K. (1958). On the periodicity of tree-growth in the tropics, with reference to the mode of branching, the leaf-fall, and formation of the resting bud. *Gard. Bull. Singapore*, *17*, 11–81.

Laibach, F. (1933). Wuchsstoffeversuche mit lebenden Orchideenpollinieen. *Ber. Dtsch. Bot. Ges.*, *51*, 386–92.

Licitis-Lindbergs, R. (1956). Branch abscission and disintegration of the female cones of *Agathis australis* Salisb. *Phytomorphology*, *6*, 151–67.

Livingston, G. A. (1950). *In vitro* tests of abscission agents. *Plant Physiol.*, *25*, 711–21.

Longman, K. A., & Coutts, M. P. (1974). Physiology of the oak tree. In *The British Oak*, ed. M. G. Morris & F. H. Perring, pp. 194–221. Faringdon: BSBI/Classey.

Longman, K. A., and Jeník, J. (1974). *Tropical Forest and Its Environment*. London: Longmans.

Merrill, E. D. (1945). *Plant Life in the Pacific World*. New York: Macmillan. 295 pp.

Millington, W. F., & Chaney, W. R. (1973). Shedding of shoots and branches. In *Shedding of Plant Parts*, ed. T. T. Kozlowski, pp. 149–204. New York: Academic Press.

von Mohl, H. (1860). Ueber die anatomischen Veränderungen des Blattgelenkes, welche das Abfallen der Blätter herbeiführen. *Bot. Ztg.*, *18*, 1–17.

Murashige, T. (1966). The deciduous behavior of a tropical plant. *Physiol. Plant.*, *19*, 348–55.

van der Pijl, L. (1952). Absciss-joints in the stems and leaves of tropical plants. *Ned. Akad. Wetensch. (Ser. C)*, *55*, 574–86.

– (1972). *Principles of Dispersal in Higher Plants,* 2nd ed. Berlin: Springer-Verlag. 162 pp.

Plumstead, E. P. (1958). The habit of growth of Glossopteridae. *Trans. Geol. Soc. S. Afr., 61,* 81–94.

Romberger, J. A. (1963). *Meristems, Growth and Development in Woody Plants.* USDA Forest Service Tech. Bull. no. 1293.

Stanley, R. G., and Kirby, E. G. (1973). Shedding of pollen and seeds. In *Shedding of Plant Parts,* ed. T. T. Kozlowski, pp. 295–340. New York: Academic Press.

Sweet, G. B. (1973). Shedding of reproductive structures in forest trees. In *Shedding of Plants Parts,* ed. T. T. Kozlowski, pp. 341–82. New York: Academic Press.

Warming, E. (1909). *Oecology of Plants,* London: Oxford University Press. 422 pp.

Wilson, B. F. (1964). Structure and growth of woody roots of *Acer rubrum* L. *Harvard Forest Paper,* no. 11, 1–14.

Part V: Organizational control

17

Formation of the trunk in woody plants

P. CHAMPAGNAT

Centre National de la Recherche Scientifique, Essone, France

Despite the diversity of morphogenetic expression in woody plants, a tree is recognized by its single trunk and a typical shrub by its inability to develop such a trunk. Certain shrublike forms, characterized by a cluster of equivalent trunks borne at or close to soil level, serve as intermediates between the two extreme conditions (Rauh, 1939). Two general statements can be made on the basis of this observation:

1. All trunks result from the dominance, in the broadest sense of the word, of one main axis (i.e., one meristem) over all others, whether or not directly developed from it. This axis is often, but not always, the epicotyledonary axis resulting from the germination of a seed. Numerous factors are involved in dominance: e.g., superior growth in length and diameter, orthotropy, ability to induce more or less permanent plagiotropy, branching, inability to form flowers or inflorescences directly.

2. When the epicotyledonary axis is destroyed (e.g., accidentally), it can be replaced rapidly and directly by another shoot originating in a distal position from the remaining part. This "positional dominance" surpasses in importance that of the terminal bud itself because even in the absence of this bud the formation of a trunk is assured. This dominance is particularly intense in very young plants. There is no branching of the young trunk; if it is broken, a single replacement axis (relay axis) is developed. Dominance is diminished in somewhat older individuals in a distal position, either in the presence or in the absence of a terminal bud, and several terminal shoots may arise. One of these then becomes dominant, and the others die (natural pruning). The appearance of the first persistent branch, which escapes this natural pruning, is a distinctive step because that

branch is the origin of the first elements of the crown, even if the trunk remains readily discernible. It is known that height of the crown depends not only on the species, but on the environment, density of planting, etc. This branching tendency is accentuated with time, more or less progressively according to species, so that in a mature tree the character and potential of all the vegetative buds are comparable.

These observations pose several problems of terminology that will be discussed before we approach the subject itself.

Apical dominance, acrotony, and trunk formation

The dominance of distal buds and axes over all others is often called "apical dominance." This term is acceptable in a descriptive sense only. It has been used in this way in the preceding paragraphs. The term "acrotony" was proposed by German morphologists to indicate apical dominance in woody plants (Troll, 1937; Rauh, 1939). Either term can be employed in the same sense. However, the question arises whether it is logical to speak of apical dominance or acrotony when one discusses causal morphology or physiology rather than descriptive morphology.

As we will show later, little is known about the factors that determine acrotony. Our present knowledge of the mechanism of apical dominance results largely from analyses of young herbaceous plants, generally annuals such as garden peas or maize, during a very short period of their growth (a few weeks at the most). Even if it is reasonable to apply the results of such work, with minor modifications, to the growth of herbaceous shoots of woody plants, it is not evident how they apply to older axes, between the buds of different branches at various distances from each other.

In the classic researches on correlations between buds of herbaceous plants, insufficient attention has been given to the fact that after removal of the terminal bud from a young individual, it is generally the axillary meristem closest to the level of decapitation that provides the continuing dominant axis. The youth of this meristem and its ability to produce diffusible auxins are cited as the reasons. Rarely, however, is a distinction made between the mechanism leading to the superiority of this distal axillary bud and the mechanism by which the new leader inhibits other axillary buds. If the latter mechanism can logically be ascribed to apical dominance in the classic sense, because of active correlation between buds, then the two mechanisms are different (Chenou, 1976).

It is, therefore, proposed that the term *apical dominance* be used to describe the effect of a parent bud on the daughter buds it produces.

Apical dominance exists equally in a young pea and in a herbaceous shoot of a woody plant. It is also the effect of one of the daughter axes on equivalent axes.

The phenomenon of release of daughter axes, which permits them to become dominant, could then be called *acrotony*. Without previous acrotony, apical dominance is not possible. Acrotony, therefore, is an essential prerequisite for the production of a trunk (Champagnat, 1969).

On the other hand, if basally located buds are released, we speak of *basitony*. This aspect is important when one compares growth in temperate and tropical regions (see Chap. 7) or observes certain species cultivated in growth chambers in a uniform climate (see Chap. 21).

Growth continuity and trunk formation

Trunk formation continues over a long period. The morphogenetic tendencies that favor a single axis must, therefore, be expressed without interruption. For example, in a temperate climate with winter dormancy, the morphology of a young shrub (*Sambucus, Berberis*) is often identical with that of a young tree (*Fraxinus, Carpinus*). Both originate from a seed and cease growth at the end of the summer. In the following spring, after a long period of apparent rest, a distinction can be made. In the tree, effects of dominance that permit the formation of a trunk are conserved with the same strength as in the preceding year. In the shrub, some of these effects, notably acrotony, disappear. The distal buds are no longer favored; numerous vigorous branches are borne in the proximal position. This different behavior of tree and shrub is repeated every spring. Knowledge of the reasons for this morphogenetic discontinuity would permit, on a comparative basis, a better understanding of the general principles that lead to a single dominant axis. It would be of particular interest to investigate the discontinuity of development in cases where it is not seasonally induced, as in tropical shrubs.

If the epicotyledonary axis of a germinated seedling continues growth for an extended period (several years) owing to its inherent functional nature, it produces a trunk by the mechanism of apical dominance in the strict sense. Whether or not it existed, acrotony would pass unobserved. Its absence would be revealed only in the case of accidental destruction of the functioning terminal meristem.

Total apical dominance, constantly maintained, results in a monoaxial form. Apical dominance, partially expressed, permits the appearance of lateral branches, which can be compared to "precocious branches" (the "sylleptic shoots" of German authors: e.g., Spath, 1912) of trees of temperate regions (Champagnat, 1954b). The latter are rarely of significant morphogenetic importance because they are more or less rapidly eliminated by natural pruning, but they can be important in the architecture of many tropical trees. In temperate climates, removal of the weather-imposed inhibition of growth coincides with periods when the growth of the principal axis is particularly rapid. As the season advances, partial dormancy is progressively more strongly imposed on the meristems. Such a process may not exist in a climate where dormancy is absent. It will be of interest to investigate whether the cause of vegetative dormancy in the tropics is different from that known in Western Europe.

Another important aspect of removal of partial apical dominance, the differential growth in thickness of the principal axis and its branches, is also poorly understood. Is it correlative? What degree of dependency is there between the two types of shoots? Is this dependence comparable to that which allows a dominant bud to inhibit a lateral one?

There are tree species that exhibit continuous growth (Hallé & Oldeman, 1970). However, they do not appear to reach exceptional dimensions and are not very common in the tropics. Some kinds of discontinuous growth do not prevent trunk formation; others do, like that of shrubs of the *Sambucus* type growing in a temperate climate. A trunk forms when acrotony persists and replaces apical dominance that is no longer imposed after the cessation of active growth.

Other kinds of discontinuous growth, more apparent than real, also fail to interrupt apical dominance. For example, when young trees are grown under uniform conditions – continuous daylight, moderate light intensity (15,000 ergs cm^{-2} sec^{-1}), constant high temperature (above 20° C) – and one measures indirectly their elongation, internode by internode, and their apparent plastochrone, one notes that their growth is never continuously regular. After being rapid, it may become very slow, even cease for several days before reestablishing a rapid rate (Lavarenne et al., 1971). These changes may occur very abruptly (Fig. 17.1). The terminal bud never aborts; acrotony is maintained. One or more shorter internodes on the mature stem are the result of such irregular growth. The mechanism of this irregular growth is poorly understood. Growth does not seem to follow any regular rhythm that would allow one to predict periods of activity or inactivity. The work of Millet (1970) on *Vicia faba* and of Melin (1973) on *Periploca graeca* seems to indicate that this irregularity may be the combined result of several rhythms.

As another example, discontinuous growth and regular rhythmic growth (by flushes) are very common in tropical climates (Hallé & Oldeman, 1970; Greathouse, Laetsch, & Phinney, 1971; Vogel, 1975).

Fig. 17.1. Irregular and discontinuous growth of woody plants cultivated in uniform conditions: *A, Sambucus nigra* L.; *B, Betula pubescens* Ehrh. In this example, death of the terminal bud does not restrict trunk development because the most distal axillary bud substitutes as a relay axis.

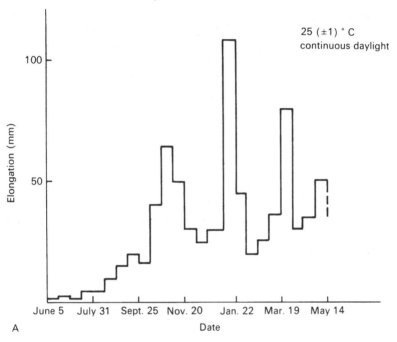

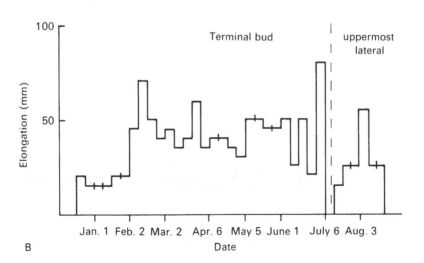

Under certain cultural conditions shoots of certain species may also show no true rest. The classic example is young oaks grown in growth chambers at 25° C, under continuous moderate light intensity (15,000 ergs cm^{-2} sec^{-1}) (Fig. 17.2). There is no interruption of apical dominance; the principal shoot apex resumes activity after periods of rest lasting 8–12 days. If vigor is quite pronounced, the axillary bud nearest the terminal bud grows out simultaneously, without substituting for it. It is noteworthy that under these conditions thickening growth is feeble and the plant becomes lianoid. Histologic study shows that rest is only apparent. At the end of the flush period, terminal bud scales are produced within which the new leaf primordia are formed during the period of apparent rest. When a certain number of primordia has been produced, organogenesis is inhibited. If the primordia are removed before they have reached a length of a few millimeters, the apex continues to generate additional primordia continuously and regularly for more than 6 months. After the cessation of organogenesis, internode elongation commences. There is no evidence for a direct causal link between the two phenomena. But as long as elongation continues, there is no resumption in activity of the apex. Only when the young shoot has reached its final dimensions is there a resumption of organogenesis. There is no proof that the two forms of growth are equally capable of maintain-

Fig. 17.2. Diagram illustrating several characteristics of pedunculate oak (*Quercus robur* L.). Note that apical organogenesis begins when preformed internodes have ceased to elongate.

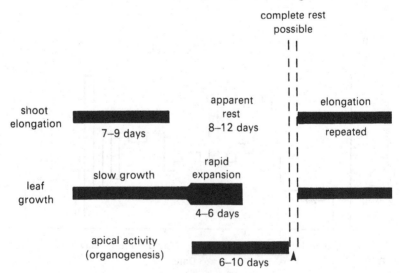

ing an inhibition on the axillary buds. The mechanism of apical dominance should be investigated in such material because it permits a separate analysis of a possible influence of the two types of growth.

We do not know whether tropical species that show a comparable developmental rhythm also have a similar antagonism between organogenesis and elongation of internodes and leaves. We do not even know whether, by artificially modifying the environmental conditions for the growth of oak, we could not suppress this incompatibility. If this were the case, the absence of a real discontinuity in growth would be more evident.

Growth discontinuity opposed to trunk formation

Young elders (*Sambucus*) maintained under uniform conditions (25° C, continuous light) continue to grow irregularly for several years. Apical dominance is total (Fig. 17.3B). Decapitation always releases the bud nearest the point of decapitation. Apical dominance is also responsible for the absence of branching close to the base of the plant. Dormancy is not imposed, because decapitation releases the buds immediately. When the plants are placed in a temperature of 12° C, the terminal bud ceases activity immediately and rest is imposed. Within 3–4 weeks growth does not resume in the terminal bud, but one to three axillary buds very close to the stock grow out (Fig. 17.3A). Their development is slow, and if the low temperature period is long (8–10 weeks) they alone will continue growth and produce vigorous branches after a return to 25° C. The growth discontinuity has upset the morphogenetic tendencies within a much shorter time than that associated with winter rest. The effect is thus not uniquely a result of length of rest period (Barnola, 1972).

If *Corylus* seedlings are placed after germination, at 25° C under conditions of long or continuous days they develop within 4–5 months the habit of young trees. They then cease to grow without evident reason. At that moment, or even as soon as apical activity diminishes (weak apical dominance), branches appear. These are not apical, but basal. The ability to form a trunk is lost, so that henceforth a cluster of axes is produced with no obligatory hierarchy. The same response is shown by *Rhamnus* (Fig. 17.4). If a young ash (*Fraxinus*) is cultivated under the same conditions, it likewise ceases development after several months. Growth renewal can be induced immediately by defoliation of the plant or can be retarded for several months if leaves are left on the plant. Growth is always apical. The basal buds of the single trunk do not develop. Knowledge of the cause of this difference in the behavior of these morphologically

Fig. 17.3. *A. Sambucus nigra* L. cultivated at 25° C in continuous daylight prior to an 8-week period of rest at 12° C. Two buds developed (to right and left of the main axis) that grew into vigorous branches after return to 25° C. *B.* Control kept continuously at 25° C. All basal buds remain dormant. They can develop within several days after decapitation of the main axis.

comparable buds would permit us better to understand why and how a trunk is formed (Barnola, 1976).

Young *Corylus* plants identical to those described above were placed in a heated greenhouse with natural light in March. The temperature was not regulated, but did not drop below 18° C. In July, after continuous, irregular elongation during 3–4 months, the plants became dormant. In the following spring, but without exposure to normally low winter temperatures, the buds close to the base produced long shoots with extended growth. Others swelled and developed into short branches. This behavior was repeated every year. In the resulting complex shrub the shoots of a single year were juxtaposed. There was no cluster of trunks as there is in this plant grown naturally in a temperate climate. This morphology can also be observed in the vicinity of Tananarive, Madagascar, where the climate is similar to these experimental conditions (Rakotondrainibe, 1974; Barnola, 1976).

In central France, where the development of vigorous basal suckers in *Corylus* is associated with an interruption in growth, the situation is very different from that in *Rubus* or *Sambucus*. In *Corylus* no important morphogenetic phenomenon is observed during autumn and winter. Between August 15 and September 16, when all active growth has ceased in young shoots (i.e., when apical dominance is

Fig. 17.4. *Rhamnus frangula* L. seedling cultivated at 25° C under continuous light. Shoots have developed at the level of the stock after the main axis has ceased to grow (cf. Fig. 17.3B).

weak but dormancy has not yet been fully imposed), short branches grow from the stock and reach a length of several centimeters or decimeters. They give rise to the monopodial slender shoots ("baguettes") characteristic of hazel. If the basal shoots are not formed in August, no baguettes appear before the end of the summer of the following year (Barnola, 1976). As in the growth chamber with artificial light, it is the weakness of apical dominance at the end of the vegetative growth period that induces the shrubby habit. Thus, in the *Corylus* type of shrub, the removal of apical dominance at a precise moment is what prevents the plant from becoming a tree. The crucial moment exists only for a short interval at the end of summer. Prior to this, following a suppression of apical dominance, the effect is the same in both cases: A renewal of acrotony contributes to the formation of a trunk.

These examples show that:

1. Continuity of growth with maintenance of apical dominance can prevent the basitonic tendency

2. If this tendency does not exist, and if acrotony is present, even the prolonged absence of apical dominance does not prevent trunk formation

3. If the basitonic tendency is intense, complete cessation of growth is not necessary; strong basitony can exist even without any variation in apical dominance

4. In *Sambucus* and *Corylus*, the basitonic ability is expressed differently and at different times of the year

These observations, though they show clearly the importance and respective roles of apical dominance, acrotony, and basitony in the branching of woody plants, permit no conclusions about the nature of these three morphogenetic principles. We restrict ourselves to a discussion of the last two.

Acrotony and basitony

Experimental analysis can be guided by either of two working hypotheses: (1) the cause(s) of acrotony and basitony lie(s) in the buds themselves or (2) the cause(s) reside(s), at least in part, outside the bud, mainly in the supporting stem and/or in the root system.

The second hypothesis favors the concepts of acrotony and basitony and distinguishes them clearly from apical dominance. Apical dominance assumes primarily correlations between buds, although the importance of the roots and the location of the bud has now become clear (Guern & Usciati, 1972; Champagnat, 1974; Usciati, 1976). Let us now discuss the two hypotheses separately.

Determinism within buds
Experimental principles

The interruption of morphogenesis of the shrubs *Sambucus* or *Corylus* between July and March of the following year (described above) may support the hypothesis of within-bud determinism.

If apical dominance of a herbaceous plant is interrupted (e.g., by decapitation), the distal axillary bud generally becomes dominant: It is the first to be no longer subjected to basipetal inhibition or to have access to previously inaccessible nutritional and hormonal factors. The released apex grows out, acquires dominance, and thus inhibits the less favored shoots.

If, in a herbaceous shoot of a perennial plant with acrotony, decapitation is not followed by outgrowth of another bud, the hypothesis of within-bud determinism is more difficult to apply. A bud may retain dominance throughout the period of rest because of its constitution, structure, or chemical composition (e.g., possessing free or bound growth substances). Inequalities in the level of dormancy may reside in the buds themselves and may be retained or amplified during winter. The simplest method of verifying this hypothesis is to isolate a bud with a portion of the adjacent internode and expose it to different environmental conditions (temperature, light), some of which must be particularly favorable. The period from the experiment to the first signs of growth in 1%, 25%, 50%, or 100% of these "nodal cuttings" can be measured, as well as the total percentage of bud release at the end of the experiment (e.g., after 30, 45, 60 days), and finally the time separating the first and the last of these bud expansions.

Results

In a large number of species the growth potential of buds along an apparently resting branch varies with the bud's position along the axis. Here one can speak of a gradient of fixed properties. In trees (*Fraxinus*, *Tilia*) and shrubs (*Sambucus*, *Rubus*) the gradient does not account for, or contrast, acrotony and basitony. In late summer and early autumn, the buds at the base of the branches or close to the root collar have the greatest growth potential, and the distal ones, particularly the terminals, are the most inert (Fig. 17.5). This tendency is obscured during winter. The disappearance of the gradient is more or less rapid; this varies from year to year and depends on the species. Several weeks before the natural resumption of growth, all buds have usually acquired the same capacity for growth (Fig. 17.5).

Examples of reversal of this tendency have been reported (Crabbé,

1968). In early spring the distal cuttings may develop most rapidly and produce the most vigorous shoots.

In *Juglans regia* the buds at the base of the stem lose a large part of their capacity for growth in March, whereas the distal buds retain, but do not increase, this capacity (Mauget, pers. comm.). This change certainly favors acrotonous branching; its specific character does not allow us to consider it a factor of basic importance.

The autumn basitonic gradient can account only for certain cases of shrubbiness. In *Rubus* and *Sambucus,* the proximal buds of the stem of the young plant can grow from October to December at low temperature (8°–12° C); the other buds cannot (Figs. 17.6, 17.7). Such temperatures are common in autumn in central France; they allow the favored buds to increase 10–100 times in volume within several weeks. This acquired precedence cannot be suppressed (Barnola, 1970).

In *Corylus avellana* no bud expansion occurs at the level of the stock, even in the presence of light, at temperatures below 20° C. Above 20° C the dominance of the proximal cutting is expressed clearly, but in nature in our climate, is not reflected in any morphologic development. As the winter progresses, growth becomes possible at lower and lower temperatures, but as the basitonic tendency is becoming obscure at the same time, it cannot have any morphogenetic consequence.

Fig. 17.5. Basitonic gradient of development of buds on isolated nodes of *Tilia* in December and January (see text). In March no differences are evident. In December terminal buds cannot grow out.

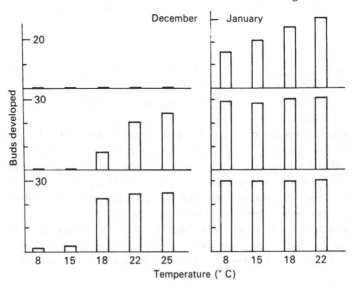

Fig. 17.6. Basitonic gradient of development of buds on isolated nodes of *Rubus*. *A*. In November, when temperatures drop, only basal buds are capable of developing (cf. Fig. 17.5). *B*. In January, difference is much less pronounced. See text.

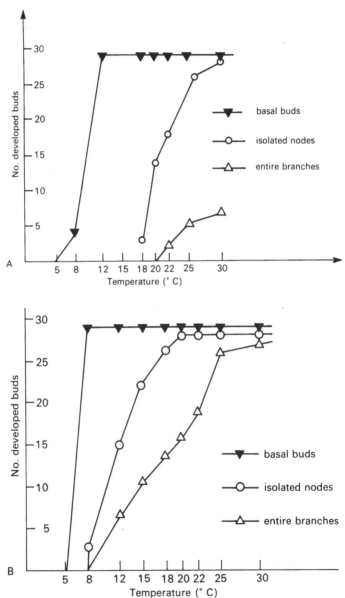

In many trees either at the level of a branch or the whole plant, one finds an identical behavior: Low temperatures are incompatible with growth when basitonic precedence exists. When growth becomes possible, this precedence no longer exists.

It would be interesting to study in tropical climates whether this basipetal gradient exists and whether it is, at least in part, responsible for a shrubby habit.

According to Vegis (1965), there is a close relation between the intensity of dormancy and the possibility of bud burst (or the germination of a seed or resting organ) in a more or less wide range of temperatures. The wider the range, the less intense is the dormancy. This shows in all cases winter development of trees and of shrubs, whatever the level of insertion of the buds. By analogy, our experiments reveal a basitonic gradient of intensity of dormancy that in certain shrubs is strong enough to control branching, at least in part. In most cases, and in a temperate climate, it is without consequence.

Determinism outside buds
Role of roots

The root system of many woody plants is the center of cytokinin synthesis (e.g., Arias & Crabbé, 1975). Roots can also export gibberellins. When dormancy begins and translocation slows, only the

Fig. 17.7. Growth of apical and basal buds of *Rubus idaeus* L. from August to March of following year. Only basal buds grow during autumn and winter. Based on mean of 75 basal buds and 275 apical buds.

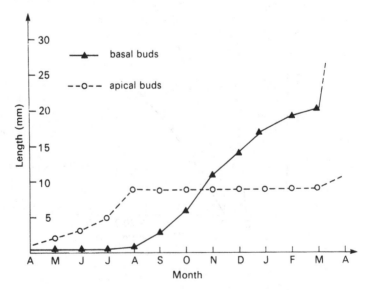

proximal buds may be sufficiently supplied with hormones. They can thus retain a growth potential the others no longer possess (Arias & Crabbé, 1975). Certain basitonic developments may be explained in this way. The terminal bud of the stolon of *Rubus fruticosus* does not root, and its developmental potential remains weak if it does not encounter the soil. If it reaches the soil, the stem ceases to elongate and spontaneous rooting takes place. The ability for bud release increases very rapidly, especially at low temperatures, and becomes identical to that of the stock bud. If the roots are suppressed as they appear, acquisition of strong potentials for growth is much retarded (Barnola, 1971). This very simple model shows how root effects lead to certain basitonies.

Role of the axis
The role of the axis can be demonstrated with the aid of experiments comparable to those using isolated nodes, but employing whole branches or young plants with or without their roots. On entire branches of acrotonic species, it is the distal and particularly the terminal bud that manifests a clear dominance. During the fall the basitonic gradient revealed by the isolated nodal cuttings is not able to express itself in intact axes. It is masked or dominated by a much more powerful acrotonic gradient along the stem.

In basitonic species of the *Rubus* and *Sambucus* type, the stock or root collar buds develop equally well attached or isolated. The presence of distal parts has no effect on them (Fig. 17.8). In the case of young trees, there is no comparable ability at the level of the stock; bud release requires the removal of distal stems. In *Corylus*, isolation from the stock is necessary for autumnal bud release. Capacity for basal bud burst weakens considerably toward winter, and it is weakest with the approach of spring.

In basitonic species of the *Rubus* and *Sambucus* type at the scale of the whole branch, the development of proximal axillary buds is initially inhibited by the acrotonic gradient. This gradient weakens rapidly during the autumn, and from December on, the basal axillary buds can develop; they can then take advantage of the release demonstrated by isolated nodal cuttings (Fig. 17.8).

In summary, acrotony results especially from a powerful "whole-branch factor" (Champagnat, Barnola, & Lavarenne, 1971) that remains effective during the whole dormancy period. In a tree, the whole-branch factor operates at the level of the whole plant and at the level of the branch. In a shrub of the *Corylus* type, the situation is the same, except that shrubbiness can be expressed only at a definite period of the year (i.e., the onset of dormancy). In a shrub of the *Rubus* and *Sambucus* type basipetal inhibition is never effective at the level of the root collar, is much weaker than in the above-mentioned species, and is canceled at the end of autumn. Thus, depending on the species and the conditions of culture, there are several possibilities for the expression of basitony.

Causal correlations of acrotony

The following experiments have been carried out. On a whole branch all buds were removed, except for one inserted in an apical, median, or basal position. The capacity for bud release was compared in the three cases, under various regimens of temperature and illumination. Similarly, the behavior of a single bud in apical, median, or basal position, attached to different axis lengths, was studied. The most interesting case was that of a median bud, attached to the same length of stem placed either apically or basally. All cases showed correlations between the axis and the buds it bears, which determine acrotony and which are thus the essential factors in the formation of the trunk (Meng Horn, Champagnat, Barnola, & Lavarenne, 1975).

A basipetal inhibition is exercised by the stem on proximal buds. It is intense and permanent, at least between September and April, and in the species studied acquires its maximum effectiveness for a

Fig. 17.8. Origin of acrotony and persistence of basitony between October and March in shoots taken from naturally grown *Rubus idaeus* plants, maintained at 25° C in continuous light. *A*, apical buds; *B*, basal buds.

Month	Oct.	Nov.	Dec.	Jan.	Feb.	Mar.
Diagram of sequence of bud burst	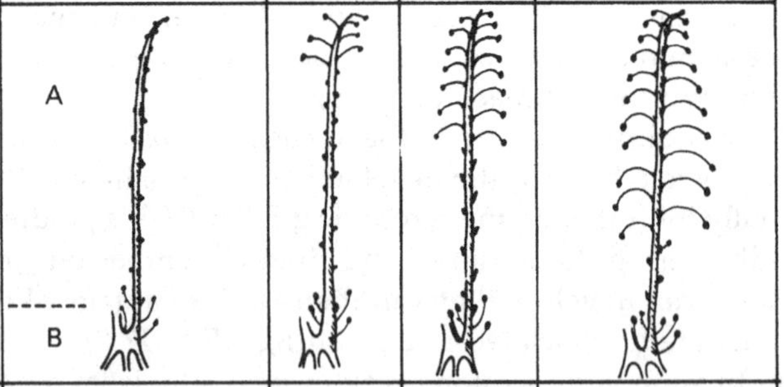 A B					
No. developing buds (B)	4		5	5	5	4
No. developing buds (A)	1		7	12	25	
Days of dormancy (B)	26–15		7–9	6–8	2–3	
Days of dormancy (A)	∞		13–15	9–12	4–6	

length of axis greater than 20 cm (Fig. 17.9). The median buds are suppressed in comparison with the apical ones, which at the same time dominate the basal ones (Fig. 17.10). The buds of the stock of trees are sensitive to this influence; those of certain shrubs are not.

An acropetal stimulation, less general and weaker than the basipetal inhibition, increases the growth potential of the terminal bud

Fig. 17.9. Basipetal inhibition exercised by the stem on bursting of basal buds in *Rhamnus frangula* (experiment of March 20, 1976). Similar results have been obtained with many other species.

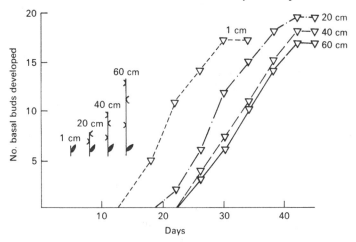

Fig. 17.10. Effect of main axis on development of buds of *Rhamnus fragula*. Bud burst occurs earlier in more distal positions. Each value is based on 12 samples (experiment of March 20, 1974).

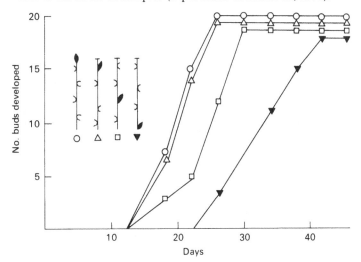

(Fig. 17.11). It appears that this stimulation is greatest if the stem is about 15 cm long. This optimal length decreases as the time of bud break approaches. In certain species it seems to disappear in March and even give way to a contrary acropetal inhibition (Lavarenne, pers. comm.).

These two stem effects, basipetal inhibition and acropetal stimulation, act in the same way and thus inhibit all vigorous branching that would disturb trunk formation. These effects cannot be discerned, at least from November on, in *Sambucus* and *Berberis* (Fig. 17.12). They cannot be opposed to the basitonic precedence. Their absence or weakness coincides with the plant's inability to form a trunk.

Discussion and conclusions

Even if the growth phenomena discussed in this chapter are not essential for trunk formation, they are important factors in the morphogenesis of woody plants. There are many kinds of growth discontinuity, and these have to be studied in tropical areas.

First, for example, there is the climate-dependent growth cessation that may be referred to as "entry into dormancy" and may interrupt morphogenetic events. More should be known about this series of events in tropical climates with one or more dry seasons. Observation and description will have to be followed by detailed experimental analysis.

Fig. 17.11. Acropetal stimulation exercised by axis on development of terminal bud in *Rhamnus frangula*. Each value is based on 20 samples (experiment of February 15, 1974).

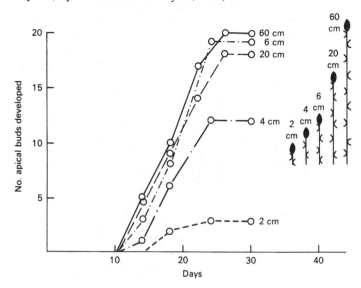

Second, there are slow and progressive endogenous changes that are independent of the environment. Certain shoots can spontaneously lose dominance; the best example is the development of plagiotropy in a principal axis (Roux, 1968; see also discussion by Hallé & Oldeman, 1970, of "mixed" axes). A shrub may result if acrotony is absent. In temperate climates entry into dormancy makes interpretation difficult. Why do trees not have to become dormant under certain uniform climatic conditions (simulated or real)?

Third, a rapid rhythm of growth (of a few weeks), which has nothing to do with true dormancy and is independent of the environment, does not interfere with trunk formation. There are many examples in the tropics where different effects of organogenesis, leaf expansion, and internodal elongation should be studied.

It is true that we know one case of true dormancy being introduced and lost without change in the environment (e.g., *Corylus* grown at 12° C in long days: Barnola et al., 1977), but the periodicity of this rhythm is about 1 year. The mechanism must be original.

There are also slower rhythms of extension growth (lasting several months) that may be synchronizable with climate; this may lead to seasonally synchronous dormancy characteristic of temperate trees. In certain bulbs organogenesis and extension are separated in time and have different induction requirements that may or may not be related to partial dormancy (Le Nard & Cohat, 1968; Le Nard, 1972).

In the previous discussion, the distinction between apical dominance, acrotony, and basitony has been made, and it is clear that acrotony occupies the prime place in trunk formation. Apical dominance must act continuously to produce a trunk alone. Examples are known in herbaceous plants where acrotony can be distinguished from apical dominance and is localized in the bud (Chenou, 1976). In

Fig. 17.12. Effect of axis on one of its buds in 1-year-old shoots. Effect is negligible in *Sambucus nigra,* but strong in *Corylus avellana,* and indicates acrotony expressed within 1 year. *P*, median bud in *proximal* position; *D*, median bud in *distal* position (see Fig. 17.10).

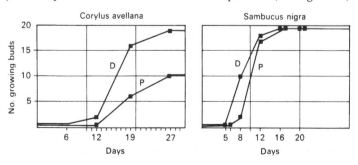

other examples (e.g., Cutter & Chiu, 1976) and in woody plants studied at Clermont-Ferrand, the relation between the axis and the buds
it bears is the result of complex correlations, and the actual mechanism is not known. In other examples significant differentiation
occurs at the moment of initiation of buds, as seems clear in the intense acrotony of a pine or in *Araucaria*.

Acrotony could even be regarded as an expression of the polarity
of all trunk axes. However, it is difficult for polarity to modify the
properties of neighboring buds that do not differ morphologically or
cytologically, but that become distinguished by such characteristics
as vigorous growth of the axes they produce. Furthermore, the clear
polarity of cambial activity, callus formation, and rhizogenesis in a
cutting of *Sambucus* are not equated with hierarchy of dominance
among buds along the entire shoot. A further aspect of trunk formation that merits detailed analysis is the differential accumulation of
wood by branch and trunk axis; this in turn is related to cambial activity and subsequent cellular differentiation (Zimmermann, 1964).
In this respect, a systematic comparison of trees and shrubs would be
of interest.

The "fixed properties" of buds (Champagnat, 1954a), i.e., certain
abilities that buds retain when they remain isolated and that are
subsequently expressed by a distinctive rate or method of growth,
require detailed examination. Numerous examples have been provided by Nozeran's school (Nozeran, Bancilhon, & Neville, 1971;
Nozeran & Neville, 1974). Distal isolated nodes removed from apple
trees in March (Crabbé, 1968) show a response that suggests a factor
for acrotonic precedence has been permanently established.

A similar phenomenon is the differences in the stump sprouts of
young *Fraxinus, Corylus,* and *Sambucus,* which show different reac-
tions or properties at certain times of the year. It may be useful to
make the distinction between "donor organs" (Champagnat, 1974),
like roots (so-called because morphogenetically correlating substances emanate from them), and "receptor organs," like buds. The
latter, having received certain correlative messages, may acquire
properties that render them insensitive to other influences. Examples
are discussed in Champagnat (1965).

The potential for growth expression of a species is often much
greater than the effect of the environment. On the other hand, the
plant may be preadapted to other conditions and under them may
develop an unrecognizable form.

Future research on tropical woody plants should build on the classic foundation of Klebs (1912, 1915) and the older literature. The early
interest in transplanting woody species from temperate to tropical
climates should be revived. The greatest need is for the establish-

ment of cooperative research, and this present symposium indicates
the increasing recognition of this need.

References

Arias, O., & Crabbé, J. (1975). Les gradients morphogénétiques du rameau
d'un an des végétaux ligneux en repos apparent: Données complé-
mentaires fournies par l'étude de *Prunus avium* L. *Physiol. Vég.*, *13*, 69–81.
Barnola, P. (1970). Recherches sur le déterminisme de la basitonie chez le
Framboisier (*Rubus idaeus* L.). *Ann. Sci. Nat. (Bot.)*, *Ser. 12*, *11*, 129–52.
– (1971). Recherches sur le déterminisme du marcottage de l'extrémité apicale des tiges de la Ronce (*Rubus fruticosus* L.). *Rev. Gén. Bot.*, *78*, 185–99.
– (1972). Etude expérimentale de la ramification basitone du Sureau noir
(*Sambucus nigra* L.). *Ann. Sci. Nat. (Bot.)*, *Ser. 12*, *13*, 369–400.
– (1976). Recherches sur le mécanisme du buissonnement chez le noisetier
(*Corylus avellana* L.). *Ann. Sci. Nat. (Bot.)*, *Ser. 12*, *17*, 223–57.
Barnola, P., Champagnat, P., & Lavarenne, S. (1977). Mise en évidence
d'une dormance rythmique chez le Noisetier (*Corylus avellana* L.) cultivé
en conditions contrôlées. *C.R. Acad. Sci. (Paris) 284*, 745–8.
Champagnat, P. (1954a). Les corrélations sur le rameau d'un an des végétaux
ligneux. *Phyton (Argentina)*, *4*, 1–104.
– (1954b). Recherches sur les "rameaux anticipes" des végétaux ligneux.
Rev. Cytol. Biol. Vég., *15*, 1–51.
– (1965). Quelques caractères de la ramification du rameau d'un an des
végétaux ligneux. *C. R. 96 Congr. Soc. Pomol. Fr.*, 9–33.
– (1969). La notion de préséances entre bourgeons. *Bull. Soc. Bot. Fr.*, *116*,
323–48.
– (1974). Introduction à l'étude des complexes de correlations. *Rev. Cytol.
Biol. Vég.*, *37*, 175–208.
Champagnat, P., Barnola, P., & Lavarenne, S. (1971). Premières recherches
sur le déterminisme de l'acrotonie des végétaux ligneux. *Ann. Sci. For.*, *28*,
5–22.
Chenou, E. (1976). Recherches sur le déterminisme de l'inhibition de croissance des bourgeons latéraux chez une fougère aquatique, le *Marsilea
drummondii*. Thesis, Université de Paris. 109 pp.
Crabbé, J. (1968). Evolution annuelle de la capacité intrinsèque de débour-
rement des bourgeons successifs de la pousse de l'année, chez le pommier
et le poirier. *Bull. Soc. Bot. Roy. Belg.*, *101*, 187–94.
Cutter, E. G., & Chiu, H. W. (1976). Differential responses of buds along the
shoot to factors involved in apical dominance. *J. Exp. Bot.*, *26*, 828–40.
Greathouse, W. C., Laetsch, W. M., & Phinney, B. O. (1971). The shoot-
growth rhythm of a tropical tree, *Theobroma cacao*. *Amer. J. Bot.*, *58*, 281–6.
Guern, J., & Usciati, M. (1972). The present status of the problem of apical
dominance. In *Hormonal Regulation in Plant Growth and Development*, ed.
H. Kaldewey & Y. Vardar, pp. 383–400. Weinheim: Verlag Chemie.
Hallé, F., & Oldeman, R. A. A. (1970). *Essai sur l'architecture et la dynamique
de croissance des arbres tropicaux*. Paris: Masson.
Klebs, G. (1912). Ueber die periodischen Erscheinungen tropischer Pflanzen. *Biol. Zbl.*, *32*, 257–85.
– (1915). Ueber Wachstum und Ruhe tropischer Baumarten. *Jb. Wiss. Bot.*,
56, 734–92.

Lavarenne, S., Champagnat, P., & Barnola, P. (1971). Croissance rythmique de quelques végétaux ligneux de regions tempérées cultivés en chambres climatisées à température élevée et constante et sous diverses photopériodes. *Bull. Soc. Bot. Fr.*, *118*, 131–62.

Le Nard, M. (1972). Incidence de séquences de hautes et basses températures sur la différenciation des bourgeons, l'enracinement et la bulbification chez la Tulipe. *Ann. Amélior. Plant.*, *22*, 39–60.

Le Nard, M., & Cohat, J. (1968). Influence des températures de conservation sur l'élongation la floraison et la bulbification de la tulipe (*Tulipa gesneriana* L.). *Ann. Amélior. Plant.*, *18*, 181–215.

Melin, D. (1973). Analyse du déterminisme du port volubile chez une espèce à rameaux polymorphes (*Periploca graeca* L.). Thesis, Université de Besançon. 289 pp.

Meng Horn, C., Champagnat, P., Barnola, P., & Lavarenne, S. (1975). L'axe caulinaire, facteur de préséances entre bourgeons sur le rameau de l'année du *Rhamnus frangula* L. *Physiol. Vég.*, *13*, 335–48.

Millet, B. (1970). Analyse des rythmes de croissance de la Fève (*Vicia faba* L.). Thesis, University de Besançon. 132 pp.

Nozeran, R., Bancilhon, L., & Neville, P. (1971). Intervention of internal correlations in the morphogenesis of higher plants. *Adv. Morphogen.*, *9*, 1–66.

Nozeran, R., & Neville, P. (1974). Morphogenèse des feuilles et des bourgeons: Résultante d'interactions multiples. *Rev. Cytol. Biol. Vég.*, *37*, 217–42.

Rakotondrainibe, F. (1974). Contribution à l'étude du déterminisme de la ramification du noisetier (*Corylus avellana* L.). Thesis, Université de Clermont-Ferrand. 2 vols., 87 pp., 15 pp.

Rauh, W. (1939). Ueber Gesetzmässigkeit der Verzweigung und deren Bedeutung für die Wuchsformen der Pflanzen. *Mitt. Dtsch. Dendr. Ges.*, *52*, 86–111.

Roux, J. (1968). Sur le comportement des axes aériens chez quelques plantes à rameaux végétatifs polymorphes: Le concept de rameau plagiotrope. *Ann. Sci. Nat. (Bot.)*, *Ser. 12*, *9*, 109–255.

Späth, H. L. (1912). *Der Johannistrieb*. Berlin: Parey.

Troll, W. (1937). *Vergleichende Morphologie der höheren Pflanzen*, Bd 1: 1. Berlin: Bornträger.

Usciati, M. (1976). Contribution à la recherche des mécanismes de la levée d'inhibition de croissance des bourgeons axillaires du *Cicer arietinum* L. Thesis, Université de Paris. 162 pp.

Vegis, A. (1965). Die Bedeutung von physikalischen und chemischen Aussenfaktoren bei der Induktion und Beendigung von Ruhezuständen bei organen und Geweben höherer Pflanzen. *Encyclop. Plant Physiol.*, *15* (2), 534–668.

Vogel, M. (1975). Croissance rythmique du Cacaoyer. Thesis, Université de Paris-Sud. 88 pp.

Zimmermann, M. H. (ed.) (1964). *The Formation of Wood in Forest Trees*. New York: Academic Press.

18

Multiple growth correlations in phanerogams

R. NOZERAN
Université de Paris-Sud, Centre d'Orsay, Orsay, France

This chapter discusses in an analytic way growth correlations that appear during the life cycle of a phanerogam, with emphasis on tropical examples. These events begin with the seed from which the plant arises and terminate with the seed it produces. Development depends more or less on environmental factors, such as the photoperiodic regimen and the temperature. In the following discussion it is assumed that external conditions permit normal development of the cycle. In this favorable context, we will attempt to identify the major internal correlations and interactions that ultimately determine morphogenesis.

The internal mechanisms implicated appear as a series of interactions among cells or groups of cells, whether or not they are located in organs. A given cell or group of cells pursuing a genetically determined activity creates around itself a particular medium that influences the activities of its neighbors. These latter cells, thus endowed with new properties, modify in turn their immediate environment, thus affecting still others, including those that originated the initial modifications. The result is a complex system of correlations. The present state of our knowledge of higher plants does not yet permit analyses at the cellular level. Nevertheless, we can analyze organs, groups of organs, and meristems.

Growth correlations of vegetative aerial parts

The normal morphogenetic progression of the vegetative aerial parts is characterized by different types of behaviors among meristems, in particular those responsible for branch differentiation. Several types of behaviors are distinguishable: temporary or permanent rest, change of direction of development, or quantitative control of growth. Although the last type of behavior is important, we will concentrate on the first two.

Behavior of vegetative meristems
Temporary rest
Several examples may be cited. Rest can be spontaneous and an integral part of a normal program: i.e., when meristems display a functional rhythmicity of endogenous origin, as in cacao (*Theobroma cacao* L.), rubber (*Hevea brasiliensis* Müll.-Arg.), and certain *Cephalotaxus* species. Rest ceases the moment internal constraints diminish or are artificially removed. This situation is found in many, probably almost all, meristems of higher plants. Rest can be terminated only when the plant is no longer exposed to certain ecologic conditions (e.g., cold, heat, drought). A good example of this is the dormancy of woody plants in various latitudes.

Permanent rest
Vegetative morphogenesis can involve the loss of function of certain meristems. In species such as *Gleditsia triacanthos* L., *Vitis* spp., and *Maclura pomifera* (Rafin.) Schneider the apex of leafy twigs aborts. In other species the meristems produce a certain number of leaves, after which the cells lose their meristematic properties, differentiate, and abort. This is the case in subtropical Apocynaceae and Euphorbiaceae (Chap. 9). This behavior may be associated with a specific morphogenetic program, such as the short shoots of *Pinus*, the thorny twigs of *Gleditsia triacanthos*, the tendriled twigs of *Vitis* spp., and the plagiotropic shoots of *Phyllanthus urinaria* L.

Change in direction of development
Good examples of a change of morphogenetic direction are plagiotropic branches: e.g., those of *Phyllanthus* spp. (*P. amarus* Schum. & Thonn., *P. odontanius* Müll.-Arg.), gymnosperms (*Araucaria* spp., *Abies* spp.), and coffee (*Coffea arabica* L.). Once induced, the meristems will retain "indefinitely" the modified program of development.

Acquisition of a new function may be less spectacular than the change from orthotropy to plagiotropy, but still may be "permanent." *Hevea brasiliensis* provides an example (Nozeran & du Plessix, 1969).

Dynamics of vegetative meristems
Internal correlations in the plant are well known; it is of particular interest to locate the elements that appear to be the most efficient in provoking a functional change in another part of the plant.

The phenomenon also has to be placed in an evolutionary and dynamic context, in regard to both the inductive and the induced units.

Inductive correlations

So far, little insight has been gained into interaction at the cellular level; experimentation has concerned whole units such as leaves and meristems. Leaves, for example, can control the inhibition of caulinary meristems as well as growth rhythm and dormancy of shoot meristems, and they may induce abortion of vegetative apices. Shoot meristem development is also affected by other parts of the plant (e.g., by other neighboring shoot meristems).

In many species, correlative inhibition is predominantly induced by the young leaves of the terminal bud (Snow, 1929; Champagnat, 1965). These leaves are practically the only factors in *Gleditsia triacanthos* (Nozeran, Bancilhon, & Neville, 1971). It is interesting to note that not only is an inhibiting effect induced in proximally located meristems at a lower level, but function and morphogenesis of the organs produced by these secondary meristems may also be affected. For example, the scaly structure of the primary leaves of the axillary buds of *G. triacanthos* is directly controlled by this type of correlative inhibition (Nozeran et al., 1971).

Consider, for example, the rhythmic growth of *Theobroma cacao*. Vogel (1975) showed that the presence of leaves at a particular developmental stage was necessary for the continuation of rhythmic growth. When young leaves were excised, rhythmic growth was eliminated (Fig. 18.1). Comparable phenomena have been observed in *Hevea* (Hallé & Martin, 1968) and in cuttings of plagiotropic shoots of *Phyllanthus distichus* Müll.-Arg. (L. Bancilhon, pers. comm.). It is, therefore, possible that the normal periodic phenomenon is due to the inducing action of leaves, not only on the terminal meristem itself, but also on the newly formed leaves of the shoot. The ontogeny of such induced leaves is reduced so that they do not develop beyond the stage of scales. Thus they are unable to pass on the effect of their preceding leaves, which by now have lost all inductive potential after their initial period of activity. Therefore, the meristem returns to normal function and produces normal leaves again. These new leaves will then be able to function as inhibitors, and so on. The result is rhythmic growth.

This type of action seems to represent a specific case of a more general phenomenon. The meristem functions, forms a product (a leaf in the above example), and this product has a negative feedback effect on the meristem. The leaves of *Dipsacus strigosus* Willd. and *Dipsacus laciniatus* (Cutter, 1964), *Gleditsia triacanthos* (Bancilhon & Neville, 1966), and *Theobroma cacao* (Vogel, 1975), and the plagiotropic twigs of *Phyllanthus distichus* (Bancilhon & Neville, 1966), all exert a feedback inhibition on the meristems that generated them. If these lateral organs are excised when young, a noticeable acceleration (at least a doubling) in the organogenic activity of their parent meristem results (Fig. 18.1).

The phenomena associated with dormancy are comparable. Young leaves and growing vegetative apices of certain varieties of *Vitis vinifera* L. act in a manner highly unfavorable for the inception of dormancy (Nigond, 1966). After a certain period of time, when they are older, the leaves appear to activate dormancy.

Leaves can cause partial abortion of the terminal meristem. For example, terminal portions of vegetative twigs of some species can be aborted under certain ecologic conditions. In *Gleditsia triacanthos* re-

Fig. 18.1. Effect of removal of young leaves of two different ages on leaf production in orthotropic shoots of *Theobroma cacao* L. In plant *a* leaves were removed when 6 mm long; in plant *b*, when 20 mm long (both figures from 30 plants). Plant *c* is a single control plant. (After M. Vogel, unpubl.)

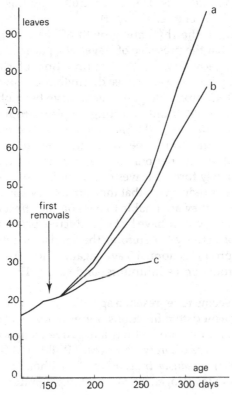

moval of young leaves delays the cessation of growth and, therefore, at least indirectly, the onset of senescence (Fig. 18.2; Neville, 1969). We can ask if comparable phenomena are responsible for the behavior of the apical meristem of the short shoots of *Pinus*, the thorny twigs of *Gleditsia triacanthos*, and the branches of Euphorbiaceae (as in *Anthostema*) or Apocynaceae (as in *Alstonia*), in which loss of the terminal meristem after a certain functional period is the rule. Premature interruption of correlations in *Pinus* and *Gleditsia* can induce the meristem of these branches of normally limited development to continue activity.

After examination of the role of leaves at different stages of devel-

Fig. 18.2. Effect of young leaves on senescence of terminal buds in *Gleditsia triacanthos* L. Ordinate, percentage of plants cultivated in short days with senescent terminal bud; abscissa, age of plants. Broken line, controls with leaves uncut; solid line, plants deprived of five leaves at 14 days, and then at 34 days, deprived once more of four to five newly formed young leaves. Removal of young leaves retards senescence. (After Neville, 1969)

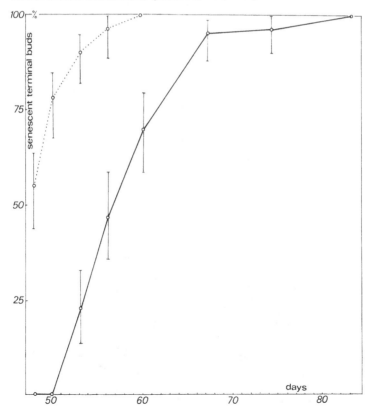

opment in the regulation of growth rate, temporary rest, and abortion of shoot meristems, we shall consider several of the conditions known to be necessary for the modification of the direction of vegetative morphogenesis.

One of the best analyzed phenomena concerns plagiotropy, where a range of different situations has been demonstrated. In *Ajuga reptans* L. (Labiatae) horizontal growth of the stolons is due to the influence of the inflorescence in the middle of the flowering cycle. If the inflorescence is removed before flowering, the stolon becomes erect and produces an orthotropic rosette (Pfirsch, 1962). The same author showed that the plagiotropic epigeal stolons of *Stachys sylvatica* L. (Labiatae) are initially under the influence of the erect axis. The nodes just formed by the terminal bud determine the plagiotropic orientation of the subsequently extending nodes of the lateral axis, and so on (Pfirsch, 1962, 1965). In this example, the plagiotropic orientation is due to the parent foliar axis.

Fig. 18.3. *Phyllanthus amarus* Schum. & Thonn. Fragment of plant showing orthotropic axis (*o*) and plagiotropic, flower-bearing branch (*pl*).

In the dimorphic *Phyllanthus* axis (Fig. 18.3) the plagiotropic function of a lateral meristem has also been established by the meristem of the parent orthotropic axis, but in this case irreversibly (Bancilhon, 1965). Removed from the plant, these lateral twigs remain plagiotropic (Fig. 18.4).

The orthotropic meristem does not always have the role of an organizer. Indeed, during the initial stages of growth following the germination of the seed, there are no plagiotropic branches. They appear only at a later stage of development. The plant, especially its orthotropic meristem, changes its mode of function. It is conceivable that the meristem itself is under the control of organs and tissues it has established during the initial ontogenetic stages.

The list of examples could be greatly expanded, but only a few additional cases will be mentioned. There are correlations within an organ, such as a leaf (Neville, 1964, 1974). There are correlations that determine the morphogenesis of leaves (Pellegrini, 1961; Nozeran & Neville, 1974) and others that determine the histologic structure of parts of the plant (Wardlaw, 1944, 1946a,b; Neville, 1968; Bancilhon, 1972). Polymorphism of roots is also the result of correlative growth (Dyanat-Nejad, 1970). The above examples are limited to those cases where our present knowledge allows the demonstration of interactions exerted in both directions modulating a morphogenetic event. The nature of these changes is a function of the stage attained by each of the causal elements.

Fig. 18.4. *Phyllanthus amarus* Schum. & Thonn. Plagiotropic shoots of same age (*A*) developed on mother plant (*B*) rooted cutting removed from parent plant when young.

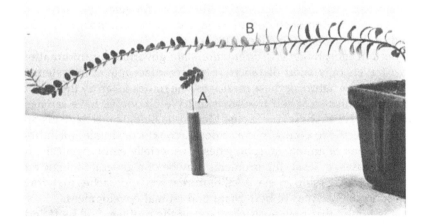

Mechanism of internal correlations

The above examples give the impression that a morphogenetic act accomplished at a determined site is the result of the integration of various signals. These circulating signals are of different polarities, their origins are in different parts of the plant, and these different parts can be at different developmental stages. Analysis of this type of phenomenon presents great difficulties: The effect can be the result of the combined action of several parts, each with a differing mode of function.

The messages emitted by the different active plant parts may sometimes travel considerable distances before reaching the target. Inhibition, for example, may be caused over a long distance or very close by. *Anthocleista nobilis* G. Don (Loganiaceae, West Africa) may serve as an example (Nozeran, 1956). This is a tree that can reach a height of 10–15 m. The trunk bears conspicuous foliar scars, with dormant axillary buds, each enveloped in a pair of prophylls modified into hard thorns. The tree remains monocaulous for a considerable time, but the upper portion begins to branch at the moment the tree flowers. Thus the terminal meristem simultaneously undergoes a transformation and then terminates its activity. If, however, the trunk is decapitated prior to flowering, dormant axillary buds grow out below the cut, regardless of the level at which the cut is made. In this case then, the terminal bud of the main axis inhibits the growth of the lateral buds (originally produced by it), and this inhibition is effective over a distance of several meters.

On the other hand, the points of emission and reception of correlative messages can be very close. A leaf may inhibit its axillary bud. Here, the two elements are in close proximity.

It should be noted that inhibition-inducing messages are not exclusively transported in a basipetal direction. Movement in acropetal direction was demonstrated in *Syringa vulgaris* (Champagnat, 1951) and in *Phyllanthus* (Bancilhon, 1969).

Certain morphogenetic orientations are governed by information that travels only short distances (e.g., organizer role of meristems). What are the nature of these messages, the routes taken by them, and the mechanisms of their transmission? We do not yet have satisfactory answers to these questions. Many physiologic mechanisms are surely specific to plants, but a wider outlook is desirable, including phenomena of animal morphogenesis, especially embryogenesis. In certain cases, at least, the problems may be of a general biologic nature, and their solution may well require novel approaches that draw on specific examples of both plant and animal development.

It is hoped that new ways of attacking the problem will be found.

A new approach has been found by Desbiez (1975), whose remarkable analysis is surely the most thorough and original concerning the correlations between cotyledon and its axillary bud. Three rather recently recognized groups of phenomena can be distinguished:

1. Some morphogenetic processes are indefinitely maintained and some of their aspects are similar to phenomena in the animal kingdom. Certain types of plagiotropy in roots or aerial parts, the behavior of certain buds of *Hedera helix* (Fig. 18.5) and *Hevea brasiliensis* (Nozeran & du Plessix, 1969), and the functional modalities of secondary meristems of *Vitis* (Favre, 1970) belong to this category. We should also mention the possible existence of developmental programs that we call "variants." These processes are determined by functional modes integrated in the genetic apparatus, but they are not expressed under normal conditions.

2. An organizing role is located in a meristem and has an effect on other groups of cells over a certain distance, similar to certain parts of animal embryos. Examples are the determination of certain leaves (Pellegrini, 1961; Nozeran & Neville, 1974), certain plagiotropic twigs (e.g., *Phyllanthus* sp., *Araucaria* sp., *Coffea arabica*), and certain plagiotropic roots (*Theobroma cacao*: Dyanat-Nejad, 1970; Dyanat-Nejad & Neville, 1972). In the last case, it has been found that the organizer messages are transported in meristematic tissue by cells that will ultimately give rise to woody elements (Dyanat-Nejad, 1970; Dyanat-Nejad & Neville, 1972). The stimulus affecting the morphogenesis of a group of cells in a given direction can arise in regions outside the actual meristematic region.

3. Intermediate forms may be produced at a morphogenetically characteristic level. By varying the time of exposure of *Phyllanthus* to the organizer, forms intermediate between plagiotropic and orthotropic twigs can be obtained (Bancilhon, 1965). The same observation has been made in the underground parts of *Theobroma cacao* (Fig. 18.6; Dyanat-Nejad, 1970). These structures, which translate a quantitative signal into a qualitative change, could perhaps be useful in the analysis of these morphogenetic events.

Conclusions
It seems clear that the entire vegetative morphogenesis of the plant is the result of multiple interactions that occur over longer or shorter distances among elements exhibiting temporally variable properties. The sum of these correlations leads to an organism whose functions are more integrated than is initially apparent. The visible result of morphogenesis may be short-lived structures, such as leaves, or structures with a function maintained indefinitely, such as plagio-

432   R. Nozeran

Fig. 18.5. *Hedera helix. A.* Plagiotropic branch axis from climbing plant: phyllotaxis is distichous, leaves are polygonal, adventitious roots (*r*) occur on lower surface. *B.* Orthotropic axis from region of plant that is flowering: phyllotaxis is spiral, leaves are obovoid, and inflorescence is terminal. Cutting of such an axis, when propagated, may produce an "ivy tree."

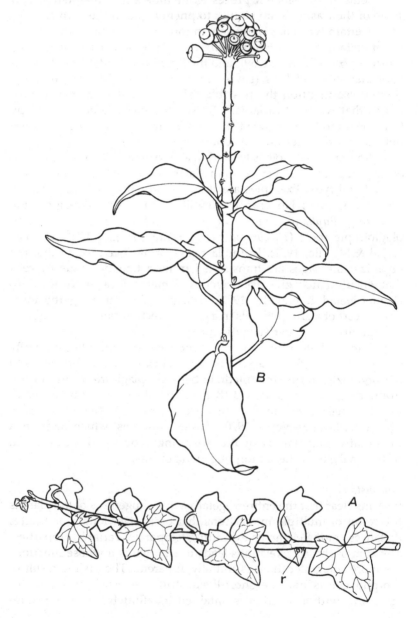

tropic shoots. However, these qualitative results should not make us lose sight of the quantitative aspects of their determination. The plant is a dynamic organism, both in its entirety and at the level of individual organs; even the "arrested" elements are participants. Indeed, in certain cases, one can show that a part of the subsequent morphogenesis of a bud is associated with the period over which it is inhibited (Fig. 18.7).

Rejuvenation

Aging and death are inherent in all plants. In addition to visible morphogenetic aspects, the life cycle is influenced from the onset of germination by modification in the function of the genetic apparatus. This may be translated into loss of function, perhaps by partial repression of hereditary information. For example, portions of branches taken from young seedlings can generate roots more easily than branches taken from adults. Tissue cultures are more easily established from fragments of young plants. In the sequence of events leading to death, certain rejuvenation processes are possible. The most important of these is flowering, but there are others that involve the vegetative portion of the life cycle.

Flowering

Flowering can be induced by ecologic conditions, but in some cases it can be the result of the culmination of correlations involving meristems, leaves, and even roots (Miginiac, 1973). It can be induced by organs that may acquire their inductive capacity gradually. It has been demonstrated that young leaves of *Phyllanthus amarus* Schum. & Thonn. inhibit flowering, whereas mature leaves promote it (Bancilhon, 1969). Similar observations were made on *Scrophularia arguta* Sol. by Chouard & Lourtioux (1959). In other plants the inductive dose can be modified by altering either the ecologic conditions or the correlations. Structures intermediate between floral and vegetative can thus be obtained (Fig. 18.8; Brulfert, 1965, 1968; Roux, 1968).

The morphogenetic sequence of flowering resembles the one described for vegetative growth. The qualitative phenomenon of flowering is a sudden and irreversible change brought about in quantitative steps; under certain conditions these steps can be recognized by the production of intermediate structures.

Once a meristem is induced to flower, senescence and death invariably follow. Nevertheless, the resulting seeds benefit from a return to the original functional mode of the hereditary apparatus that takes place when a new young individual is produced. It is important that there is not a complete return to "zero" because a strict

434   R. Nozeran

Fig. 18.6. *Theobroma cacao. A.* Experimental plan: Embryo is detached and basal extensions of cotyledons are removed to expose hypocotyl radicle, about 6 mm long. Marks are made every millimeter. Controls (c) are left in this state. On other plants, apical millimeter is removed at different developmental stages between 0 and 5 days (0–5). Hypocotyl-radicle elongation, indicated by dotted line (e), was not accounted for. Dashed line, average development of part suppressed. *B.* Radicle system of these plants 15 days later (o, principal orthotropic root; o₂, secondary orthotropic roots; pp, precocious plagiotropic roots formed on portion of principal root existing before germination; pt, retarded plagiotropic roots).

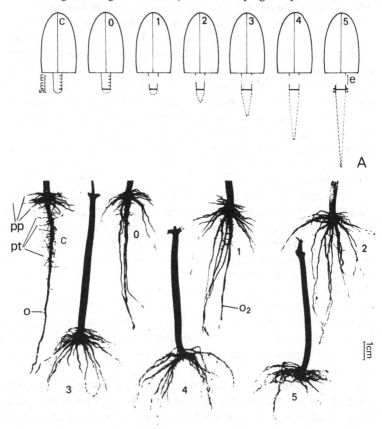

identity between seeds of successive generations is possible only for purely homozygous plants, grown under identical ecologic conditions. It can safely be said that such conditions are never realized in nature.

This momentary reversion is not the result only of sexual reproduction, because plants arising from apomictic embryos, as in *Panicum maximum* Jacq., are indistinguishable from those of the same species arising from fertilized embryos, at least as far as morphogenetic behavior is concerned. Such rejuvenation processes occur also in vegetative parts near the flowering portion.

Fig. 18.6. *Theobroma cacao. C.* Schematic representation of result of experiment (retarded plagiotropic roots are not shown; only two of the six vertical series of precocious roots are represented). Precocious lateral roots closest to wound are more clearly orthotropic the earlier cut was made. Until third day, operation leads to increase in number of precocious roots in region of collar. (After Dyanat-Nejad, 1968)

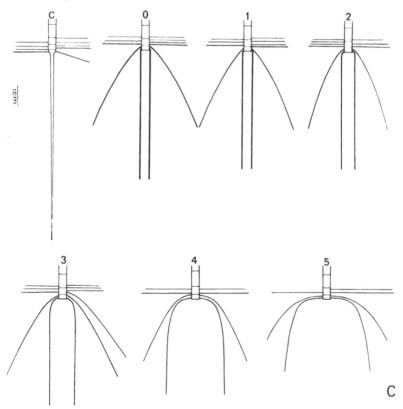

Vegetative rejuvenation

Propagules (e.g., runners, tubers, aerial or underground bulblets, and suckers) give rise to new plants whose characteristics are more or less the same as those of plants arising from seeds. Apomictic seeds represent an extreme case. Vegetative reproduction is often from roots or from other parts close to the ground. These regions retain juvenile-type functions throughout the development of the plant. In other cases, vegetative reproduction takes place near flowers or in their place (i.e., where reversion mechanisms occur).

Fig. 18.7. Modification of secondary orthotropic axis formed in axil of third foliage leaf as function of developmental stage of plant. *A*, *B*. Intact individuals of *Phyllanthus amarus* representing two extreme stages of development on which experiments were conducted (*A*, two plagiotropic shoots; *B*, six plagiotropic shoots). *A'*, *B'*: Individuals corresponding to *A* and *B*, respectively, whose main orthotropic axis (*o*) has been removed above third foliage leaf (l_3) and latent orthotropic buds situated below this leaf have also been removed (*ob* replaced by a cross). For simplification and clarity, leaves borne by plagiotropic twigs are not shown. *c*, cotyledon; l_1 to l_4, assimilating leaves; *o*, principal orthotropic axis; *ob*, latent orthotropic bud; *pl*, plagiotropic twig; *sl*, scale leaf. (Data from Bancilhon, pers. comm.)

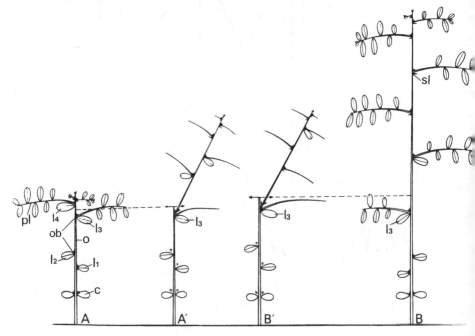

Artificial vegetative propagation makes use of these properties. Parts taken from these two regions of the plant possess the greatest potential (i.e., they are of juvenile character comparable to seedling tissue). They are most useful for tissue culture. Cuttings are best taken from juvenile parts, and adventitious buds arise most easily on them. Numerous cases are known where in vitro cultures are best started with fragments taken from the region of the floral apparatus of a plant (inflorescence axis, floral peduncle, parts of flowers). They have the greatest ability to form buds. This has been observed in the carrot, cabbage, beet (Margara, 1969, 1970), gladiolus (Ziv, Halevy, & Shilo, 1970), narcissus (McChesney, 1972), hybrids of *Alstroemeria* (Amaryllidaceae; Ziv, Kanterovitz, & Shilo, 1973), and tulip (Bancilhon, 1975).

Other parts of the plant exhibit similar phenomena. It has been shown, for example, that in a number of herbaceous *Phyllanthus* species with dimorphic axes, a cutting from a plagiotropic branch

Fig. 18.8. *Anagallis arvensis.* Proliferating flower formed after insufficient induction of flowering. *l*, chlorophyllous leaf; *p*, petal; *pp*, petaloid structure; *s*, sepal; *st*, stamen. (After Brulfert, 1965)

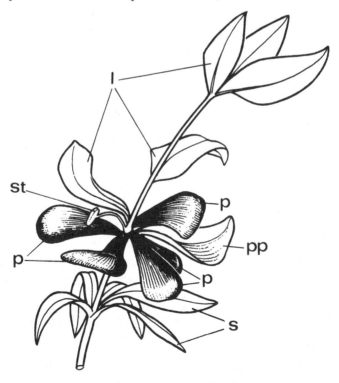

leads to an individual that remains indefinitely plagiotropic (Bancilhon, Nozeran, & Roux, 1963). In *Hevea brasiliensis* (Nozeran & du Plessix, 1969), *Lolium* (Bresse, Hayward, & Thomas, 1965), and *Panicum maximum* (Pernes, Combes, & Rene-Chaume, 1970) there exists in vegetative offshoots a "memory" of parental physiologic processes. In other words, a part of the expression of hereditary infor-

Fig. 18.9. *Vitis vinifera.* Cultured in vitro, clones of this species acquire a morphology identical to that of plants grown from seeds. This organization has been maintained throughout successive cutting for over 15 years. *A,* individual cultivated in a tube; *B,* same removed from tube.

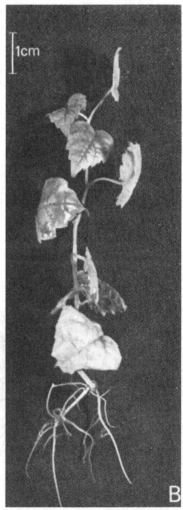

mation contained in the region of vegetative multiplication of the mother plant persists in the progeny.

It is now possible to obtain experimentally individual plants with juvenile features from portions of older plants. New individuals can be grown in vitro from a very small number of cells or, indeed, from one cell. The particular conditions under which buds are regenerated lead to a miniaturization of the meristem that will yield the stem. This meristem then functions in a juvenile state that is expressed morphologically. The resulting shoots are morphologically indistinguishable from those arising from seeds in several species: e.g., in orchids (Morel, 1963, 1964; Champagnat, Morel, Chabut, & Cognet, 1966), in *Solanum tuberosum,* and in *Vitis* (Fig. 18.9). The juvenile state is also expressed by physiologic responses. Several *Citrus* species that propagate as cuttings very poorly in nature acquire root-forming capabilities in vitro quite comparable to those of individuals arising from seeds (Bouzid & Lasram, 1971). These characteristics can be retained for a long time in vitro. A clone of *Vitis* has been cultivated in our laboratory in its juvenile form for over 15 years (Fig. 18.9).

Rejuvenation achieves its ultimate form with embryoids, somatic embryos whose structure and ontogenesis recall those of normal embryos in the seed (Halperin & Wetherell, 1964, 1965). Embryoids have been obtained in calluses from various regions of the plant and from a wide diversity of plants. Others have been produced from cultures of isolated cells.

Comparisons with similar phenomena in mushrooms (Marcou, 1961; Chevaugeon, 1968) suggest that at least a part of the phenomenon of rejuvenation is associated with the decrease in the number of cells participating in the construction of a new individual.

These processes show how rejuvenation may be effected from adult, even old, plant parts entirely by vegetative multiplication. Thus, the plant's life cycle can be artificially regenerated in the absence of sexual reproduction. The discovery of different means, natural and artificial, of inducing the renewal of a plant will perhaps open the door to studies of a similar nature in higher organisms. This will be possible if research concentrates on the fundamental principles that must exist in different organisms.

General discussion

Stein: Dr. Nozeran, you said that defoliation eliminates rhythmic growth in *Theobroma cacao.* This argues against the diversion of materials from the leaves into the meristem. If I understand the system, there is vigorous leaf growth and little activity of the meristem up to a point, and then there is resumption of growth by the meristem because of a redirection of materials into the meristem.

Nozeran: I have never used the word "substance" because there may well be different circulations in different directions; the transmitted message may not be a particular "substance."

Alvim: I believe that there must be a substance, because one has to remove the very small leaves in defoliation experiments. If they are not removed when they are very small, terminal growth will stop. This suggests that when whole, the small leaves expand and remove something that should go to the bud, whose growth is consequently arrested.

Nozeran: There are different circulations, and the messages may be in another form than that of a particular substance. In the particular case of the organizational role of the meristem, I am convinced that the diffusion of a particular substance from the meristem is not the cause of organization or specialization. This is in agreement with work on animal embryology. This is one of the reasons why the chapter, though incomplete, ended with an allusion to animal embryology.

Givnish: You said that in *Ajuga* the orthotropic flowering spike controls the lateral plagiotropic shoots. When the spike was removed, the plagiotropic shoots reverted back to orthotropic spikes. What does this mean to *Ajuga* ecologically? What happens to it in a lawn or pasture when it is mown repeatedly?

Nozeran: Such experiments were done by Madame Pfirsch in Strasburg. In the natural cycle of the plant, when the fruits are dispersed and the inflorescences dry out, the lateral axes grow orthotropically. This happens prematurely when you mow your lawn. The more you mow, the more orthotropy is expressed. Sexual reproduction could not occur if cutting is excessive.

References

Bancilhon, L. (1965). Sur la mise en évidence d'un rôle "organisateur" du méristème apical de l'axe orthotrope de *Phyllanthus*. *C.R. Acad. Sci.*, *260*, 5327–9.

– (1969). Etude expérimentale de la morphogenèse et plus specialement de la floraison d'un groupe de *Phyllanthus* (Euphorbiacées) à rameaux dimorphes. *Ann. Sci. Nat. (Bot.)*, Ser. 12, 10, 127–224.

– (1972). Action des feuilles sur le système vasculaire des rameaux plagiotropes au travers de quelques espèces de *Phyllanthus* à rameaux dimorphes. *Phytomorphology*, 22, 181–94.

– (1975). Premiers essais de multiplication végétative en culture *in vitro* chez *Tulipa gesneriana* L., variété *Paul Richter* (Liliacée). *C.R. Acad. Sci.*, *279*, 983–6.

Bancilhon, L., & Neville, P. (1966). Action régulatrice des jeunes organes latéraux à rôle assimilateur sur l'activité du méristème de la tige principale chez *Phyllanthus distichus* Müll.-Arg. et *Gleditsia triacanthos* L. *C.R. Acad. Sci.*, *263*, 1830–3.

Bancilhon, L., Nozeran, R., & Roux, J. (1963). Observations sur la morphogenèse de l'appareil végétatif de *Phyllanthus* herbacés. *Nat. Monsp.*, *15*, 5–12.

Bouzid, S., & Lasram, M. (1971). Utilisation de cultures *in vitro* pour l'obtention de clones de *Citrus* homogènes et de bon état sanitaire. *Sème Congr. Intern. Agrum. Méditer. C.L.A.M. (Madrid)*, 2, 1–6.

Bresse, E. L., Hayward, M. D., & Thomas, A. C. (1965). Somatic selection in perennial rye-grass. *Heredity*, *20*, 367–9.

Brulfert, J. (1965). Etude expérimentale du développement végétatif et floral chez *Anagallis arvensis* L., ssp. *phoenicea* Scop.: Formation de fleurs prolifères chez cette même espèce. *Rev. Gén. Bot.*, *72*, 641–94.

– (1968). Deux modalités de floraison en position non axillaire chez l'*Anagallis arvensis* L., ssp. *phoenicae* Scop. à la suite d'ébourgeonnements successifs. *C.R. Acad. Sci.*, *267*, 1957–60.

Champagnat, P. (1951). Les corrélations d'inhibition sur la pousse herbacée du Lilas II. Un curieux exemple d'inhibition acropète. *Bull. Assoc. Philomat. Alsace Lorraine*, *2*, 54–6.

– (1965). Physiologie de la croissance et de l'inhibition des bourgeons: Dominance apicale et phénomènes analogues. *Handb. Pflanzenphysiol.*, *15*, 1106–64.

Champagnat, P., Morel, C., Chabut, P., & Cognet, A. M. (1966). Recherches morphologiques et histologiques sur la multiplication végétative de quelques Orchidées du genre *Cymbidium*. *Rev. Gén. Bot.*, *73*, 706–46.

Chevaugeon, J. (1968). Etude expérimentale d'une étape du développement du *Pestalozzia annulata* B. et C. *Ann. Sci. Nat.* (*Bot.*), Ser. 12, *9*, 417–32.

Chouard, P., & Lourtioux, A. (1959). Corrélations et réversions de croissance et de mise à fleurs chez la plante amphicarpique *Scrophularia arguta* Sol. *C.R. Acad. Sci.*, *249*, 889–91.

Cutter, E. C. (1964). Phyllotaxis and apical growth. *New Phytol.*, *63*, 39–46.

Desbiez, M. O. (1975). Base expérimentale d'une interprétation des correlations entre le cotylédon et son bourgeon axillaire. Thesis, Université de Clermont-Ferrand. 174 pp.

Dyanat-Nejad, H. (1970). Contrôle de la plagiotropie des racines latérales chez *Theobroma cacao* L. *Bull. Soc. Bot. Fr.* (*Mém.*),117, 183–92.

Dyanat-Nejad, H., & Neville, P. (1972). Etude du mode d'action du méristème radical orthotrope dans le contrôle de la plagiotropie des racines latérales chez *Theobroma cacao* L. *Rev. Gén. Bot.*, *79*, 319–40.

Favre, J. M. (1970). Influence de facteurs internes sur la cinétique d'enracinement de la vigne cultivée *in vitro*. *Rev. Gén. Bot.*, *77*, 519–62.

Hallé, F. & Martin, R. (1968). Etude de la croissance rythmique chez l'Hévéa (*Hevea brasiliensis* Müll..-Arg. Euphorbiacées-Crotonoïdées). *Adansonia*, *8*, 475–503.

Halperin, W., & Wetherell, P. F. (1964). Adventive embryony in tissue cultures of the wild carrot *Daucus carota* L. *Amer. J. Bot.*, *51*, 274–83.

– (1965). Ontogeny of adventive embryos of wild carrot. *Science*, *147*, 756–8.

Marcou, D. (1961). Notion de longévité et nature cytoplasmique du déterminant de la sénescence chez quelques champignons. *Ann. Sci. nat.*, Ser. 12, *2*, 653–764.

Margara, J. (1969). Sur l'aptitude des tissus au bourgeonnement et à l'initiation florale *in vitro*. *C. R. Acad. Sci.*, *268*, 803–5.

– (1970). Néoformation de bourgeons *in vitro* chez la betterave sucrièré, *Beta vulgaris* L. *C. R. Acad. Sci.*, *269*, 698–701.

McChesney, J. D. (1972). *In vitro* culture of the monocot *Narcissus triandrus* L., cv. "Thalia". *Trans. Kans. Acad. Sci.*, *74*, 44–7.

Miginiac, E. D. (1973). Mise à fleurs et corrélations entre organes chez le *Scrophularia arguta* Sol. Thesis, Université de Paris. 172 pp.

Morel, G. (1963). Le culture *in vitro* du méristème apical de certaines Orchidées. *C. R. Acad. Sci.*, *256*, 4955–7.

– (1964). La culture *in vitro* du méristème apical. *Rev. Cytol. Biol. Vég.*, *27*, 307–14.

442   R. Nozeran

Neville, P. (1964). Corrélations morphogènes entre les différentes parties de la feuille de Gleditsia triacanthos L. Ann. Sci. Nat. (Bot.), Ser. 12, 5, 785–98.

– (1968). Morphogenèse chez Gleditsia triacanthos L. I. Mise en évidence expérimentale de corrélations jouant un rôle dans la morphogenèse et la croissance des bourgeons et des tiges. Ann. Sci. Nat. (Bot.), Ser. 12, 9, 433–510.

– (1969). Morphogenèse chez Gleditsia triacanthos L. III. Etude histologique et expérimentale de la sénescence des bourgeons.Ann. Sci. Nat. (Bot.), Ser. 12, 10, 301–24.

– (1974). Morphogenèse chez Gleditsia triacanthos L. VII. Rôle des ébauches de folioles dans les corrélations morphogènes intrafoliaires. Rev. Cytol. Biol. Vég., 37, 419–28.

Nigond, J. (1966). Recherches sur la dormance des bourgeons de la Vigne. Thesis, Faculté des Sciences, Orsay. 170 pp.

Nozeran, R. (1956). Sur la structure des épines et des bourgeons dormants d'Anthocleista nobilis Don. Nat. Monsp., 8, 167–75.

Nozeran, R., Bancilhon, L., & Neville, P. (1971). Intervention of internal correlations in the morphogenesis of higher plants. Adv. Morphogen., 9, 1–66.

Nozeran, R., & du Plessix, J. (1969). Amélioration de la productivité multiplication végétative et morphogenèse de l'Hevea brasiliensis: Trois aspects d'un même problème. R.G.C.P., 46, 821–7.

Nozeran, R., & Neville, P. (1974). Morphogenèse des feuilles et des bourgeons: Résultante d'interactions multiples. Rev. Cytol. Biol. Vég., 37, 217–42.

Pellegrini, O. (1961). Modificazione delle prospettive morfogenetiche in promordi fogliari chirurgicamente isolati dal meristema apicale del germogiol. Delpinoa N. S., 3, 1–12.

Pernes, J., Combes, D., & Rene-Chaume, R. (1970). Différenciation de populations naturelles du Panicum maximum Jacq. en Côte d'Ivoire par acquisition de modifications transmissibles les unes par graines apomictiques, d'autres par multiplication végétative. C. R. Acad. Sci., 270, 199–205.

Pfirsch, E. (1962). Recherches sur le conditionement plagiotropique chez quelques plantes à stolons. Thesis, Université de Strasbourg. 121 pp.

– (1965). Déterminisme de la croissance plagiotropique chez les stolons epigés Stachys silvatica L. Ann. Sci. Nat. (Bot.), Ser. 12, 6, 339–60.

Roux, J. (1968). Sur le comportement des axes aériens chez quelques plantes à rameaux végétatifs polymorphes: Le concept de rameau plagiotrope. Ann. Sci. Nat. (Bot.), Ser. 12, 9, 109–255.

Snow, R. (1929). The young leaf as the inhibiting organ. New Phytol., 28, 345–58.

Vogel, M. (1975). Recherche du déterminisme du rythme de croissance du Cacaoyer. Café Cacao Thé, 19, 265–90.

Wardlaw, C. W. (1944). Experimental and analytical studies of Pteridophytes. 4. Stelar morphology: Experimental observations on the relation between leaf development and stelar morphology in species of Dryopteris and Onoclea. Ann. Bot. Lond. N. S., 8, 387–99.

– (1946a). Experimental and analytical studies of Pteridophytes. 7. Stelar morphology: The effect of defoliation on the stele of Osmunda and Todea. Ann. Bot. Lond. N. S., 9, 97–107.

– (1946b). Experimental and analytical studies of Pteridophytes. 9. The effect of removing leaf primordia on the development of Angiopteris evicta Hoffm. Ann. Bot. Lond. N. S., 10, 223–35.

Ziv, M., Halevy, A. H., & Shilo, R. (1970). Organs and plantlets regeneration of *Gladiolus* through tissue culture. *Ann. Bot. Lond. N.S.*, *34*, 671–6.

Ziv, M., Kanterovitz, R., & Halevy, A. H. (1973). Vegetative propagation of *Alstroemeria in vitro. Sci. Hort.*, *1*, 271–7.

19

Relation of climate to
growth periodicity in tropical trees

PAULO DE T. ALVIM AND RONALD ALVIM
Centro de Pesquisas do Cacau, Itabuna, Bahia, Brazil

In a review paper on growth rhythms of tropical plants presented at the Second Cabot Symposium (Alvim, 1964), several examples were given of rain forest trees showing periodic changes in leaf production, leaf fall, flowering, and fruiting. In many cases, the periodic phenomena could not be attributed to endogenous control, as suggested by some authors, but rather to changes in external conditions. It was recognized that much experimental work was needed in order to define the regulatory mechanism of growth rhythms in the tropics.

Several papers on the subject have been published in recent years (Janzen, 1967; Daubenmire, 1972; Tomlinson & Gill, 1973; Huxley & Van Eck, 1974; and Frankie, Baker, & Opler, 1974). In regard to the controlling mechanism of periodic growth, reference should be made to the papers by Hallé & Martin (1968), Borchert (1973), Alvim, Machado, & Vello (1972), and Vogel (1975). From a study with young cacao plants growing under controlled environmental conditions, Greathouse, Laetsch, & Phinney (1971) concluded that flushing in this species was under the control of an endogenous mechanism. This conclusion, however, does not seem applicable to mature plants grown under field conditions. Tropical species exhibiting rhythmic growth are always asynchronous with regard to growth flushes when they are young, but invariably show close synchronism in growth behavior when mature (Alvim et al., 1972). This seems to indicate the presence of a triggering stimulus coming from outside. Borchert (1975) explains the difference between juvenile and adult trees in terms of water economy, suggesting that when the plant increases in size, the internal resistance to water movement also increases, and this reduces the plant's capacity to maintain or rapidly restore a balanced water economy. Internal water potential, then, appears to be a factor directly responsible for rhythmic growth, moisture stress

445

being the primary cause of vegetative rest. Alvim et al. (1972) produced experimental evidence indicating that the internal water balance is indeed important, but seems to operate indirectly on the mechanism of growth periodicity, through its influence on the hormonal balance in the plant. This hypothesis has recently received further support from the work of Alvim, Alvim, Lorenzi, & Saunders (1974), who established some provisional relations between bud growth and internal changes in water potential and in the abscisic acid/cytokinin balance. Contrary to the widely postulated concept that bud rest results from moisture deficiency, a theory was advanced suggesting that a period of moisture stress is actually a prerequisite for breaking bud rest, as clearly demonstrated in coffee (Alvim, 1960). A recent study by Magalhães and Angelocci (1976) indicated that the internal water potential of coffee must reach a critical threshold value of at least – 12 bars in order to break bud dormancy. Once this stress requirement is met, the buds will open in response to rain.

The present chapter summarizes the results of some studies on the physiology of growth and flowering of cacao (*Theobroma cacao* L.) and several trees commonly found in the rain forest region of southern Bahia, Brazil. These studies were designed to look at a possible relation between climatic factors and rhythmic phenomena in a typical rain forest region.

Material and methods
The region where the studies were conducted is located near the coastal area of Bahia at a latitude of 12°–16° S. It has a mean annual temperature of about 23.5° C and a total rainfall of around 1700 mm, with no pronounced dry season. There are, however, variations in the rainfall pattern from year to year.

The studies with cacao included greenhouse experiments at the Cacao Research Center, near Itabuna (lat. 14°43′ S), and phenologic observations on mature plants under field conditions in two localities: Juçari (lat. 15°25′ S) and São Francisco do Conde, near Salvador (lat. 12°37′ S). Normally, rainfall is more evenly distributed during the year in Juçari than in São Francisco do Conde.

The phenologic studies on forest trees were carried out in a 40-ha biologic reserve located at the Cacao Research Center.

The experimental procedures described below were adopted.

Greenhouse experiments with cacao
Cacao trees aged 3 years were grown in 250-liter pots and subjected to different water regimens. In five such pots, soil moisture was monitored by means of Bouyoucos gypsum blocks placed at

depths of 6, 12, 18, and 24 in. Data were taken on the frequency of flushing and leaf shedding, and on diurnal changes in stem circumference. For the latter, the liquid displacement dendrometer recently developed by Alvim (1975) was used, and daily readings were taken at 7 A.M. and 3 P.M.

Phenologic studies on cacao

In Juçari, the observations were made on 16 8-year-old plants, growing without overhead shade. In São Francisco do Conde the observations were made on 10 trees about 5 years old, which were partially shaded by bananas and *Erythrina glauca*. In both areas the experimental procedures adopted to study growth and flowering rhythms were the same as described by Alvim et al. (1972). Growth flushes were recorded by counting each week the newly formed leaves on 10 selected shoots per plant. The weekly leaf fall was estimated by counting the number of leaves shed over 10 1-m² collectors, randomly placed under the trees. Flowering intensity was assessed by counting, at weekly intervals, the number of nonfertilized flowers that had dropped on a plastic net, about 2 by 2 m, placed around the trunks of 10 plants. Trunk growth was measured by daily readings of liquid displacement dendrometers (Alvim, 1975) at 7–8 A.M.

Cumulative soil moisture deficits and excesses have been used in previous work to establish correlations between the rhythmic shoot growth pattern of cacao and the alternations of periods of moisture stress and rain (Alvim et al., 1972). A simplified method of representing rainfall distribution was used in the experiment described here. It depicts the alternation of relatively dry and wet periods (hydroperiodicity) in a clearer way. The new system consists in discarding daily rainfall data below 4 mm, which is approximately equivalent to the mean daily evapotranspiration in the region. Rainfall is represented only when it exceeds 5 mm/day (by short, thin columns) or 10 mm/day (by longer, thin columns). Relatively heavy rain (more than 30 mm/day) or continuous rain averaging more than 10 mm/day over 3–5 days is represented by thicker columns.

Phenologic studies on forest trees

A group of 120 trees of 30 different species (4 representatives of each species) were included in these studies. They are listed in Table 19.1. The plants were observed at 2-weeks intervals for 3 calendar years (i.e., from January 1973 through December 1975). Data on flowering, fruit ripening, leaf fall, and flushing were obtained by simply observing the tree canopies with binoculars. Trunk growth was measured by means of a dial gauge dendrometer, similar to the

Table 19.1. *Trees observed fortnightly for flowering, leaf shedding, leaf formation, fruit ripening, and trunk growth: January 1973 through December 1975*

Latin name	Family, subfamily	Common name
Acanthosyris paulo-alvinii Barroso	Santalaceae	Mata-cacau
Andira stipulacea Benth.	Leguminosae, Papilionoideae	Angelin-côco
Artocarpus integrifolia Linn.	Moraceae	Jaqueira
Bombax macrophyllum K. Schum.	Bombacaceae	Imbiruçu
Cariniana estrellensis (Raddi.) O.Ktze	Lecythidaceae	Jequitibá branco
Cassia verrucosa Vog.	Leguminosae, Caesalpinioideae	Cobi preto
Cedrela glaziovii D.C.	Meliaceae	Cedro rosa
Erythrina glauca Willd.	Leguminosae, Papilionoideae	Eritrina
Erythrina velutina Willd.	Leguminosae, Papilionoideae	Eritrina
Ficus doliaraia Mart.	Moraceae	Gameleira branca
Ficus eximia Schott.	Moraceae	Gameleira preta
Gallesia scorododendrum Casar.	Phytolacaceae	Paul d'alho
Genipa americana Linn.	Rubiaceae	Jenipapeiro
Guarea rosea D.C.	Meliaceae	Rosa branca
Inga affinis Benth.	Leguminosae, Mimosoideae	Ingá cipó
Lecythis pisonis Camb.	Lecythidaceae	Sapucaia
Lonchocarpus glabrescens Benth.	Leguminosae, Papilionoideae	Cabelouro
Lucuma sp.	Sapotaceae	Mucuri
Machaerium angustifolium Vog.	Leguminosae, Papilionoideae	Sete capotes
Manilkara elata (Fr. Allem.) Monac.	Sapotaceae	Maçaranduba
Poecilanthe sp.	Leguminosae, Papilionoideae	Mucitaíba branca
Quararibea floribunda Schum.	Bombacaceae	Virote verdadeiro
Simarouba amara Aubl.	Simarubaceae	Pau-paraíba
Spondias lutea Linn.	Anacardiaceae	Cajazeira
Sterculia sp.	Sterculiaceae	Samuma preta
Swartzia macrostachya Benth.	Leguminosae, Caesalpinioideae	Jacarandá branco
Tabebuia sp.	Bignoniaceae	Peroba branca
Tapirira guianensis Aubl.	Anacardiaceae	Pau-pombo
Tecoma sp.	Bignoniaceae	Pau d'arco
Terminalia brasiliensis Eichl.	Combretaceae	Araçá d'agua

one described by Daubenmire (1945). Climatic data were obtained from a meteorologic station located at a distance of about 2 km from the study area. Changes in soil moisture availability during the study period were expressed for 10-day periods and estimated from rainfall data and potential evapotranspiration calculated according to Thornthwait's method.

Results and discussion
Greenhouse experiments with cacao

Figure 19.1 shows typical data on leaf fall, flushing, and diurnal changes in stem diameter, as affected by changes in soil moisture. The daily shrinkage and swelling in stem diameter (difference between morning and afternoon readings) became more pronounced when soil moisture approached the wilting percentage in the upper layers (6–12 in.). A daily net gain in trunk diameter was observed as

Fig. 19.1. Effect of soil moisture availability on trunk growth, leaf fall, and flushing of cacao plants growing in 250-liter containers under greenhouse conditions.

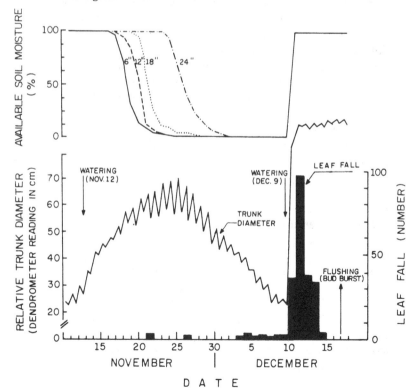

long as soil moisture remained within the available range for at least a part of the root system. When soil moisture dropped below 50% of the available range at the lower depth, a progressive net loss in diameter was observed.

Rewatering the soil (December 9) caused a very rapid swelling of the stem, bringing about resumption of trunk growth in less than 24 hours. The new growth curve followed approximately a trend similar to that found by extrapolating the curve for the period before the plant was stressed. This appears to indicate that cambial activity was not completely arrested during the period of moisture stress and that progressive shrinkage following soil moisture deficiency was mainly caused by dehydration of the bark. Rewatering the wilted plants also caused an abrupt increase in leaf shedding. This peculiar phenomenon is probably the result of the rupture of the abscission layers in consequence of the rapid swelling of the bark. About 1 week after rewatering, many terminal buds started swelling, initiating a new flushing cycle. Such flushing cycles could be repeatedly reproduced under greenhouse conditions whenever the plants were alternatively subjected to dry and wet periods. Control plants, kept under similar conditions, but frequently irrigated, never showed simultaneous or synchronous bud burst, although a few isolated branches (and as a rule only the ones located in the upper part of the canopy) occasionally showed new leaves. This fact seems to indicate that changes in internal moisture status associated with flushing can be brought about not only by changes in soil moisture availability but also by changes in transpiration losses of individual branches.

Phenologic studies on cacao

Figures 19.2 and 19.3 show data on flushing, leaf fall, flowering, and changes in stem circumference in cacao, together with rainfall, mean air temperature, and the new method for representing seasonal dry and wet periods (hydroperiodicity). These results provide further evidence that synchronous bud break in the field is triggered by alternating dry and wet periods: Peaks of flushing were always preceded by a sequence of dryness and wetness. In both areas, increase in trunk circumference was markedly reduced during dry periods. In Juçari (Fig. 19.2) actual trunk shrinkage occurred during the dry spells of May-June and September-October 1975. In São Francisco do Conde (Fig. 19.3) stem shrinkage was particularly evident in October-November, coinciding with a drought period of about 50 days. Trunk growth was temporarily resumed following heavy rains in early December, but shrinkage was again observed during the subsequent dry spells in late December and January.

As shown in previous studies (Alvim, 1967; Alvim et al., 1972), flowering of mature cacao plants in Bahia usually follows a clearly seasonal pattern. As a rule, there is practically no flowering from July to September, the controlling factors apparently being mainly the competition between growing fruits and flower buds, and to a lesser extent, the relatively low temperature prevailing during that period. From October to June flowering intensity always shows much variation, and the main factor responsible for this variation is obviously the rainfall pattern or the hydroperiodicity of the environment. As shown in Fig. 19.2, flowering in Juçari started in October in

Fig. 19.2. Flushing, flowering, and trunk growth of mature cacao plants growing under field conditions in Juçari, Bahia, together with data on rainfall, temperature, and hydroperiodicity of environment. See text for explanation.

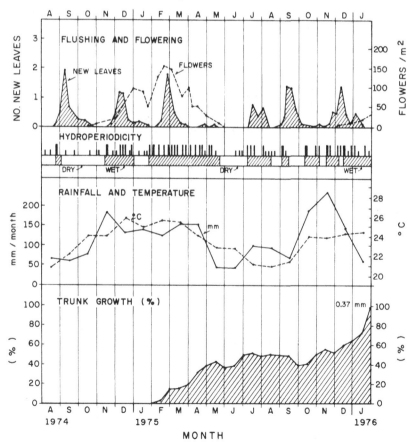

452 *Paulo de T. & Ronald Alvim*

1974 and in December in 1975. The delay observed in 1975 possibly resulted from excessive rainfall during October-November. In São Francisco do Conde, on the other hand, the October-November 1975 were quite dry, and flowering was relatively intensive during these months.

Fig. 19.3. Flushing, flowering, and trunk growth of mature cacao plants growing under field conditions in São Francisco do Conde, Bahia, together with data on rainfall, temperature, and hydroperiodicity of environment. See text for explanation.

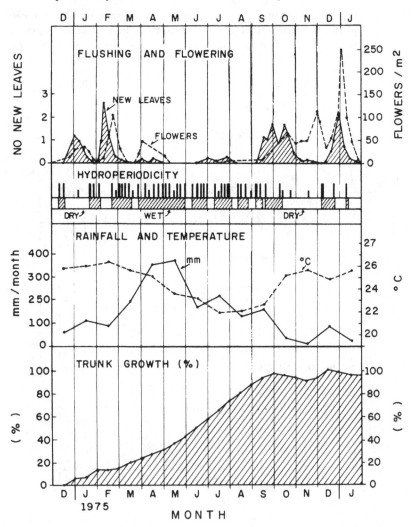

Phenologic studies on forest trees

It is not the intention of this chapter to report and analyze in detail the enormous amount of data obtained from all phenologic events observed at biweekly intervals (trunk growth, flowering, leaf shedding, flushing, and fruit ripening). This study is still in progress, and additional data are obviously needed to answer many questions, especially those concerning the growth and flowering patterns of the species included in the study. However, a preliminary attempt is made here to establish possible relations between climatic factors and some periodic phenomena, such as flowering and trunk growth.

As a general observation, it should be pointed out that complete synchrony for a specific phenologic event (e.g., flowering) was seldom recorded in all four representatives of a given species. It seems obvious, therefore, that internal factors or slight differences in microclimatic factors play an important role in the phenologic pattern of most species included in this study.

Data on rainfall and evapotranspiration are given in Fig. 19.4. The percentage of species flowering throughout the experimental period is given in Fig. 19.5, which also depicts rainfall data for the 3 years as well as the mean monthly data for temperature and day length. These figures reveal that the percentage of species flowering in a given month varied from year to year. Complete inhibition of flowering was observed in all species from January to mid-March 1973, probably because of the dry conditions occurring during the period.

Comparing the flowering pattern of individual species with climatic factors shows clearly that some species flowered at about the same time every year, showing little or no response to variation in rainfall pattern from year to year. This is true for *Andira stipulacea, Cassia verrucosa, Erythrina glauca, Erythrina velutina, Lecythis pisonis, Lonchocarpus glabrescens, Simarouba amara, Tabebuia* sp., and *Tapirira guianensis* (Table 19.2). In these species flowering appears to be strongly affected by environmental factors like photoperiod and/or temperature. This is indicated by the fact that they flowered conspicuously at approximately the same period of the year; the best examples are *Erythrina glauca* and *Cassia verrucosa,* which seem to be short-day plants. However, two species included in this group (*Andira stipulacea* and *Lonchocarpus glabrescens*) failed to flower in 1973; one (*Lecythis pisonis*) did not flower in 1974.

Another group of plants seemed to flower in response to a hydroperiodic stimulus (i.e., rain following an extended dry period), as previously reported for coffee (Alvim, 1960). This seems to be the case for *Gallesia scorododendrum* and *Cariniana estrellensis* (Table 19.3).

Flowering in a third group of trees (Table 19.4) showed no clear relation with external factors. Among these species were some in which flowering was observed almost every month of the year, although following a different pattern each year (*Acanthosyris paulo-alvinii* and *Artocarpus integrifolia*). Others showed scarce flowering (*Bombax macrophyllum, Cedrela glaziovii, Machaerium angustifolium,* and *Manilkara elata*), and one did not flower at all during the study period (*Tecoma* sp.).

Fig. 19.6 shows trunk growth, flowering, fruit setting, leaf fall, and

Fig. 19.4. Seasonal variation in soil moisture availability, January 1973 to December 1975, as estimated by rainfall and potential evapotranspiration.

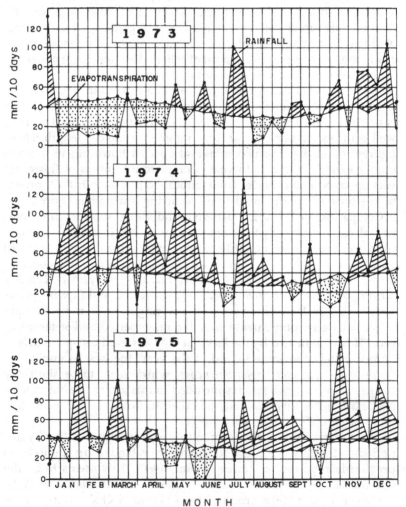

flushing in two typically evergreen trees (*Artocarpus integrifolia* and *Acanthosyris paulo-alvinii*) and four deciduous trees (*Erythrina glauca, Tabebuia* sp., *Lecythis pisonis,* and *Simarouba amara*) during 1974. Among the species studied, *A. integrifolia* was the only one that showed no seasonality as regards flowering, trunk growth, and fruit setting, in contrast to the deciduous species, which were all grouped among strongly seasonal species in regard to flowering (Table 19.2). Leaf shedding in the deciduous trees was probably induced by short days and always preceded leaf flushing. *Erythrina glauca* showed the

Fig. 19.5. Percentage of forest trees flowering during experimental period and data on monthly rainfall, day length, and 3-year mean monthly temperature.

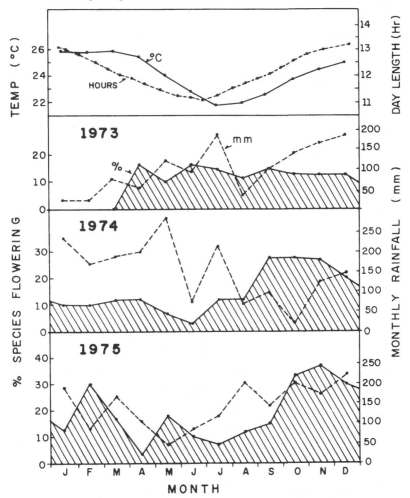

Table 19.2. *Species with flowering affected by photoperiod and/or temperature*

Species	Year	J	F	M	A	M	J	J	A	S	O	N	D
Andira stipulacea	1973												
	1974										1	1	
	1975											1	1
Cassia verrucosa	1973				1	3	1						
	1974			1	3	1							
	1975				3	1							
Erythrina glauca	1973						3	4					
	1974						4	4	3				
	1975						4	4	2	1			
Erythrina velutina	1973								1	1			1
	1974								1	1			
	1975						2	2		2			
Lecythis pisonis	1973									1	1		
	1974												
	1975	1									1		
Lonchocarpus glabrescens	1973												
	1974											1	2
	1975	3											2
Simarouba amara	1973							1					
	1974						3	4					
	1975						2	3					
Tabebuia sp.	1973								1				
	1974								3				
	1975									3			
Tapirira guianensis	1973								2				
	1974								3	4	2	1	
	1975									3	2		

Table 19.3. *Species with flowering affected by hydroperiodic stimulus*

Species	Year	J	F	M	A	M	J	J	A	S	O	N	D
Gallesia scorododendrum	1973				4								
	1974												
	1975												
Cariniana estrellensis	1973				3	3	1	1	1				
	1974												
	1975		4	4									

longest leafless period and was the only species in which all four plants observed showed simultaneous flowering at the same month in the 3 years studied (see Table 19.2). Differences in periodicity of cambium growth are well illustrated in Fig. 19.6: *Artocarpus integrifolia* and *Simarouba amara* showed continuous trunk growth; in the other species a period of rest seemed to occur in association with short days. This period of dormancy is quite obvious in *Erythrina glauca* and *Lecythis pisonis*, but is not so well defined in *Simarouba amara, Acanthosyris paulo-alvinii,* and *Tabebuia* sp. The sharp decrease in stem diameter in *Lecythis pisonis, Acanthosyris paulo-alvinii,* and *Tabebuia* sp. from October to November, was apparently due to trunk shrinking in response to water shortage in the period (see Fig. 19.4). Absolute annual radial growth varied from a very low value of 0.84 mm in *A. paulo-alvinii* to 7.9 mm in *S. amara.*

Fig. 19.6. Trunk growth, flowering (o), fruit setting (o), leaf fall (↓), and flushing (↑) in two evergreen (*Artocarpus integrifolia* and *Acanthosyris paulo-alvinii*) and four deciduous trees during 1974. Numbers in the top left corners represent annual increments in trunk radius.

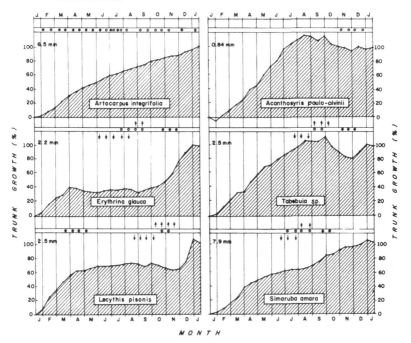

Table 19.4. *Species with flowering showing no clear relation with external factors*

Species	Year	No. of plants in flower in each month											
		J	F	M	A	M	J	J	A	S	O	N	D
Acanthosyris paulo-alvinii	1973						1	1				2	3
	1974				1	1					1	2	2
	1975	1	2	1	1							4	3
Artocarpus integrifolia	1973									1	3	3	1
	1974						1	1		2	3	2	
	1975	3	1	1		1	1			3	2	4	4
Ficus eximia	1973			1						1		2	
	1974		1						1	2	1	1	
	1975		2		1				1			2	
Ficus doliaria	1973									1			
	1974					2					1	2	
	1975		2	1			1		3				
Genipa americana	1973						1	1		1			
	1974		1					1		1			1
	1975	3	3										1
Poecilanthe sp.	1973												
	1974												
	1975			1		1	2			1	1		
Guarea rosea	1973			1									
	1974		1										
	1975		1		1							2	4
Quararibea floribunda	1973											1	1
	1974			1									
	1975											1	2
Inga affinis	1973												2
	1974			2	1						1	2	2
	1975										3	4	4
Lucuma sp.	1973									1	1		
	1974								1	1	1		
	1975	1	1			1	1				1	1	1
Swartzia macrostachya	1973			3	2								
	1974			1	2		1						
	1975		3			2				2			
Spondias lutea	1973												2
	1974	2	1	1							1	3	3
	1975										1	3	2
Sterculia sp.	1973								1			1	
	1974	1										1	
	1975		1								1	1	1
Terminalia brasiliensis	1973						3	4	4	1	1		

Table 19.4 (*cont.*)

Species	Year	No. of plants in flower in each month											
		J	F	M	A	M	J	J	A	S	O	N	D
	1974											1	1
	1975								2	3	3		
Bombax macrophyllum	1973												
	1974	1											
	1975												
Cedrela glaziovii	1973												
	1974								1				
	1975		1		1				1				
Machaerium angustifolium	1973												
	1974												
	1975		1	1									
Manilkara elata	1973												
	1974	1	2						1	3	2	1	
	1975										1	1	
Tecoma sp.	1973												
	1974												
	1975												

Conclusions

The results obtained with cacao in the present study support a previously proposed theory (Alvim et al., 1972) according to which the flushing mechanism of this species is triggered by a sequence of dryness and wetness. It would, therefore, appear that the vegetative buds of cacao behave similarly to the flower buds of coffee with respect to their response to moisture availability. In other words, a period of moisture stress appears to be a necessary prerequisite for synchronous vegetative flushing to take place in mature plants. As pointed out by Sale (1970), the flower buds of cacao, unlike those of coffee, do not seem to pass through a state of rest imposed by high water potential. This is indicated by the fact that some flowering usually occurs even under conditions of frequent watering or constantly high humidity. However, it has been clearly demonstrated (Alvim, 1967; Sale, 1970; Alvim et al., 1972) that flowering of cacao is greatly enhanced when rain follows a period of moisture stress.

The phenologic studies of the various tropical trees described here show clearly that some species are strongly responsive to photoperiodic or/and thermoperiodic stimulus (e.g., *Erythrina glauca*, *Cassia verrucosa*, and *Simarouba amara*), whereas others appear to flower in response to a hydroperiodic stimulus (e.g., *Gallesia scorodo-dendrum* and *Cariniana estrellensis*). Special attention should be called to the fact that the four representative trees of *G. scorodondendrum* flowered simultaneously only once, in April 1973, when rain (in mid-

Fig. 19.7. Monthly mean over 3 years of percentage of species flowering in Itabuna, Bahia, and in Manaus, Amazonas, together with data on rainfall and temperature.

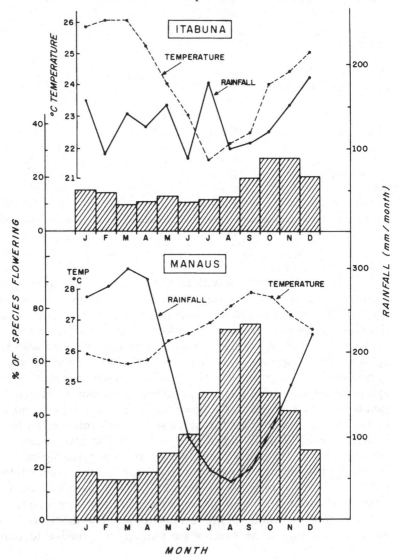

March) followed an extended dry period that lasted more than 2 months.

Based on the results of this study the mean percentage of species in flower during the various months of the year were calculated and plotted, together with the mean monthly values for temperature and rainfall during the study period (Fig. 19.7). For comparative purposes, similar data for Manaus (lat. 3°01′ S), Amazonas, are presented in the same figure. Data on the flowering pattern in Manaus were extracted from the phenologic observations reported by Prance & da Silva (1975). It can be observed that in Itabuna, where no typical dry season occurs, some 10%–20% of the species flowered throughout the year, with no significant differences from month to month. In Manaus, on the other hand, where a well-defined dry season occurs from June to October, flowering is much more seasonal, with over 70% of the species flowering during the peak of the dry season (August to September) and only 17%–19% during the wettest months (January to April). These figures indicate that in rain forest regions with pronounced dry season, as in Manaus, flowering appears to be predominantly controlled by a hydroperiodic stimulus.

General discussion

Fisher: Did you make any observations on periodicity of growth in *Terminalia catappa*? Is periodicity of leaf drop related to day length or dry season?

Alvim: To day length. In the state of Bahia, Brazil, at a latitude of about 15° S, *T. catappa* grows commonly in the gardens and along the beaches. It drops its leaves almost completely in July or August, which happens to be the wettest period of the year, when the days are shorter.

Ng: In Malaysia I think *T. catappa* responds to drought because I have seen large numbers of deciduous trees only within the last 4 months when we had a bad drought. Normally it is not deciduous for more than a few days. In this recent drought the trees remained deciduous for a month or so.

Fisher: Corner (1952) cites that in Singapore *Terminalia* loses its leaves twice a year – in the winter season and the summer season – and it is interesting that you also note this. In Florida, of course, the shortest days coincide with the middle of the dry season, so there is only one leafless period.

Alvim: I said short days, but it could be a slightly lower temperature because both conditions coincide. I am quite sure that the effect is not due to the dry season, because every year leaf loss occurs at the same time and we don't have a dry season. The leaf fall occurs during July or August when there are short days and low temperatures. Dry season could have an additional effect, of course, as there may be two mechanisms. I am quite sure that where I live the mechanism is not related to rainfall.

Browning: In the experiments Sale (1970) did, which were in controlled environments, was there any relation between flushing and leaf loss? Is flushing in the field invariably associated with leaf loss?

Ng: In Pasoh, Malaysia, I have been working on leaf flushing where Dr. Kira's group has been working on leaf loss. I found a peak in the flushing in the forest at the same time his team found a peak in leaf litter formation. Among many of our so-called evergreen trees in Malaysia, the evergreen condition is due to slightly overlapping phases of leaf loss and flushing, producing a periodic thinning of the crown. I think it is flushing that brings on the leaf loss by inducing the stem to thicken and thereby inducing abscission.

Alvim: Leaf fall is closely related to dry season where this occurs. Cacao is no exception. In the proposed hypothesis I am assuming that even without leaf fall the abscission layer is formed and blocks transport of growth substances. However, actual leaf fall, as I have showed in my chapter, quite often follows rather than precedes rain, primarily because of presumed swelling of the stem. However, in some cases flushing does not show a very close correlation with leaf fall. The correlation is best when a relatively dry period is followed by a relatively wet period.

Browning: We know that abscisic acid is produced in leaves; we are led to believe that abscisic acid may inhibit stem growth, and it follows then that if the leaves are removed, the source of abscisic acid is removed, and growth results. But this can be a completely artificial situation. It may well be that the natural mechanisms that, for example, might relieve abscisic acid inhibition of the apex, could be completely different. Experiments where one defoliates and then stimulates growth have to be interpreted with caution.

Oldeman: You spoke about leaf fall in the wet season. I encountered a similar situation in French Guiana. When one flies over the country during the rainy period, one sees many leafless trees in the valleys. At the height of the dry season, many trees of higher elevations are without leaves. I suggest that the rainy season is accompanied by waterlogging of the soil, and the root system's activity is then reduced because of insufficient aeration. Photosynthesis may be adversely affected by a lowered supply of root assimilates, and leaves may therefore drop. New leaves are formed as soon as the soil is sufficiently aerated for the roots to resume their normal activity. Is this explanation reasonable?

Alvim: In an experiment carried out near Manaus, where the terrain is flooded every year by the river, we simultaneously measured the level of the water table, trunk growths, and flushing in cacao plants, for a period of 2 years. Between May and July in both years the plants remained with the root system underwater for 2–3 months. Both trunk growth and flushing stopped completely during the flooded period and were resumed soon after the water table went down, usually in August. So the cycle of growth of the plants under these conditions seem to be primarily controlled by soil aeration.

Givnish: In connection with Dr. Alvim's paper, I feel challenged to produce an example of maintenance rather than growth. Bjorkman at the Carnegie Institution of Washington at Stanford has been looking at the various physiologic factors that allow plants growing in hot environments to survive. It was found that the principal adaptation of plants growing in hot environments is their ability to quickly repair and maintain a high turnover rate of the thermally unstable photosynthetic enzymes. The turnover is less in plants growing at lower temperatures. This is an example of maintenance physiology. It demonstrates the necessity to produce and maintain enzyme pools, as enzymes are not static, they can be destroyed chemically, they have half-lives, they can fall apart thermally with certain probabilities. I think we need to know a lot more about maintenance versus growth, respiration and its variation as a function of temperature.

Borchert: As long as a tree is growing, I see no possibility of distinguishing between metabolic activities that support continued growth and those of "maintenance metabolism" (i.e., that keep alive the cells already there).

Givnish: This discussion arose in connection with the "cold night" effect, where it is claimed that plants make greater dry weight gain because they do not have to maintain metabolism at a high level during the night. They can throw most of their metabolic weight into productive growth rather than maintaining machinery.

Zimmermann: This is a very complex question. One important factor here is translocation. Normally plants photosynthesize more carbohydrates during the day than they can use and export, and they export the excess during the night. If this is done efficiently, there is plenty of material available during the night for growth. Night may be the best time for growth because water stress is lower than during the day.

Givnish: Is anything known about the relative translocation abilities of arctic and temperate or tropical plants, since the former are often exposed to 24-hour days?

Zimmermann: I don't know anything about arctic plants, but we know that some plants can adapt to low temperature conditions. Translocation decreases when the temperature is lowered. In cold-hardy plants adaptation is rapid (i.e., translocation rates return to normal).

Baker: I want to point out something that everybody must have had in mind, but I don't think anybody has expressed yet: the difference between proximate and ultimate causes. In these phenologic considerations we have been talking about the triggering of flowering. But we must also keep in mind what is the ultimate benefit to the plant of flowering at a particular time. The ultimate cause is to have flowering at the time when there is an abundance of the appropriate pollinators. We must not lose sight of this fact when we start making our syntheses about tropical ecology.

Alvim: In the specific case of cacao the flowers are pollinated only by a very small insect of the genus *Forcipomyia* (Diptera, Ceratopogonidae), as Dr. Baker mentioned. I doubt if flowering in cacao would have any relation to the presence or absence of the insect because *Forcipomyia* appears to be present throughout the year.

Baker: I don't think *Theobroma cacao* is the ideal example of pollinator relation. Dr. Janzen has shown that insect activity is much greater in Costa Rica in the dry season; this correlates with the flowering of a lot of trees in the dry season and makes their triggering by the onset of the dry season very appropriate.

Alvim: Yes, this is probably true. Also in the dry season defoliation occurs, and according to teleologic theories this should make the flowers more visible and accessible to pollinators.

References

Alvim, P. de T. (1960). Moisture stress as a requirement for flowering of coffee. *Science, 132,* 354.

– (1964). Tree growth periodicity in tropical climates. In *Formation of Wood in Forest Trees,* ed. M. H. Zimmermann, pp. 479–95. New York: Academic Press.

– (1967). Eco-physiology of the cacao tree. In *Conférence Internationale sur les Recherches Agronomiques Cacaoyeres, Abidjan,* 15–20 Nov., 1965, pp. 23–35. Paris: Jouve, Institut Français du Café et du Cacao.

464   *Paulo de T. & Ronald Alvim*

- (1975). A new dendrometer for monitoring cambium activity and changes in the internal water status of plants. *Turrialba, 25*, 445–7.
Alvim, P. de T., Machado, A. D., & Vello, F. (1972). Physiological responses of cacao to environmental factors. In *Fourth International Cocoa Research Conference*, Trinidad and Tobago, January 9–18, pp. 17.
Alvim, R., Alvim, P. de T., Lorenzi, R., & Saunders, P. F. (1974). The possible role of abscisic acid and cytokinins in growth rhythms of *Theobroma cacao* L. *Rev. Theobroma, 4*, 3–12.
Borchert, R. (1973). Simulation of rhythmic tree growth under constant conditions. *Physiol. Plant., 29*, 173–80.
- (1975). Endogenous shoot growth rhythms and indeterminate shoot growth in oak. *Physiol. Plant., 35*, 152–7.
Corner, E. J. H. (1952). *Wayside Trees of Malaya*, 2nd ed. Singapore: Government Printers. 2 vols.
Daubenmire, R. F. (1945). An improved type of precision dendrometer. *Ecology, 26*, 97–8.
- (1972). Phenology and other characteristics of tropical semi-deciduous forest in North-Western Costa Rica. *J. Ecol., 60*, 147–70.
Frankie, G. W., Baker, H. G., & Opler, P. A. (1974). Comparative phenological studies of trees in tropical wet and dry forests in the lowlands of Costa Rica. *J. Ecol., 62*, 881–919.
Greathouse, D. C., Laetsch, W. M., & Phinney, B. O. (1971). The shoot-growth rhythm of a tropical tree, *Theobroma cacao*. *Amer. J. Bot., 58*, 281–6.
Hallé, F., & Martin, R. (1968). Etude de la croissance rythmique chez l'Hévéa (*Hevea brasiliensis* Müll.-Arg. Euphorbiacées-Crotonoidées). *Adansonia, 8*, 475–503.
Huxley, P. A., & Van Eck, W. A. (1974). Seasonal changes in growth and development of some woody perennials near Kampala, Uganda. *J. Ecol., 62*, 579–92.
Janzen, D. H. (1967). Synchronization of sexual reproduction of trees within the dry season in Central America. *Evolution, 21*, 620–37.
Magalhães, A. C., & Angelocci, L. R. (1976). Sudden alterations in water balance associated with flower bud opening in coffee plants. *J. Hort. Sci. 51*, 419–23.
Prance, G. T., & da Silva, M. F. (1975). *Arvores de Manaus*. Manaus: INPA. 312 pp.
Sale, P. J. M. (1970). Growth, flowering and fruiting of cacao under controlled soil moisture conditions. *J. Hort. Sci., 45*, 99–118.
Tomlinson, P. B., & Gill, A. M. (1973). Growth habits of tropical trees: Some guiding principles. In *Tropical Forest Ecosystems in Africa and South America: A Comparative Review*, ed. B. J. Meggers, E. S. Ayensu, & W. D. Duckworth, pp. 129–43. Washington, D.C.: Smithsonian Institution Press.
Vogel, M. (1975). Recherche du determinisme du rythme de croissance du cacaoyer. *Café Cacao Thé, 19*, 265–90.

20

Control of shoot extension and dormancy: external and internal factors

K. A. LONGMAN

Institute of Terrestrial Ecology, Penicuik, Midlothian, Scotland

It is often imagined that shoot extension of tropical trees is rapid and continuous and hardly affected by environment. Many examples certainly occur of height growth exceeding 2m a year, but these are usually found with saplings and coppice shoots along forest margins and in large gaps, or (as gaps occur in the soil as well as the canopy) in young plantations (Richards, 1952; Longman & Jeník, 1974). If the growth potential of buds throughout the life of a tree is considered, there must clearly be a great array of factors that reduces growth rates far below the maximum attainable in that species. Indeed, many buds never become active at all unless the tree is damaged, and the remainder generally alternate between growth and inactivity. Thus periodicity of shoot growth is predominant after the first few years (Schimper, 1898; Zimmermann & Brown, 1971), with less than 20% of woody species still exhibiting continuous growth (Coster, 1923; Koriba, 1958). These include a number of small dicotyledonous trees and shrubs and some palms and conifers, but virtually none of the dominant and codominant plants of the lowland tropical forests.

What are the many factors that influence and control the rate, duration, and distribution of bud activity? It is hoped that the examples discussed in the following pages will convey an impression of the magnitude and diversity of effects that have been discovered in a variety of tropical woody genera. Naturally, there are many other factors involved: Some are referred to by other contributors to this volume; others are discussed in reviews and books (Romberger, 1963; Alvim, 1964; Kozlowski, 1964, 1971; Longman & Jeník, 1974). The available evidence suggests that both the external conditions surrounding the tree and various internal changes and competitive effects are involved, and that the response of a particular bud is complicated still further by interactions between environmental and inner factors (see Chap. 23), and by substantial genetic variation within species as well as amongst them.

The approach taken here towards unravelling the complex situation in the tropical forest is to consider one or two factors experimentally in controlled environments, keeping other conditions as uniform as possible. It should be understood at the outset that the primary aim of these studies in growth physiology is to investigate if the particular species is sensitive to these factors and their interactions. Whether or not the factors play a role in the forest setting is much more difficult to show, but at least results obtained in growth rooms cannot be ignored when the ecologic picture is pieced together. It is also logical to measure the variability of environmental factors at different levels in the forest canopy before assuming that observed fluctuations in growth rate or flushing behaviour are necessarily controlled endogenously (Chap. 21).

The effects of two environmental factors, photoperiod and temperature, are discussed in the first section, where it is shown that, contrary to the general impression, large differences in the rate of growth can occur when these factors are altered. Several facets of shoot growth are affected in both pioneer and climax species representative of the upper tree layer. The same external factors are also referred to in the following two sections, in which internal considerations assume prominence. These cover dormancy, the cessation and restarting of active growth, and some of the physiologic changes that occur with increasing age in perennial woody plants. The interaction of internal and external factors is again the subject of the fourth section, in which the phenomenon of apical dominance is considered in relation to the effects of the orientation of the shoot relative to the gravitational field and of the position of single buds in the absence of competition. Finally, some examples are given of genetic variation under both glasshouse and field conditions.

Materials and methods

The responses of species of tropical forest trees and other woody plants were studied in a wide range of experiments conducted in West Africa and in tropical environments in Scotland. For eight of these, details of the plants used (*Terminalia superba*, *Chlorophora excelsa*, *Ceiba pentandra*, *Gmelina arborea*, *Triplochiton scleroxylon*, and the shrub *Manihot esculenta*) are given in Table 20.1. For the sake of brevity, the genus alone is cited subsequently.

Measurements of initial height and numbers of nodes on the day treatments commenced were used to derive the gain in stem height and in new leaves and to calculate mean internode length. The macroscopic appearance of these organs provided the criterion for es-

Table 20.1. *Experimental plant material*

Experiment no.	Species	Type of material	Origin	Location of experiment	Studying effect of	At start of experiment				
						Date	Approx. age (mo)	Approx. height (m)	No. of replicates[a]	Condition of buds
1	Terminalia superba	Seedlings	Forest Res. Inst., Kumasi, Ghana	Accra, Ghana (5.5° N)	Day length	Mar. 2	9	0.7–0.8	13	Terminals very active
2	Chlorophora excelsa	Seedlings	Sunyani, Ghana	Accra	Day length	Mar. 4	10	0.65	16	Terminals very active
3	Ceiba pentandra	Seedlings	Legon, Ghana (spineless variety)	Accra	Day length, night temperature	Aug. 5	1	0.1	15	Terminals active
4	Gmelina arborea	Seedlings (wildings)	Kpong, Ghana	Accra	Day length, night temperature	Aug. 10	c.3	0.15	14	Terminals active
5	Triplochiton scleroxylon	Rooted cuttings (1 clone)	Nigeria Ilugun	Edinburgh, Scotland (56° N)	Day length, temperature	Feb. 5	33 from sowing; 4–8 from striking	0.15	12	Inactive (decapitated, defoliated)
6	Triplochiton scleroxylon	Rooted cuttings (5 clones)	Nigeria Ilugun	Edinburgh	Orientation	June 18–20	25.5 from sowing; 12 from striking	0.9–1.0	7	Mainly inactive (decapitated)
7	Manihot esculenta	Unrooted stem sections	Ghana (unknown clones)	Accra	Bud position, orientation		Unknown from seed; fresh cuttings	0.18	12	Inactive (leafless)
8	Manihot esculenta	Unrooted stem sections	Ghana (unknown clones)	Accra	Bud position, orientation		Unknown from seed; fresh cuttings	0.18	18	Inactive (leafless)

[a] In each treatment.

timating rates of apical activity; no account was taken of the possibility that some were preformed in an inactive bud. The absence of activity in an apex was taken to be demonstrated when no new leaves appeared over two or three successive measurements and the gain in stem length was slight or nil. Except where dominance relations were studied, measurements were made on the terminal shoot (or in its absence the most distal lateral). Leaf lengths were taken from comparable leaves that had completed their enlargement during the period of treatment and were of the lamina only, either the whole blade or the longest lobe or leaflet.

Experiments 1 and 2

Effects of day length on shoot extension growth (*Terminalia* and *Chlorophora*). Potted plants were grown in a lightly shaded glasshouse for approximately 7 hours in the middle of the day, at temperatures normally between 30° and 35° C. For the remaining 17 hours they were in growth rooms at 32.5° C ($\pm$ 1°), where all plants received approximately 2 hours 10 minutes of fluorescent light of 2000–3000 lx. Day length was extended by white incandescent light of 60–120 lx. Light breaks were provided by 1 hour fluorescent followed immediately by 1 hour incandescent in the middle of the dark period. All plants received approximately the same total light energy.

Treatments: 1, 17 hours 10 minutes; 2, 13 hours 10 minutes (= June 21 in Accra); 3, 9 hours 10 minutes; 4, 9 hours 10 minutes + light break

Experiments 3 and 4

Effects of day length and night temperature on shoot extension growth (*Ceiba* and *Gmelina*). As in experiments 1 and 2, potted plants were moved daily to and from the glasshouse, where the average day temperature was about 31° C ($\pm$ 3°). Lighting regimens were obtained in the same way in the growth rooms, with temperatures maintained at 31° C during the period under fluorescent lighting.

Treatments: Three day lengths were used (9 hours 10 minutes, 13 hours 10 minutes, 17 hours 10 minutes) and three night temperatures (26°, 31°, 36° C), giving a total of nine treatments

The growth rooms for two regimens (9 hours 10 minutes at 31° C and 17 hours 10 minutes at 36° C) were interchanged from September 11 to October 22, as a check on inherent cabinet differences. No appreciable modifications of growth were found. In experiment 3 some wilting occurred during the first 12 days before repotting to larger pots.

Experiment 5

Effects of day length and temperature on shoot extension growth and dominance relations (*Triplochiton*). Rooted cuttings, raised under the conditions described by Leakey & Longman (1976), were decapitated and defoliated, and then grown continuously in modified Fison's cabinets at constant temperatures. Light intensities of approximately 16,000–18,000 lx were maintained at the top of the plants, consisting of about 83% warm white fluorescent and 17% white incandescent. Day length extension was provided by white incandescent light of approximately 150–200 lx. Vapour pressure deficits were maintained approximately the same in all cabinets by adjusting the water level in a trough across which the air circulated. Air circulation rate and fresh air intake were also closely similar.

Treatments: Three day lengths (11, 13, 15 hours) and two constant temperatures (20°, 25° C) were used, giving six treatments. Except when cabinets were opened, the Telemax controllers maintained air temperatures ±0.75° C or better. Some of the plants grew poorly, probably owing to the complete defoliation and to previous overcrowding. Between three and five plants per cabinet were therefore discarded.

Experiment 6

Effects of orientation on dominance relations (*Triplochiton*). Rooted cuttings were decapitated and repotted at the appropriate orientation in plastic pots of 190 mm diameter, which were cut where necessary. Additional high-intensity lighting was provided in the glasshouse by 400-W MBFR/U mercury/fluorescent lamps. Day length was 19½ hours, temperature approximately 25°–30° C.

Treatments: 1, 0° (upright); *2*, 45° (angled); *3*, 90° (horizontal).

Experiments 7 and 8

Effects of orientation and bud position on rates of sprouting (*Manihot*). Leafless stem segments were prepared from the main stems (order 0); material with very long or short internodes was rejected. The apical and basal bud positions were equidistant from the ends and on the same side of the segment; the midposition was in the middle. The material was soaked in water for several hours and then treated with hypochlorite. All buds except those required were removed, together with some of the adjacent bark to discourage bud regeneration. Complete bark ringing was done down to the xylem, a few millimeters proximal to apical buds and distal to basal buds. Cut surfaces were treated with petroleum jelly or paraffin wax.

Cuttings were placed at the appropriate orientation in moist river sand in the dark at 27°–30° C in a growth room. When orientation was horizontal, the buds were on the upper side (see Fig. 20.7).

Treatments (*experiment 7*): Three orientations: 0° (upright), 90° (horizontal), 180° (inverted); bud positions: 1 bud (apical), 2 buds (apical & basal), giving 6 treatments.

Treatments (*experiment 8*): Three orientations: 0°, 90°, 180°; three bud positions: 1 bud at apical, mid, and basal positions; plus 1 bud at apical position (ringed) and 1 bud at basal position (ringed), giving 15 treatments in all.

Photoperiod, temperature, and rate of shoot growth

Because the length of the day varies much less in equatorial regions than at higher latitudes, this environmental factor has often been dismissed as being of little or no importance. However, it has become clear in recent years that many tropical herbaceous plants are sensitive to photoperiod in their flowering (Mathon, 1975). Not all are short-day plants, as has sometimes been suggested, and some seem to respond to differences of only 5 minutes (Njoku, 1959; Roberts, 1963). It is important to distinguish between true photoperiodic responses, demonstrated by low-intensity supplementary illumination or light breaks during the dark period, and experiments in which the effects of day length and light intensity are confounded.

Photoperiodic effects on shoot growth have been demonstrated for a number of tropical trees (see Bünning, 1948; Njoku, 1964; Longman 1966, 1969, 1972), and the overall effects on shoot extension of day length and other factors have been summarised by Longman & Jeník (1974). In general, increasing the photoperiod increased shoot growth rates, usually more in one part of the range of day lengths used. Some of the effects were very large, with fourfold and fivefold differences in *Terminalia* and *Chlorophora excelsa*. One instance was recorded in *Chlorophora regia* of somewhat reduced shoot growth in long days, but this could not be confirmed because of shortage of seed.

It has also been found that differences in temperature of 5° (or even less) often have a striking effect on overall shoot extension rates of tropical trees, even when they are varied only during the night. Optimal temperatures can lie above 30° C, maxima above 40° C, and minima around 15° C (Kwakwa, 1964). Damage often occurs at 10° (Went, 1962), and even at 20° shoot growth rates may be much less than at higher temperatures. It is also not uncommon for a warm or hot night to offset in part the reduction in growth rate caused by a cool day, even when nights of 30° C and above are used, which seldom occur under natural conditions in the humid tropics. In some species optimal conditions involve cooler nights than days; others appear to grow best in constant temperatures (Kwakwa, 1964).

Thus it has been firmly established that overall shoot extension can be profoundly modified by photoperiod and temperature, and it is also clear that the two factors often show strong interactions. In the following sections, the various components of shoot growth will be discussed in more detail and examples drawn from a number of experiments. It should be noted that screened air temperatures were maintained at the stated values and that the extending internodes, shoot apices, and expanding leaves may have been at higher or lower temperatures. Moreover, root growth was no doubt influenced also by the soil temperature differences that existed.

Internode elongation

The amount of shoot growth made by a bud is partly determined by the final length achieved by the internodes, and the distribution of these stem units influences the height and crown diameter of the tree and the leaf mosaic in the forest. Mean internode lengths can be substantially modified by day length, and in *Terminalia*, for example, were about four times as long when the plants received long days or a light break as in short days (Fig. 20.1*A*). A quantitative effect of photoperiod is suggested in this instance, as the internode length at 13 hours 10 minutes was intermediate between that at long days and short days. On the other hand, a threshold effect between short and intermediate day lengths appears more likely in *Ceiba* (Fig. 20.1*B*), with a large and highly significant effect of increasing to 13 hours 10 minutes, but little or no effect of further lengthening of the day.

Increasing the night temperature from 26° to 31° C also increased internode lengths significantly in *Ceiba*, but only by about 20%. Interactions between photoperiod and temperature were not marked in *Ceiba*, but were very pronounced in *Triplochiton* (Fig. 20.1*C*). Day and night temperatures were the same in this experiment, and a 5° C rise had a large effect at 13 hours and 15 hours, but not at 11 hours. Similarly, increasing day length had a large effect at 25° C but not at 20° C. The presence of such interactions indicates that it would be unwise to assume that an environmental factor has no effect upon a particular species unless several trials have been conducted under various regimens.

Less information is available on the rate and duration of growth of a given internode under controlled conditions. Measurements of this kind were not made in these experiments, but the response to changed environment can be very rapid. The *Terminalia* plants, for example, exhibited photoperiodic effects (significant at the 1% level) on overall shoot extension growth after only 3 days' treatment (Long-man, 1966). In *Chlorophora, Ceiba,* and *Gmelina,* shoot growth rates in some regimens were more than double those in others within 3–6 days from the start of treatments. It seems likely, therefore, that the responses may occur directly to the current environment, but cumu-lative effects also appear to be present. In *Terminalia* and *Chlorophora* there was evidence that mean internode length increased in impor-tance during the second and third month as a component of overall shoot extension.

Production of new leaves

Combining with internode elongation to determine the growth rate of an active bud is the number of internodes that extend during a period of time. Most or all the nodes that are formed and separate during this process may bear one or more leaves, so that both overall stem length and foliation are affected. Striking effects of day length on numbers of new leaves were found in *Chlorophora* (Fig. 20.2*A*), with 4.5 times as many leaves being formed during the first month in long days as in short days. These very large differences were not maintained, although after 3 months the long-day treat-ment had still produced twice as many leaves. The intermediate day

Fig. 20.1. Effect of day length and temperature on mean final inter-node lengths in (*A*) *Terminalia superba* at 13.5 weeks, (*B*) *Ceiba pen-tandra* at 21 weeks, and (*C*) *Triplochiton scleroxylon* at 7 weeks. 25° / 25° indicates controlled day and night temperatures; /26°, con-trolled night temperature. Vertical bars show least significant dif-ferences for individual treatments, at 5% (*), 1% (**), and 0.1% (***) levels of probability, using minimum numbers of replicates. *LB*, light break.

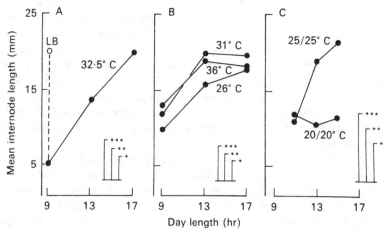

length did not increase leaf production significantly over the short day, and in this instance the light break was not as effective as a long photoperiod. A smaller, but highly significant reduction under short days during the first month in *Terminalia*, associated with temporary cessation of extension growth, had largely disappeared by the end of the experiment.

Persistent and highly significant effects of night temperature were shown in *Gmelina* (Fig. 20.2B). After 5 months, plants experiencing very hot nights had produced more than twice the number of leaves formed by those receiving warm nights, with the data for hot nights intermediate. This species is clearly very responsive to temperature and shows a quantitative increase even where night temperatures were in excess of those experienced in nature; there was little evidence of an influence of day length. In *Ceiba*, however, there were significant effects and interactions of the two factors (Fig. 20.2C). Night temperatures of at least 31° C appeared to be needed for rapid leaf production, and short days reduced the rate. The effect of night temperature seemed also to be of greatest importance at the intermediate day length.

Leaf expansion

The final size attained by the leaves was also considerably affected by the different regimens, both at early and later stages in the experiments. For example, a 50% increase in lamina length with increased day length occurred in *Chlorophora* (Fig. 20.3A), and as this species also produced a greater number of leaves under long days (see Fig. 20.2A), the increase in total leaf area per plant must have been considerable. In *Gmelina* there were small but highly significant effects of treatment on lamina length (Fig. 20.3B), which were closely paralleled by the data on lamina width. Thus it is clear that the combined area of the two leaves at each node would have shown larger differences.

Some evidence of interaction between day length and night temperature was shown by *Gmelina*, and this was more marked in *Ceiba* (Fig. 20.3C). When the latter was grown in warm nights, final leaf length increased with increasing day length, but with very hot nights the reverse occurred. As the reduction was highly significant, this interesting result may perhaps be taken to support the view that there is no intrinsic difference between long-day and short-day plants, but as the data are for the seventh leaf produced, cumulative rather than direct effects may be involved. Competition between various growth centres is likely to have become very different by this time, for leaf production rates and internode lengths had been influenced by treatment, and some temporary dormancy was occurring

Fig. 20.2. Effect of day length and night temperature on mean gain in numbers of leaves in (A) *Chlorophora excelsa* at 3.5 weeks, (B) *Gmelina arborea* at 6 weeks, and (C) *Ceiba pentandra* at 21 weeks. 25° / 25° indicates controlled day and night temperatures; /26°, controlled night temperature. Vertical bars show least significant differences for individual treatments, at 5% (*), 1% (**), and 0.1% (***) levels of probability, using minimum numbers of replicates. *LB*, light break.

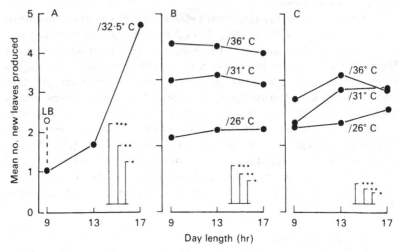

Fig. 20.3. Effect of day length and night temperature on mean final leaf lengths in (A) *Chlorophora excelsa*, (B) *Gmelina arborea*, and (C) *Ceiba pentandra*, based on leaf 1, leaf 1, and leaf 7, respectively. 25° / 25° indicates controlled day and night temperatures; /26°, controlled night temperature. Vertical bars show least significant differences for individual treatments at 5% (*), 1% (**), and 0.1% (***) levels of probability, using minimum numbers of replicates. *LB*, light break.

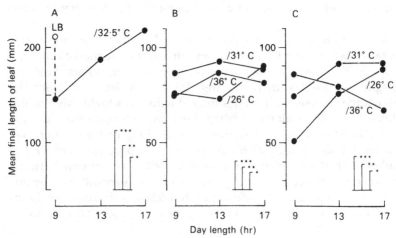

under short days. In *Terminalia*, leaf length was also slightly greater under short days, but the difference from those in long days was not significant. An example of the influence of competition on final leaf size is also given in Table 20.4.

In *Triplochiton* there are pronounced effects on final leaf size. Leaves were smaller when plants received a combination of 11-hour days and 20°-nights than when days were longer and/or nights warmer (Longman, 1966; Longman & Jeník, 1974). Further recent studies with young, rooted cuttings of this species have emphasised the importance of interactions. In experiment 5, total leaf area per plant after 10 weeks was estimated to be approximately 100% greater with 15- or 13-hour days than with 11-hour days. This result was obtained at a constant temperature of 25° C, but at 20° C there was little or no effect of day length. The promoting effect of high night temperatures on leaf length was confirmed in another experiment in which longer leaves were formed with a 20°-day/30°-night regimen than at 20°/20°, 20°/30°, or even 30°/30°. However, the large leaves formed under this unusual cool-day/hit-night treatment tended to be a pale, yellow green colour, and a similar colour effect was also produced in *Terminalia ivorensis* and *Terminalia superba*.

Although temperature is known to play an important role in determining final leaf size (Milthorpe, 1956, 1976), further detailed studies are needed on these tropical trees. Knowledge of the interacting effects of temperature and photoperiod may, for instance, be gained from investigating the rates and duration of growth of individual leaves under various regimens. It is possible that the powerful effect of night temperature in *Triplochiton* may be related to greater night-time leaf expansion. In species such as this, with a large lamina whose position is altered by leaf movements, the close relation between petiole and lamina lengths found in experiment 5 may also be of significance.

Dry weight accumulation

Because the amount of leaf and stem tissue produced by a bud has been shown to be strongly influenced by day length and temperature, it is pertinent to enquire whether there are appreciable effects on the accumulation or distribution of dry matter. The important role of photoperiod in tropical forest trees is underlined by results for *Chlorophora* (Table 20.2). Shoot dry weights of the plants at the end of the experiment were 60% greater when they had been subjected to long days than when they had received short days. As no treatment provided more total light energy than another, this is clearly a true day-length effect, which is probably related to the greater numbers and lengths of leaves under long days (Figs. 20.2*A*, 20.3*A*). Total plant dry weights were also significantly affected, but the root dry weights were similar. As a result, the mean shoot/root ratio changed from 1.11 in short days to 1.88 in long days. Similar but smaller significant trends were found in *Terminalia* (experiment 1). Effects of temperature on dry weight were not studied in experiments 3–5, but they have been demonstrated for *Ceiba* (Kwakwa, 1964) and are presumably of widespread occurrence.

Dormancy and rest

Terminal buds can be said to be in a state of rest if they are not making any active growth, apart from the accumulation of initials at the apex. They are sometimes described as being "dormant," using the word in its wide sense, but it may be more satisfactory to restrict the latter term to its precise meaning: that the bud itself is unable to grow actively, although placed in conditions normally favouring growth (Doorenbos, 1953; Vegis, 1964; Wareing, 1969a,b). There is then a close analogy with the usage of the same word in connection with seed dormancy. Other types of rest are referred to as "predormancy" and "postdormancy."

True dormancy appears to exist in tropical trees, though it has been rather little studied. Coster (1923), in his pioneer observations and experiments on trees in Southeast Asia, showed that the buds on cut shoots of *Bombax malabaricum* were more "reluctant" to flush when taken in the middle part of the leafless period than when taken earlier or later. He drew a parallel with the "midrest" (true dormancy) often found in temperate trees. Here dormancy is broken and buds rendered postdormant by several weeks of temperatures a little above freezing point. Other tropical trees have been reported as showing true dormancy (Njoku, 1963), but it is not known whether temperature conditions or other factors are involved in breaking it.

However, it should not be assumed that true dormancy is involved in every instance of a shoot system with its terminal buds in a state of rest. Cut leafless shoots of *Tectona grandis* standing in water generally

Table 20.2. *Effect of day length on dry weight of* Chlorophora: *Expt. 2, 13 weeks*

Dry weight (gm.)	Day length				Significance (P)
	9 hr. 10 min	13 hr 10 min	Light break	17 hr 10 min	
Shoot	9.93	10.30	11.55	15.86	0.1%
Root	9.18	7.18	8.37	8.68	n.s.
Total	19.12	17.49	19.92	24.54	1.0%
Shoot/root	1.11	1.46	1.41	1.88	0.1%

flushed within 10 days, irrespective of the conditions under which they were placed (Coster, 1923). In such cases it is likely that the buds are postdormant (Wareing, 1969a; Kozlowski, 1971), perhaps prevented from growing only by shortage of water (Longman & Jeník, 1974).

Predormancy may well prove the commonest type of rest in tropical trees. Here the environment may be favourable, and the bud is not intrinsically dormant, but it is prevented from growing out by the presence of the mature leaves. The common tropical habit of shoot growth by repeated intermittent flushes (Borchert, 1973) may be due to the periodic development of predormant terminal buds. *Couroupita guianensis* was stimulated to grow continuously for a period of several months by successive removal of mature leaves (Klebs, 1926). Temperate species of the genus *Quercus*, which has a number of tropical representatives, have also been converted from intermittent to continuous growth by removal of mature leaves (Longman & Coutts, 1974). On the other hand, periodic cessation of outgrowth in other instances (e.g., *Camellia* spp.) may be due to a rapid rate of extension of internodes exceeding their rate of formation (Bond, 1945; Kulasegaram, 1969), which could thus represent a type of true bud dormancy.

Predormancy can be induced in some tropical tree seedlings that normally grow continuously. Short days brought about temporary predormancy in *Terminalia* (experiment 1) and *Ceiba* (experiment 3), which did not occur with intermediate or long days or light breaks (Longman, 1969). *Cedrela odorata* plants showed very rapid onset of predormancy within 2 weeks in short days (Fig. 20.4). The importance of the leaves was demonstrated here, for when plants had few or no mature leaves extension growth did not cease, and it restarted if it had previously stopped. Similarly in *Hildegardia barteri*, a species showing an intermittent growth habit even in the first year, the presence of mature leaves prevented terminal bud outgrowth under short days (Longman, 1969), and differences in bud activity were found in response to 30-minute changes in day length (Njoku, 1964).

Lower night temperatures also brought about the temporary cessation of terminal shoot growth in *Gmelina* (experiment 4) and in *Bombax buonopozense*, particularly under short days (Longman, 1969). Conversely, higher temperatures have been reported to stop extension growth in the subtropical conifer *Pinus elliottii* var. *elliottii* (Slee & Shepherd, 1972). In a recent experiment with *Triplochiton*, shortage of mineral nutrients caused some terminal bud rest when the plants were also growing in small pots, but this did not occur in larger pots or when nutrients were added to the soil twice weekly.

A combination of internal and external factors, therefore, appears to control rest and dormancy in terminal buds, though they are as yet imperfectly understood. Lateral buds often remain inactive because they are inhibited by the actively growing terminal shoots; these dominance relations will be considered separately in a later section. Terminal bud rest has important effects on the annual growth rate of a shoot system, for in many trees the rate of extension may be rapid, but the period of activity confined to 1 or 2 months in the year (Njoku, 1963; Longman & Jeník, 1974). The timing of bud break, with the subsequent renewal of the photosynthetic surfaces at a higher level in the forest, may well be vital to the competitive success of a tree during most of its life. In some cases flushing peaks occur irregularly, or regularly but not related to seasonal changes (Holttum, 1953), but in many instances, especially where there is a definite dry season, distinct seasonality is often found (Taylor, 1960; Hopkins, 1970). A surprising feature is the frequency of prerain or equinoctial flushing (Bünning, 1948; Alvim, 1964; Frankie, Baker, & Opler, 1974), which may perhaps be related to the display of the newly expanded leaves at the time of maximum solar irradiation (Longman & Jeník, 1974). The nature of the physiologic triggers for seasonal and nonseasonal budbreak remains one of the most intriguing of tropical problems.

Fig. 20.4. Effect of day length on duration of shoot growth in *Cedrela odorata* seedlings after 5 weeks of treatment. *Left.* Plant grown under long-day conditions (17 hours 10 minutes at 31° C) showing active terminal bud producing new leaves. *Right.* Plant grown under short-day conditions (9 hours 10 minutes at 31° C), with inactive (predormant) terminal bud and all leaves fully expanded except single arrested leaf at tip. Plants were decapitated prior to treatment for comparisons with "mature" grafted shoots.

Changes with age

As a seedling grows older, patterns of shoot growth and rest generally change, and a single, short flush often replaces intermittent or continuous growth, leaves become smaller and sometimes simpler, branching patterns become complex, and reproductive activity commences (Bünning, 1952). However, it is important to distinguish among three types of physiologic change, some of which are easily reversible, whereas others involve subtler, more permanent modifications to the growth potential (Longman, 1976b).

First, differences can exist early in the life cycle. Young *Triplochiton* plants that have been grown normally and then cut up into single-node, leafy cuttings, will root easily and grow into ordinary plants with a vertical main stem. However, if trees 1.5 m tall are decapitated and grown at an orientation of about 45° from the vertical (experiment 6), lateral shoots produced near the previous apex on the main stem develop as plagiotropic branches with a distichous leaf arrangement. Nearer to the roots vertical shoots are formed, with leaves spirally arranged (Leakey, Chapman, & Longman, 1975; Longman, 1976c). This type of response to gravimorphic treatments has been described in temperate trees (Wareing & Nasr, 1961; Smith & Wareing, 1964a; Champagnat, 1965) and is probably widespread amongst woody plants.

However, it has been found in *Triplochiton* that cuttings from the apical type of branch root less easily and continue to grow plagiotropically, at least for a time, whereas material from the basal type root freely and grow vertically (Fig. 20.5). Thus, buds less than 1 m

Fig. 20.5. Within-clone variation in growth habit of *Triplochiton* cuttings about 5 months from striking, 29 months from germination of seed. *Left*. Basal cutting, showing orthotropic growth and radial symmetry. *Right*. Apical cutting, showing plagiotropic growth and dorsiventral symmetry.

apart on the same tree can have clearly different growth potentials, and it will be important in the vegetative propagation of this valuable timber tree to determine whether these are permanent or temporary carryover effects from the stock plant (Longman, 1976a,b). It is possible that the explanation will prove to be that the apical cuttings are morphologically branches and the basal cuttings are main stems, but in *Terminalia* spp. cuttings from branches often grow vertically, though in *Chlorophora* there is some tendency towards plagiotropy.

A second class of differences exists between a young plant with one or a few rapidly growing apices and an older tree with an aged shoot system, consisting of a large number of slowly growing buds (Wareing, 1959, 1970; Moorby & Wareing, 1963). Implicit in the concept of ageing is that the reduced growth can be easily reversed, because it results from competition rather than a permanent loss of growth potential. A short-lived tropical woody plant in which the reversibility of an aged shoot system was tested is *Manihot esculenta*, which in a year or so produces (through its sympodial habit and repeated equal branching) an aged shoot system of the Leeuwenberg or umbel-disk type (Hallé & Oldeman, 1970; Brunig, 1976). As the number of apices increases (Table 20.3), there is a rapid decline in the rate of stem elongation and in the amount of shoot produced in each successive order of branching.

In one experiment, plants growing in the field in southern Ghana were paired, and two equivalent active apices matched. One plant was left untreated, and in the second all other branches were re-

Table 20.3. *Changes with age in shoot growth of untreated* Manihot *in field conditions*

Characteristic	0 (main stem)	1	2	3	4	5	6
Number of active apices	1	3.1	7.1	13.0	17.0	15.1	10.0
Shoot extension rate per apex (mm/week)	78	50	16	15	9	4	2
Final length of branch order (mm)	1460	465	285	200	125	65	50
Number of nodes on branch order	37	20	12	9	8	7	

Order of branching spans columns 0–6.

Data collected from farms in Ghana and Sierra Leone.
Source: After Damptey (1964).

moved, together with any shoots that emerged later from apices inactive at the outset (Damptey, 1964). Within a week, highly significant differences had developed, with shoot extension rate of the singled apex already twice that of the control (Table 20.4). Reversal of slow growth rates was also obtained by rooting cuttings from each branch order. Once the effect of the initial differences in size and rate of establishment of the cuttings had passed, growth rates tended to become similar in all plants.

It is usually found when grafting buds from old trees onto small root stocks that their vigour is enhanced by removing them from a competitive situation. However, such scion material also tends to retain strongly certain other features of the mature tree (e.g., the lack of thorniness in *Citrus* spp.). In contrast, buds propagated vegetatively from young seedlings or coppice shoots generally exhibit juvenile characteristics. It was found that grafts from mature *Cedrela odorata* trees grew intermittently, whereas pruned seedlings of the same size grew continuously. The mature tissue was also apparently more sensitive to day length (Longman, 1969). The contrasting branching habits often seen in seedlings and mature grafts confirm the permanent retention of characteristics of the old tree by the graft.

A third type of difference is the greater tendency of mature tissue to flower. The singled apices of *Manihot* were situated as terminal growing points on branch orders 2–4 (i.e., the plants had already formed a terminal inflorescence and branched two, three, or four times before treatment). Although, as has been seen, their growth rate was enhanced, most pruned plants actually flowered *more* frequently than the controls (Table 20.4), suggesting strongly that they had retained a mature tendency achieved earlier. In contrast, the same experiment performed by singling apices of order 1 produced a significant increase neither of growth rate nor of flowering (Damptey, 1964).

The unlimited potential of a vegetative bud changes to a finite capacity as its meristem becomes reproductive and determinate, except in those cases where the floral organs are all produced on lateral axes. Little is known yet about the factors that cause a bud to initiate flowers, because of the technical problems of experimenting with large trees and perhaps because of the many factors that may be involved. A large number of herbaceous tropical plants have now been recorded as photoperiodically sensitive (Mathon, 1975), and also some woody shrubs (Longman & Jeník, 1974). There are short-day, day-neutral, and long-day plants, and quantitative as well as threshold effects. The possibility of studying flower initiation by tropical forest trees in controlled conditions has recently been increased by the discovery that in *Triplochiton* the specialised reproductive

branches (Fig. 20.6) can be induced in a glasshouse on plants not yet
3 years old (Leakey & Longman, 1976). Flower buds are formed on
these laterals under certain conditions, and so significant research
developments in morphogenesis, floral biology, and breeding may
in time become possible.

Dominance, gravimorphism, and bud position

In preceding sections, emphasis has been placed on the responses of single buds. In all genera except the unbranched palms,
tree ferns, and dicotyledonous shrubs, there will, of course, be many
shoots growing at the same time, with complex dominance relations
between them. Other lateral buds do not grow out into shoots; it is
quite likely that most or all of these are capable of growth, but are
prevented owing to correlative inhibition by active apices. Generally
speaking, any lateral bud can be stimulated into active growth by
pruning or complete ringing in the internode distal to it, unless it is
in a state of rest. Similarly if cuttings are made, the top bud or buds
usually grow out, and all the buds in the case of single-node cuttings.

In *Cedrela odorata*, which often grows without branching for many
months, placing the plant at an orientation of about 45° or 90° from
the vertical leads to the outgrowth of some lateral buds. This is also
true of decapitated *Triplochiton*, as mentioned in the previous section. Here the buds on the top are much more likely to grow out than

Table 20.4. *Effect of removing all competing shoot apices from
branched* Manihot *plants in field conditions*

Characteristic	Control (untreated)	Singled apex	Significance of difference (P)
Shoot extension rate (mm/week) (first week)	15.5	33.7	0.1%
Final leaf length (mm) (sample leaf-longest leaflet)	197	289	0.1%
Percent of plants flowering once	29	64	
Percent of plants flowering twice	0	21	
Percent of plants flowering three times	0	21	

Source: Plants had initiated terminal inflorescences and branched
two to four times before beginning of experiment, which ran for 12
weeks from January 15. After Damptey (1964).

those on the sides; those underneath are the least likely to sprout (Table 20.5). Similar responses have been known for many years in temperate trees (Vöchting, 1878) and in *Citrus* (Halma, 1923).

Another aspect of gravimorphic treatment is found after a few weeks, when dominance sets in again amongst the many active shoots. In a vertical plant the uppermost or the second bud is nearly always the one that reasserts dominance, unless further treatments are applied (Wareing & Nasr, 1961; Wareing, 1970). However, in horizontally grown plants there is a distinct tendency for the most vigorous shoots to be those near the roots (Fig. 20.7), though the basal dominance is generally not so pronounced as the apical dominance reimposed in the vertical plants. At 45°, weak apical dominance was found in *Triplochiton*. At all three orientations, strong evidence that correlative inhibition is being imposed again is provided by the fact that many of the smaller shoots grow very slowly or stop. In *Triplochiton* and *Cedrela odorata*, the apex or even the whole of a suppressed lateral axis is often abscised (Table 20.6), and in the latter species it is not uncommon for all shoots except one to disappear or die back.

Many other environmental factors besides the force of gravity can influence dominance relations, including light, water relations, and temperature (Zimmermann & Brown, 1971). In experiment 5, for ex-

Fig. 20.6. Characteristic reproductive type of branching, induced on a *Triplochiton* cutting less than 3 years from germination of seed. These laterals have short internodes, small, pale leaves, and branch frequently.

ample, the uppermost lateral shoots in the decapitated *Triplochiton* had mostly become orthotropic after 2.5 months at 25° C. In contrast, few of those on plants at 20° C had curved up appreciably (mean orientations 19.4° and 35.5° from the vertical, respectively). Decapitated *Terminalia ivorensis* showed a different response to higher temperatures, with greater production of both orthotropic and plagiotropic branches, so that at 30°/30° C plants had nearly four times as many active apices as at 20°/20° C.

Table 20.5. *Gravimorphic effect on bud outgrowth in decapitated* Triplochiton *cuttings (numbers of inactive buds on top, sides, and bottom of horizontal and angled shoots): Expt. 6, 9 weeks*

Clone No.	90° quadrants in which buds were situated	Shoot orientation		
		90° (horizontal)	45° (angled)	0° (upright for comparison)
8036	Top	4.0	4.0	11.0
	Side/2	12.0	13.5	13.0
	Bottom	23.0	25.0	18.0
8038	Top	0.0	0.0	7.0
	Side/2	5.5	3.0	13.5
	Bottom	7.0	7.0	10.0

Each figure represents total for 7 plants, which in 90° and 45° treatments were potted so that most distal lateral pointed downwards (middle of bottom quadrant).

Table 20.6. *Reassertion of dominance in decapitated* Triplochiton *cuttings (percent sprouted buds that later abscised apex or whole shoot): Expt. 6, 16 weeks*

Clone no.	Orientation			Clone means
	0° (vertical)	45° (angled)	90° (horizontal)	
8036	60	43	42	49
8038	7	14	19	14
Orientation means	31	26	29	

Source: After Leakey & Longman (1976).

A simpler system in which to study dominance is in the outgrowth of inactive buds from a length of leafless *Manihot* stem. Such segments will root easily from the basal end, and it has been demonstrated that they exhibit pronounced effects of orientation, with strong apical dominance in upright and near-upright cuttings, basal dominance in cuttings at orientations from 90° to 150°, and weak apical dominance in completely inverted cuttings (Longman, 1968). An even simpler system was subsequently used, in which short segments, restricted to two buds on the same side of the stem, were allowed to flush for 1–2 weeks in the dark, positioned either upright, horizontal, or inverted in moist sand at constant temperature (Fig. 20.8).

At first sight it would appear that there were no appreciable effects of orientation (Fig. 20.9A), with the apical bud growing twice as fast as the basal, a highly significant difference overall. However, it was felt that in order to understand the interactions of two buds it was necessary to know how a bud on its own would grow; so a series of cuttings was also included with a single bud in the apical position. Here there were pronounced effects of orientation, as has been found in *Populus* and *Malus* (Borkowska, 1969), with greatly reduced growth in the inverted segments. This result, which was obtained in several experiments (see Fig. 20.8), leads to the unexpected conclusion that the presence of a basal bud in inverted cuttings actually

Fig. 20.7. Gravimorphic effect on distribution of lateral growth in *Triplochiton*. Decapitated main stems, placed at three different orientations, were subdivided into five-bud segments. Each point on graphs represents mean length of lateral stem produced per segment in 15 weeks. Symbols indicate five clones originating from same seed lot. (After Leakey & Longman, 1976)

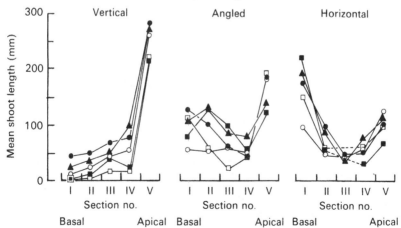

promotes the growth of the apical bud, whereas at other orientations it does not.

In Fig. 20.9B an experiment is shown that provided a second unexpected result. Single bud segments were used throughout, and the inhibiting effect of inverting the cutting was found to be marked only when the bud was in the apical position. There was much less effect when it was basal, and little or none when the bud was situated in the middle. Thus the growth potential of a bud can be influenced even by its position on a piece of stem, and this is true even when the two loci lie a few centimeters apart on the upper side of a horizontal cutting. Clearly, it will be even more difficult than appeared likely to build up a clear picture of the interrelations of several buds.

When a complete ring was made distal to a basal bud, growth was hardly affected, but a ring placed proximal to an apical bud had a highly significant reducing effect. It appears likely, therefore, that it is the stem proximal to the bud that "locates" it and is important in the orientation response. These results could be interpreted in terms of polar auxin transport together with the existence of a promoting factor, synthesised in the roots and rising to the highest point of the stem (Smith & Wareing, 1964b; Zimmermann & Brown, 1971), for only when a bud is both apically placed and at the highest point on the stem does it grow rapidly. A root factor effect is also suggested by the finding that ordinary cuttings of *Manihot* growing in the light show very strong apical dominance and little or no effects of orientation when they are rooted at the apical ends (Longman, 1968).

Fig. 20.8. One-bud stem segments of *Manihot* after 13 days in another experiment in same series as experiments 7 and 8. Note inhibition of growth when bud is at apical end and cutting is inverted. A few nodal and many basal-end roots were formed.

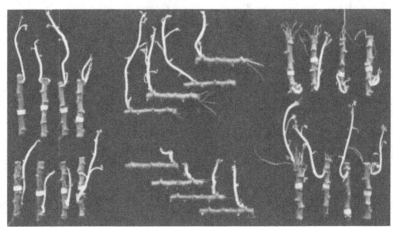

Fig. 20.9. Effect of orientation and bud position on amount of shoot growth made by buds of *Manihot esculenta* at 7 days. *A,* experiment 7; *B,* experiment 8. Vertical bars show least significant differences for individual treatments at 5% (*), 1% (**), and 0.1% (***) levels of probability, using minimum numbers of replicates.

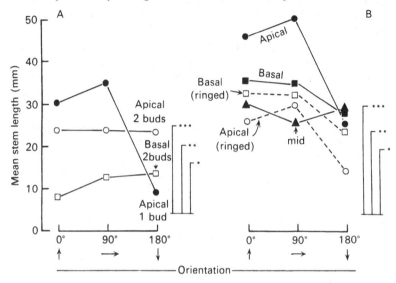

Fig. 20.10. Genetic variation in bud activity in *Triplochiton* seedlings, decapitated to 10 nodes at 4-months. Six examples are illustrated, with contrasting proportions of buds showing active growth. Seed lot 1: Igbo-Ora, Nigeria; seed lot 2: Mile 19, Iwo-Ibadan Road, Nigeria. (After Leakey & Longman, 1976)

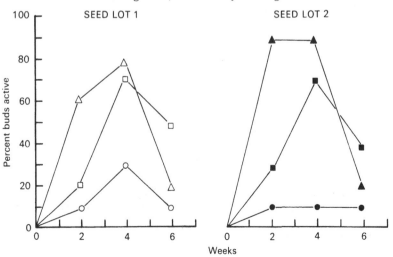

Inherent differences

This account of some of the diverse factors that can affect the shoot growth made by a bud will be concluded with a brief reference to genetic variability. This is very marked in many forest tree species, and it provides an added dimension to the richness of the tropical forests. Unfortunately, however, most forests are in the process of dwindling to small, isolated fragments (Gómez-Pompa, Vázquez-Yanes, & Guevara, 1972). New methods and approaches will be needed if conservation of gene resources is to succeed in retaining sufficient examples of the diversity of a species (Frankel & Bennett, 1970; Burley & Styles, 1976).

It may also be appropriate for botanists to avoid rigidly categorising patterns of shoot extension growth in forest trees, because the range of variation within a species can be so great. Figure 20.10 shows a striking example in variation in response to decapitation in *Triplochiton*. Even within a seed lot, individual seedlings displayed widely different responses. In some only one new shoot was stimulated; in others many branches were produced quickly and dominance was then rapidly reasserted; in a third situation, branches were formed and suppressed more slowly. When clonal cuttings were made from these seedlings, it was confirmed that they demonstrated the strong degree of inherent variation that has been found amongst the clones at present in use (see, for instance, Fig. 20.4, Tables 20.5, 20.6).

Planting of clonal trials of *Triplochiton* in Nigeria has already taken place, and early assessments show large variation amongst clones both in means and in variances (Tables 20.7, 20.8). When the initial

Table 20.7. *Variation of clonal means in field-grown* Triplochiton *cuttings*[a]

	Clone no.					Significance of clonal effect (P)
Characteristic	166/8	175/5	142/10	175/1	177/10	
Number of branches	4.5	4.6	5.0	5.9	8.0	0.1%
Mean length of branches (mm)	102.0	92.0	79.0	74.0	57.0	1.0%
Total branch length (mm)	441.0	427.0	403.0	450.0	453.0	n.s.

[a]Cuttings planted at Gambari, near Ibadan, Nigeria, in June-July 1975 and assessed in October 1975. Clones 175/1 and 175/5 are derived from the same seed lot.
Source: Data from M. R. Bowen.

Table 20.8. *Within-clone coefficients of variation (percent) in field-grown* Triplochiton *cuttings* [a]

Characteristic	Clone no. (no. of replicates)					Mean for 5 clones
	166/8 (40)	175/5 (22)	142/10 (31)	175/1 (26)	177/10 (22)	(141)
Number of branches	49.8	40.7	40.6	35.7	21.8	36.6
Mean length of branches	61.1	61.1	47.0	52.1	40.9	57.1
Total branch length	67.6	70.7	69.0	64.8	49.7	65.3

[a]Cuttings planted at Gambari, near Ibadan, Nigeria, in June-July 1975 and assessed in October 1975. Clones 175/1 and 175/5 are derived from the same seed lot.
Source: Data from M. R. Bowen.

effects of establishment and previous conditions have been super-seded, the idiosyncrasies of some of these new forest tree clones may become as familiar to research scientists as a particular clone of banana now is to the farmer, the shipper, and the consumer. At that stage, it may become a little easier to elucidate the part played by environmental, internal, and genetic factors in controlling rest and activity in buds, and thus the rates and patterns of shoot growth in tropical forest trees.

Acknowledgments

I am very grateful to my many colleagues in West Africa and Scotland who helped and stimulated me in various ways. Dr. R. R. B. Leakey had a major part in experiments 5 and 6, and Miss M. N. Cole in experiment 7. Professor F. T. Last made invaluable contributions. Thanks are also due the United Kingdom Overseas Development Ministry, who are sponsoring the tropical studies in Edinburgh and Ibadan, and the Forest Research Departments of Nigeria, Ghana, and Sierra Leone.

General discussion

Ashton: I measured tree growth in forests on different soils and at different altitudes for some years in Borneo. The mean increment in diameter does not vary in relation to the soils, which was a surprise to us. We did find there was a decline of increment of girth (or diameter) with increase in alti-tude over quite a narrow range, from about 50 to 500 m altitude above sea level. This seems to fit in with your observations, but is in conflict with the so-called cool-nights theory of tropical foresters that explains (Went, 1957), some observations from Africa, where trees grow more rapidly at higher altitudes. The explanation offered is that the respiration rate at night in the lower altitudes is so much greater that it leads to a lower net rate of production. Do you have any comment?

Longman: It is always a large jump from controlled experiments to field circumstances. Obviously, there will be many other factors that influence the rates and the periodicity of shoot extension growth and cambial activity. Went's cool-night theory has been misinterpreted, for he did also provide examples of plants that grew and flowered better with a warm night: e.g., the African violet, *Saintpaulia* (Went, 1957). Other research has shown that if the day is a little cooler than the optimum for the species, compensation may be afforded by very warm nights (Hussey, 1965), and we have data supporting this (Longman, 1972). One explanation, at least as far as dry weight increase is concerned, is that during the warm night more and bigger leaves are made, which can increase the photosynthetic capacity of the plants. Thus the so-called wastage of photosynthates by rapid respiration during warm nights can actually be geared to productive effort that increases the plant's photosynthetic surface.

Givnish: The cold-night hypothesis rests on the notion, as you indicate, that energy used at night is wasted rather than diverted to the production of further organs. What is known about the relative importance of maintenance versus productive metabolism?

Longman: Perhaps Dr. Borchert will comment on this, as I do not know anything specifically about the balance of metabolic activity within the plant. I do know that species vary and that different trees show different responses to temperature changes. In Ghana, Dr. Kwakwa and I found, for instance, certain species that grew better with a cooler night, but others that grew better with a constant temperature and responded unfavorably to fluctuating temperatures. In some species growth increased even when the day temperature was suboptimal if the night temperature was raised (Kwakwa, 1964). I have given a similar example concerning leaf growth of *Triplochiton*.

Borchert: The term "maintenance metabolism" does not make sense in a tree, because as long as a tree is alive (and not dormant) it grows. Therefore, one cannot separate maintenance metabolism, as we think of it in a mature animal, and metabolism needed for sustaining growth. Furthermore, it is well documented that many trees do much, if not most, of their growing during the night.

Fisher: Regarding the interaction of water stress with bud growth and bud release, is there any evidence that the inhibitory effect of a branch on its parent terminal axis also could be related to water stress? Eliminating lateral branches may release water stress for the apical bud and, therefore, stimulate main stem axis growth, and there may be no hormonal or nutritional relation. Is that a possibility?

Longman: I don't know about the particular case of *Terminalia*, and I don't, myself, subscribe to the view that this intermittent stopping of the terminal bud and then restarting is a water stress phenomenon. A relevant experiment was done by the eminent Czech experimental morphologist, Dostál (1967). He showed that if small cuttings of European oak, *Quercus petraea*, are rooted, they go on flushing repeatedly as quite small shoots with the root system in water. He suggested that this may have something to do with the relative size of root and shoot system. If the root system is big, as with the coppice shoot or a small plant put in a large pot, this promotes repeated flushing.

Ng: There are many seedlings in which flushing starts with germination (e.g., *Agathis, Barringtonia*). Continuous growth until first branching may be found in some species or families, but it is not general.

Givnish: As an ecologist I thought that the controversy about whether growth response is determined by internal control or by the differential use of water and nutrients is a classic argument that has never and, in most cases, can never be resolved. Is that correct?

Fisher: I would agree. That's why I was curious whether Dr. Longman had any further evidence that growth control might be due to hormones or water balance, in relation to this one example of branch inhibition of terminal growth.

Longman: No.

Lee: What limits the growth of bonsai trees?

Longman: Professor Wareing did some work on this when he was in Manchester (U.K.) He felt that part of the restriction on the bonsai tree was the small pot and thus a restriction in water and nutrients; but there was also an element of genetic selection.

Borchert: It appears from your answer, Dr. Longman, that you agree that the root/shoot ratio plays a role in rhythmic shoot growth, but you do not accept the proposed role of temporary water deficits (Chap. 21). I do not say that the arrest of shoot growth is always due to a water deficit, I just used that as a convenient example. However, we do not get around the fact that the morphologic relation between root and shoot that you are alluding to must somehow be translated into a functional relation involving the availability of hormones, water, or nutrients to the growing shoot.

I am not quite sure whether Dostál's work with small oak cuttings is comparable to work with 2- or 3-three-year-old seedlings. In such small cuttings there is virtually no carbohydrate reserve. Richardson (1953) showed that in 1-year-old, very small seedlings current growth of both root and shoot is completely dependent upon current photosynthesis, whereas with every year that passes young trees acquire a larger buffer of stored carbohydrates, and spring flushing occurs largely at the expense of these reserves. In Dostál's example I would have said off-hand that the limiting factor for shoot growth was most likely not water but carbohydrates, which during flushing are usually used up faster than they are produced.

Jeník: The interrelations between the seasonality of the above-ground and underground parts has been studied in the tropics only with young plants and tree crops, and apparently not with mature trees. Work on temperate trees is quite extensive. In reviewing the results on the periodicity of underground parts, we realized that there are two views and two types of evidence. One says that the root system of European trees grows early in the spring, before flushing of the above-ground parts, and again late in the fall. Another view is that underground roots grow whenever there are favorable conditions – particularly moisture. The prevailing evidence favors the first view.

Nozeran: Phyllanthus distichus Müll.-Arg. is a tree from Southeast Asia that flowers only when it is a tree, even in the greenhouse. If the first plagiotropic branches are taken after the seedling stage has passed, and cuttings are made of them, a large percentage will flower. The experiments show a progressively diminishing probability of flowering of plagiotropic branches. Can Dr. Longman interpret this phenomenon? We think the seed that has been formed on the tree retains a kind of "memory" of the floral phenomenon and the expression of this memory as a flower or inflorescence is inhibited by the juvenile leaves; it is known that they can inhibit flower-

ing. Only when adult leaves are produced and probably under certain ecologic conditions is a new flowering phase possible. Initially there is a progressive loss of the "memory" and then another gradient of reflowering
when this genetic "memory" can be expressed. Here an explanation is offered that depends not on ecologic phenomena but on internal gradients.
Longman: This is extremely interesting. One generally thinks of the youngest
seedling as being the one which cannot flower and as it grows older it passes
from the juvenile condition to the mature condition where it can flower. I
have been arguing recently that what is really happening is that one can
make a young seedling flower if it is influenced strongly enough, but it is
easier with an older tree (Longman, 1976b). In the example of *Phyllanthus*, I
do not know whether it is really a memory from the previous conditions, but
we certainly understand in vegetative reproduction that there is for some
time a memory of the previous growing conditions.

I know only one other example of a reverse trend: Occasionally a grape-
fruit flowers precociously in the first year, but the plant usually dies subsequently.

Nozeran: Other observations relevant to this discussion are the transmission
from one generation to another of dormancy, which is now being experi-
mented on in *Panicum maximum* in Abidjan. In general, this is shown by
apomictic seeds, but some may be sexually produced. This suggests that the
transfer of factors from one generation to another is cytoplasmic. An analogy
is found in young mammals that lactate when newly born.

References

Alvim, P. de T. (1964). Tree growth periodicity in tropical climates. In *Formation of Wood in Forest Trees*, ed. M. H. Zimmermann, pp. 479–95. New
York: Academic Press.

Bond, T. E. T. (1945). Studies in the vegetative growth and anatomy of the
tea plant (*Camellia thea* Link.) with special reference to the phloem. II. Further analysis of flushing behaviour. *Ann. Bot. Lond., 9,* 183–216.

Borchert, R. (1973). Simulation of rhythmic tree growth under constant conditions. *Physiol. Plant., 29,* 173–80.

Borkowska, B. (1969). The influence of gravity on the growth of buds cultured with a piece of stem in sterile conditions. *Bull. Acad. Pol. Sci. Cl. V,
Ser. Sci. Biol., 17*(8), 503–5.

Brunig, E. F. (1976). Interpretation of tree forms in terms of environmental
conditions. In *Tree Physiology and Yield Improvement*, ed. M. G. R. Cannell
& F. T. Last, pp. 139–56. London: Academic Press.

Bünning, E. (1948). Studien über Photoperiodizität in den Tropen. In *Verna-
lisation and Photoperiodism*, ed. A. E. Murneek & R. O. White, pp. 161–6.
Waltham, Mass: Chronica Botanica.

– (1952). Ueber die Ursachen der Blühreife und Blühperiodizität. *Ztschr.
Bot., 40,* 293–306.

Burley, J., & Styles, B. T. (eds.) (1976). *Tropical Trees: Variation, Breeding and
Conservation.* London: Academic Press.

Champagnat, P. (1965). Physiologie de la croissance et de l'inhibition des
bourgeons: Dominance apicale et phénomènes analogues. In *Handb.
Pflanzenphysiol., 15* (1), 1106–64.

Coster, C. (1923). Lauberneuerung und andere periodische Lebensprozesse in dem trockenen Monsunggebiet Ost-Javas. *Ann. Jard. Bot. Buitenzorg,* 33, 117–89.

Damptey, H. B. (1964). *Studies of apical dominance in woody plants.* M.Sc. thesis, University of Ghana.

Doorenbos, J. (1953). Review of the literature on dormancy in buds of woody plants. *Mededelingen van het Laboratorium voor truinbouwplantenteelt, Landbouwhoogeschool te Wageningen,* 53, 1–24.

Dostál, R. (1967). *On Integration in Plants,* Cambridge, Mass.: Harvard University Press.

Frankel, O. H., & Bennett, E. (eds.) (1970). *Genetic Resources in Plants: Their Exploration and Conservation.* Oxford: Blackwell.

Frankie, G. W., Baker, H. G., & Opler, P. A. (1974). Comparative phenological studies of trees in tropical wet and dry forests in the lowlands of Costa Rica. *J. Ecol.,* 62, 881–919.

Gómez-Pompa, A., Vázquez-Yanes, C., & Guevara, S. (1972). The tropical rain forest: A nonrenewable resource. *Science,* 177, 762–5.

Hallé, F., & Oldeman, R. A. A. (1970). *Essai sur l'architecture et la dynamique de croissance des arbres tropicaux.* Paris: Masson.

Halma, F. F. (1923). Influence of position on production of laterals by branches. *California Citrog.,* 8, 146–80.

Holttum, R. E. (1953). Evolutionary trends in an equatorial climate. *Symp. Soc. Exp. Biol.,* 7, 159–73.

Hopkins, B. (1970). Vegetation of the Olokemeji Forest Reserve, Nigeria. VI. The plants on the forest site, with special reference to their seasonal growth. *J. Ecol.,* 58, 765–93.

Hussey, G. (1965). Growth and development in the young tomato. III. The effect of night and day temperatures on vegetative growth. *J. Exp. Bot.,* 48, 373–85.

Klebs, G. (1926). Ueber periodisch wachsende tropische Baumarten. *Sitz. Heidelb. Akad. Wiss. Kl. Math.-Naturwiss.,* 2, 1–31.

Koriba, K. (1958). On the periodicity of tree-growth in the tropics, with reference to the mode of branching, the leaf-fall, and the formation of the resting bud. *Gard. Bull. Singapore,* 17, 11–81.

Kozlowski, T. T. (1964). Shoot growth in woody plants. *Bot. Rev.,* 30, 335–92.

– (1971). *Growth and Development of Trees,* vol. I. New York: Academic Press.

Kulasegaram, S. (1969). Studies on the dormancy of tea shoots. 1. Hormonal stimulation of the growth of dormant buds. *Tea Quart.,* 40, 31–46.

Kwakwa, R. S. (1964). The effects of temperature and day-length on growth and flowering in woody plants, M.Sc. thesis, University of Ghana.

Leakey, R. R. B., Chapman, V. R., & Longman, K. A. (1975). Studies on root initiation and bud outgrowth in nine clones of *Triplochiton scleroxylon* K. Schum. In *Proceedings of Symposium on Variation and Breeding Systems of Triplochiton scleroxylon K. Schum,* pp. 86–92. Ibadan, Nigeria: Federal Department of Forest Research.

Leakey, R. R. B., & Longman, K. A. (1976). *Root and Bud Formation in West African Trees.* Project 248, 2nd ann. rep. Institute of Terrestrial Ecology.

Longman, K. A. (1966). Effects of the length of the day on growth of West African trees. *J. W. Afr. Sci. Assoc.,* 11, 3–10.

– (1968). Effects of orientation and root position on apical dominance in a tropical woody plant. *Ann. Bot. Lond.,* 32, 553–66.

– (1969). The dormancy and survival of plants in the humid tropics. *Symp. Soc., Exp. Biol.,* 23, 471–88.

- (1972). Environmental control of shoot growth in tropical trees. In *Essays in Forest Meteorology: An Aberystwyth Symposium*, ed. J. A. Taylor, pp. 157–67. Aberystwyth: J. A. Taylor, Dept. of Geography, University of Wales.
- (1976a). Conservation and utilization of gene resources by vegetative multiplication of tropical trees. In *Tropical Trees: Variation, Breeding and Conservation*, ed. J. Burley & B. T. Styles, pp. 19–24. London: Academic Press.
- (1976b). Some experimental approaches to the problem of phase-change in trees. *Acta Hort.*, 56, 81–90.
- (1976c). Tree biology research and plant propagation. *Combined Proceedings of the International Plant Propagators' Society*, 25, 219–36.
Longman, K. A., & Coutts, M. P. (1974). Physiology of the oak tree. In *The British Oak*, ed. M. G. Morris & F. H. Perring, pp. 194–221. Faringdon: BSBI/Classey.
Longman, K. A., & Jeník, J. (1974). *Tropical Forest and Its Environment*. London: Longmans.
Mathon, C. C. (1975). *Ecologie du développement et phytogéographie*. Poitiers: Facultés des Sciences de l'Université.
Milthorpe, F. L. (1956). *The Growth of Leaves*. London: Butterworth.
- (1976). Quantitative aspects of leaf growth. In *Perspectives in Experimental Biology*, vol. 1, *Botany*, ed. N. Sunderland, pp. 33–40. Oxford: Pergamon Press.
Moorby, J., & Wareing, P. F. (1963). Ageing in woody plants. *Ann. Bot. Lond.*, 27, 291–308.
Njoku, E. (1959). Response of rice to small differences in day-length. *Nature Lond.*, 183, 1598.
- (1963). Seasonal periodicity in the growth and development of some forest trees in Nigeria. I. Observations on mature trees. *J. Ecol.*, 51, 617–24.
- (1964). Seasonal periodicity in the growth and development of some forest trees in Nigeria. II. Observations on seedlings. *J. Ecol.*, 52, 19–26.
Richards, P. W. (1952). *The Tropical Rain Forest*. Cambridge: Cambridge University Press.
Richardson, S. D. (1953). Studies of root growth in *Acer saccharinum* L. II. Factors affecting root growth when photosynthesis is curtailed. *Proc. Koninkl. Akad. Wetenschap. Amsterdam*, C56, 186–93.
Roberts, E. H. (1963). *Photoperiodism of Rice*. Bull. No. 7, West African Rice Research Station, Rokupr, Sierra Leone.
Romberger, J. A. (1963). *Meristems, Growth and Development in Woody Plants*. USDA Forest Service Techn. Bull. No. 1293.
Schimper, A. F. W. (1898). *Pflanzengeographie auf Physiologischer Grundlage*. Jena. English edition (1903), *Plant Geography on a Physiological Basis*, tr. W. R. Fisher; ed. P. Groom & I. B. Balfour. London: Oxford University Press.
Slee, M. U., & Shepherd, K. R. (1972). The importance of high temperatures in the induction of the resting phase of *Pinus elliottii*. *Aust. J. Biol. Sci.*, 25, 1351–4.
Smith, H., & Wareing, P. F. (1964a). Gravimorphism in trees. 2. The effect of gravity on bud-break in osier willow. *Ann. Bot. Lond.*, 28, 283–95.
- (1964b). Gravimorphism in trees. 3. The possible implication of a root factor in the growth and dominance relationships of the shoots. *Ann. Bot. Lond.*, 28, 297–309.
Taylor, C. J. (1960). *Synecology and Silviculture in Ghana*. Edinburgh: Nelson.

Vegis, A. (1964). Dormancy in higher plants. *Ann. Rev. Plant Physiol.*, 15, 185–224.

Vöchting, H. (1878). *Ueber Organbildung in Pflanzenreich.* Bonn: Cohen.

Wareing, P. F. (1959). Problems of juvenility and flowering in trees. *J. Linn. Soc. (Bot.)*, 56, 282–9.

– (1969a). Control of bud dormancy in seed plants. *Symp. Soc. Exp. Biol.*, 23, 241–62.

– (1969b). Germination and dormancy. In *The Physiology of Plant Growth and Development*, ed. M. B. Wilkins, pp. 605–43. New York: McGraw-Hill.

– (1970). Growth and its co-ordination in trees. In *Physiology of Tree Crops*, ed. L. C. Luckwill & C. V. Cutting, pp. 1–21. London: Academic Press.

Wareing, P. F., & Nasr, T. A. A. (1961). Gravimorphism in trees. 1. Effects of gravity on growth and apical dominance in fruit trees. *Ann. Bot. Lond.*, 25, 321–40.

Went, F. W. (1957). *The Experimental Control of Plant Growth.* Waltham, Mass.: Chronica Botanica.

– (1962). Some problems of plant physiology in the tropics. *Assoc. Trop. Biol. Bull.*, 1, 90–1.

Zimmermann, M. H., & Brown, C. L. (1971). *Trees: Structure and Function.* New York: Springer Verlag.

21

Feedback control and age-related changes of shoot growth in seasonal and nonseasonal climates

ROLF BORCHERT

Department of Botany, Physiology, and Cell Biology, University of Kansas, Lawrence, Kansas

One of the most characteristic properties of trees as long-lived growing systems is the dramatic annual changes in activity displayed in the temperate zone and elsewhere. It is now well established that these changes in activity are not only correlated with, but are responses to, seasonal variations in temperature, photoperiod, moisture availability, and other environmental factors. The less dramatic, yet nevertheless clearly notable, variations in activity that commonly occur in tropical trees, even those growing in uniform climatic conditions such as prevail in Singapore (Koriba, 1958), have been likewise ascribed to the more subtle variations occurring in such tropical environments. However, good correlations between fluctuations in environmental conditions and activity changes in tropical trees have been only rarely established, and experimental evidence for the causal effect of certain environmental cues is scarce (see Chap. 19). Rhythmic growth – the regular alternation between periods of active shoot growth (flush) and periods when no active growth is shown (rest) – displayed by many trees (e.g., cacao, rubber, mango) in their juvenile stages, has been shown to occur not only in nonseasonal climatic conditions, but also in constant, controlled environments (Hallé & Martin, 1968; Greathouse, Laetsch, & Phinney, 1971). The mechanism underlying these rhythmic changes in the activity of apical meristems – and the corresponding hormonal changes – may, therefore, reside within the tree itself, rather than in its environment. In sum, the relative importance of exogenous vs. endogenous controls determining shoot growth patterns in trees appears to vary widely.

As a first step in the analysis of the dynamics of trees as growing systems I showed earlier that rhythmic growth under constant conditions can result from feedback interaction between the subsystems of a tree (Borchert, 1973). I will now discuss how variations in environmental conditions and an increase in tree size may interact with the internal controls of shoot growth in the determination of shoot growth patterns in trees. This analysis will comprise the following steps: (1) integration into a dynamic model of knowledge obtained from relatively simple experimental systems and concerned with the structure-function relations and the interaction between tree and environment, (2) comparison between experimentally induced variations in shoot growth patterns and changes predicted by the model, and (3) extrapolation from the model to predict the functional changes that are likely to result from increased structural complexity of a tree. I shall then discuss observed differences in shoot growth patterns between young and old trees and hope to show that these age-dependent changes in shoot growth are the consequences of the structural changes that trees as growing systems of extraordinary size, complexity, and longevity undergo with age.

Model of rhythmic growth and its implications

In the model of rhythmic growth (Fig. 21.1; Borchert, 1973), a tree has been considered as a system consisting of two major subsystems, root and shoot, the latter being in turn composed of stems and leaves. In view of the interdependence of these subsystems, it can be assumed that continuous growth of the entire tree is possible only if all its subsystems are in a functional equilibrium. If equilibrium is disturbed, growth of some subsystem must be temporarily arrested until a functional balance is restored. For instance, if the shoot system of a tree grows faster than its root system for any length of time, transpiration will increase faster than the capacity to absorb water, and a water deficit will result in the shoot. This water deficit may slow or stop shoot growth until a functional equilibrium between root and shoot (a favorable root/shoot ratio) has been restored. Any feedback system in which one subsystem is rate-limiting for another one can cause endogenous growth rhythms under constant environmental conditions, and feedback interaction between root and shoot has been considered the basis for rhythmic growth (Fig. 21.1).

Basically, the temporary arrest of shoot growth may depend on deficiencies in any one of the various substances roots supply to shoots. But in my view, available indirect evidence supports the notion that arrest of shoot growth is caused mostly by water stress, whose inhibitory effect on the apical meristem might be mediated by hormonal changes (Alvim et al., 1974; Borchert, 1975).

The model has the following implications:

Fig. 21.1. Computer-simulated time course of rhythmic shoot growth. During first four time intervals (days), shoot growth rate increases from zero to its maximum value; root growth takes place at constant rate; and a new leaf is initiated every second day. Growth rate for each leaf is increased from zero to its maximum value, and leaf growth is arrested once leaf has attained its genetically determined maximum length. Leaf area is calculated from leaf length; rate of transpiration is proportional to total leaf area and is reduced by existing water deficit. Rate of water absorption is proportional to root length. As more and more leaves attain maturity, transpiration exceeds water absorption by root, and water deficit begins to develop at day 14. At day 17, water deficit surpasses critical value, and shoot and leaf growth rates are gradually reduced to zero. As flush ends, development of immature leaves is terminated before they have reached their maximum size. After arrest of shoot and leaf growth, water deficit gradually disappears because of continuation of root growth; transpiration reaches its maximum value for given leaf area; and new flush begins. Continuous shoot growth will result if transpiration and root elongation increase at same rate. For details, see Borchert (1973).

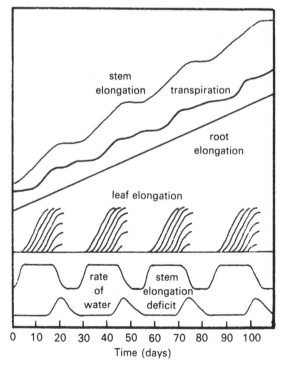

1. Growth rhythms occurring under constant conditions are endogenous (i.e., the changes causing the temporary arrest of shoot growth under favorable environmental conditions arise within the tree).

2. The period length of the rhythm depends on genetic factors (e.g., those determining the rates of shoot, root, and leaf growth) as well as on the set of environmental conditions under which the tree is growing. Period length will thus be altered if the tree is transferred from one set of constant environmental conditions to another, and in a variable environment a secondary, exogenous periodicity will be imposed over the basic endogenous rhythm.

3. There are no basic – only gradual – differences between rhythmic (determinate) and continuous (indeterminate) shoot growth. For instance, during "continuous" shoot growth in seedlings of avocado (*Persea americana*; Fig. 21.2B) the rate of shoot elongation varies at least as much as during rhythmic growth in pin oak (Fig. 21.2A), *Hevea*, or cacao (Hallé & Martin, 1968; Greathouse et al., 1971), but it never becomes zero. The increase in leaf area is, therefore, more continuous than during rhythmic growth. Similar variations in the rate of shoot elongation have been observed during continuous growth of the tropical tree *Tabernaemontana crassa* (Prévost, 1972) and of such temperate zone trees as *Pinus taeda* (Kramer & Kozlowski, 1960) and

Fig. 21.2. Time course of shoot and leaf growth in (A) vigorous single-shoot pin oak seedling and (B) seedling of avocado (*Persea americana*). s, shoot elongation during week preceding indicated date; l, total leaf area of tree. (A, from Borchert, 1975; B, data supplied by Dr. J. Kummerow, San Diego State University, San Diego, California)

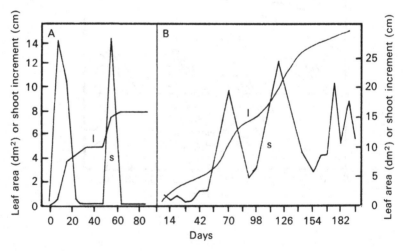

Populus deltoides (Borchert, 1976b), suggesting that feedback oscillations occur whose effect is to maintain an adequate root/shoot ratio during continuous growth. Rhythmic or continuous growth might result from differences in the sensitivity of apical meristems to feedback inhibition of a certain intensity, which could induce formation of a resting bud in oak, cacao, or rubber, cause shoot tip abortion in *Tilia* and similar species, or simply slow shoot growth in continuously growing trees.

Experimental modifications of shoot growth patterns
In agreement with the model, experimental *reduction of each maturing leaf* to about one-third normal size (i.e., slowing the rate of increase in total leaf area) caused a transition from rhythmic to continuous shoot growth in *Hevea* (Fig. 21.3A, D; Hallé & Martin, 1968;

Fig. 21.3. Computer simulation of time course of shoot growth rate under constant environmental conditions, as function of decreasing maximum leaf length. Maximum leaf length was (A) 24, (B) 18, (C) 14, and (D) 12. For details, see Borchert (1973).

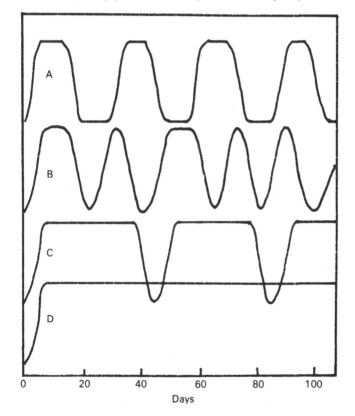

Borchert, 1973) but not in pin oak (*Quercus palustris:* Borchert, 1973). In the latter species the same treatment resulted in a series of short, consecutive flushes: i.e., a pattern predicted by the model for the transition from rhythmic to continuous growth (Table 21.1, Fig. 21.3*B*). The coincidence between this intuitively unexpected experimental result and a prediction derived from the model is an excellent illustration of the power of modeling, which permits predictions of a more detailed nature than those made on the basis of a more general working hypothesis. Similarly, such findings increase confidence in the correctness of the basic assumptions incorporated into the model. In oak, continuous shoot growth for an entire season is observed only in stump sprouts, which possess an extremely favorable root/shoot ratio. Similarly, vigorous shoots in tea may grow continuously for rather long periods and form up to 20 leaves, whereas normally only 4 leaves are formed in each flush (Bond, 1942).

In order to study the effect of *enhanced leaf growth* on rhythmicity, shoot growth patterns of single-shoot oak saplings were compared with those of trees in which several shoots were permitted to grow. Necessarily, the total leaf area, and thus transpiration, will increase faster in the multiple-shoot trees than in the control group. Hence (assuming the root growth rate remains the same), the functional bal-

Table 21.1. *Means of selected variables characterizing seasonal shoot growth in potted saplings of pin oak.*

Variable	Experimental group[a]			
	A	B	C	D
Number of flushes	1.9	1.0	2.9	2.0
Days of active shoot growth	25.4	8.0	48.2	19.1
Total leaf area (dm²)	10.5	12.9	0.9	2.9
Number of leaves	26.7	64.4	46.0	18.8
Percent of leaves in first flush	66.0	97.0	60.0	55.0
Area/leaf (cm²)	39.4	20.0	2.0	15.7
Total shoot growth (cm)	33.7	16.3	30.2	8.6
Maximum shoot growth rate (cm/day)	2.7	1.9	1.7	0.7
Number of shoots per tree	1.0	7.3	1.0	1.0
Number of trees per group	7.0	7.0	9.0	8.0

[a]Experimental treatments: A, one shoot per tree allowed to grow; B, all shoots allowed to grow (total shoot growth and maximum shoot growth rate data obtained from one vigorous shoot per plant); C, each leaf experimentally reduced to about one-fifth its normal length; D, trees growing in sand instead of soil.

ance between root and shoot system should be lost faster and should require longer for restoration after shoot growth has been arrested. Growth flushes should thus become shorter and less frequent. Experimental results showed that – in agreement with these predictions – shoot growth in multishoot trees lasted only for a very short period, and there was only one seasonal flush of shoot growth compared with two in the single-shoot controls (Table 21.1A, B; the limited number of consecutive flushes is ascribed to limited pot size, as discussed below).

The effect of *variations in environmental conditions* on shoot growth patterns of cacao has been extensively documented by Sale (1970a). Young cacao trees were grown in containers in a glasshouse and kept under three different soil moisture regimens. Soil water was restored to field capacity when available water had been depleted to approximately 85% in the wet treatment (W), to 50% in the medium treatment (M), and to 15% in the dry treatment (D). Frequency of watering in the three treatments was about three times a week, once a week, and once every 3 weeks, respectively. In addition, to determine the effect of atmospheric humidity on plant responses to soil moisture, the watering treatments were combined factorially with the two ranges of atmospheric humidity that occur naturally during the day in the dry season (50% relative humidity at midday, 100% at night) and in the wet season (70%–80% relative humidity at midday, 100% at night) in Trinidad, where the work was performed.

The total leaf area attained by experimental plants at the end of the first 6-month experimental period (dry season) was directly proportional to the available soil moisture (Fig. 21.4). During the second period (wet season), net leaf area increased only if soil moisture was equal to or higher than during the previous period (e.g., treatments WW, MW, DW, MM), but remained the same or declined if soil moisture was lower than before (e.g., WM, WD, MD). During the third period, net leaf area increased where a higher soil moisture followed a lower one (e.g., WDW, MDM), remained unchanged where soil moisture was kept the same (WWW, MMM), and declined where soil moisture was decreased (e.g., MWM, DWD). This means that under identical soil moisture conditions (e.g., M) the net leaf area increased (MDM), remained the same (MMM), or declined (MWM), depending only on the previous moisture regimen and thus on the functional state of the tree (root/shoot ratio) at the beginning of the period. If trees entered a W period with a high root/shoot ratio, as established by relative drought, there followed a marked increase in leaf area. If they entered a D period with a low root/shoot ratio, leaf abscission occurred (DWD) until the foliage was reduced to an amount that could be adequately supplied by the root.

Fig. 21.4. Leaf growth and abscission in saplings of cacao (*Theobroma cacao*) as function of varying soil moisture regimens (W, wet; M, medium; D, dry, as described in text) during three consecutive 6-month periods. Total length of each bar indicates total leaf area formed to date; lower (striped) part of bar, net leaf area present at end of respective 6-month period; upper (dotted) part of bar, reduction of total leaf area by abscission; horizontal line through bars, leaf area present at beginning of experiment. (Data from Table 1 and Fig. 2 in Sale, 1970a)

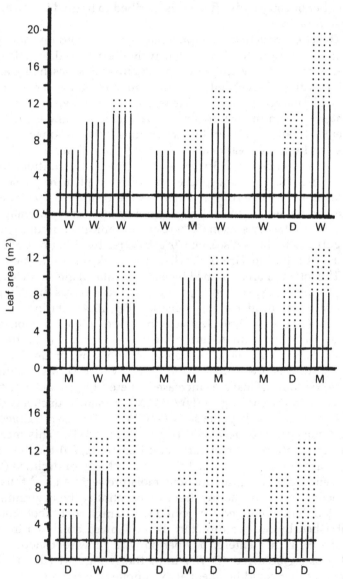

Sale's data also show that differential increases in leaf area under W and M treatments were achieved mainly by variations in the size of mature leaves, which ranged from 110 to 217 cm², and the number of simultaneous growth flushes, but not by variations in the frequency of flushing (Fig. 21.5). Comparisons between pin oak saplings growing in soil or sand, respectively, gave similar results (Table 21.1D; Borchert, 1975). Under D conditions, however, the 3-weekly watering-drying cycle imposed a higher than normal frequency of flushing upon the trees: Whereas under W and M conditions trees passed through four to five flushes during the third period, nine flushes in 26 weeks or one flush every 3 weeks occurred under the D regimen. As under these conditions many axillary buds were released from correlative inhibition, the plants were extensively branched and produced large numbers of simultaneous flushes (Fig. 21.5). However, newly expanded leaves mostly dried up and ab-

Fig. 21.5. Net leaf area (*l*) and number of flushes (*f*) in saplings of cacao during three consecutive 6-month periods under varying soil moisture regimens (*W*, wet; *M*, medium; *D*, dry, as described in text). (Redrawn from Figs. 1 and 2 in Sale, 1970a)

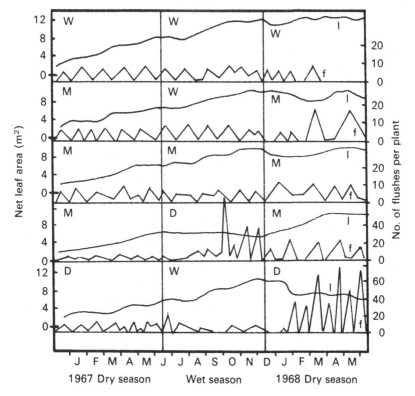

scised during the second, drier half of the watering cycle, thus restoring a high root/shoot ratio favorable for the initiation of new shoot growth after the next watering. The readjustment of the root/shoot ratio after each seasonal change in soil moisture regimen had another consequence: The total leaf growth over three periods was highest in those groups that spent one or two periods under D conditions (Fig. 21.4), and à correspondingly higher fraction of their total dry matter production was allocated to the formation of leaves (e.g., 53% in WDW vs. 32% in WWW). Thus, the internal allocation of resources was profoundly affected. Finally, the different soil moisture regimens exerted a marked influence on the architecture of the experimental trees. Almost no branching occurred in the continuously W plants, which developed a few long, thick branches carrying very large leaves. Plants under the M treatment formed more branches and resembled field-grown trees. Under D treatment, the trees became excessively branched as a result of repeated axillary flushing; yet the plants were stunted in appearance and their thin branches carried only small leaves.

The effect of *restricted root growth* on shoot growth patterns can be assessed experimentally in container-grown plants. The size of the container determines the maximum extension and functional capacity of the root system, and thus, indirectly, the maximum size of the shoot system. Seasonal shoot growth of potted, well-watered trees should end when they have attained this maximum size (trees are pot-bound). In agreement with these considerations, among different groups of pin oak saplings grown in containers of the same size, mean total leaf area per tree was found to be in the same range, whereas other components of the shoot growth pattern varied widely. Mean leaf area was 12.9 dm^2 in multishoot trees (Table 21.1B), 10.5 dm^2 in single-shoot trees having passed through two consecutive flushes (Table 21.1A), and 9.5 dm^2 in less vigorous single-shoot trees having completed three growth flushes. When rhythmic growth patterns of the last two groups were computer-simulated by appropriate combinations of root and leaf growth rates, limitation of root size to a maximum value likewise restricted the number of flushes to two or three, respectively (Borchert, 1975). Renewed flushing of "quiescent" (pot-bound) trees can be induced either by transferring the trees to a larger container or by partial defoliation. This means that the frequently observed "correlative inhibition of shoot growth by leaves" does not depend simply on the presence of a certain leaf area per shoot, but on the ratio between leaf area and size of the root system (Borchert, 1975).

The same phenomenon was observed in the work with cacao discussed above (Sale, 1970a). Under each soil moisture regimen a certain leaf area could be supported by the root system (approximately 12 m² in W, 8.5 m² in M, and 4 m² in D). Once this leaf area was reached, shoot growth ceased if leaves persisted (WWW) or continued if leaf abscission occurred. Under constant soil moisture regimens, the decline in relative humidity associated with the transition from the wet to the dry season (third period) also caused a decline in the maximum leaf area (Fig. 21.5). Finally, trees grown under W conditions in smaller containers reached about half the final leaf area of the trees described above (Sale, 1970b).

The examples discussed so far have shown that experimental modification of the functional equilibrium between root and shoot (root/shoot ratio) – whether by changing the structure of the tree, by varying environmental conditions, or by imposing limits to growth of a subsystem – generally caused the changes in shoot growth patterns predicted by the model of rhythmic growth. More specifically, even in the relatively small experimental trees, the increase in size and complexity, which results from the very nature of the tree as a growing system, was found to enhance the problem of maintaining the functional balance between the subsystems of a tree. Based on these observations, one would expect that in older trees the amplitude of feedback oscillations should be even larger, and the response to environmental conditions more dramatic.

Periodicity of primary shoot growth in large trees

Saplings of many tropical trees have been reported to remain evergreen and show continuous, monopodial growth for 2–3 years or until they reach a height of several meters. With increasing size, trees branch, may shed their leaves during the dry season, and become dormant for increasingly longer periods. There is a transition from continuous to intermittent or periodic shoot growth (e.g., *Bombax, Ceiba, Pithecellobium, Tectona, Terminalia:* Klebs, 1912; Coster, 1923; *Cussonia, Lophira:* Hallé & Martin, 1968; *Persea:* Kummerow, pers. comm.; *Bombax* and other species: Njoku, 1963, 1964). A temporary arrest of shoot growth accompanied by distinct morphologic changes in apical buds or leaves has been observed in the majority of trees growing under the rather uniform climatic conditions of Singapore (Koriba, 1958). As many descriptions of shoot growth patterns in tropical trees found in the literature are not accompanied by adequate climatic records, the degree of seasonal variation and its effect on growth patterns often cannot be assessed. The situation is complicated further by the fact that a good number of trees range from permanently moist to seasonally dry regions (e.g., *Hevea, Tectona*), where they may vary from an evergreen to a deciduous habit and probably display significant differences in shoot growth patterns as well.

In tropical trees showing rhythmic growth of seedlings, the interval between consecutive growth flushes tends to lengthen as trees age (e.g., *Hevea brasiliensis:* Huber, 1898; *Theobroma cacao:* Greenwood & Posnette, 1950; *Mangifera indica:* Holdsworth, 1963). For example, the number of annual growth flushes declines from 8 to 10 in many young trees to about 4 in mature cacao and tea and to 1–2 in many other trees. Whereas flushing of young trees is apparently rather independent of variations in environmental factors, in adult trees it is usually considered to be well correlated with seasonal changes in climate.

In the temperate zone, equivalent age-related changes in shoot growth patterns occur. Under favorable conditions, seedlings of many trees grow continuously throughout the growing season and terminate shoot growth in fall by formation of a resting bud or by shoot tip abortion. Shoot elongation in adult trees of the same species is usually arrested earlier in the season and at a time when environmental conditions permit continued growth of seedlings (e.g., *Ailanthus, Betula, Liriodendron, Populus, Robinia:* Klebs, 1917; Critchfield, 1960; Kozlowski, 1971). In several species of the genera *Acer, Fraxinus,* and *Prunus,* seedlings grow continuously, but saplings may grow rapidly for a while, then undergo a marked decline in shoot growth rate accompanied by a partial transformation of leaves into cataphylls, resume rapid growth again, and finally form a resting bud (concealed lammas shoots: Späth, 1912). In adult trees of these species seasonal shoot growth is usually limited to one rather short flush. The same phenomenon was described for tea (Bond, 1942). In oak, the only temperate zone tree commonly displaying rhythmic shoot growth of seedlings, the number of seasonal flushes declines from three to four in seedlings to one or two in mature trees.

In sum, it appears that under a variety of climatic conditions shoot growth in older trees is arrested earlier in the growing season than that in small, young trees. Consequently, the number of annual recurrent flushes or the duration of continuous shoot growth, and hence annual shoot growth increments, are reduced as trees become older. Ultimately this trend results in a single, rather short period of annual shoot growth, as observed in many trees growing under strongly seasonal climates.

Essentially, these observations and conclusions agree with the model predictions based on the described experimental modifications of shoot growth patterns in small trees. However, in view of the scarcity of detailed observations on mature trees, the following interpretation of the age-related changes in shoot growth patterns must remain tentative.

Growing trees increase in size as well as complexity. The increase in complexity of the shoot system is due to the dramatic rise in the number of growing shoots, which necessarily leads to a high degree of positional differentiation among individual shoots. Shoots may be long (terminal) or short (lateral) shoots, they may grow in the upper or lower part of the crown, and angles of growth may vary widely. Consequently, the network of internal, physiologic correlations, which constitutes one of the least understood aspects of organismic plant biology, must become highly complex. A progressively larger fraction of the shoots and buds present will be subject to some kind of correlative inhibition and will not realize their full growth potential (Borchert, 1976a; Chap. 20). For instance, growth of many lateral shoots is known to be inhibited by an actively growing terminal bud (apical dominance). If the latter is removed, lateral shoots are released from inhibition and resume normal, unrestricted growth. This type of correlative inhibition depends primarily on the specific geometric relation between apical and lateral buds and only indirectly on the environmental factors affecting growth of the tree as a whole. The correlative inhibition is thus endogenous in nature. The apical bud is thought to exert its inhibitory influence mainly through the action of plant hormones, but the mechanism of inhibition is poorly understood (Phillips, 1975). As lateral (short) shoots may account for more than 90% of all growing shoots in some mature trees (Wilson, 1966), this rather localized correlative inhibition is likely to represent a major cause of the reduced vigor of shoot growth in mature trees.

However, apical dominance is not sufficient by itself to explain the observed cessation of growth in vigorous terminal shoots under apparently favorable environmental conditions. In view of the differences between shoot growth in single-shoot and multishoot pin oak saplings described earlier, can it be that this arrest of shoot growth results from the type of feedback inhibition postulated earlier as the basis of rhythmic growth? Such feedback controls have low sensitivity, and considerable overshoot has been shown to occur even in the regulation of shoot growth of small trees. As there is no evidence that large trees possess different or better control systems than small ones, it must be expected that feedback inhibition of shoot growth becomes stronger and occurs more frequently as trees increase in size and complexity. The observed age-related transition from continuous to periodic growth or the reduced frequency of recurrent flushing, as observed even under uniform environmental conditions, is therefore considered to be primarily a consequence of the tree's increased complexity (i.e., of the age-related changes in the structure of the system "tree"). The same type of feedback inhibition is thus thought to be responsible for the termination of each flush during rhythmic growth of seedlings and for the eventual cessation of continuous growth in mature trees. This feedback inhibition imposes a state of quiescence or "summer dormancy" – not true dormancy – on all growing shoots of a tree alike. As shown earlier, in small trees this inhibition can be readily overcome by appropriate changes in tree structure or environmental conditions. A few similar observations exist for mature trees. If trees in the temperate zone are defoliated in early summer by excessive drought or insect attack and are subsequently supplied with adequate soil moisture, they will initiate a second flush of shoot growth at a time when they are normally quiescent. Likewise, severe pruning of a large tree, which reduces its complexity and establishes a favorable root/shoot ratio, greatly enhances the growth vigor of the remaining shoots.

If, as postulated here, the arrest of shoot growth under favorable environmental conditions is indeed the result of an unbalanced root/shoot ratio, the question of the adaptive significance of such rest periods or of rhythmic growth becomes irrelevant. Instead, the adaptive value of slow indeterminate compared with rapid determinate shoot growth should be discussed. For instance, Huber (1935) pointed out that rapid determinate growth, accompanied by the formation of ring-porous wood, is common in areas characterized by cold winters and dry summers, where trees must make optimum use of the rather short period favorable for shoot growth.

In contrast to the limitations on shoot growth resulting from correlative and feedback inhibition among the subsystems of the tree, the expanding root system faces progressively deteriorating growing conditions that are the direct consequence of the physical nature of the soil. With increasing depth, soil structure becomes less favorable for root penetration, and root metabolism becomes increasingly restricted by inefficient gas exchange. In most trees the great majority of absorbing roots are therefore found in the upper 60–100 cm of the soil. Expansion of the root system is often limited by competing root systems of neighboring trees. Thus, sooner or later in the life of a tree, the root system approaches its physical limits, root growth must slow, and – in the absence of comparable restrictions on shoot growth – the root/shoot ratio declines. Such a reduction in the ratio between below-ground and above-ground biomass as a function of increasing age has indeed been observed in a mixed deciduous forest (Harris, Goldstein, & Henderson, 1973) and is likely to play a significant role in the reduction of annual shoot growth.

Moderate seasonal fluctuations in climatic conditions may affect large trees by enhancing the periodically occurring functional disequilibrium and thus tend to synchronize endogenous growth periodicity in populations of adult trees. The problem of maintaining a functional equilibrium in a large tree is compounded if seasonal variations in climatic conditions are strong enough to cause deciduousness and regularly impose a rest period. During the burst of intensive shoot growth, which usually follows rest periods, a large supply of nutrients must be provided by mobilizing the tree's organic reserve materials and absorbing minerals from the soil, an efficient transport system capable of distributing water and nutrients to thousands of competing shoots must be formed, and the rapidly increasing leaf area must be adequately supplied with water by the root system. It appears virtually impossible that the demands of many rapidly growing shoots can be met during long periods, particularly if the growing system is subject to additional stress (e.g., from rising temperatures or declining soil moisture). For instance, within a period of 8 weeks from May to July, the leaf area of a young tree increased by 68%, transpirational water loss per unit leaf area by 45%, and the total transpirational water loss of the tree by 142% (Chapman & Parker, 1942). Under such conditions, a small tree still able to expand its root system and thus its capacity to absorb water and nutrients from the soil may be able to maintain the balanced water economy necessary for continued shoot growth. However, a large tree whose root system has already occupied all available soil space when the current flush of shoot growth was initiated, soon develops a water deficit and ceases to grow. Depletion of soil moisture has long been considered the most important factor limiting extension growth in trees (Zahner, 1968).

Size and complexity of a tree have been shown to affect both the internal correlations between its structural components and the interaction with its environment. If the role of increasing complexity is taken into account, age-related changes in shoot growth patterns can be predicted on the basis of the assumptions that underlie the model of rhythmic growth. This approach implies that any basic concept that adequately considers the behavior of whole trees as large, complex, and long-lived growing systems should be applicable to trees of all climatic zones alike. If tropical trees express a greater range of the developmental potentials inherent in the life form of a tree, and, therefore, display a greater structural diversity than temperate trees (Tomlinson & Gill, 1973), this is most likely due to the favorable tropical environment, which imposes fewer exogenous constraints on tree growth than temperate climatic conditions, and not to fundamental differences in basic mechanisms controlling tree development.

Acknowledgments

Support of this work by the University of Kansas General Research Fund is acknowledged. I am indebted to Dr. J. Kummerow (Department of Botany, San Diego State University) for permission to use his unpublished data on growth of *Persea americana*, and to Dr. G. Browning (Department of Botany, University College of Wales) for critically reading the manuscript.

General discussion

Givnish: Your model, as currently stated, depends on more or less constant root growth, yet you did not present any figures for root growth. In relatively nonseasonal climates do trees tend to show rhythmic root as well as shoot growth? In such climates is rhythmicity or periodicity in shoot growth more marked and the amplitude greater than in root growth?

Borchert: Both have been observed. Dr. Hallé has shown that root growth in seedlings of *Hevea* is continuous; but Dr. Nozeran has data, and I have heard this from other people, that root growth in cacao and *Citrus* is rhythmic, yet the main periods of shoot growth alternate with those of root growth. In other words, a flush of root growth precedes one of shoot growth, even under constant conditions.

Alvim: The model is very interesting, but it does not fit some of the observed facts. For instance, in Sale's work you showed some figures that indicate that after 2.5 years if plants are given plenty of water, they stop flushing. Sale, in spite of finding that he could induce flushing by manipulating soil moisture, concluded that moisture stress was needed to induce flushing. In a sense, he agreed with my previous suggestion. Simple pruning of rubber has been shown to change growth from intermittent to continuous. If this were related to moisture, then it would be very simple to maintain a rubber or a cacao tree growing continuously by keeping it in a very moist atmosphere or high humidity. This has been tried and its effect is just the contrary: it stops flushing completely. I think pruning operates, not through an effect on water stress, but through upsetting some hormonal equilibrium, as other authors have suggested. The model could be reorganized with something else other than water in mind. Water is important, but it operates through some more critical internal factor that is acting on growth as a stress effect. There has been discussion whether this stress is a cold shock, but cold factor has to be demonstrated more clearly. It is mentioned in the literature quite often, but it has never been clearly demonstrated. I think whenever you have a shower causing growth, it seems to be related to some other factor, but not to temperature. We have tried to reproduce the temperature effect, but without success.

Borchert: With reference to your comment on growth and humid conditions, let me first reiterate that I have not claimed that shoot growth is always arrested by water deficiency. You refer to this observation (Fig. 21.5, top graph third period), and I ascribe this effect to the fact that in a containerized plant obviously the expansion of the root system is limited. Once the tree is potbound, there is no possibility for additional increase in the absorbing capacity of the root system; hence, growth of the shoot system is limited as well. In wet conditions, as shown here, where the plant was adequately supplied with water, there was no leaf abscission when the maximum possible leaf area was reached. Shoot growth remained arrested. I can support this view with the example of the multiple-shoot oak seedlings I have shown. One can induce renewed flushing in two ways: (1) by defoliation and (2) by transplanting such a multiple-shoot tree into a larger container, which permits renewed growth of the roots. In both cases renewed flushing occurs. The same set of leaves that prevents additional shoot growth as long as the tree is potbound is no longer inhibitory if the tree is transplanted into a larger pot. I might add that Dr. Sale wrote me that he finds my interpretation of his data convincing.

Hallé: I have in my greenhouse in Montpellier a lot of young rubber trees; they are well supplied with water and grow continuously.

Borchert: Plants in greenhouses often grow much longer and much more continuously than they do outside. It is well known that the greenhouse atmosphere is quite different from the outside environment. Here you have the same effect that you produced in your earlier work by reduction of leaf area; namely, a reduction of transpiration stress. Would you agree with this interpretation?

Hallé: Yes. I would like to offer one objection to your model, taking rubber as an example, because I have studied this tree for a long time. When the apical meristem is beginning to rest, the leaves are expanding, and this expansion demands a large amount of water. In my opinion, the problem is not the absorption of water by roots, but a balance between the water used by the apical meristem and the water used by the expanding leaves.

Borchert: I completely agree; Dr. Browning and I have discussed this phenomenon (see also Chap. 22). I would never have expected my simplified approach on the organism level to explain everything. What we are talking about here is a similar feedback mechanism, but on the level of the growing shoot. Last year, at the Leningrad Botanical Congress, Dr. Lundegård from Sweden presented data that I consider quite relevant. He showed that the decline in water potential measured in growing pea plants has to be ascribed more to growth of the shoot apex than to transpiration occurring through the leaves. In other words, a growing shoot apex, or in your case a growing leaf, can be a greater sink for water than a transpiring surface.

References

Alvim, R., Alvim, P. de T., Lorenzi, R., & Saunders, P. F. (1974). The possible role of abscisic acid and cytokinin in growth rhythms of *Theobroma cacao* L. *Plant Physiol.* (*Suppl.*), p. 69.

Bond, T. E. T. (1942). Studies in the vegetative growth and anatomy of the tea plant (*Camellia thea* Link.) with special references to the phloem. I. The flush shoot. *Ann. Bot., N.S., 6,* 607–30.

Borchert, R. (1973). Simulation of rhythmic tree growth under constant conditions. *Physiol. Plant., 29,* 173–80.

– (1975). Endogenous shoot growth rhythms and indeterminate shoot growth in oak. *Physiol. Plant., 35,* 152–7.

– (1976a). The concept of juvenility in wood plants. *Acta Hort., 56,* 21–36.

– (1976b). Computer-aided evaluation of shoot growth patterns in trees. *Univ. Kansas Sci. Bull., 51* (4), 129–40.

Chapman, H. D., & Parker, E. R. (1942). Weekly absorption of nitrate by young, bearing orange trees growing out of doors in solution cultures. *Plant Physiol., 17,* 366–76.

514   *Rolf Borchert*

Coster, C. (1923). Lauberneuerung und andere periodische Lebensprozesse in dem trockenen Monsungebiet Ost-Javas. *Ann.Jard.Bot.Buitenzorg, 33,* 117–89.

Critchfield, W. B. (1960). Leaf dimorphism in *Populus trichocarpa. Amer. J. Bot., 47,* 699–711.

Greathouse, D. C., Laetsch, W. M., & Phinney, B. T. (1971). The shoot-growth rhythm of a tropical tree, *Theobroma cacao. Amer. J. Bot., 58,* 281–6.

Greenwood, M., & Posnette, A. F. (1950). The growth flushes of cacao. *J. Hort. Sci., 25,* 164–74.

Hallé, F., & Martin, R. (1968). Etude de la croissance rythmique chez l' Hévéa (*Hevea brasiliensis* Müll.-Arg.). *Adansonia, 8,* 475–503.

Harris, W. F., Goldstein, R. A., & Henderson, G. S. (1973). Analysis of forest biomass pools, annual primary production and turnover of biomass for a mixed deciduous forest watershed. In *Proceedings of the Working Party on Forest Biomass of IUFRO,* ed. H. Young, pp. 41–64. Orono: University of Maine Press.

Holdsworth, M. (1963). Intermittent growth of the mango tree. *J. W. Afr. Sci. Assoc., 7,* 163–71.

Huber, B. (1935). Die physiologische Bedeutung der Ring- und Zerstreut-porigkeit. *Ber. Dtsch. Bot. Ges., 53,* 711–9. English translation available from: Translation Center, John Crear Library, 35 West 33rd St., Chicago, Ill. 60616.

Huber, J. (1898). Beitrag zur Kenntnis der periodischen Wachstumser-scheinungen bei *Hevea brasiliensis* Müll.-Arg. *Bot. Centralbl., 76,* 258–64.

Klebs, G. (1912). Ueber die periodischen Erscheinungen tropischer Pflan-zen. *Biol. Centralbl., 32,* 257–85.

– (1917). Ueber das Verhältnis von Wachstum und Ruhe bei den Pflanzen. *Biol. Centralbl., 37,* 373–415.

Koriba, K. (1958). On the periodicity of tree growth in the tropics. *Gard. Bull. Singapore, 17,* 11–81.

Kozlowski, T. T. (1971). *Growth and Development of Trees,* vol. 1. New York: Academic Press. 443 pp.

Kramer, P. J., & Kozlowski, T. T. (1960). *Physiology of Trees.* New York: McGraw Hill. 642 pp.

Njoku, E. (1963). Seasonal periodicity in the growth and development of some forest trees in Nigeria. I. Observations on mature trees. *J. Ecol., 51,* 617–24.

– (1964). Seasonal periodicity in the growth and development of some forest trees in Nigeria. II. Observations on seedlings. *J. Ecol., 52,* 19–26.

Phillips, I. D. J. (1975). Apical dominance. *Annu. Rev. Plant Physiol. 26,* 341–67.

Prévost, M.-F. (1972). Rythme d'allongement des articles de *Tabernaemon-tana crassa* Benth. (Apocynacées). *Candollea, 27,* 219–27.

Sale, P. J. M. (1970a). Growth, flowering and fruiting of cacao under con-trolled soil moisture conditions. *J. Hort. Sci., 45,* 99–118.

– (1970b). Growth and flowering of cacao under controlled atmospheric rela-tive humidities. *J. Hort. Sci., 45,* 119–32.

Späth, H. L. (1912). *Der Johannistrieb.* Berlin: Parey. 91 pp.

Tomlinson, P. B., & Gill, A. M. (1973). Growth habits of tropical trees: Some guiding principles. In *Tropical Forest Ecosystems in Africa and South Africa: A Comparative Review,* ed. B. J. Meggers, E. S. Ayensu, & W. D. Duckworth. Washington, D.C.: Smithsonian Institution Press.

Wilson, B. F. (1966). Development of the shoot system of *Acer rubrum* L. *Harvard Forest Paper*, no. 14, pp. 1–21.

Zahner, R. (1968). Water deficits and growth of trees. In *Water Deficits and Plant Growth*, ed. T. T. Kozlowski, vol. 2., pp. 191–254. New York: Academic Press.

22

Structural requirements for optimal water conduction in tree stems

MARTIN H. ZIMMERMANN
Harvard University, Harvard Forest, Petersham, Massachusetts

Stems of higher plants are generally considered to serve four major functions: (1) the mechanical function of supporting the plant in an upright position, (2) conduction of water and soil nutrients from roots to leaves, (3) conduction of carbohydrates and other assimilates from leaves to stem and roots, and (4) storage of assimilates, usually in the form of starch, for future use. In arborescent monocotyledons, these four functions are spread more or less evenly throughout the transverse section of the stem with the qualification that the mechanical function is primarily at the stem periphery in the area of greatest stress, where the fibrous sheaths of vascular bundles are most highly developed. Xylem and phloem serve the two respective transport functions, and the ground parenchyma is storage area. In conifers and most dicotyledonous trees phloem and xylem are morphologically separate. Water transport and the mechanical function are located in the woody cylinder, the xylem. Storage takes place in vertical and radial parenchyma of xylem and phloem.

An evolutionary view of the mechanical and conduction functions of the xylem has been illustrated by Bailey & Tupper (1918) and is shown in the plant anatomy book by Esau (1965, Fig. 11.1, p. 228). In primitive wood, tracheids serve as mechanical as well as water-conducting elements. During angiosperm evolution the two functions became increasingly separated and assigned to specialized cells: files of tracheary elements (vessels) for the conduction of water; fibers for mechanical support. The fact that angiosperm fibers are shorter than conifer tracheids shows that the great length of tracheids does not serve a mechanical function, but rather an increase of hydraulic conductivity, a point that will be discussed in more detail later in this chapter. I shall not deal with mechanical properties of tree stems; readers interested in this topic are referred to the paper by McMahon (1975).

The water-conducting unit

In conifers and a few primitive angiosperms most of the cells visible in a single transverse section are tracheids. The lengths of these are considerable, around 3 mm in most present-day conifers, up to almost 8 mm in *Agathis* (Carlquist, 1975, p. 87), and even longer in some extinct species. In most angiosperms water moves primarily through vessels; i.e., series of water-conducting cells (vessel elements) arranged end-to-end, with their end walls partly or entirely dissolved. From a functional point of view it is these cell rows, the vessels, that concern us (Fig. 22.1). The length of vessel elements is functionally meaningless, as far as we know.

Fig. 22.1. Vessel elements and vessels in angiosperm xylem. (From Zimmermann, 1976)

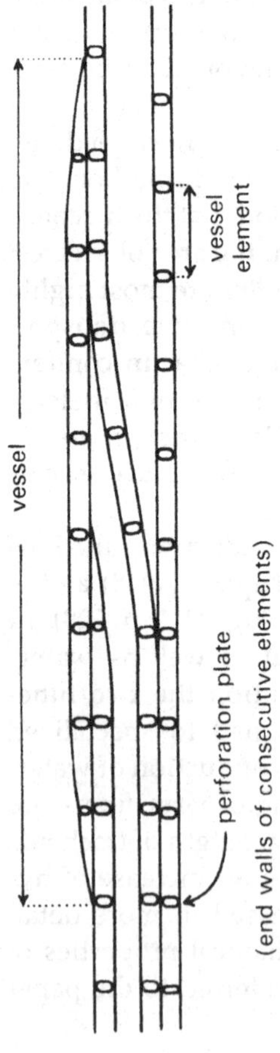

Vessels are of quite variable length. There are indications that wide vessels are longer than narrow ones, but there are relatively few measurements of vessel length in the literature. In older reports measurements concern either the longest vessels (see the citations in Esau, 1965) or some measure of average length of the shorter ones (e.g., Scholander, 1958). It was Skene & Balodis (1968) who, for the first time, presented a measure of length distribution in a tree species. In *Eucalyptus obliqua* they found a grading of vessel length from very short (0–10 cm) to a maximum of about 3.5 m. Their results have been critically reviewed by Milburn & Covey-Crump (1971), who pointed out that they could have been dealing with four length groups rather than a continuous grading of lengths. Skene & Balodis are probably more nearly correct, as their conclusions have been backed up by cinematographic analysis (Skene, unpubl.). At any rate, the discussion is somewhat irrelevant in this connection, interesting as it is from a theoretic point of view.

Water moves up the tree through the lumen of the conducting unit, and from unit to unit via pits. This is easily comprehended in the case of coniferous wood, as tracheids are closely packed and tracheid-to-tracheid pits are visible on radial walls. It is less obvious in angiosperm wood containing vessels, as vessel-to-vessel contacts, and especially vessel ends, are less easily recognized. Cinematographic analysis of xylem reveals the three-dimensional arrangement of vessels (Zimmermann, 1976). In all species investigated (with rare exceptions so far only seen in a shrub) individual vessels always end in pairs or in groups as shown diagrammatically in Fig. 22.1 (Zimmermann, 1971). In arborescent monocotyledons individual vessels also end always in pairs or groups (Fig. 22.3) (Zimmermann, 1973). This is rather significant, albeit obvious, as water has to move from vessel to vessel every few centimeters or meters, whatever the vessel length, and blind endings would be functionally inappropriate in a stem.

A very important function of the conducting unit is confinement of accidental embolism. Xylem water is usually under tension (i.e., in a metastable condition). If the conducting unit is injured in any way, or if a gas nucleus appears (e.g., as branches are tossed about by wind), air will enter the injury or the gas nucleus will probably expand, until the gas (air in the first case, water vapor in the second) has filled the whole unit. The surface tension of water prevents the air-water interface from penetrating the vessel-to-vessel pits because pit pores are very small (Zimmermann & Brown, 1971, p. 207). Therefore, gas embolism remains confined within the conducting unit. If tension persists for a longer period, as it usually does in trees, the damage becomes permanent. If root pressure renders xylem sap tension periodically positive, as in some herbs overnight, embolism can be repaired if the vessel is not physically injured (Milburn & McLaughlin, 1974).

The water supply of the tree depends a great deal on the dimensions of the conducting units. Dimensions have been modified in various ways during the course of evolution, and it is obvious that the optimal structure of xylem depends very much upon environmental requirements. It is reasonable to assume that the two main factors influencing evolution are efficiency and safety. The two factors oppose each other: Evolution of a more efficient water-conducting capacity involves greater risk. Let us first discuss the efficiency of water conduction.

Hydraulic conductivity

There are various ways of looking at the efficiency of the water-conducting system. The hydraulic conductivity (L_P) determines the volume flow of water (dV/dt) at a given pressure gradient (dP/dl), according to the Poiseuille equation, $dV/dt = L_p \, dP/dl$, L_p is a constant containing the reciprocal value of liquid viscosity and r^4. That is, hydraulic conductivity, and therefore volume flow, is directly proportional to the fourth power of the capillary radius.

This is extremely important, because it tells us that a very small increase in diameter of tracheary elements during evolution causes a greatly increased efficiency of water transport. For example, all other parameters being equal or properly scaled, a vessel of a diameter of 200 μm moves, at a given pressure gradient, as much water as 16 vessels 100 μm in diameter or 256 vessels 50 μm in diameter. This means that when two vessels run parallel, one 200 μm and the other 50 μm wide, the wider vessel carries 99.6% of the total volume and the narrower one only 0.4%! If we compare equal transverse sectional conducting areas instead of diameters, flow volumes through the capillaries of diameters 1, 2, and 4 are still proportional to the square (i.e., 1:4:16).

Tracheids and vessels are not ideal capillaries; the hydraulic conductivity, therefore, does not depend upon the diameter alone. It is also influenced by conducting unit length and unit-to-unit contact. Vessel length can be measured, but it is difficult to assess quantitatively its effect on efficiency. The resistance to flow from tracheid to tracheid or vessel to vessel is also difficult to measure experimentally. However, it is possible to measure resistance to flow in wood and compare this with the theoretical resistance of capillaries of the same diameter. Such measurements have been made by Berger (1931), Riedl (1937), and Münch (1943) and summarized in Zimmermann & Brown (1971, p. 199). There are two methods: Either the volume of water flowing through the wood at a given pressure gradient is measured, or peak velocity is estimated by some sort of titration procedure detecting the first arrival of a displacement solution. Both these methods have their drawbacks. In the case of volume flow, it is necessary to measure the diameters of all conducting vessels. It is not easy to know how many of the visible vessels may have been embolized during the experiment. The second method avoids this problem, because it is necessary to measure the diameters of only the widest vessels. But it is difficult to obtain a good measure for peak velocity because the diffusion of solute blunts the tip of the flow paraboloid (Taylor, 1953).

Efficiency of wood as a conductor can be expressed as percent hydraulic conductivity of ideal capillaries of the same diameter. Measurements reported by the three authors listed above range from 100% in lianes to about 12% in various shrubs. This percentage value is a measure of the combined effects of limited vessel length and of resistance to flow from vessel to vessel via pits. The 100% value means that vessels are very long and vessel-to-vessel resistance is insignificant. It is interesting to note, for example, that the values obtained for *Abies pectinata* (fir) ranged from 26% to 43%. In other words, the resistance to flow through tracheids of fir is only about three times that through capillaries of equal diameter.

M. T. Tyree and later D. R. Lee, while working at the Harvard Forest, made additional measurements of the value of percent efficiency. Both found that the scatter of results is considerable. In addition to the difficulties mentioned above, the greatest problem is the measurement of vessel diameter. The diameter of all vessels of the piece of wood must be measured on a single (!) transverse section, or some must be measured and the rest estimated. A small error in measurement results in a large error of the calculated hydraulic conductivity. If we consider that $1.1^4 = 1.4641$ and $0.9^4 = 0.6561$, we realize that a 10% error in diameter measurement would give us a scatter of efficiency values between 66% and 146% for wood with the same efficiency as ideal capillaries (100%).

In summary, we can say that in dicotyledonous trees the limited vessel length and vessel-to-vessel contact reduces the hydraulic conductivy to about half that of ideal capillaries. In other words, the hydraulic conductivity of vessels is of the order of capillaries 15% more narrow. We may, therefore, conclude that the diameter of tracheids or vessels is by far the most important factor of conduction efficiency.

Efficiency vs. safety

If a small diameter increase makes vessels so much more efficient, why have species with small-diameter tracheary elements not become replaced by the "advanced" large-porous species? This question can be expressed more dramatically by asking: How is it possible that even within a single genus, such as *Quercus*, there are species with very large and others with small vessel diameters? The answer is that small pores are very much safer water conductors than large pores. We have seen that for every large vessel, a tree needs tens, hundreds, or thousands of small ones (considering that small-diameter vessels are also shorter) to move comparable amounts of water.

Let us assume that a certain small-porous tree species has 100 vessels in a given volume of wood and another species with larger vessels has only 1 within the same volume. Any accidental embolism occurring at a given point will be 100 times more damaging in the large-porous species. It is, therefore, not surprising that trees living in stress areas such as deserts are small-porous. In the genus *Quercus*, the evergreen live oaks of dry habitats are small-porous, and oaks of temperate regions are large-porous.

Both small- and large-porousness have been successful during evolution if stature of trees is an indicator of success. The two tallest tree species in the world (up to 100 m high) are *Sequoiadendron giganteum*, a conifer of the North American west coast, and *Eucalyptus regnans*, an Australian dicotyledon.

Among the largest vessel diameters we find in nature are those of the temperate ring-porous trees whose survival strategy has been discussed elsewhere (Zimmermann & Brown, 1971, p. 211), certain tropical trees, and tropical lianes. The absolute maximal vessel diameter seems to be about 0.5 mm; above this, the risk may be too great. One could, of course, argue that evolution has not had time to produce larger vessels. This is very unlikely because the largest vessels are of comparable diameter in many unrelated plant families.

Conductivity differences within an individual tree
Structural difference of tracheary elements
in top and bottom of trees

It was long ago recognized by plant anatomists that if one compares tracheary elements of the secondary xylem of different parts of an individual tree, one finds progressively more "advanced" forms as one moves basipetally from branches to stem and roots. Vessel diameters tend to be greater in roots than in the stem, and greater in the stem than in branches. Similarly, perforation plates may be simpler further down (see citations in Carlquist, 1975).

There is a relatively simple functional explanation of this. No matter whether the absolute pressures in a tree are positive or negative, there is always a gradient of decreasing pressure acropetally. For a "standing" or "hanging" water column at rest, this is 0.1 bar/m height. If the water is moving, the gradient is steeper owing to flow resistance; it may be as much as 0.2 bar/m under conditions of high transpiration (Zimmermann & Brown, 1971). This means that the top of the tree always lives under conditions of lower absolute water pressure than the basal part of the tree. The safety restriction on evolution of vessels in the direction of greater diameter and length is less severe at the base than at the top of the tree. In view of the efficiency vs. safety discussed in the previous section, we would expect more efficient vessels in the roots and safer ones at the top of the tree. This is, indeed, the case.

Leaf-specific conductivity

Hydraulic conductivity can be calculated or measured for single capillaries, or it can be expressed per transverse sectional area of xylem (Farmer, 1918). But such values do not really tell us how well the crown is supplied by xylem channels. We need to relate water conduction to the number of supplied leaves. Several hundred years ago, Leonardo da Vinci, who was a very keen observer of nature, wrote in his notebook: "All the branches of a tree at every stage of its height when put together are equal in thickness to the trunk (below them). – All the branches of a water (course) at every stage of its course, if they are of equal rapidity, are equal to the body of the main stream. – Every year when the boughs of a plant (or tree) have made an end of maturing their growth, they will have made, when put together, a thickness equal to that of the main stem," (Richter, 1970, Leonardo's notes nos. 394, 395). Jaccard (1913, 1919) made very careful measurements of the annual growth increments of several tree species and concluded that the trees produce xylem of equal conductivity along the length of their branches and stems. Japanese workers, investigating the matter from an ecologic point of view, called this simple situation the "pipe model" (Shinozaki et al., 1964).

Huber (1928) refined this concept by measuring total transverse sectional xylem area at different points in trees and expressing it per fresh weight of supplied leaves. For the plants of the forests of Western Europe, he found values of the order of 0.5 mm^2/gm. Such measurements become interesting as soon as they are made comparative. Huber showed for *Abies* that whereas stem and side branches have values of the order of 0.5 mm^2/gm, the terminal shoot has a value of 2–4 mm^2/gm. This is an interesting expression of apical dominance: In terms of water conduction, the terminal shoot is greatly favored, and in case of drought it will suffer last.

Comparisons of plants of different habitats are equally interesting. Huber (1928) found a range of values from 0.3 mm^2/gm in plants of humid habitats to 3.4 mm^2/gm in (nonsucculent) desert plants. One could wonder why desert plants have more xylem than plants of mesic areas. The answer can be found again in terms of efficiency vs. safety: If desert plants must have smaller vessels for safety, the very much lowered efficiency of these requires larger transverse sectional xylem areas to accommodate more vessels.

Huber's value tells us only part of the story, because it takes into account neither vessel diameter (i.e., the efficiency of xylem) nor nonfunctional xylem of older trees. What we really need is information about the hydraulic conductivity of the xylem portion supplying a given leaf mass. A simple measurement of this is flow rate of water through a stem section, expressed per fresh weight of supplied leaves. We have begun to make such measurements, but can give only some very tentative values at this point. They are expressed as microliters of water flowing through xylem, per hour, at a gradient of 0.1 bar/m (gravity flow), per gram fresh weight of supplied leaves. We call this "leaf-specific conductivity." The few plants we have looked at so far gave values of 20–50 for stems and 1–10 for twigs.

Flow-rate measurements are easy to make, but usable values are difficult to obtain, as erratic results will quickly indicate to the experimenter. Embolized vessels at the cut surface can decrease the conductivity enormously and make measurements worthless. It is necessary to cut all plant parts under water and to store the segments to be measured for several hours in water to get them fully relaxed. Details of the procedure will eventually be published. The following examples are meant merely to point out the principle (Fig. 22.2). The *Tsuga canadensis* was a forest-grown sapling about 2 m in height (several persons spent many hours defoliating it!). As can be seen from the figure, the leaf-specific conductivity of the main stem is considerably larger than that of the side branches. It is interesting to note that the terminal shoot does not have the characteristics of a leader: Its conductivity is comparable to that of the lateral branches. For this tree, Huber's values (square millimeters of transverse sectional xylem area per gram fresh weight of leaves supplied) were also measured. They ranged between 0.66 and 0.99 mm²/gm throughout the tree without showing the apical dominance Huber found in *Abies*. This lack of strong apical dominance (from the point of view of leaf-specific conductivity) does indeed agree with the observed growth habit of *Tsuga*.

The *Myrsine floridana* was a greenhouse-grown seedling. Its leaf-specific conductivity is similar to that of *Tsuga*: considerably larger in the stem than in the branches. The leader shows again a relatively low conductivity. In this case, however, we were dealing with immature vascular tissue.

Fig. 22.2. Leaf-specific conductivity. Hydraulic conductivity of xylem of axes of three small trees: *Tsuga canadensis* (2 m tall), *Myrsine floridana* (90 cm tall), and a small specimen of palm *Rhapis excelsa* (50 cm tall from ground to apex).

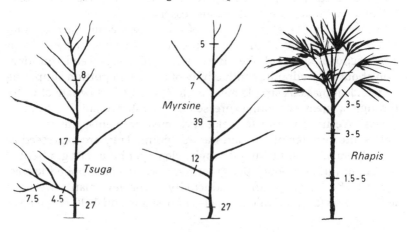

There is no question that these values are causally related to the tree architecture described by Hallé & Oldeman (1970; see also Chaps. 7–10), but we do not yet know how they are related. It is perhaps most likely that the initial events leading to a specific tree morphology are of a hormonal nature, but the resulting hydraulic characteristics have, no doubt, a further formative effect.

It will be most interesting to make measurements on larger trees. We can speculate that, as in the seedling, leaf-specific conductivity will be much higher in the main stem than in lateral branches. The leaves at the top of the tree are in a more unfavorable environment than those at the base of the tree, because xylem pressure decreases at least 0.1 bar with every meter of height. During times of water shortage, the survival of the tree hinges on its ability to prevent the lower leaves from taking most available water, thus causing the leaves at the top to wilt. Resistance to flow in a lower lateral branch must be higher than in the main stem so that the leaves at the top of the tree can favorably compete for a fair share of available water.

Palms and other arborescent monocotyledons pose a rather interesting problem, because the stem, once formed, has a permanent complement of xylem vessels. If any change in hydraulic conductivity takes place over a period of time at a given point in the stem, it can only be a decrease caused by embolism of some of the vessels. A few initial measurements with the small palm *Rhapis excelsa* indicate that stem and petioles have comparable leaf-specific conductivities of the order of 2–5 (Fig. 22.2). Our extensive knowledge of the vascular anatomy indicates that the leaf insertion must represent a hydraulic bottleneck (Zimmermann & Tomlinson, 1965). A leaf trace consists of basically two components: the leaf trace proper, leading into the petiole, and the vascular tissue providing continuity up the stem via bridges and an axial bundle branch (Fig. 22.3). The xylem leading into the petiole consists of narrow protoxylem elements (Fig. 22.3, *LT* at point *A* and at *B*), the continuing axial components of wide metaxylem vessels (Fig. 22.3, vessels *1, 2, 3* at *A*). Comparative estimates of hydraulic conductivity from stem to leaf and axially along the stem can be obtained as follows. All vessel radii of a leaf trace at *A* are measured (diameters in ocular scale units may be used to simplify matters, as we are not interested in absolute values). The *average* of the fourth power of the large vessel diameters (*1, 2, 3* in Fig. 22.3) is taken because only one axial bundle per leaf trace is continuous up the stem. The *sum* of the fourth powers of all small vessels (*LT* in Fig. 22.3) is taken, because all the small vessels go into the petiole. A comparison of the two results indicates that the hydraulic conductivity up the stem is two to five times greater than that from stem to petiole. We have not yet tested this calculation with water flow experiments, but plan to do so in the future. The result is rather important because it means that the pressure gradient between stem and petiole is considerably steeper than the gradient up the stem or along the petiole. The pressure in the leaf must, therefore, be lower than in the stem above the leaf insertion, the difference being greater, the greater is transpiration. Under stress conditions, embolism must take place first in the leaf. The accumulating effects of repeated embolisms in an aging leaf drop the pressure further. Eventually, an old leaf is deprived of water. This may be the sequence of events in palms whose leaves are not abscised: Lower leaves simply dry out.

Possible significance of scalariform perforation plates

It is generally assumed that scalariform perforation plates are remnants of scalariform pits whose membranes have disappeared during the course of evolution. This transition from tracheid to vessel reduced resistance to flow, and the lateral bordered pits were strengthened to take over the mechanical function (Bailey, 1953). The next evolutionary step was the complete dissolution of the plate (i.e., the disappearance of the bars).

Fig. 22.3. Leaf trace departure in palm *Rhapis excelsa: left,* longitudinal view; *right,* transverse sections of conducting bundles at different points (*A–D*). At *A,* a departing leaf trace breaks up into four components: leaf trace proper (*LT*), continuing axial bundle (*1*), and two bridges (*2, 3*). Axial water movement is via large metaxylem vessels (*1, 2, 3*); movement of water into petiole is via small protoxylem vessels or tracheids (*LT*). Note that distribution of conducting units prevents embolism from spreading when leaf drops.

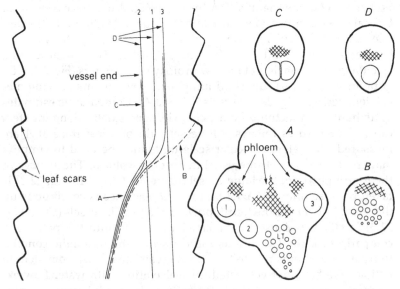

Scalariform perforation plates are relatively common. The question, therefore, arises whether there is any functional advantage gained by maintaining them. Carlquist (1975) assigns them a mechanical function. I doubt this explanation, because the plates are too far apart and investment of the same amount of cellulose in the wall proper would probably be as effective. But, of course, we are both speculating.

At the present time, I can think of only one functional benefit due to the presence of scalariform perforation plates: It concerns trees growing in cold climates. When the xylem water melts in late winter, vessels must be riddled with bubbles (Scholander, 1958). The larger the vessels, the greater the danger that bubbles within a single vessel will rise and combine to form bigger ones that will not be able to redissolve before tension is called for. It seems probable that the movement of bubbles from one vessel element to the next is prevented by the scalariform perforation plates. Indeed, when thick, fresh, longitudinal sections of wood with scalariform perforation plates (such as birch) are frozen and, during subsequent thawing, are observed with a microscope while in an upright position (vessels vertically oriented), it can be seen that the tiny rising bubbles easily get caught on the slightest irregularity along their path, such as the bars of a scalariform perforation plate. This means that bubbles are kept separate, remain very small, and, therefore, redissolve more easily.

If the conclusion from this observation is correct, we have an explanation of why woody plants colonizing the frontiers of cold areas are (besides conifers) willows and birches, both genera with scalariform perforation plates. Indeed, near Umiat, in northern Alaska, north of the Brooks Range and well north of the timberline, in the region of the tundra, "forests" of several-meter-tall willow trees can be found along the Colville River.

Scalariform perforation plates may be "useful" in cold climates, but this explanation would apply to tropical trees only if there are species in which positive xylem pressures occur periodically. Other than this, we have to assume that the plates are true relics, with possibly little selection for their disappearance as their interference with hydraulic conductivity is not very serious. But alternate functions may yet be found.

Concluding remarks

The preceding discussion is not artificially restricted to tropical trees, because the mechanisms of water movement are similar in all tall woody plants regardless of whether they grow in tropical or temperate regions.

Unfortunately, the mechanisms of water conduction have been studied very little in tropical trees. Many questions of considerable importance are still unanswered. For example, in temperate regions large-porousness is always correlated with ring-porousness, because all large vessels embolize during winter and a new set must be produced quickly before the new leaves come out and large quantities of water have to be moved. What is the significance of ring-porousness in a tropical tree such as *Tectona grandis*? A comparison of the rate of embolism in large- and small-porous tropical trees would be of a particular interest. It would permit us perhaps to relate the rate of embolism to vessel volume, undisturbed by the complicating factor of freezing temperatures. It would be highly desirable to measure xylem pressures and pressure gradients in tropical trees. One often hears rumors about guttating tropical trees or trees whose xylem shows vigorous bleeding when cut. How common are positive pressures? Are they continuous or periodic? How does the tree protect itself from excessive losses by leakage – a problem automatically solved with below-atmospheric pressures, as described in the second section of this chapter?

Arborescent monocotyledons seem to be restricted to the tropics and subtropics, though a few hardy species have ventured into areas where they have to endure occasional frost. However, nobody has measured actual stem temperatures, and it is unlikely that the xylem water in the stem, which is well protected by the fibrous layers of old leaf bases, actually freezes when the air temperature drops briefly below freezing.

We still know very little about hydraulic conductivity distribution in different types of tropical trees, including arborescent monocotyledons where the situation is so different from that in dicotyledons. Most important, we have to make efforts to study mature trees and not merely seedlings grown in a greenhouse.

A better knowledge of transport processes in trees is not of academic interest only. It is imperative for economically important species, because we cannot hope to grow trees in monocultures without inviting diseases, and we cannot understand or combat diseases without understanding how the healthy organism functions. A good, and at the same time rather tragic, example is lethal yellowing of coconut palms, a disease that has already killed millions of trees in the Caribbean area. Lethal yellowing is a translocation disease, and yet we are still sadly ignorant about some fundamental questions of translocation in coconut.

Lastly, it is important to realize that by studying a phenomenon in a wide variety of plants, a much better understanding can be achieved than if only a narrow range of species is investigated. Thus the study of tropical trees will benefit botany as a whole, even botany of temperate regions.

General discussion

Oldeman: What happens when a tree is suddenly presented with a larger amount of energy, such as increased illumination or fertilizer in the soil? It then has to transmit a greater amount of nutrients either from upper to lower or from lower to upper parts. This could be accomplished by increasing the diameter of tracheary elements.

Zimmermann: It would be easier to do it by moving solutions at a higher concentration. When the soil is fertilized, for example, the roots may increase secretion into the xylem.

Oldeman: Does this not assume a change of the whole process of synthesis in the roots? The change in transport capacity is most effective by the development of wide vessels.

Zimmermann: The answer is probably no. When vessel diameter is increased there is a higher risk involved, and a mechanism can be effective only if the risk is not detrimental to the tree. There are other possibilities: e.g., the regulation of transpiration. Cambial activity also needs to satisfy the requirements of proper structural proportions, so that a branch does not break off or bend excessively under its own weight (see McMahon, 1975). I have concerned myself with only one among many factors.

Jeník. There are interesting phenomena in roots that relate to the efficiency of water conduction in large vessels. An anatomic analysis of roots shows that the taproots and single roots have mostly small vessels, whereas horizontal roots, penetrating the upper horizons of the soil, have extremely large vessels. The maximum is even larger than you mentioned, 0.6 – 0.7 mm in European oaks. On the other hand, the taproot has few large vessels. Can you visualize this in relation to the function of the taproot as a pumping organ?

Zimmermann: At a given pressure in the base of the trunk, and the vessel diameter being so much larger in the lateral roots, the bulk of the water probably comes from the lateral roots. It is interesting that you found very large vessels in the roots, because the roots are the least endangered place in the whole tree. The highest pressures are here and under conditions of even incipient root pressure (i.e., when root pressure is manifested only as far as the stump), these large vessels could refill.

Givnish: Has anyone done an experiment using, say, grape vines having different xylem diameters to measure how the risk of embolism per unit time depends on water potential for xylem of different diameters?

Zimmermann: I don't think anybody has actually measured that. I visualize it as a function of pressure in the xylem: the lower the pressure, the higher the risk. I assume it is also a function of volume: the larger the volume, the higher the probability of embolism.

Ashton: Can the steepened pressure gradient at the leaf insertion of the palm be ascribed to the basipetal growth of the leaf? In the dicot, has it any connection with the physical characteristics of the abscission layer of the leaf? If it does, is it of any relevance to what Dr. Givnish was telling us the other day?

Zimmermann: The answer to your first question is yes: Extended growth of the leaf base produces small-diameter tracheary elements. I am not familiar enough with the dicotyledonous leaf insertion to answer your second question, but I do not believe it has anything to do with the abscission layer.

Hartshorn: I am intrigued by your findings of a reduced hydraulic conductivity in the terminal shoot of *Tsuga*. Might this be a consistent relation with plagiotropic terminal shoots? If so, it has very significant evolutionary implications.

Zimmermann: Yes, but I do not know which is the causal factor. *Tsuga* may be unable to make tracheids of bigger diameter near the top and, therefore, fail mechanically or it may be the other way around. But it is certainly striking that *Tsuga* is not like *Abies*, which has a very strong apical dominance. Unfortunately, we have not yet done experiments with *Abies*. It would be interesting to do experiments with tall trees, but since one has to pick off all the leaves, the task would be difficult.

Givnish: How do the diameters of xylem elements at the top of a *Eucalyptus regnans* tree compare with those of the tracheids at the top of a *Sequoia*?

Zimmermann: Stem tracheids of *Sequoia* are 50 – 65 μm (and up to 80 μm) in diameter; the vessels of *Eucalyptus regnans* are of the order of 100–200 μm. However, I do not know if anyone has systematically looked into the question of diameters at different heights.

Fisher: Petch (1930) reported on the poisoning of lateral roots in buttressed trees, which suggested that one of the possible causes of root buttressing and the fluting of trunks of trees was the existence of preferential sites directly above roots where water and food conduction is enhanced or more efficient, thus providing more nutrition to the cambium and resulting in eccentric growth. Do you know of any structural evidence for longitudinal zones related to lateral roots that would indicate more efficient transport?

Zimmermann: This would suggest straighter grain in buttressed trees. Water conduction normally spreads out in the growth ring even over short vertical distances. I do not have a translocation answer for buttressing. For fluted trunks, there probably is an answer. Münch (1937) described longitudinal grooves in the stem below heavily shaded branches in red oaks. This indicates that even shaded branches control the formation of vascular tissue in the stem below, but growth is not as fast as in neighboring tissue, controlled by fully illuminated crown parts. When the branch is cut, the groove fills out because the area is "invaded" by neighboring tissue. One may postulate that in strongly fluted trees the tissue below a dead branch is not invaded by tissue controlled by living parts of the crown. Hormonal control of cambial activity seems to be much more linear than conduction in either xylem or phloem.

Jeník: Chalk & Akpalu (1963) in West African rain forest trees found a positive relation between the structure of the wood and the possibilities of lateral transport and buttress formation.

References

Bailey, I. W. (1953). Evolution of the tracheary tissue of land plants. *Amer. J. Bot.*, 40, 4–8.

Bailey, I. W., & Tupper, W. W. (1918). Size variation in tracheary cells. I. A comparison between the secondary xylems of vascular cryptogams, gymnosperms and angiosperms. *Amer. Acad. Arts Sci. Proc.*, 54, 149–204.

Berger, W. (1931). Das Wasserleitungssystem von krautigen Pflanzen,

Zwergsträuchern und Lianen in quantitativer Betrachtung. *Beih. Bot. Centralbl., 48,* 363–90.

Carlquist, S. (1975). *Ecological Strategies of Xylem Evolution.* Berkeley: University of California Press.

Chalk, L., & Akpalu, J. D. (1963). Possible relation between the anatomy of the wood and buttressing. *Commonw. For. Rev. Lond., 42,* 53–8.

Esau, K. (1965). *Plant Anatomy,* 2nd ed. New York: Wiley.

Farmer, J. B. (1918). On the quantitative differences in the water-conductivity of the wood in trees and shrubs. I. The evergreens. II. The deciduous plants. *Proc. Roy. Soc. Lond., B, 90,* 218–50.

Hallé, F., & Oldeman, R. A. A. (1970). *Essai sur l'architecture et la dynamique de croissance des arbres tropicaux.* Paris: Masson.

Huber, B. (1928). Weitere quantitative Untersuchungen über das Wasserleitungssystem der Pflanzen. *Jb. Wiss. Bot., 67,* 877–959.

Jaccard, P. (1913). Eine neue Auffassung über die Ursachen des Dickenwachstums. *Naturw. Forst, 11,* 241–79.

– (1919). *Nouvelles recherches sur l'accroissement en épaisseur des arbres.* Zurich: Foundation Schnyder von Wartensee.

McMahon, T. A. (1975). The mechanical design of trees. *Sci. Amer., 233*(1), 93–102.

Milburn, J. A., & Covey-Crump, P. A. K. (1971). A simple method for the determination of conduit length and distribution in stems. *New Phytol., 70,* 427–34.

Milburn, J. A., & McLaughlin, M. E. (1974). Studies of cavitation in isolated vascular bundles and whole leaves of *Plantago major* L. *New Phytol., 73,* 861–71.

Münch, E. (1937). Versuche über Richtungen und Wege der Stoffbewegungen im Baum. *Forstwiss. Centralb.* 59, 305–24, 337–51.

– (1943). Durchlässigkeit der Siebröhren für Druckströmungen. *Flora, Jena, 136,* 223–62.

Petch, T. (1930). Buttress roots. *Ann. Roy. Bot. Gard. Peradeniya, 9,* 101–17.

Richter, J. P. (1970). *The Notebooks of Leonardo da Vinci (1452–1519), Compiled and Edited from the Original Manuscripts.* New York: Dover. Reprint of a work originally published by Sampson, Low, Marston, Searle and Rivington, London, 1883.

Riedl, H. (1937). Bau und Leistungen des Wurzelholzes. *Jb. Wiss. Bot., 85,* 1–72.

Scholander, P. F. (1958). The rise of sap in lianas. In *The Physiology of Forest Trees,* ed. K. V. Thimann, pp. 3–17. New York: Ronald Press.

Shinozaki, K., Yoda, K., Hozumi, K., & Kira, T. (1964). A quantitative analysis of plant form: The pipe model theory. I. Basic analyses. II. Further evidence of the theory and its application in forest ecology. *Jap. J. Ecol., 14,* 97–105, 133–9.

Skene, D. S., & Balodis, V. (1968). A study of vessel length in *Eucalyptus obliqua* L'Hérit. *J. Exp. Bot., 19,* 825–30.

Taylor, G. (1953). Dispersion of soluble matter flowing slowly through a tube. *Proc. Roy. Soc. Lond., A, 219,* 186–203.

Zimmermann, M. H. (1971). Dicotyledonous wood structure made apparent by sequential sections. Film E 1735. (Film data and summary available as a reprint.) Institut für den wissenschaftlichen Film, Göttingen, Germany.

– (1973). Transport problems in arborescent monocotyledons. *Quart. Rev. Biol., 48,* 314–21.

– (1976). The study of vascular patterns in higher plants. In *Transport and Transfer Processes in Plants*, ed. I. F. Wardlaw & J. B. Passioura, pp. 221–35. New York: Academic Press.

Zimmermann, M. H., & Brown, C. L. (1971). *Trees: Structure and Function*. New York: Springer-Verlag.

Zimmermann, M. H., & Tomlinson, P. B. (1965). Anatomy of the palm *Rhapis excelsa*. I. Mature vegetative axis. *J. Arnold Arbor.*, 46, 160–80.

Part VI: Community interactions

23

Architecture and energy exchange of dicotyledonous trees in the forest

ROELOF A. A. OLDEMAN

Mission O.R.S.T.O.M. en Equateur, Quito, Ecuador

Scientific publications give the impression that the forest is a structured object in which processes like regeneration, tree fall, or seed distribution are happening. The principal constituents of this object, the trees, generally are considered as standard elements. For a functional classification of trees, one must look primarily to publications on forestry. By virtue of his profession, the forester knows that function and structure of the forest are liable to change together. Hence he is bound to regard the forest as a process, though sometimes with a certain bias toward wood production dynamics. The theories of Clements (see Allred & Clements, 1949) at first sight seem to form a bridge between science and its application.

The essence of Clements's succession concept is the replacement of one forest block, characterized by a distribution of species numbers, by another similarly defined block. Species are taxonomically distinguished by rather minute organs (e.g., leaves, twigs, flowers, and fruits), whereas generally the overall architecture of plants is not considered. The model of a successional sere (Clements) or of a vegetation analyzed by the methods of Braun Blanquet (1932) thus becomes a set of various clouds of flowers, fruits, leaves, and twigs. This image is useful because of its bearing on reproduction strategies and population dynamics, but it does not provide a framework for total forest biology or its energetic principles.

The impact of floristically oriented approaches on scientific thought has been tremendous and was certainly one of the early driving forces in tropical forest research. The distribution of organs in the forest is, for instance, the foundation of most work on photosynthesis in whole canopies (Chaps. 24, 25) and in single trees

Present address: Agr. University, "Hinkeloord," Box 342, Wageningen, Netherlands.

535

(Horn, 1971) and has led to basic concepts such as leaf area index (LAI) and leaf area density (LAD), single-layered and multilayered crowns, and crown-depth diagrams (Ogawa et al., 1965).

Floristic and Clementsian succession does not conflict with anything that can be said about vertical vegetation structure; neither does Raunkiaer's life form classification, which is another method based on organ distribution (i.e., of meristems), but going further toward their integration as whole organisms. The stratified forest theory, therefore, has remained unchallenged for a long time. Richards's discussion of it (1952) leaves the impression of a concept in which some people believe and others do not. The origin of the idea of stratification looks like one of the many genial intuitive notions of the late nineteenth century. Some scientist or forester has probably *seen* it in the forest, and its existence has been shown to be possible, though difficult to observe directly (Oldeman, 1974a).

More recently, skepticism regarding universal forest stratification has grown. Richards (1969) and Godron (1971) tend to view the idea as convenient for vegetation analysis rather than as a general biologic or ecologic reality. Rollet (1969, 1974) proved by a statistical study that seems to be unparalleled in scope, as it concerns an inventory of 5000 km^2 in Venezuela, that the generalization "the tropical forest is stratified" is false, even if limited to the climax forest. His results appear unassailable from every point of view. Still, the existence of layered forests was demonstrated by Oldeman (1974a). The contrasting results of these authors can be reconciled if it is assumed that layers exist in the forest, but that they are neither permanent nor even preponderant.

The present symposium considers tropical trees as living systems. In this chapter, energy exchanges between these systems and the environment, and energy distribution within their bodies, will be linked to their architecture and growth model. The integration of trees into a collective living system – the forest – allows us to check existing hypotheses on forest layering and to draw an image of forest architecture and sylvigenesis. Overall morphology of whole organisms or vegetation (i.e., architecture) will be used as a pointer for function, which in its turn causes architecture to change. This basic approach should be borne in mind when reading the next sections, of which form and behavior are the two principal components.

The tree
Architecture
There are many observations on stereotypical development patterns during tree growth (see Hallé & Oldeman, 1970; Chaps. 7, 9–13). Such patterns can be recognized by the form of individual

plants observed at successive moments, which leads to a chronologic series of architectural phases. The pattern that emerges when one follows trees from germination onward proves species-constant (but see Chap. 8). The spatial and temporal organization of organs and organ complexes and their morphologic characteristics, which are the result of a meristematic production program, follow a regular pattern. Quantitatively, however, there is a marked plasticity in number (branches, internodes, flowers, fruits) and size (leaves, branches) of the organs produced (Chap. 20). This depends on the energy input, which determines the quantity of living matter that can be formed. If a tree were relying upon a very precise environmental energy level, it could not survive, for there are always fluctuations of this level.

After observation of numerous species, Hallé & Oldeman (1970) described a number of architectural models among the myriad tree species. A limited number of basic organization patterns among trees is in agreement with the work of Stevens (1974), who suggests, on theoretic grounds, that the spectrum of basic natural forms also must be limited. The different nomenclature of models or morphologic types (see Oldeman, 1975, p. 19) is irrelevant to this fundamental insight: Any one may be adopted according to its usefulness.

Because of my familiarity with the architectural models, these will here be used as units of morphogenesis, but other systems may suggest themselves and be applied in the same way. Jeník's work (Chap. 14) on subterranenn architecture should ultimately be incorporated, although it is too recent to be included here. The architectural tree models have been discussed elsewhere (e.g., Hallé & Oldeman, 1970; Hallé, 1974; Oldeman, 1974a,b, 1975) and will not be considered here in further detail (see Chap. 7). In the illustrations every aerial part of a tree that conforms to a model will be represented by a symbol resembling an upright spoon (Fig. 23.2).

Energy exchange and distribution

In the model, certain active meristems produce well-defined organs at specific times. Other meristems remain at rest. The whole organism has a surface through which energy exchange with the environment takes place and a corresponding volume through which energy is distributed internally. Surface can also serve other purposes (e.g., isolation), as can volume (e.g., mechanical strength: McMahon, 1975), but in a developmental context their function in energy distribution can be considered separately. Energy comes in through leaf surfaces as radiation and goes out as chemical bonds in gaseous molecules. It enters roots, perhaps mostly through root hairs, as chemical energy in solution and leaves them mainly bound in gases. Finally, it circulates inside the tree bound in chemical substances in aqueous solutions and propelled by physical pressure gradients (Chap. 22).

Figure 23.1 gives a generalized diagram of these energetic currents. Their circulation is regulated by the system of physiologic correlations through phytohormones, which establish priorities and pass information, acting as a system of traffic lights upon the traffic currents in a town (see Chaps. 17, 20). When energy is directed toward certain meristems, these are activated and, if they were not differentiated earlier for a special activity, are incited to produce well-defined organ complexes. Resting or dormant meristems remain outside the main currents of energy. In dicotyledonous trees, part of the stream is directed toward the cambium, which makes transport tissue to keep energy transfer at the necessary level. Cambial production may be seen, in part, as a race against the embolism of old vessels.

Three main production systems can be distinguished in a dicotyledonous tree. The activity of two of these is dictated by the environmental energy level, because they transform various kinds of external energy into biochemical energy needed for growth. These two systems are (1) photosynthesis, depending on light intensity and its periodicity, and (2) rhizosynthesis, relying on soil fertility, humid-

Fig. 23.1. (*facing page*) Energy distribution in a dicotyledonous tree. *A*. General integration of photoproduction system (*pps*), cambial production system (*cps*), and root production system (*rps*). Shaded parts are metabolically inactive. Leaves depend on input from climate, roots on input from soil; cambium lacks direct communication with environment. Single-headed black arrows, environmental input; double-headed black arrows, output into environment; simple arrows, liquid energy transport; double-headed white arrows, production of conductive tissue (cambial output). *B*. Detail of energetic pathways in branch, following source-sink pattern. Data are lacking for analogous root diagram. Volume function: liquid energy transport (simple black arrows). Surface function: assimilation (leaves: source). Dotted area, volume and surface expansion (sink); hatched area, meristematic zone (sink, surface, and volume production: three-headed white arrow). When pathways are opened to lateral meristems (1–10), priority among them is established by physiologic correlations; result is reiteration. Double-headed white arrows, cambial production. *C*. Interaction between inputs and outputs of environment and production systems. Increased or decreased environmental input, fed through the three systems, causes more or less general vigor. When one system reaches its inherent upper production limit per unit time, and cannot increase in size (e.g., cambium), general vigor cannot increase further.

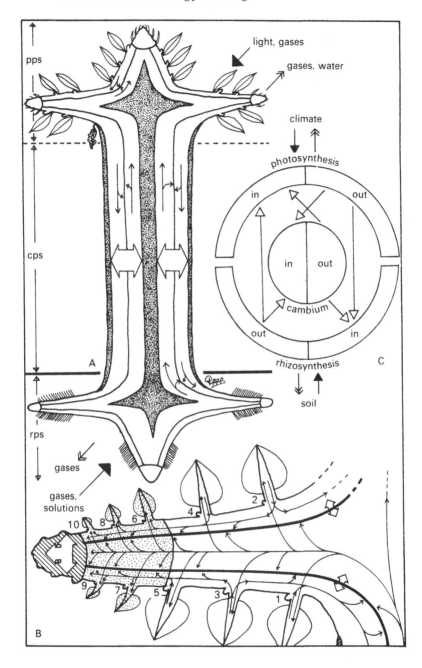

ity, and structure. The cambial system does not seem to exchange energy directly with the environment (Philipson, Ward, & Butterfield, 1971). It produces transport tissue in quantities that are determined by the intensity of energy transfer between roots and leaves and toward itself. The cambium produces other important tissues, but they do not interest us primarily in the context of the internal energy budget.

The plasticity of photosynthesis and rhizosynthesis in assimilating the environmental energy offered is secured by quantitative variations in leaf and root production, possible within the model. The corresponding plasticity in cambial production mainly depends on the kinds of vessels produced. Zimmermann (Chap. 22) stresses the relation between transport capacity and safety in vessels. The energetic factor linked to these variables is reaction speed: Formation of large but less secure vessels permits a rapidly increased transport capacity when photosynthesis or rhizosynthesis rises fast. The adaptive significance of large vessels is not their volume as such, but the possibility they offer of rapidly modifying total transport capacity, even at some risk of security. Hence, cambium production can be added to feedback control between shoots and roots (Chap. 21). The three systems are balanced: When one is active, the activity of the other two is amplified; when one is less active, the others also slow (Fig. 23.1C; for details, see Oldeman, 1974a). The cambium yields only secondary volume adjustment, but meristems add in the aerial part volume (internodes) and surface (leaves) to the photosynthetic system. Probably the same is true for subterranean meristems. Hence, the architectural model is a standard solution for harmonization of volume and surface inside fairly narrow limits of energy availability from the environment.

When this energy level remains rather constant and at the same value in the seedling biotope, a tree species adopts no other architectures except the chronologic series dictated by its model. Indeed, this is clearly the case with pioneer species (e.g., tropical *Cecropia* spp., *Musanga cecropioides*, *Ochroma* spp., and temperate *Pinus* spp.). It is also the case with species that spend their whole lives in the lower reaches of the forest: e.g., tropical *Diospyros matherana* (Ebenaceae), *Tabernaemontana* spp., *Pourouma minor* (Moraceae), and temperate *Cornus* spp. The Asian *Strobilanthes* possibly belongs to the same category of trees, but its architecture and model conformity should be checked.

Changing energy levels and reiteration

The majority of trees encounter important environmental changes as they develop. These can be absolute (e.g., an increase of sunlight, caloric energy, dry air, water, or even chemical substances like natural or artificial fertilizers) or relative (e.g., half the volume of a tree crown is shorn off by the fall of a branch, so that the leaf surface is halved, but still remains exposed to about the same light intensity). The reaction to increasing energy offer is positive morphogenesis: All three production systems (Fig. 23.1) enlarge. A new equilibrium is established when one of them reaches a productivity ceiling. For instance, on poor and shallow soil the root system can be the limiting agent. Tree expansion in this case is halted by exogenous influences.

When the environment does not impose limits, the productive capacity of the organs implied in the model (meristems) and of the cambium have an inherent overhead. One has to distinguish between the ecologic production threshold and the absolute production threshold, which can usually be reached only under laboratory conditions. The forest tree will never be exposed to the artificial day lengths or unnaturally high night temperatures that can be generated in a growth chamber. When light intensity exceeds the value that can still promote photosynthesis (gm CO_2 cm^{-2} min^{-1}) and a well-structured soil furnishes all necessary elements, the meristems of all parts of the tree are active at their ecologic maximum (cm^3/min volume, and cm^2/min leaf or root surface), and the cambium may also produce the highest volume of tissue it can build (cm^3 cm^{-2} min^{-1}).

If, at such a moment, energy is still offered in excess of these maximum possibilities (i.e., photosynthesis and rhizosynthesis continue to intensify while meristematic and cambial activities are at their upper limit), energy is first stored in specialized tissues (parenchyma roots). Soon, however, the energetic currents (Fig. 23.1*B*) inside the tree become overcharged and, guided by the physiologic correlations system, are directed toward resting meristems, which become activated. Champagnat (1954) accounts for the activation of such meristems by "circumstantial influences, linked to the vigor of growth," which is another way of saying the same thing.

These meristems, mobilized during vigorous growth, do not belong to the group that functions within the architectural model and consequently function beyond it. Some 3 years of checking numerous trees has shown that such meristems do not produce branches simply within the limits of model conformity; their growth pattern repeats the architectural model itself, completely or partially. Processes of repetition are not new in biology; they bear distinctive names (e.g., cell division or replication). Therefore, Oldeman (1974a) coined the term *"reiteration"* to designate the repetition of any architectural model from a meristem instead of from a seed. Reiteration imples more than regeneration, as it may involve a change in the level of differentiation of an existing functional meristem. Reiteration is an opportunistic process determined by environmental influ-

ences, whereas the model itself is a hereditary standard organization. Reiteration complicates the network of energetic pathways
within the tree (Fig. 23.1*B*) and increases the number of points
among which the total energy supply from below must be divided
and from which photosynthates must be exported. The same generalization can be applied to roots. This is why the first reiterated
complexes in a tree, though still displaying the size of trees themselves, are much smaller than the whole tree. The number of arborescent reiteration complexes is higher, and the complexes more chao-
tically arranged, in free-standing than in forest trees, to which this
text is limited. In the forest, a wave of arborescent reiteration is followed by a wave of more numerous and smaller reiterations (frutescent reiteration) if an excess of energy is still offered. These may then
bear suffrutescent organ complexes, which in their turn carry herbaceous reiteration. From the suffrutescent complexes on, the model
is usually imperfectly followed, and parts of the organoge-
netic sequence cease to become visible in the form of organs. According to Roux (pers. comm.) such parts are "telescoped" and can be
brought back into visible existence by experimental means, such as
the techniques of vegetative multiplication.

Figure 23.2 diagrammatically represents the life of a forest tree that
germinates in the energy-poor environment of the undergrowth and
reacts by reiteration every time a change in the forest canopy allows
increased irradiation or when the death of a neighboring tree liber-
ates inorganic or organic soil resources. The age-related changes in
the expression of the model, from arborescent to herbaceous, may be
compared with analogous changes in shoot growth (Chap. 21). Once
it reaches the phase of herbaceous reiteration, the tree lacks further
capability for building large structures and, therefore, has attained
its ultimate size.

So far, the behavior of the tree has been in response to an ecologic
sequence of stepwise increases in energy offered: Its reaction has
been positive morphogenesis. Often, however, the environmental
energy level decreases. Northern winters are an example of this, as
are the overtopping of a tree by a faster growing competitor or climatic seasons during which the soil dries out or is waterlogged. In such
cases trees react by what might be called "negative morphogenesis,"
what Addicott (Chap. 16) cites as examples of abscission strategies.
Figure 23.2 shows that every time a forest tree has to cope with unfavorable years it sheds structural parts and becomes reduced to the
status of a suppressed tree, awaiting the opportunity to resume expansion.

At every moment of its life the tree architecturally expresses a balance between positive and negative morphogenesis. This starts

when, through intervention of seedling predators (Chaps. 4, 5), surviving but damaged seedlings react to the relative increase in energy offer by reiteration, which regenerates the plant. Observations of 50 *Mouriri crassifolia* seedlings in the Guianese rain forest showed 48 cases of this early reiteration. The tree subsequently grows according to the initial, seed-originated model or an early reiterated one. These cannot be distinguished, as all traces of reiteration are rapidly obliterated by secondary thickening. The earliest examples of this are possibly adventitious hypocotyl meristems (Chap. 12). So far, the life of a forest tree runs parallel to that of a pioneer species growing in the open (Fig. 23.2).

For the tree growing in shade, there is no energy to continue this initial growth period. Branches are shed, and the crown becomes shallow, often with the physiognomy of a paint brush (van Steenis, 1956). This suppressed state often ends with death, but as far as I know there are no survivorship curves for trees in this phase. Such data would come from the research by Sarukhán (Chap. 6), in particular from the study of fluctuations in sapling populations. If the canopy above the tree opens, growth is resumed, first still conforming to the model, then by reiteration. Here, reiteration has nothing to do with regeneration, only with expansion.

During this stage, the tree may be overtopped and survive again in a suppressed state after shedding all or most of the existing reiterated complexes, even returning to a model-like architecture. If it does not die, growth can start again if the state of the canopy once more allows it. Finally the tree may expand its crown into the canopy itself, exposed to the macroclimate, which is the highest energy level of the biotope. Examples of trees in all growth phases here described can be found in the 22 profile diagrams of forests from many continents, which have been reproduced by Oldeman (1974a).

When the tree is young and slender, potential cambium production far exceeds the requirements to integrate leaf and root assimilation. However, the cambium production is gradually driven to its maximum, because the number of aerial and subterranean meristems grows and with them the tree's transport requirements. More fundamentally, the cambium is a surface that has to produce volume. The elementary mathematic comparison of a second-degree with a third-degree function shows that the increase of surface lags increasingly behind the enlarging volume of growing cylinders or cones (Stevens, 1974, p. 19). When maximum cambium production is reached, the only possibility for further harmonization between volume and surface is change in the form of the cambial surface, by fluting. This indeed happens, as can be seen in the buttresses and flanged trunks of many tropical trees. Preliminary observations near

Fig. 23.2. Architecture and energy exchange of growing tree. *Upper level.* Development of pioneer (P) and forest (F) tree as tree of future (P_f, F_f), in transition (P_t, F_t), of present (P_p, F_p), and in senescence (F_s). Pioneer dies after some 15 years (P:D): forest tree alternately expands and is suppressed in its phase of future. Reiteration first is regenerative (RR), then adjustive and arborescent (AR), frutescent (FR), suffrutescent (SR), and finally herbaceous (HR). Death is possible after each suppressed period (pD). Age of 5–6 centuries is hypothetic. Hatched trees overtopping forest tree (*if F*) mean decreased light availability: this increases as indicated by sun symbol (first two for pioneer only: *if P*). *Second level.* Changes in energy availability or distribution, diagramed as in Fig. 23.1C. Circled $+$ or $-$ indicates relative change (energy level same, not tree); $+$ or $-$ in triangle indicates absolute change. Between diagrams, situation is stable, does not change. *Third level.* Symbols used in Figs. 23.3, 23.4, and 23.6 for corresponding trees of first level. *Fourth level.* Values of parameter H/d measure cambial activity compared with other growth processes. *Fifth (lowest) level.* Morphologic inversion point, expressed as value H_l/H_t (free trunk height/total height), measures blaance between positive and negative morphogenesis.

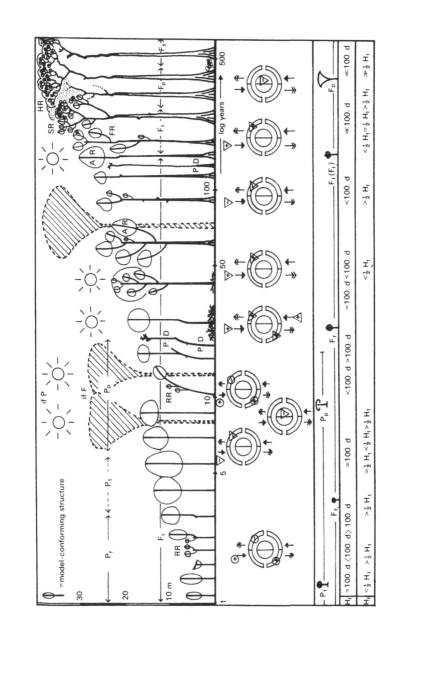

Saül (French Guiana) showed that on 32 of 35 buttressed trees reiteration had begun; the other 3 were understory *Pourouma minor*, in which reiteration is rare. Comparison with not-yet-reiterated individuals of the buttressed species suggests that no buttresses occurred before pronounced reiteration.

Trees can start reiterating at different heights (Fig 23.2), so these observations do not mean that buttresses are formed only in trees of a particular height class. Richards's data (1952) show that buttresses are bigger and more common on more fertile ground. In that situation, high rhizosynthesis is furthered, so that the whole metabolism of the tree can function at a high level, cambium production included. This agrees with the suggested mechanism of buttress formation.

One may prudently presume that the ultimate endogenous brake on tree growth is deficient cambium production. This handicaps leaf and root assimilation, whose deceleration in turn slows cambium production still further. A metabolic vicious cycle is the result, and the end is the "natural" death of the tree, which dries out. That this is the case is suggested by the vernacular names for dead erect trees: "*bois sec*" by the Creoles of French Guiana; "*arbol seco*" by other forest-dwelling Latin Americans. This occurrence implies that senescence and death of trees are not programmed but, on the contrary, are attributable to the failure of growth programs when biovolume increases beyond certain thresholds.

The forest
Trees as building elements
The image of the growth of a forest tree, as given above, allows a functional division among the trees that form a forest. "Trees of the future" architecturally conform to their model or have not yet shown prolific reiteration. These trees have at least a potential future, if they do not die during periods of suppression. "Trees of the present" have already reached their maximum expansion and can be recognized by a prolific reiteration of various kinds, from arborescent to herbaceous. Architecturally, they dominate the existing forest. "Trees of the past" have been irreparably damaged, show a chaotic reiteration, and often exhibit a spectacular disproportion between a massive broken trunk and a multitude of reiterated trunklets (Fig. 23.3A). They are no more than a nuisance for the growth of other trees, and their destiny is rapid elimination.

It is difficult to extend these circumscriptions to pioneer and understory species that always conform to their model and reiterate only for purposes of regeneration. Nevertheless, it is evident that, like all trees, they first have a potential future, then reach a stage of maximal expansion, maintain this for some time, and finally decay. It is possible to recognize these phases when examining the forest as a whole.

An indicator of the general vigor of a tree is the position of the lowest living branch or reiterated trunk on the main trunk. Using these points, Ogawa et al. (1965) established "crown depth diagrams" of forest plots. Oldeman (1974b) showed that they can be related to a reference line that indicates half the total height of each tree, in order to judge the maturity and degree of architectural homogeneity of a forest plot. A lowest living branch below this middle indicates predominant positive morphogenesis in an ecologic situation of increasing energy offer. When this point lies in the upper half of the trunk, negative morphogenesis (branch shedding) predominates, indicating either a diminishing environmental energy level or, in trees of the present, a deficient interior energy distribution (Fig. 23.2). The point of insertion of the lowest living branch is called the "morpholigic inversion point," and the point at half total height is the "ecologic inversion point" (Oldeman, 1974a,b).

Another parameter of the architectural state of the tree is the relation between diameter above the roots and total height (H/d relation). It was empirically found that in model-conforming trees, $H = 100 \cdot d$. When reiteration has occurred as a regeneration mechanism, the relation becomes $H > 100 \cdot d$, and when reiteration has taken place as an energetic adjustment mechanism, $H < 100d$. The total height of trees of the future equals or exceeds 100 times their diameter; it is less than this value in trees of the present (Fig. 23.2). This development is a symptom of the trends in cambium production during the life of a tree, as suggested above. Statistics and more detailed treatment of these questions can be found in Oldeman (1974a).

It is undoubtedly possible to express these relations in a more refined mathematic way (McMahon, 1975), but so far their formulation as simple rules of thumb has been sufficient for our purposes. For the same reason, no efforts have been made to express mathematically the architectural tree models, though this is unquestionably feasible (Hallé, pers. comm.). It is probable that refined mathematic treatment will be requisite for designing artificial landscapes in ecologic equilibrium.

Dynamic and homeostatic phases in sylvigenesis

Many of the definitions belonging to the Clementsian succession theory (see the glossary in Allred & Clements, 1949) do not or only imperfectly apply to the tropical forests, and the meaning of those terms, when used by tropical foresters and botanists, becomes rather vague. Current use of the word "climax," for instance, has

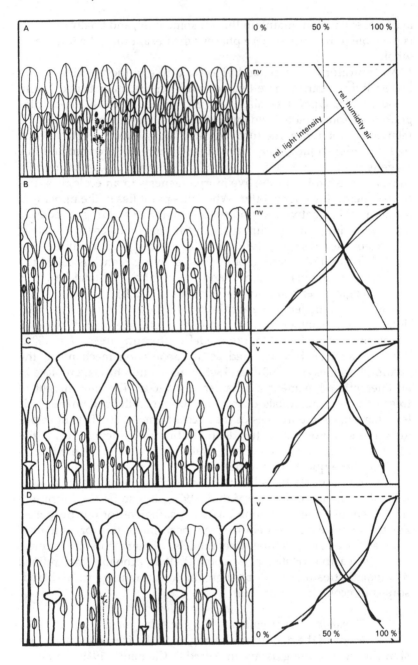

come to include all tropical forest except the pioneer phase. Therefore, the term "succession" will not be used here. The word "regeneration" is also excluded, because forest development is a cyclic process in which *each* phase can be regenerated. Moreover, the term is
ambiguous because it also refers to seedlings and saplings that are
the agents of forest establishment. The word "sylvigenesis" here indicates the process by which the forest develops and includes many
partial processes continuing through successive phases in shorter or
longer cycles.

During sylvigenesis two states alternate. The first is dynamic and
consists of growth toward the second (i.e., it has no constant architecture, for it changes incessantly). The second is the conclusion of
the first (i.e., with a distinctly layered architecture induced by a
severe hierarchy among its constituent trees). As long as the trees
remain organized this way, the forest cannot change and hence is
homeostatic; this steady state can last for centuries. To be thermody-
namically exact, it displays the homeostasis of an open system, as it
exchanges energy with its environment.

In Fig. 23.3, the aerial parts of the forest are symbolized as indicated in Fig. 23.2, with the subterranean parts omitted. The mechanism that reinforces the hierarchy among trees is energetic and depends on an increasing modulation of vertical microclimatologic
gradients. At the right side of Fig. 23.3, the establishment of this
modulation is diagrammatically represented for the factors light and
relative humidity of the air (Oldeman, 1974a,b). Direct measurement
of the vertical light gradient in a mature rain forest near Cayenne
(French Guiana) by Bonhomme (unpubl. ms., 1973) confirmed one
part of this image. The other part is supported by relative air humidity measurements by Cachan & Duval (1963) in Ivory Coast.

Fig. 23.3. (*facing page*) Sylvigenesis during one phase, in its dynamic (*A, B*) and steady (*C, D*) states. Fast, competitive growth
without forest structure (*A*) and with unmodulated vertical gradients (*right*), leads to transitional stage (*B*), when some trees overtop others and modulate vertical gradients (*right*). Modulation
creates layer with relatively high light availability, in relation to
average trend, where smaller crowns can expand. In these stages
(*A, B*), gradients have not been verified (*nv*). Forest then becomes
homeostatic, stable in time (*C*), with structural ensembles formed by
canopy and lower privileged trees and suppressed set of future;
structure strongly modulates vertical gradients (*right*). Senescence
(*D*) by exhaustion of set of present furnishes more light to set of future and prepares a following sylvigenetic phase. Light intensity
expressed as percentage of total macroclimatic light; relative humidity of air as usual. See text.

The terms "set of the future," "set of the present," and "set of the past" are used when trees in such phases participate as groups in sylvigenesis. The set of the present is divided into "structural ensembles" at different height levels (Fig. 23.3C). The dynamic state of each sylvigenetic phase has only a set of the future, whose trees grow as fast as reciprocal competition permits (Fig. 23.3A). No clear-cut structure or layers can be distinguished until, during a continuing dynamic transition, the fastest growing (reiterating) trees start to overtop the others (Fig. 23.3B).

The stragglers left behind in shade and high humidity are provided with less energy and react as described above. Their crowns diminish in size, transpire more, but photosynthesize less. This adjustment exerts an influence on lower parts of the forest, where light intensity is now higher, and relative air humidity lower, than the values of the average regression curve at this level. Other species that are adapted to reach maximal photosynthetic production at lower light intensities (represented by the low asymptote of the photosynthetic curve) here find an energetic environment where positive morphogenesis can dominate and form a structural ensemble. This layer is not continuous and includes crowns smaller than the upper ones, though fully expanded in terms of the tree's physiognomy. Modulations of gradients may extend downward and cause the formation of lower structural ensembles made up of treelets or shrubs.

Figure 23.2 shows the architecture of the trees of the present in structural ensembles (corresponding to Fig. 23.3C) and of suppressed trees of the future filling the gaps between these ensembles, but still without themselves determining forest architecture. Energy distribution inside trees of the present is such that ultimately their numerous growing points (Fig. 23.2) and deficient cambium production (Fig. 23.1) become handicaps for survival. Gradually, negative morphogenesis comes to dominate, the morphologic inversion point ascends, and the crown becomes shallow. This steady state ends with senescence (Fig. 23.3D), when trees of the present are exhausted and those of the future suppressed for a long time, both of them showing morphologic inversion points well above half total height. The end of senescence comes with the trees of the present dying while erect or falling down. In both cases, the situation becomes suitable for resumption of growth by surviving trees of the future, initiating the dynamic part of another sylvigenetic phase.

Sylvigenesis and its cycles
Negative sylvigenesis: the chablis
Just as there is positive and negative tree morphogenesis, the forest annihilates parts of itself (negative sylvigenesis) and develops anew (positive sylvigenesis). The agents of destruction are falling

trees. The fall of a tree, its impact on the forest, the fallen tree itself, and the resulting destruction are together described by the medieval French word *"chablis."* The word will be used in this sense here, because as far as I know no other term exists with this meaning.

Figure 23.4 shows the configuration of a chablis (after Oldeman, 1974b), but modified because the gap in the canopy, which was formerly filled by the crown of the fallen tree, has now also been taken into account. In different parts of a fresh chablis, different energetic conditions prevail. In the crown gap, the canopy has disappeared, and lower layers have been slightly damaged by the falling trunk. The set of the future is here exposed to the macroclimate, whereas the underlying modulation of the vertical light gradient (maybe also that of relative humidity) has remained approximately the same. The general light intensity has, however, increased; the whole curve (Fig. 23.3D) is displaced to the right.

Where the crown has fallen, there is an epicenter of destruction under the larger limbs, and the soil level is exposed to the macrocli-

Fig. 23.4. The chablis. Tree fall produces distinct energetic environments: crown gap with intact lower storys, epicenter of crown fall (macroclimatic), and its periphery with gradual transition from macroclimate to forest microclimate. In each, sylvigenesis is different. In diagram, intact forest is hatched, and periphery extends beyond destruction because of lateral light offer.

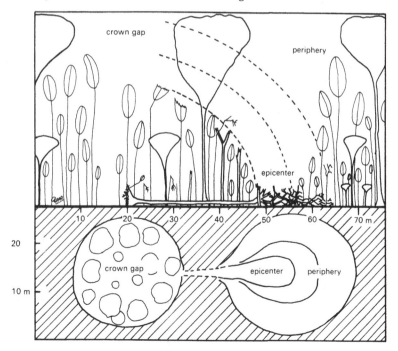

mate. Around this area, the chablis is more or less funnel-shaped. The transition from total destruction to architectural and energetic forest conditions is gradual, and sylvigenesis occurs according to different modalities on the epicenter, in its periphery, and under the crown gap.

The size of a one-tree chablis depends on the size of the fallen tree, between 10 m by 20 m and 20 m by 40 m. An area about 10 m by 10 m of this surface is covered by the epicenter and an equivalent area by the crown gap. Often, however, one falling tree knocks down others, and the chablis may thus be larger (Schulz, 1960). Comparable observations for Southeast Asia are given by Poore (1968). The frequency of chablis rises with the rigor of the climate. In the occidental equatorial zone of Ecuador, visible chablis number 15/ha on flat land and 18/ha on hillsides. At an altitude of 650 m above sea level, these figures are respectively 16/ha and 23/ha, and at 1900 m, their average is 33/ha. The interval of total decomposition and the size of the trees also change with climate (i.e., with floristics), and the image of negative sylvigenesis is modified accordingly; as yet there are no quantitative data.

Trees dying erect represent another form of negative sylvigenesis. Their death causes the same energetic changes as those under the crown gap of a chablis, but more gradually, so that damage to forest architecture can be repaired as fast as it occurs by slowly growing trees of the future.

Positive sylvigenesis

In the epicenter of a chablis, sylvigenesis starts from scratch: Pioneer species soon overtop herbaceous vegetation, which grows during a preliminary period. The dynamic state of this pioneer phase often contains trees (*Cecropia, Ochroma*) that remain unbranched for some years. Other trees (*Inga, Trema, Guazuma*) branch early. The following transition to the steady state is rather short in the first case, longer in the second. Budowski's data (1961) and my own observations indicate a dynamic state of 5–10 years and homeostasis for 10–20 years in pioneer forest. A single structural ensemble with rather thin foliage induces only a slight modulation of vertical gradients during homeostasis, and the trees of the future are likely to be in shadow cast more densely by the liane-rich thicket that includes them than by the pioneer trees. This phase in short, but very important ecologically and economically.

Trees recruited from the shrubby set of the future build another set of the present after the death of the pioneers. Generally their reiteration remains modest and they conform to their models for most of their lives: e.g., *Rollinia* (Annonaceae); *Dyalianthera, Iryanthera, Pyc-*

nanthus, Virola (Myristicaceae); *Pourouma* spp. (Moraceae); *Symphonia globulifera* (Guttiferae). Chronologic data are few. In Surinam, the cutting cycle for *Virola surinamensis* in natural forest is 40 years, and this period should cover approximately the transition from the set of the future to that of the present. There are often two structural ensembles: the canopy and layer of small trees 8–10 m high, the latter markedly model-conforming (e.g., understory *Diospyros* in tropical Africa and America).

When the set of the present of this second, postpioneer phase dies, the dynamic state of a third phase appears. The successful trees belong to another floristic complex with more prolific reiteration, slower growth (which may be keyed to slower decay of their predecessors), a longer life span than their predecessors, and generally more slender branches and harder wood (Budowski, 1961). The third phase may include the elimination of palms, many of which are light-loving and live less than 2 centuries. In steady state, this phase is elaborately organized, with two arborescent and one subarborescent structural ensembles at least (i.e., vertical gradients show a higher frequency of modulation).

In the humid tropics, there are at least four phases and probably a

Fig. 23.5. Species-area curves. Statistical points of *"levantamientos estructurales"* (*le 1, le 4, le 6*) of Venezuelan rain forest. Floristics change stepwise from one sylvigenetic phase to next (*sp 1, sp 2, sp 3, sp 4*), as illustrated by my general curve (thick line). Currently published graphs of this type only show a regression curve. (Based on Finol Urdaneta, 1972)

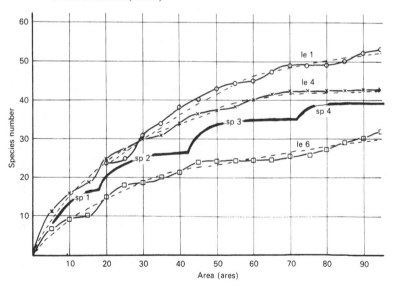

Fig. 23.6. The forest mosaic. *A*. Reduced and simplified architectural transect of 480 m by 20 m in Ecuador (corredor Castro, km 120 via Puerto Quito: alt. 650 m; lat. 0°02′11″N, long. 79°00′00″W). Facets of mosaic (sylvigenetic phases) evaluated by floristics and with a plan. *, chablis. Phases in dynamic (*1d*, *2d*, *3d*, *4d*) or steady (*1h*, *2h*, *3h*, *4h*) state. *B*. Facets of corredor Castro, schematic and chronologically arrayed. Percentages are result of balance between forest destruction and regrowth. *C*. Sylvigenetic cycles following a chablis. As there are four phases, 13 different cycles following a chablis are possible. Regression from older phase x to younger one y ($r\ x-y$) or to beginning (*rc*) is most general. Regression to dynamic state of same phase causes autoregeneration (*ar*), the only process that strictly deserves name regeneration. Sometimes, enough trees of future of a following phase can be activated in anticipated sylvigenesis (*a*). As all phases are in equilibrium with macroclimate, there is no evident climax, and in different cycles, same phase may represent a sere of different rank.

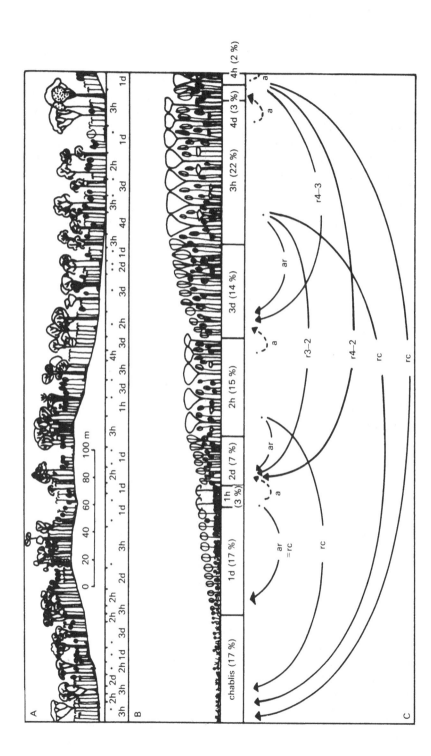

rare fifth one, which succeed one another in the above-described manner. The most convenient way to distinguish these phases is by floristics (i.e., by a rapid assessment of the most common species in the canopy), so as to avoid the time-consuming architectural analysis of the lower layers. In more advanced phases individuals of species are more scattered and less grouped. Data presented in Fig. 23.5 show that species-area curves, generally represented by their regression curves, in reality contain an alternation of rapid increases and near constancy. This can also be observed in the species-area graphs for species-rich vegetation published by Holdridge, et al. (1971), whose "synthetic forest profiles" hide this. Species-area curves are composed of cumulative graphs for each sylvigenetic phase.

The forest mosaic

The tropical forest owes its configuration at any moment to the balance between destruction and regrowth. Destructive processes are not synchronized by ecologic factors, such as the temperate hurricanes, which affect whole regions. Therefore, tropical sylvigenetic phases are distributed more or less at random, except where edaphic factors such as soil or slope increase the likelihood of chablis.

The longer positive sylvigenesis goes on, the greater becomes the risk that negative sylvigenesis (chablis) will throw the forest back into an earlier phase (Fig. 23.6). At a certain stage or age this risk becomes 100%, and no further sylvigenesis is possible. In reality, the process halts when survival chances become too small for a viable genetic density of a tree population (i.e., before the risk of destruction is total). A long lifespan can here compensate for a sparse distribution in space.

The tropical forest is a mosaic of coexisting facets, each of which represents a chronologic point in positive sylvigenesis (Fig. 23.6), showing its proper interaction between environmental energetics, which are more or less intensely regulated by the architecture of the vegetation, and plant energetics, which are largely a function of plant architecture.

Spatial distribution of the forest mosaic

In Fig. 23.6B, the coexisting facets of the forest mosaic are chronologically arrayed, as far as this can be done with a transect representing generalized architectural characters.

Nature is always more subtle than analysis, and, therefore, it is impossible to go beyond a certain precision in the description of a tropical forest mosaic. Figure 23.4 shows that in the epicenter of a chablis positive sylvigenesis starts again, beginning with the pioneer phase. Under the crown gap, development depends on the set of the future that is activated: If this contains strong and numerous trees of a following phase, anticipated development may occur; otherwise, the phase of the fallen tree, or some prior one, repeats itself in sylvigenetic regression. The periphery of the chablis ideally shows some concentric bands, each of which is microclimatically (i.e., energetically) suitable for the regressed or anticipated growth of some phase and has or has not in its set of the future the elements to build it.

The forest mosaic is composed of more or less circular spots and rings or horseshoe-shaped areas (see Chap. 27). Schulz (1960, Fig. 58) published maps of the individual distribution of *Dicorynia guianensis* (Papilionaceae) and *Ocotea rubra* (Lauraceae), both over 20 cm DBH, in the rain forest of Surinam. These corroborate the image of sylvigenesis given here. I evaluate *D. guianensis* as belonging to the third sylvigenetic phase, and *O. rubra* to the fourth, according to my observations in French Guiana. Both species show a distribution in horseshoe-shaped groups, the first narrower than the second. On the maps, the epicenters of the chablis within the groups of each species can often be seen to be the same. Moreover, there are isolated trees of each species, which would have grown in an understocked set of the future, in a crown gap, or under a big tree dying erect. Schulz's maps are, of course, no proof of the energetic-architectural forest model here described, but this model could explain them fully.

The intricate cohesion of different energetic processes in space and time of one and the same tropical forest, distinct modes of which can be found in other climates, allows us to see layering as a slowly transient phenomenon in a more complicated process. Constancy of the forest does not lie in age-resistant structures, but in repetitive sylvigenetic processes. Except during their very beginning, forest dynamics are extremely slow at the human time scale, but this should not obscure the fact that the forest is a living system by virtue of change.

General discussion

Ashton: The areas taken up by the stages you recognized in your diagram can be very different from one another. I mention this because in conversation with you I learned that you had been working these ideas in a forest that is periodically subjected to rather strong winds. Consequently, you drew the picture of the chablis at one end and the big tree crown at the other end, with the epicenter placed eccentrically. In Southeast Asia, in forests such as Professor Kira and I have experience of, we must bear in mind that a tree in what Dr. Whitmore calls the mature phase will reach a maximum height and then continue to increase in bole diameter for many years. This so-called mature phase can thus occupy a very considerable percentage of the total area of the forest. Furthermore, as this is usually associated with a forest where there is not much wind, the patterns by which the trees die, as Dr. Whitmore has described in his book (see also discussion, Chap. 27), is rather different from those in the Guianese forest.

Oldeman: Quantitative differences can be accounted for in my examples. My model, moreover, comes from regions where there is no appreciable wind. Chablis are not primarily formed by wind, though perhaps triggered by wind; they are formed by soaking of the soil in the rainy season.

Whitmore: The horseshoe-shaped gaps caused by the fall of a single tree are probably the kind produced in areas without strong winds. In Southeast Asia, in gaps as small as those created by a one-tree fall, pioneers do not invade either the dipterocarp rain forest or, in fact, the Asian rain forest in general. Existing shade-tolerant seedlings are "released." For pioneers like the Southeast Asian equivalents of *Cecropia* to come in, there must be a really big gap. An experiment in the 1920s showed that gaps had to be 0.2–0.3 ha for a switch from existing seedling growth to pioneer invasion (Kramer, 1926, 1933).

Oldeman: Similar experiments were done in the Republic of Zaire. When pioneers establish in a gap, there is a complete regression. However, in many cases there is a complete regeneration of the same phase or of an earlier one, or even anticipation of the next phase. This may effectively explain the fact that there are not more pioneers and few major phases in the forest. Evidence for this is that in many chablis there are saplings rather than pioneers in the epicenter.

Whitmore: In the Southeast Asian forest I do not think that anybody has ever managed to recognize a series of floristic phases in successional relation as shown in your diagram. Pioneers and climax forest both can be recognized. It would be very difficult to draw a successional series with discrete stages connecting these two extremes.

Oldeman: This forest (Fig. 23.6) is at an altitude 650 m above sea level and floristically less rich than at lower levels. It is easier to make analyses in floristically poorer vegetation. There are also certain architectural criteria. Pioneers generally conform to their models, and reiteration plays a more important role as the phases progress. Consequently, the more strongly reiterated trees belong to the oldest phases. In general, gregariousness of species diminishes from early to late phases. Phases are mixed, but monospecific groups of trees indicate earlier phases in correlation with architectural criteria. There is also a tendency to find the thickest lateral axes in the early phase and more slender twigs in the later phases. I could, however, cite a lot of exceptions to each of these rules. We have to keep in mind that they must be considered together, floristics included.

Fisher: In relation to your very interesting hypothesis about the effect of energy and the relation of energy input and changing form with respect to reiteration, is there a quantitative correlation with solar radiation or much experimental data that would support the hypothesis?

Oldeman: To give one quantitative element, I calculated the probable shape of the modulated gradients in the forest by planimetering the surface of the trees in the set of the present and the set of the future. The results have been substantiated by direct measurements of light intensity with a fish-eye lens in the forest near Cayenne in French Guiana, by Bonhomme (1973). Vertical moisture gradients have been measured along towers, much like those used by Dr. Kira, in Ivory Coast as early as 1963 by Cachan & Duval. They are tabulated in the original publication. I applied these values, calculated from crown surface, to the probable form of these gradients, and they coincided with certain architectural phenomena (e.g., prolific reiteration, rare reiteration, model conformity) in the manner I explained.

May I continue with another question to our colleagues who have worked in Asia. Dr. Hallé saw in Africa and I in South America that when a pioneer tree dies upright, its branches seem to curl up, but it is not known whether orthotropic or plagiotropic. Is this true in Asia also?

Ashton: Anthocephalus chinensis does that.

Whitmore: Branches of some species of *Macaranga* do it also.

References

Allred, B. W., & Clements, E. S. (1949). *Dynamics of Vegetation: Selections from the Writings of Frederic E. Clements.* New York: Wilson.

Bonhomme, R. (1973). *Mesures préliminaires des rayonnements solaires au dessus et à l'intérieur d'une forêt guyanaise.* Guadeloupe, French Antilles: I.N.R.A.

Braun Blanquet, J. (1932). *Plantsociology: The Study of Plant Communities.* New York: McGraw-Hill.

Budowski, G. (1961). Studies on forest succession in Costa Rica and Panama. Ph.D. thesis, Yale University.

Cachan, P., & Duval, J. (1963). Variations microclimatiques verticales et saisonnières dans la forêt sempervirente de basse Côte d'Ivoire. *Ann. Fac. Sci. Dakar, 8,* 5–87.

Champagnat, P. (1954). Les corrélations sur le rameau d'un an des végétaux ligneux. *Phyton (Argentina), 4,* 1–104.

Finol Urdaneta, H. (1972). *Estudio fitosociológico de las unidades II y III de la reserva forestal de Caparo, estado Barinas.* Venezuela: Universidad de los Andes, Facultad de Sciences Forestales, 81 pp.

Godron, M. (1971). Essai sur une approche probabiliste de l'ecologie des végétaux. Thesis, Montpellier, No. CNRS AO 2820. 247 pp.

Hallé, F. (1974). Architecture of trees in the rain forest of Morobe district New Guinea. *Biotropica, 6*(1), 43–50.

Hallé, F., & Oldeman, R.A.A. (1970). *Essai sur l'architecture et la dynamique de croissance des arbres tropicaux.* Paris: Masson.

Holdridge, L. R., Grenke, W. C., Hatheway, W. H., Liang, T., & Tosi, J. A. (1971). *Forest Environments in Tropical Life Zones.* Oxford: Pergamon Press.

Horn, H. (1971). *The Adaptive Geometry of Trees.* Princeton: Princeton University Press.

Kramer, K. (1926). Onderzoek naar de natuurlijke verjonging in den uitkap in Preanger gebergte-bosch. *Med. Proefst. Boschu. Bogor, 14.*

– (1933). Die natuurlijke verjonging in het Goenoeng Gedeh complex. *Tectona, 26,* 156–85.

McMahon, T. (1975). The mechanical design of trees. *Sci. Amer., 233*(1), 93–102.

Ogawa, H., Yoda, K., Kira, T., Ogino, K., Ratanowongse, D., & Apasutaya, C. (1965). Comparative ecological study on three main types of forest vegetation in Thailand. I. Structure and floristic composition. II. Plant biomass. *Nat. Life S.E. Asia, 4,* 13–48, 49–80.

Oldeman, R.A.A. (1974a). L'architecture de la forêt guyanaise. *Mém. O.R.S.T.O.M.*, no. 73.

– (1974b). Ecotopes des arbres et gradients écologiques verticaux en forêt guyanaise. *Terre Vie, 28*(4), 487–520.

– (1975). *Bioarquitectura de las vegetaciones y método práctico para su observación*. Nota técnica no. E.1, 21 pp. Quito, Ecuador: Min. Agr. y Ganad., Dpto. de Regionalización.

Philipson, W. R., Ward, J. M., & Butterfield, B. G. (1971). *The Vascular Cambium: Its Development and Activity*. London: Chapman & Hall.

Poore, M. E. D. (1968). Studies on Malaysian rain forest. 1. The forest on Triassic sediments in Jengka Forest Reserve. *J. Ecol., 56*, 143–96.

Richards, P. W. (1952). *The Tropical Rain Forest*. Cambridge: Cambridge University Press.

– (1969). Speciation in the tropical rain forest and the concept of the niche. *Biol. J. Linn. Soc. Lond., 1*, 149–53.

Rollet, B. (1969). Etudes quantitatives d'une forêt dense humide sempervirente de la Guyana vénézuélienne. Thesis, Toulouse, No. CNRS 2969. 404 pp.

– (1974). *L'architecture des forêts denses humides sempervirentes de plaine*. Nogent-sur-Marne: Centre Technique Forestier Tropical. 298 pp.

Schulz, J. P. (1960). Ecological studies on rain forest in Northern Surinam. *Verh. Kon. Nederl. Akad. Wet. (Afd. Natuurk.) Amsterdam, Reeks 2, 53*(1), 1–267.

van Steenis, C. G. G. J. (1956). Basic principles of rain forest sociology. *UNESCO, Actes Coll. Kandy*, 159–65.

Stevens, P. S. (1974). *Patterns in Nature*. Boston: Little, Brown.

24

Community architecture and organic matter dynamics in tropical lowland rain forests of Southeast Asia with special reference to Pasoh Forest, West Malaysia

TATUO KIRA

Department of Biology, Faculty of Science, Osaka City University, Osaka, Japan

Our knowledge of the floristics, physiognomy, and structure of humid tropical forests has been summarized by Richards (1952), but their properties as a peculiar type of ecosystem remain little investigated. Even after 10 years of International Biological Program (IBP, 1965–74), for instance, relatively few papers offer a reasonable estimate of primary productivity of humid tropical forests (e.g., Müller & Nielsen, 1965; Kira, Ogawa, Yoda, & Ogino, 1967; Ogino, Ratanawongse, Tsutsumi, & Shidei, 1967; Hozumi, Yoda, & Kira, 1969; Odum & Pigeon, 1970; Golley & Golley, 1972; see also the review by Murphy, 1975).

The integrated ecosystem research at Pasoh Forest Reserve, Negeri Sembilan, West Malaysia (1970–74), by Malaysian, Japanese, and British ecologists was one of the efforts to fill this gap (Soepadmo, 1973). Ten Japanese plant ecologists were responsible for the study of primary production and organic matter dynamics within this IBP project. This chapter is a preliminary summary of their studies, drafted on my own responsibility. Additional information on Southeast Asian rain forest ecosystems is also incorporated as far as possible from the experience of the Osaka City University group on tropical ecology.

Community architecture
Tree height
Tropical rain forests of Southeast Asia seem to include the tallest of all similar plant formations in the world. Richards (1974) stated that in African and Latin American tropics he had so far met with only a single tree taller than 60 m, whereas big trees of the same height class are not uncommon in Southeast Asia, especially in the rain forests of eastern Borneo. An example from eastern Kalimantan

(Fig. 24.1) demonstrates that tall emergent trees in this forest are generally over 60 m in above-ground height. Big trees of *Koompassia excelsa*, *Shorea* spp., etc. are often taller than 70 m.

The above-ground structure of a forest stand can well be described by the hyperbolic relation between height and stem diameter (DBH) of trees characteristic of each stand (Ogawa, 1969; Kira & Ogawa, 1971; Fig. 24.1). Figures 24.2 and 24.3 relate diameter-height curves from several selected localities in Southeast Asia to differences in climate. Proceeding northward from eastern Kalimantan through the Malay Peninsula to northern Thailand, typical rain forests gradually give way to evergreen seasonal forests and then to deciduous seasonal forests. The transition is associated with a decrease in precipitation and increasing length of dry season and is accompanied by a decrease in tree height. The height of a tree with 1-m DBH is expected to be 61 m at Sebulu (1), 47 m at Pasoh (2), 46 m at Khao Chong (3), 39 m at Sakaerat (4), and 32 m at Ping Kong (5). The tree height in tropical forests seems to depend largely on the amount and seasonal distribution of rainfall, unless edaphic conditions are very unfavorable. The most luxuriant development of lowland rain forest in eastern Borneo may be ascribed to the abundant rainfall, which is extremely evenly distributed throughout the year.

Fig. 24.1. Hyperbolic relation between DBH and height of trees in equatorial rain forest, Sebulu, eastern Kalimantan, Indonesia.

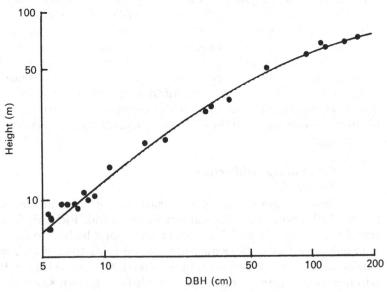

Fig. 24.2. Location and climate of five selected forests in Southeast Asia where tree DBH-height curves of Fig. 24.3 were obtained. Climate diagrams, drawn according to Walter's (1955) method, show annual distribution of monthly mean temperature and precipitation. Dotted areas indicate dry season.

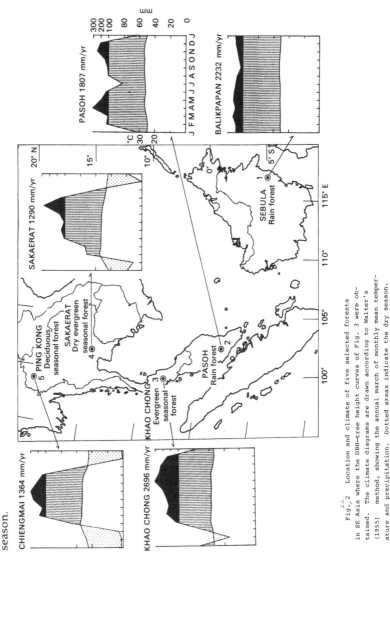

Fig. 2   Location and climate of five selected forests in SE Asia where the DBH-tree height curves of Fig. 3 were obtained. The climate diagrams are drawn according to Walter's (1955) method, showing the annual march of monthly mean temperature and precipitation. Dotted areas indicate the dry season.

Stratification and profile structure

Richards (1952) considered three stories to be the basic structure of tropical rain forests: (1) a layer of more or less isolated crowns
of giant trees overtopping the forest; (2) a layer of the crowns of large
trees, which may be dense and continuous in sufficiently humid dis-
tricts but may be open and discontinuous where the climate is less
moist; and (3) a layer of the crowns of small trees, which is not dense
and discontinuous in the former case but may become a continuous
main canopy in the latter case.

This general principle is applicable to any type of evergreen lowland forest of Southeast Asia except where the soil is very poor or is
waterlogged.

Another aspect of forest structure is revealed by examination of the
vertical distribution of foliage in a forest profile by means of destruc-

Fig. 24.3. DBH-height curves of trees in five tropical forests of
Southeast Asia (see Fig. 24.2). *1*, tropical rain forest, Sebulu, eastern
Kalimantan; *2*, tropical rain forest, Pasoh, western Malaysia; *3*,
evergreen seasonal forest, Khao Chong, southern Thailand; *4*, dry
seasonal evergreen forest, Sakaerat, central Thailand; *5*, deciduous
seasonal forest, Ping Kong, northern Thailand.

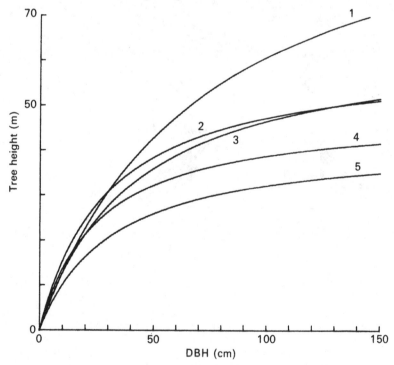

tive sampling (Kira, Shinozaki, & Hozumi, 1969). The profile diagram in Fig. 24.4, showing the vertical distribution of leaf area, leaf
biomass, and woody organ biomass, was obtained by clear felling a
20-m by 100-m strip of undisturbed rain forest at Pasoh. All trees on
the clear-felled plot were separated into main stem, branches, and
leaves according to the so-called stratified clip technique (Monsi &
Saeki, 1953) and weighed. The pattern of leaf distribution in Fig. 24.4
indicates the existence of a main canopy at 20–35 m above the ground
formed by the dense assemblage of crowns of large trees, corresponding to the second layer of Richards. The leaf area density (LAD)
decreases somewhat abruptly above 35 m, where big isolated crowns
of giant emergent trees constitute the top layer. The transition from
the main canopy to the third layer of small trees is gradual, showing
that leaves are distributed more or less evenly between 35 and 10 m.
High LAD in the ground layer below 1.3 m and in the overlying layer
between 1.3 and 5 m are due, respectively, to the abundant occur-

Fig. 24.4. Profile structure of Pasoh Forest, based on clear felling of
20-m by 100-m strip. (From Kato et al., 1974)

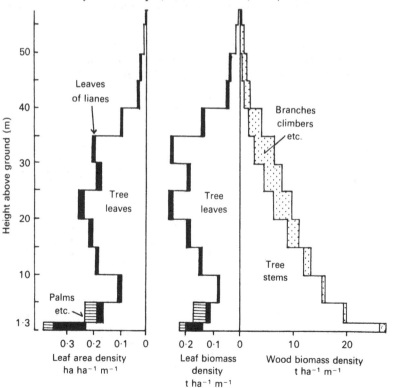

rence of tree and liane seedlings and to the development of a shrub layer including many small palms. A gap between shrub layer and small tree layer does exist, but seems specifically exaggerated in this destructive sampling plot.

Light profile

The three-dimensional distribution of light intensity and photosynthetically active radiation was thoroughly investigated by Yoda (1974c) in and near the destructive sampling plot. Figure 24.5 illustrates how light is successively intercepted by layers of leaf canopy. As the downward attenuation of light through a canopy space is expected to be exponential if LAD remains uniform (Monsi & Saeki, 1953), a linear regression between log illuminance and above-ground height in Fig. 24.5 suggests that LAD is almost homogeneous between 10- and 30-m levels. The wide space of the main canopy is fairly evenly filled with leaves despite the apparent stratification among component trees.

The mean relative illuminance on the ground surface was only 0.3%–0.4% of incident daylight over the forest, and that at breast height less than 0.5%. Such low levels of light at the bottom of

Fig. 24.5. Vertical distribution of mean relative illuminance in Pasoh Forest. Black dots indicate observations in February 1973; crosses, observations in July 1971. (From Yoda, 1974c)

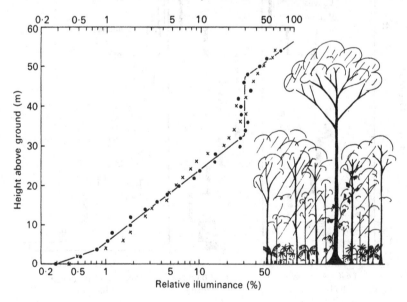

the canopy reflect a large leaf area index (LAI) of Pasoh Forest (about 7 ha/ha for all trees over 4.5 cm DBH).

Considering, however, the enormous height of the forest canopy, the mean LAD is exceedingly small: only 0.12–0.14 m²/m³. Even the maximum LAD in the 20- to 25-m layer barely reaches 0.25 m²/m³. These values are nearly half as small as the corresponding LAD's in temperate forests and almost 10 times smaller than those in herbaceous communities (Kira et al., 1969). Aoki, Yabuki, & Koyama (1975) pointed out that such a small LAD resulted in greater relative wind velocity inside the forest than in ordinary crop canopies with greater concentration of foliage. This may, in turn, favor the photosynthetic activity of the forest by preventing CO_2 starvation from which herbaceous communities tend to suffer in daylight on fine days.

Primary production rates

Methods

Rates of organic matter production by a plant community can be estimated on different principles. The methods described in the following section intend to estimate the total amount of CO_2 assimilated by the community, whereas the so-called summation method or harvest method arrives at the estimate of production rate by separately measuring the amounts of assimilated organic matter allotted to various community processes, such as community biomass increment, compensation of biomass losses due to animal grazing, death of plants, shedding of plant parts, and respiratory consumption by component plants (Kira et al., 1967; Newbould, 1967).

Plant biomass and biomass increment

The above-ground biomass of Pasoh Forest was estimated by two destructive samplings, in 1971 and 1973, including the clear felling of a 20-m by 100-m plot. Some 160 trees (DBH range from 4.5 cm to 102 cm) were felled and weighed using the stratified clip technique.

A series of allometric correlations among various dimensions of sample trees was empirically established, and a procedure was developed to estimate oven-dry weights of stem, branches, and leaves as well as total leaf area per tree from measured values of DBH. The steps of calculation were:

1. Estimating tree height (H) in meters from DBH (D) in centimeters

$$\frac{1}{H} = \frac{1}{2.0}\frac{1}{D} + \frac{1}{61}$$

2. Estimating stem dry weight (w_S) in kilograms from D and H

$$w_S = 0.313(D^2H)^{0.9733} \qquad (D^2H \text{ in dm}^3)$$

3. Estimating branch dry weight (w_B) in kilograms from stem weight

$$w_B = 0.316\, w_S^{1.070}$$

4. Estimating leaf dry weight (w_L) in kilograms from stem weight

$$\frac{1}{w_L} = \frac{1}{0.124 w_S^{0.794}} + \frac{1}{125}$$

5. Estimating total leaf area (u) in square meters from leaf weight

$$u = 11.4\, w_L^{0.900}$$

The total above-ground weight was calculated as the sum of w_S, w_B, and w_L.

The total biomass and LAI in a given forest stand could thus be calculated if the DBH of all trees was measured. The biomass of all trees ($D \geqslant 4.5$ cm) on a 0.8-ha area including the clear-felled plot in Pasoh Forest consisted of 346 tons per hectare of stem, 77.9 t/ha of branch, and 7.77 t/ha of leaf, totaling 431 t/ha. The estimate of LAI was 6.87 ha/ha. The above-ground biomass and LAI of smaller undergrowth plants were very variable from place to place, ranging from 0.5 to 8.8 t/ha and 0.23 to 0.95 ha/ha, respectively. The comparison of actually harvested and calculated biomass on the clear-felled plot is shown in Table 24.1. The relative error of estimation was mostly within 10%.

The density of dry organic matter per unit space occupied by the forest is approximately 10.9 kg/m^3 if the average height of very uneven canopy surface is presumed to be 40 m. As the corresponding biomass density is known to range from 1 to 1.5 kg/m^3 in most closed forests (Kira & Shidei, 1967), the above-ground biomass stock in this rain forest seems moderate in proportion to its height.

No attempt was made to estimate total root biomass. Only the amount of fine roots less than 1 cm in diameter was measured by digging out soil cores (Yoda, 1974a). The fine root biomass was fairly evenly distributed everywhere in the forest, amounting to about 20 t/ha.

The LAI estimate of about 7 ha/ha (8 ha/ha on the clear-felled plot) was not as large as expected. The most reliable estimates of LAI, obtained in temperate forests of Japan, are as follows: larch plantation (39 years old), 6.7 ha/ha; secondary beech forest, 6.4 ha/ha; secondary evergreen oak forest, 6.6 ha/ha. All include the leaf area of undergrowth vegetation. The LAI estimate at Pasoh is no less accurate than these values, because all these figures were obtained by taking leaf area samples from every height level of forest canopy. If this is not done, estimates may be subject to a serious bias owing to the remarkable increase in ratio of leaf area to leaf weight from the surface to the bottom of the canopy (Kira et al., 1969; Kira, 1975). Ogawa, Yoda, Ogino, & Kira (1965) and Golley (pers. comm.; see also Golley, 1972) reported LAIs of 12.3 and 20 ha/ha, respectively, from the rain forests of southern Thailand and Panama; however, their method of leaf area sampling was obviously inadequate, so these extraordinarily large values are very doubtful.

The increment of biomass could be estimated by repeating DBH census on the same plot at certain time intervals. DBH increments during a 1.9-year period ranged from nearly zero to 1.3 cm and tended to be greatest in moderate-sized trees of 20–50 cm DBH (Fig. 24.6). Calculated biomass increment rates on the 0.8-ha plot are shown in Table 24.2. These are the result of balance between the growth of surviving trees and the death of a considerable number of small trees during the period.

The annual increment rate of above-ground tree biomass, 5.3 t ha^{-1}

Table 24.1. *Dry weight and leaf area of actually harvested above-ground biomass of trees (DBH ≥ 4.5 cm) and estimates calculated with allometric relations*

	Stem (kg)	Branch (kg)	Stem + branch (kg)	Leaf (kg)	Total (kg)	Leaf area (m²)
Harvested						
A. Trees	56,852	14,288	71,140	853	71,993	6,940
B. Lianes			1,285	45	1,330	463
C. A + B			72,425	898	73,323	7,403
Calculated						
D[a]	53,586	12,550	66,136	924	67,060	7,659
Relative error						
100 (D − A)/A	−5.7	−12.2	−7.0	+8.3	−6.9	+10.4
100 (D − C)/C			−8.7	+2.9	−8.5	+ 3.5

[a]The biomass of lianes was treated as part of their host trees' biomass in the analyses of allometric relation (see Kira & Ogawa, 1971).
Source: Kato, Tadaki, & Ogawa (1974).

Fig. 24.6. DBH increments in relation to initial DBH in trees of clear-felled plot. Period of observation was 1.9 years, from April 1971 to March 1973. (From Kato et al., 1974)

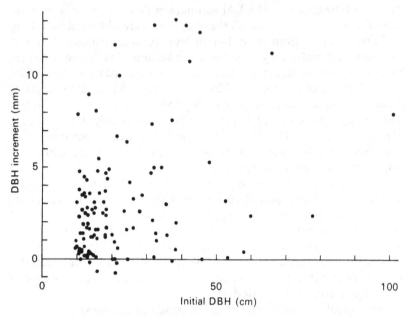

Table 24.2. *Above-ground biomass increment on 0.8-ha plot in Pasoh Forest, calculated from census of DBH of all trees (DBH ≥ 4.5 cm) in April 1971 and March 1973 (time interval: 695 days)*

	Stem	Branch	Stem + branch	Leaf	Total	LAI (ha/ha)
Biomass in 1971 (t/ha⁻¹)	337.74	75.98	413.7	7.61	421.3	6.73
Biomass in 1973 (t/ha⁻¹)	345.81	77.85	423.7	7.77	431.4	6.87
Rate of increment (t ha⁻¹ yr⁻¹)	4.238	0.982	5.221	0.084	5.305	0.074
Relative increment rate (per yr)	0.0125	0.0129	0.0126	0.0110	0.0126	0.0109

yr^{-1}, was so small it could have been easily canceled by the additional death of a tree about 60 cm in DBH per hectare and per year. The plant biomass in this climax rain forest is apparently in an almost stationary state.

Litterfall

Leaves, small branches (diameter $\leq$ 10 cm), and other fine litter falling to the ground were collected with 30 round funnels (mouth area 1 m^2) placed regularly around a 2-ha plot (including the 0.8-ha plot mentioned above). The rates of fall of branch and bark did not show any appreciable seasonal trend, though the fluctuation from week to week was considerable. The rate of leaf fall tended to reach the maximum in February to April immediately after the minimum of monthly rainfall in January, and another less pronounced peak seemed to occur following a slight drought in midsummer. The fall of flowers, bract, and caterpillar frass was more or less synchronized with that of leaves. Statistical tests suggested that the number (30) of litter receptacles was barely enough to estimate the rate of total litterfall within the relative error of 10%.

The mean fall rates of fine litter are summarized in Table 24.3. The annual amount of total litterfall (11.1 t ha^{-1} yr^{-1}) was nearly twice or three times as large as the litterfall rates observed in temperate forests (Bray & Gorham, 1964; Tadaki & Kagawa, 1968), but was similar to the rates obtained so far in other tropical rain forests (Klinge,

Table 24.3. *Mean fall rates of fine litter on 2-ha plot, including clear-felled plot and 0.8-ha plot (Tables 24.2, 24.6), in Pasoh Forest*

Type of litter	Mean daily rate (gm m^{-2} day^{-1})	Mean annual rate (t ha^{-1} yr^{-1})
Leaf	1.927	7.03
Branch (diameter < 10 cm)	0.666	2.43
Flower	0.022	0.082
Fruit, seed	0.104	0.381
Bark	0.073	0.267
Bud scale, bract, etc.	0.082	0.298
Insect body	0.003	0.011
Insect frass	0.073	0.268
Others	0.090	0.327
Nonleaf litter total	1.114	4.07
Plant litter total	2.964	10.82
Total	3.041	11.10

1974). More than 97% of this amount was of plant origin, and leaves accounted for about 63%. Dead branches tend to remain on trees for a certain period and are subject to decomposition before falling to the ground, so that the observed amount of branch fall has to be corrected for this prefall decomposition to know the real loss of biomass due to the death of branches.

As for big wood litter (diameter > 10 cm), all pieces of such big wood, either lying on the ground or standing dead, on five 2-ha areas were marked and numbered, the grade of decay recorded in terms of five arbitrary classes established visually, and their length and diameter measured to estimate their volume. The mean bulk density of wood of each decay class was also determined by appropriate sampling. The standing crop of big dead wood ranged between 42 and 65 t/ha and was much greater than the accumulation of other fine dead plant materials on the forest floor (3–5 t/ha). Similar observations repeated semiannually showed that the annual rate of supply of big dead wood to the ground amounted to 3.3–20.5 t ha^{-1} yr^{-1} with an average of 9.3 t ha^{-1} yr^{-1}. It is worthy of note that the rates of fall of fine litter and big wood litter are more or less similar in amount in such a fully mature forest community.

Table 24.4 illustrates the dynamics of big wood litter in the 2-ha plot, where the annual fall rate was the least among the five areas observed. The annual loss of biomass due to the death of big wood was estimated to be 3.7 t ha^{-1} yr^{-1}, considering the prefall decomposition.

Grazing consumption

Among the consumption of net plant production by heterotrophic organisms, only the consumption of leaves by caterpillars was estimated from the amount of frass caught by litter funnels (0.268 t ha^{-1} yr^{-1}; Table 24.3). As the mean efficiency of assimilation by forest caterpillars is around 13%, the rate of their leaf consumption is estimated at 0.31 t ha^{-1} yr^{-1}. Many other animals and heterotrophic plants may utilize live organic matter of plants, and a considerable amount of frass might have been lost without reaching litter funnels. Therefore, this amount, which equals 4% of leaf biomass or annual leaf production, is obviously a minimum estimate of biomass consumption by heterotrophs.

Net production rate

As summarized in Table 24.5, the sum of the rates of biomass increment, litterfall, and grazing consumption gives the first approximation of net production rate at 26.7 t ha^{-1} yr^{-1}. The mean turnover time of above-ground tree biomass is fairly long, amounting to

Table 24.4. *Dynamics of big wood litter (diameter > 10 cm) in same 2-ha plot in Pasoh Forest*

Grade of decay	Mean bulk density (gm/cm^3)	Accumulation on ground (t/ha)	Fall rate (t ha^{-1} yr^{-1})	CO$_2$ evolution[a] (mg m^{-2} hr^{-1})	Weight loss due to decomposition[a] (t ha^{-1} yr^{-1})
0	0.58	0.3	0.6	1	0.1
1	0.56	0.8	1.2	4	0.2
2	0.54	6.9	0.1	37	1.8
3	0.48	19.0	1.1	125	6.1
4	0.48	11.2	0.3	74	3.6
5	0.39	8.9		75	3.7
Total		47.1	3.3	316	15.5

[a]See text for explanation.
Source: Calculated from data of Yoneda, Yoda, & Kira (1974).

Table 24.5. *Estimation of net production (in t ha^{-1} yr^{-1}) on 0.8-ha plot in Pasoh Forest for 2-year period, April 1971 to March 1973, based on trees over 4.5 cm DBH*

	Stem	Big branch	Small branch	Other plant litter	Leaf	Above-ground total	Root	Total
Biomass increment (Δy)	4.238	0.982			0.084	5.305	0.530[a]	5.83
Litterfall (L')	3.30	2.43	1.36		7.03	14.12		
Estimated biomass loss (L)	3.67[b]	3.47[c]	1.36		7.03	15.53	5.0[d]	20.53
Grazing consumption (G)					0.31	0.31		0.31
Net production ($\Delta y + L + G$)		13.72			7.42	21.14	5.53	26.67

[a]Assumed to be 10% of above-ground biomass increase.
[b]Prefall loss of dry weight assumed to be 10%.
[c]Prefall loss of dry weight assumed to be 30%.
[d]Assumed to be 25% of fine root biomass.

574   Tatuo Kira

426/26.7 = 16 years. Kira (1976) obtained the following mean values of above-ground net production rate based on the studies at 316 forest stands in Japan: 11.15 ± 3.75 t ha^{-1} yr^{-1} in boreal conifer forests, 8.74 ± 3.47 t ha^{-1} yr^{-1} in cool temperate deciduous broadleaf forests, 14.25 ± 5.78 t ha^{-1} yr^{-1} in temperate conifer (except pine) forests, 13.64 ± 5.00 t ha^{-1} yr^{-1} in temperate pine forests, and 20.65 ± 7.21 t ha^{-1} yr^{-1} in warm temperate evergreen broadleaf forests. Pasoh Forest is thus considerably more productive than average temperate forests.

However, even temperate forests have been known to produce more than 30 tons of dry matter per hectare annually if the forests are dense enough and in early stages of development (Kira, 1975). Since the net productivity of a forest stand depends greatly on its age or tree size composition (Kira & Shidei, 1967; Kira, 1975), younger forests in the humid tropics may possibly be more productive than the mature forest at Pasoh, as already demonstrated with young plantations of rubber (Templeton, 1968) and oil palm (Ng, Thamboo, & de Souza, 1968).

Community respiration

Estimating the total respiration by a forest community involves serious technical difficulties. In the Pasoh Forest study, the estimate was made on essentially the same principle as first proposed by Yoda (Yoda et al., 1965; Yoda, 1967; Kira, 1968).

Leaf, branch, stem, and root samples of various component plants of Pasoh Forest, including a first-layer tree and 13 other tree species, were enclosed in plastic containers with KOH solution for a few hours under shade on the forest floor to determine the amount of absorbed CO_2 by titration with HCl. An effort was made to minimize (to within 30 minutes) the time from the felling of the sample tree to the start of measurement, because the detachment of samples from the felled tree enhanced their respiration rate to a considerable extent with the lapse of time. The negative effect of enclosing too much sample in a container, due to insufficient air circulation, was also carefully corrected for (Yoda, 1967).

The dark respiration rate of leaves decreased remarkably with the height from which samples were taken or with the mean intensity of light to which they were exposed at their original position in the canopy. As a result, the mean leaf respiration rate per tree increased with tree size or DBH (Fig. 24.7). The mean respiration rate (CO_2 production) in big trees (DBH > 20 cm) was 260–500 mg kg^{-1} hr^{-1} on a leaf fresh weight basis and 1.1–1.7 mg dm^{-2} hr^{-1} on a leaf area basis. These rates were consistent with the results of measurements with an infrared gas analyzer (Koyama, 1974).

Fig. 24.7. Dependence of mean dark respiration rate of leaves per tree on size of tree. Respiration rate expressed on fresh weight basis. (From Yoda, 1974b)

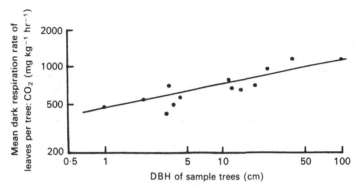

Fig. 24.8. Hyperbolic relations between wood respiration rate (fresh weight basis) and diameter of stem, branch, and root samples, corresponding to equations in text. A. *Shorea globifera*, DBH 32.4 cm. B. Mixed sample of small trees (DBH 3.1–4.2 cm) including *Hopea* sp., *Ochanostachys amentacea*, *Shorea pauciflora*, and *Shorea macroptera*. Cross, root samples. (From Yoda, 1974b)

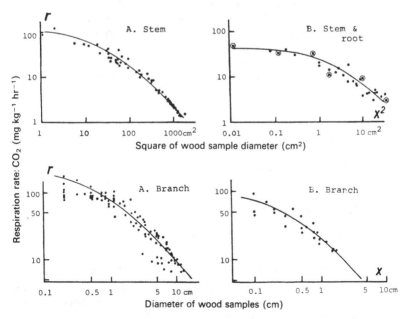

576   *Tatuo Kira*

Respiration rates of stem, branch, and root varied widely depending on their diameter, following the empiric formulation:

$$1/r = Ax + B \qquad \text{for branches}$$
$$1/r = Ax^2 + B \qquad \text{for stem and roots}$$

where r and x, respectively, refer to the respiration rate on a fresh weight basis and the diameter of woody organs (Fig. 24.8). Two coefficients, A and B, differed depending on the size (DBH) of the tree from which samples were collected. Yoda (1974b) calculated the total respiration in woody organs of a tree based on these formulations, considering the effect of tree size on A and B and the frequency distribution of x in the tree, which has been shown to follow a certain simple law (Shinozaki, Yoda, Hozumi, & Kira, 1964).

Combining all these results, the relation between DBH and total respiration per tree was worked out as shown in Fig. 24.9. The whole community respiration was then calculated for the same plot where the net production rate in Table 24.5 was obtained on the basis of DBH census and the relation of Figs. 24.7 and 24.9. The result given in Table 24.6 shows that leaves are responsible for about 53% of the total community respiration, although there is some evidence that the root biomass and root respiration were considerably underestimated because of the lack of actual biomass measurement.

Gross production rate

The gross production is here defined as the sum of net production and dark community respiration, disregarding the photorespiration, which is difficult to measure in situ. The gross produc-

Table 24.6. *Tentative estimates of community respiration and gross production rates in 0.8-ha plot in Pasoh Forest, given in t ha^{-1} yr^{-1}*

	Stem	Branch	Root	Leaf	Total
Respiration rate					
Trees (DBH > 4.5 cm)	5.2	13.1	6.1	26.1	50.5
Smaller trees and undergrowth		0.5	0.5	3.0	4.0
Total		18.8	6.6	29.1	54.5
Net production rate (trees only)		13.72	5.53	7.42	26.7
Gross production rate (trees only)		32.0	11.6	33.52	77.2

Respiration rates (at 25° C) were converted into corresponding rates of organic matter consumption, assuming a conversion factor, 0.61 gm dry matter/gm CO_2 respired.

Fig. 24.9. Respiration rates per tree of different organs in relation to tree size. (From Yoda, 1974b)

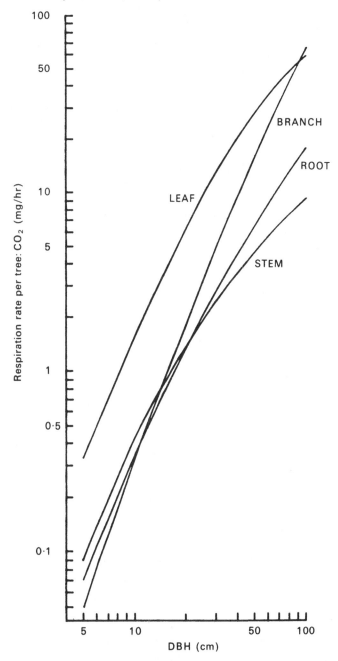

tion rate in the tree components was estimated at 77.2 t ha^{-1} yr^{-1}. Although the gross production estimates so far reported are not so reliable, particularly with respect to the assessment of respiratory consumption, that of Pasoh Forest is greater than most of the estimates in warm temperate (40–80 t ha^{-1} yr^{-1}), cool temperate (20–40 t ha^{-1} yr^{-1}), and boreal forests (20–50 t ha^{-1} yr^{-1}) (Kira, 1975).

The ratio of net production to gross production was as small as 0.35. Even smaller ratios were reported from a subhumid forest in Ivory Coast (0.26) and an evergreen seasonal forest of southern Thailand by Müller & Nielsen (1965) and Kira et al. (1967), but it is highly probable that they overestimated respiration rates owing to the delay of measurement after the felling of sample trees.

As reviewed by Kira (Kira et al., 1969; Kira, 1975), the gross production rate as well as LAI tend to be greater in forests than in herbaceous communities growing under similar environmental conditions. However, the ratio of net production to gross production may be as large as 0.5–0.6 in the latter, resulting in more or less similar net productivity in both types of plant community. The age and size distribution of trees in a forest stand are additional important factors affecting this ratio, as pointed out above (Kira & Shidei, 1967).

The very small net production/gross production ratio is not inherent in tropical rain forests, but is a common property of forest communities in general, in particular of mature climax forests dominated by a number of big, old trees.

Gross canopy photosynthesis
Net photosynthesis in detached leaves

Koyama (1974) obtained light-photosynthesis curves with detached sun leaves of several tree species at Pasoh. One of the tallest layer species, *Shorea leprosula*, showed a very high rate of net photosynthesis (CO$_2$: 20–30 mg dm^{-2} hr^{-1}) at light saturation. This species is known as a light-demanding pioneer in secondary succession, but is also common in undisturbed rain forests. It is interesting to find that other typical secondary forest species such as *Mallotus*, *Macaranga*, and *Glochidion* also exhibited fairly high photosynthetic rates ranging from 20 to 25 mg dm^{-2} hr^{-1} of CO$_2$. Shade-tolerant species growing up to the emergent layer (e.g., *Shorea macroptera*, *Shorea pauciflora*, *Dipterocarpus crinitus*, *Dipterocarpus sublamellatus*) were not different from temperate hardwoods in their photosynthetic capacity (CO$_2$: 10–15 mg dm^{-2} hr^{-1}).

Canopy photosynthesis model

Of various canopy photosynthesis models that aim at calculating total canopy photosynthesis from the rate of photosynthesis of single leaves, Monsi-Saeki's original model (Monsi & Saeki, 1953) was chosen because of its simplicity. It is based on two assumptions:

1. The vertical distribution of light flux density inside a leaf canopy is expressed by an exponential equation
2. All leaves have the same light-response curve given by a hyperbolic equation

The light distribution curve derived from Yoda's (1974c) observation is:

$$I = I_0 e^{-0.696F}$$

where I is light flux density at a given height in the canopy
I_0 is light flux density over the canopy surface
F is cumulative LAI between the height and the canopy surface

The light photosynthesis curve for *Dipterocarpus crinitus*

$$p = 7.422 \times 10^{-4} I'/(1 + 2.412 \times 10^{-5} I')$$

where p is rate of gross photosynthesis on a leaf area basis (mg dm^{-2} hr^{-1} of CO_2)
I' is light flux density on the leaf surface (lux)

is tentatively adopted as the basic equation of item 2. Then, Monsi-Saeki's model gives the rate of gross canopy photosynthesis (P) as:

$$P = \frac{7.422 \times 10^{-4}}{1.679 \times 10^{-5}} \ln \frac{1 + 1.679 \times 10^{-5} I_0}{1 + 1.679 \times 10^{-5} I_0 \exp[-0.696F^*]}$$

where F^* is the total LAI of the forest and assumed to be 8.0 ha ha^{-1}, including undergrowth leaves. A daily curve of I_0 obtained on a fine day in February 1973 (Yoda, 1974c) is utilized to estimate hourly means of I_0, which is then put into the equation to obtain hourly averages of canopy photosynthesis rate.

Assuming a dry matter/CO_2 conversion factor of 0.614, the daily gross photosynthesis on this particular day is estimated at 0.237 t ha^{-1} day^{-1}, which is equivalent to 86.5 t ha^{-1} yr^{-1}. More elaborate calculations would be possible, if more data on photosynthetic rates of leaves from various parts of forest canopy were available, using modified forms of Monsi-Saeki's original model proposed by Saeki (1960), Hozumi, Kirita, & Nishioka (1972), etc. For the moment, however, I can point out that this value is close to the estimate of gross production rate by the summation method, which would be something over 80 t ha^{-1} yr^{-1} if the production rates by smaller trees and undergrowth vegetation were taken into consideration (see Table 24.6).

Micrometeorologic methods

Aoki et al. (1975) made continuous 24-hour observations of micrometeorologic conditions in Pasoh Forest on November 7–8, 14–15, and 21–22, 1973. Anemometer, dry and wet bulb thermometer, and solarimeter were installed at six to eight height levels (0–53 m above the ground) on a walk-up tower constructed near the center of the IBP Research Area at a distance of about 800 m from the primary production study site. Air samples were also led from respective height levels into an infrared gas analyzer on the ground through a system of plastic tubes. The uppermost position of the instruments and air inlet (53 m) was well above the crown of the emergent tree (47 m tall) around which the tower was built. The net flux of CO_2 from the free air to the forest canopy was computed based on the energy balance method.

The authors obtained a CO_2 flux–incident radiation curve from the result of observations and thereby estimated the net daytime flux of CO_2 from the air to the canopy on a day receiving an average amount of radiation (347 cal cm^{-2} day^{-1}, annual mean for 1973 at Pasoh) at 36 kg ha^{-1} day^{-1}, assuming a sine curve for the daily change of solar radiation. This is equivalent to CO_2 flux of about 13 t ha^{-1} yr^{-1}. The negative flux of CO_2 during the night was 15 kg ha^{-1} hr^{-1}. In addition, there is another flux from ground surface to the air as soil respiration, which consisted of the CO_2 flux from the ground covered by fine litter (6.0 kg ha^{-1} hr^{-1}) and that from decaying big wood litter (2.4 kg ha^{-1} hr^{-1}) as stated later. The total annual soil respiration thus amounts to about 74 t ha of CO_2.

Considering the CO_2 budget of the forest, in which the horizontal flux due to advection was neglected because of the sufficient length of fetch (about 1 km), the rate of net canopy photosynthesis is expected to be the sum of the net daytime CO_2 flux and the soil respiration flux: $13 + 74 = 87$ t ha^{-1} yr^{-1} of CO_2. As the mean carbon content of wood was $49.1 \pm 1.0\%$, the dry matter/CO_2 conversion factor is 0.56 gm/gm. The net production rate in terms of dry matter is therefore estimated at 49 t ha^{-1} yr^{-1}.

The dark respiration rate of the above-ground parts of the forest is equal to the difference (CO_2 flux from the canopy to the air at night) minus (soil respiration), i.e., $15 - 6.0 - 2.4 = 6.6$ kg ha^{-1} hr^{-1} of CO_2. The annual rate becomes $6.6 \times 24 \times 365 \times 10^{-3} = 58$ t ha^{-1} yr^{-1} of CO_2. If the top/root ratio in respiration rate is assumed to be 7.3, according to Table 24.6, the total community respiration rate may amount to 66 t ha^{-1} yr^{-1} of CO_2, or to 40 t ha^{-1} yr^{-1} of dry matter consumption if the conversion factor 0.61 gm dry matter/gm CO_2 is applied to this case. This community respiration rate is considerably smaller than the estimate of Table 24.6 (54.5 t ha^{-1} yr^{-1}), whereas the micrometeorologically estimated net production rate (49 t ha^{-1} yr^{-1}) is nearly twice as large as the result of the summation method (Table 24.5: 24.7 t ha^{-1} yr^{-1}). An overestimate of soil respiration rate might be a possible cause for this inconsistency. If the mean soil respiration rate were smaller by one-third, the calculated rates of net production and community respiration would be about 35 and 57 t ha^{-1} yr^{-1} in dry matter. Further examination and improvement of the methodology of soil respiration measurement seem necessary in this connection.

Decomposition and carbon cycling
Accumulation of dead organic matter

The accumulation of dead organic matter as litter layer on the ground surface (A_0 layer) and in the mineral soil layer down to 1 m depth is shown in Table 24.7. The accumulation of big wood litter

Table 24.7. *Accumulation (tons per hectare) of dead organic matter on forest floor (A_0 layer) and in mineral soil of Pasoh Forest*

	Primary production study site (2 ha)	Mean of four 2-ha plots
Dry matter in A_0 layer		
Leaf litter	1.72	1.65
Small branch litter	2.18	1.90
Big wood litter	47.1	49.0
Other components	0.58	0.79
Total	51.6	53.3
Carbon in A_0 layer		
Leaf litter	0.79	0.76
Small branch litter	1.04	0.91
Big wood litter	23.7	24.6
Other components	0.14	0.19
Total	25.7	26.5
Carbon in mineral soil (1 m depth)	66.2	68.7
Total	91.9	95.2

Source: Yoda (1974a).

(diameter $>$ 10 cm) either lying on the ground or standing dead was vast (about 50 t ha^{-1} on average), whereas the amount of A_0 layer, consisting of fine litter only, was small (4.3 t ha^{-1}). The amount of organic matter in the mineral soil layer was about 69 t/ha^{-1} in terms of carbon, being 2.6 times as large as the carbon content of the A_0 layer.

If the linear reaction model proposed by Ogawa, Yoda, & Kira (1961) and Olson (1963) is applied to the dynamics of ground surface organic matter, the relative rate of disappearance of fine litter from the A_0 layer is calculated as (11.1 t ha^{-1} yr^{-1}) $\div$ (4.34 t ha^{-1}) = 2.56 yr^{-1} and that of big wood litter as (9.3 t ha^{-1} yr^{-1}) $\div$ (49.0 t ha^{-1}) = 0.190 yr^{-1}. The corresponding 95% disappearance time amounts to 3/2.56 = 1.2 years for fine litter and 3/0.190 = 15.8 years for big wood litter. Fresh, fine litter is expected to disappear almost completely from the ground surface within 14–15 months; newly fallen big wood litter may remain on the ground for 15–16 years before it disappears.

Soil respiration

Weekly observations of soil respiration were made with 30 pairs of Kirita's (1971) apparatus placed along the outer border of the 2-ha plot for primary production study. One of the pairs (an inverted cylindric tin container) was used for measuring the CO_2 output from undisturbed forest floor covered with fine litter; the other was used for determining the output from bare ground surface, where the A_0 layer had been completely removed. The difference between the two paired cylinders gave the amount of CO_2 evolved from the A_0 layer.

The method employed is a modification of Walter's (1952) apparatus, which consisted of an inverted box containing a certain amount of KOH solution placed on the ground surface. The original method was found to be subject to large error owing to the size of the box, the size of the KOH container, and the concentration of KOH used. The modified method used a piece of plastic sponge as the KOH holder to assure efficient absorption of CO_2. The apparatus was laid both on the undisturbed ground surface covered by litter and on the bare mineral soil surface after removal of all organic materials, in order to assess the CO_2 evolution from the ground surface litter layer. Removal of litter could increase the diffusion rate to and from the mineral soil, but no attempt was made to compensate for this.

The mean soil respiration rate of undisturbed ground surface based on 30 repetitions fluctuated mostly between CO_2 values of 500 and 700 mg m^{-2} hr^{-1} with a mean value of 595 mg m^{-2} hr^{-1} (Fig. 24.10). No apparent seasonal trend was recognized either in soil respiration rate or in soil surface temperature (24°–26° C), but there was some evidence that the moisture content of the A_0 layer and soil affected the rate of soil respiration.

On an average, about 76% of the total soil respiration was evolved from the surface of mineral soil. The annual totals were therefore, 52 t $ha^{-1} yr^{-1}$ of CO_2 for total soil respiration and 39.5 t $ha^{-1} yr^{-1}$ of CO_2 for mineral soil.

Beside this CO_2 flux from the open forest floor, an additional flux is supplied to the air by the decomposition of big wood litter. Yoneda (1975a,b) recently found that the rate of CO_2 evolution from decaying wood litter under natural conditions on the forest floor was more or less linearly correlated with the bulk density of the sample wood. Determining this relation experimentally and combining it with the amounts of accumulation of big wood litter of different grades of decay (see Table 24.4), Yoneda, Yuda, & Kira (1974) calculated the dynamics of big wood litter in Pasoh Forest as shown in Table 24.4 and Fig. 24.11.

The estimated mean rate of CO_2 evolution from big wood litter amounted to 317 mg $m^{-2} hr^{-1}$ or 28 t $ha^{-1} yr^{-1}$, which corresponded to the decomposition of dead wood of 15.5 t $ha^{-1} yr^{-1}$. The actually observed rate of supply of dead big wood to the ground (9.3 t $ha^{-1} yr^{-1}$) was much smaller than the estimated rate of decomposition. However, it is probable that the supply and decomposition are more or less balanced if the observation is made on a sufficiently wide area and over a sufficiently long period. The latter may have been too short in this study. In addition, our measurements may have overestimated the CO_2 flux from big wood litter to a certain extent, as only smaller branch wood (diameter 3-4 cm) was used for the measurements for technical reasons. Therefore, it was tentatively assumed that the real rate of decomposition of accumulating big wood litter

Fig. 24.10. Fluctuation of soil respiration rate in Pasoh Forest over period of 1 year. (From Ogawa, 1974)

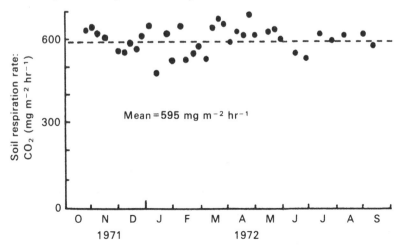

was about 12 t ha^{-1} yr^{-1}, a rate equivalent to a mean CO_2 evolution rate of 240 mg m^{-2} hr^{-1}.

The sum total of CO_2 supply from decomposing dead organic matter to the air is thus estimated at $595 + 240 = 835$ mg m^{-2} hr^{-1} or 74 t ha^{-1} yr^{-1}.

Carbon cycling in the soil

The dynamics of organic matter in the soil system in terms of carbon pools and flows can be approximated by the model shown in Fig. 24.12. Of the pools and flows that make up the model, the pools in the A_0 layer (M_0) and as humus in mineral soil (M) as well as the fluxes as fine litterfall (L), soil respiration (SR), and CO_2 evolution from mineral soil (SR_M) were actually determined at Pasoh. In the primary production study plot, these were:

fine litterfall (L) = 5.5 t C ha^{-1} yr^{-1}
soil respiration rate (SR) = 14.2 t C ha^{-1} yr^{-1}
CO_2 flux from mineral soil (SR_M) = 10.8 t C ha^{-1} yr^{-1}
carbon pool in A_0 layer (M_0) = 2.0 t C ha^{-1}
carbon pool in mineral soil (M) = 66.2 t C ha^{-1}

Fig. 24.11. Accumulation, rate of supply, and calculated rate of decomposition of big wood litter (diameter > 10 cm) in Pasoh Forest. Figures are averages of five 2-ha plots. (Recalculated from data of Yoneda et al., 1974)

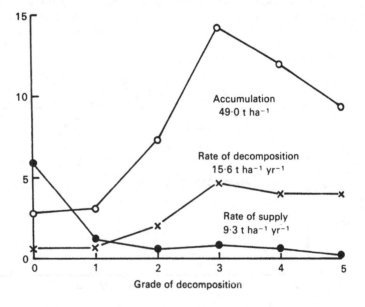

If a completely stationary state is assumed for this forest stand,

CO_2 flux from A_0 layer $(SR_A) = SR - SR_M$
$$= 14.2 - 10.8 = 3.4 \text{ t C ha}^{-1}\text{ yr}^{-1}$$

Downward transport of organic matter from A_0 layer to mineral soil $(V) = L - SR_A = 5.5 - 3.4$
$$= 2.1 \text{ t C ha}^{-1}\text{ yr}^{-1}$$

$k = V/M_0 = 2.1/2.0 = 1.050 \text{ yr}^{-1}$

Since $SR = L + L_R + R$,

$L_R + R = SR - L = 14.2 - 5.5 = 8.7 \text{ t C ha}^{-1}\text{ yr}^{-1}$

Fig. 24.12. A carbon cycling model in forest soils. (From Nakane, 1975)

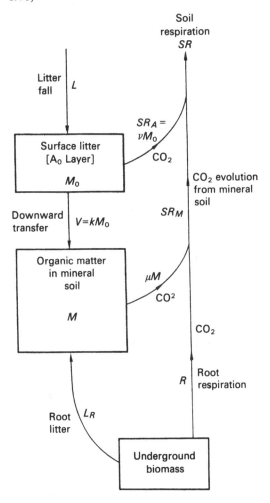

586   *Tatuo Kira*

If one of the two terms, L_R and R, is known, all parts of the model can be described quantitatively.

According to Tables 24.5 and 24.6, however, the carbon equivalents of root respiration (R) and root litter (L_R) are 2.9 t C ha^{-1} yr^{-1} and 2.5 t C ha^{-1} yr^{-1}, respectively, so that the sum (5.4 t C ha^{-1} yr^{-1}) is much smaller than the above figure. Some previous studies (Kucera & Kirkham, 1971; Edwards & Sollins, 1973) estimated the relative share of root respiration in the total soil respiration at 35%–40%. It is, therefore, tentatively assumed in the following treatments that the estimates of R and L_R were both underestimated and that the R/SR ratio equals 0.4.

Fig. 24.13. Carbon cycling in soils of Pasoh Forest and in warm temperate evergreen oak forest at Nara, central Japan. (Data for Nara Forest from Nakane, 1975)

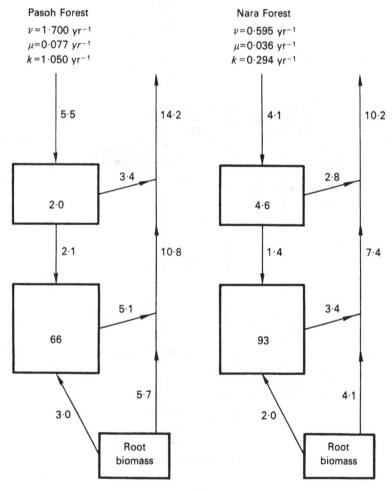

Figure 24.13 illustrates the carbon cycling system in Pasoh Forest soils and in soils of a warm temperate evergreen oak forest at Nara, central Japan. Both forests receive sufficient rainfall, but the annual mean temperature is higher at Pasoh (25° C) than at Nara (13.5° C). With the rise of temperature, the flows increase and the pools diminish in size. (Table 24.8). It is noteworthy that, in response to the temperature rise, all the pools and flows change at more or less the same ratio (1.2–1.5) except for the A_0 pool, which is apparently more sensitive than other parts of the system (2.3 in ratio). Nakane (1975) also found the same relation in the comparison of carbon cycling in the soils on different parts of a slope covered by evergreeen oak forest.

The 95% disappearance time in the A_0 layer, $3/(\nu + k)$, is 1.1 years at Pasoh and 3.4 years at Nara on a carbon basis, while the 95% decomposition time, $3/\nu$, is 1.8 years (Pasoh) and 5.0 years (Nara), respectively. That for humus in the mineral soil, $3/\mu$, is much longer, amounting to 39 years at Pasoh and 83 years at Nara.

The decomposition of big wood litter was neglected in the above analyses, because its influence on soil organic matter content seemed very local. However, considerable amounts of dead wood and leaf litter are carried by termites into their mounds (Abe, 1974; Matsumoto, 1974, 1976) and may result in another local cycling of carbon and nutrients characteristic of tropical ecosystems.

Table 24.8. *Effect of temperature on carbon cycling system in terms of ratios in carbon flux and carbon pools in Pasoh Forest (annual mean temperature 25° C) and Nara Forest (annual mean temperature 13.5° C)*

	Pasoh/Nara ratio in flux	Nara/Pasoh ratio in pool
Fine litterfall (L)	1.34	
Soil respiration rate (SR)	1.39	
CO_2 flux from A_0 layer (SR_A)	1.21	
CO_2 flux from mineral soil layer (SR_M)	1.46	
Transport from A_0 to mineral soil layer (V)	1.50	
CO_2 flux from organic matter in mineral soil (μM)	1.50	
Root litter (L_R)	1.50	
Root respiration (R)	1.39	
Carbon pool in A_0 layer (M_o)		2.30
Carbon pool in mineral soil layer (M)		1.41

Source: Data for Nara Forest from Nakane (1975).

Acknowledgments

This paper is Japanese Contribution No. 10 to the IBP Pasoh Forest Project, supported by the Japan Society for the Promotion of Science, involving the following team members: T. Kira (project leader), H. Ogawa, K. Hozumi, K. Yabuki, K. Yoda, R. Kato, Y. Tadaki, H. Sato, H. Kirita, H. Koyama, M. Aoki, and T. Yoneda. I am grateful to the other members of the Pasoh Forest Project for free use of their original data and for useful discussions.

References

Abe, T. (1974). The role of termites in the breakdown of dead wood on the forest floor of Pasoh Study Area. Malaysian IBP Synthesis Meeting, Kuala Lumpur, August 1974 (unpublished).

Aoki, M., Yabuki, K., & Koyama, H. (1975). Micrometeorology and assessment of primary production of a tropical rain forest in West Malaysia. *J. Agric. Meteor. (Tokyo)*, *31*, 115–24.

Bray, J. R., & Gorham, E. (1964). Litter production in forests of the world. *Adv. Ecol. Res.*, *2*, 101–57.

Edwards, N. T., & Sollins, P. (1973). Continuous measurement of carbon dioxide evolution from partitioned forest floor components. *Ecology, 54*, 406–12.

Golley, F. B. (1972). Energy flux in ecosystems. In *Ecosystem Structure and Function*, ed. J. A. Wiens, pp. 69–90. Corvallis: Oregon State University Press.

Golley, P. M., & Golley, F. B. (eds.) (1972). *Tropical Ecology with an Emphasis on Organic Production*. Athens: Institute of Ecology, University of Georgia.

Hozumi, K., Kirita, H., & Nishioka, M. (1972). Estimation of canopy photosynthesis and its seasonal change in a warm-temperate evergreen oak forest at Minamata (Japan). *Photosynthetica, 6*, 158–68.

Hozumi, K., Yoda, K., & Kira, T. (1969). Production ecology of tropical rain forests in southwestern Cambodia. II. Photosynthetic production in an evergreen seasonal forest. *Nat. Life S.E. Asia, Tokyo, 6*, 57–81.

Kato, R., Tadaki, Y., & Ogawa, H. (1974). Plant biomass and growth increment studies in Pasoh Forest. Malaysian IBP Synthesis Meeting, Kuala Lumpur, August 1974 (unpublished).

Kira, T. (1968). A rational method for estimating total respiration of trees and forest stands. In *Functioning of Terrestrial Ecosystems at the Primary Production Level*, ed. F. E. Eckardt, pp. 399–407. Paris: UNESCO.

– (1975). Primary production of forests. In *Photosynthesis and Productivity in Different Environments*, ed. J. P. Cooper, pp. 5–40. Cambridge: Cambridge University Press.

– (1976).*Introduction to Terrestrial Ecosystems, Handbook of Ecology*, vol. 2. Tokyo: Kyoritsu Shuppan. In Japanese.

Kira, T., & Ogawa, H. (1971). Assessment of primary production in tropical and equatorial forests. In *Productivity of Forest Ecosystems, Proceedings of the Brussels Symposium, 1969*, ed. P. Duvigneaud, pp. 309–21. Paris: UNESCO.

Kira, T., Ogawa, H., Yoda, K., & Ogino, K. (1967). Comparative ecological studies on three main types of forest vegetation in Thailand. IV. Dry matter production, with special reference to the Khao Chong rain forest. *Nat. Life S.E. Asia, Kyoto, 6*, 149–74.

Kira, T., & Shidei, T. (1967). Primary production and turnover of organic matter in different forest ecosystems of the Western Pacific. *Jap. J. Ecol., 17,* 70–87.

Kira, T., Shinozaki, K., & Hozumi, K. (1969). Structure of forest canopies as related to their primary productivity. *Plant & Cell Physiol., 10,* 129–42.

Kirita, H. (1971). Re-examination of the absorption method of measuring soil respiration under field conditions. IV. An improved absorption method using a disc of plastic sponge as absorbent holder. *Jap. J. Ecol., 21,* 119–27. In Japanese with English summary.

Klinge, H. (1974). Litter production of tropical ecosystems. Malaysian IBP Synthesis Meeting, Kuala Lumpur, August 1974 (unpublished).

Koyama, H. (1974). Photosynthesis studies in Pasoh Forest. Malaysian IBP Synthesis Meeting, Kuala Lumpur, August 1974 (unpublished).

Kucera, C. L., & Kirkham, D. R. (1971). Soil respiration studies in tallgrass prairie in Missouri. *Ecology, 52,* 912–15.

Matsumoto, T. (1974). The role of termites in the decomposition of leaf litter on the forest floor of Pasoh Study Area. Malaysian IBP Synthesis Meeting, Kuala Lumpur, August 1974 (unpublished).

– (1976). The role of termites in an equatorial rain forest ecosystem of West Malaysia. I. Population density, biomass, carbon, nitrogen and calorific content and respiration rate. *Oecologia (Berlin), 22,* 153–78.

Monsi, M., & Saeki, T. (1953). Ueber den Lichtfaktor in den Pflanzengesellschaften und ihre Bedeutung für die Stoffproduktion. *Jap. J. Bot., 14,* 22–52.

Müller, D., & Nielsen, J. (1965). Production brute, pertes par respiration et production nette dans la forêt ombrophile tropicale. *Det Forst. Fors. Danmark, 29,* 69–110.

Murphy, P. G. (1975). Net primary productivity in tropical terrestrial ecosystems. In *Primary Productivity of the Biosphere,* eds. H. Lieth & R. H. Whittaker, pp. 217–31. New York: Springer-Verlag.

Nakane, K. (1975). Dynamics of soil organic matter in different parts of a slope under evergreen oak forest. *Jap. J. Ecol., 25,* 206–16. In Japanese with English summary.

Newbould, P. J. (1967). *Methods for Estimating the Primary Production of Forests: IBP Handbook 2.* Oxford: Blackwell.

Ng, S. K., Thamboo, S., & de Souza, P. (1968). Nutrient contents of oil palms in Malaya. II. Nutrients in vegetative tissues. *Malaysian Agric. J., 46,* 332.

Odum, T. T., & Pigeon, R. F. (eds.) (1970). *A Tropical Rain Forest: A Study of Irradiation and Ecology at El Verde.* Oak Ridge, Tenn.: U.S. Atomic Energy Commission. 3 vols.

Ogawa, H. (1969). An attempt at classifying forest types based on the tree height-DBH relationship. In Interim Report of JIBP-PT-F for 1968, ed. T. Kira, pp. 3–17. In Japanese, mimeographed.

– (1974). Litter production and carbon cycling in Pasoh forest. Malaysian IBP Synthesis Meeting, Kuala Lumpur, August 1974 (unpublished).

Ogawa, H., Yoda, K., & Kira, T. (1961). A preliminary survey on the vegetation of Thailand. *Nat. Life S.E. Asia, Kyoto, 1,* 21–157.

Ogawa, H., Yoda, K., Ogino, K., & Kira, T. (1965). Comparative ecological studies on three main types of forest vegetation in Thailand. II. Plant biomass. *Nat. Life S.E. Asia, Kyoto, 4,* 49–80.

Ogino, K., Ratanawongse, D., Tsutsumi, T., & Shidei, T. (1967). The primary production of tropical forests in Thailand. *Southeast Asian Studies, Kyoto, 5,* 121–54. In Japanese.

Olson, J. S. (1963). Energy storage and the balance of producers and decomposers in ecological systems. *Ecology, 44,* 322–31.

Richards, P. W. (1952). *The Tropical Rain Forest.* Cambridge: Cambridge University Press.

— (1974). Pasoh rainforest in perspective. Malaysian IBP Synthesis Meeting, Kuala Lumpur, August 1974 (unpublished).

Saeki, T. (1960). Relationship between leaf amount, light distribution and total photosynthesis in a plant community. *Bot. Mag. (Tokyo), 73,* 55–63.

Shinozaki, K., Yoda, K., Hozumi, K., & Kira, T. (1964). A quantitative analysis of plant form: The pipe model theory. *Jap. J. Ecol., 14,* 97–105, 133–39.

Soepadmo, E. (1973). Progress report (1970–1972) on IBP-PT Project at Pasoh, Negri Sembilan, Malaysia. In *Proceedings of the East Asian Regional Seminar for the I.B.P., Kyoto,* ed. S. Mori & T. Kira, pp. 29–39. Kyoto: Japanese Natl. Committee for the IBP.

Tadaki, Y., & Kagawa, T. (1968). Studies on the production structure of forest. XIII: Seasonal change of litterfall in some evergreen stands. *J. Jap. For. Soc., 50,* 7–13.

Templeton, J. K. (1968). Growth studies in *Hevea brasiliensis.* I. Growth analysis up to seven years after bud-grafting. *J. Rubber Res. Inst. Malaya, 20,* 136.

Walter, H. (1952). Eine einfache Methode zur ökologischen Erfassung des CO_2-Faktors am Standort. *Ber. Dtsch. Bot. Ges., 65,* 175–82.

— (1955). Klimagramme als Mittel zur Beurteilung der Klimaverhältnisse für ökologische, vegetationskundliche und landwirtschaftliche Zwecke. *Ber. Dtsch. Bot. Ges., 68,* 331–44.

Yoda, K. (1967). Comparative ecological studies on three main types of forest vegetation in Thailand. III. Community respiration. *Nat. Life S.E. Asia, Kyoto, 5,* 83–148.

— (1974a). Carbon, nitrogen and mineral nutrients stock in the soils of Pasoh Forest area. Malaysian IBP Synthesis Meeting, Kuala Lumpur, August 1974 (unpublished).

— (1974b). Respiration studies in Pasoh Forest plants. Malaysian IBP Synthesis Meeting, Kuala Lumpur, August 1974 (unpublished).

— (1974c). Three-dimensional distribution of light intensity in a tropical rain forest of West Malaysia. *Jap. J. Ecol., 24,* 247–54.

Yoda, K., Shinozaki, K., Ogawa, H., Hozumi, K., & Kira, T. (1965). Estimation of the total amount of respiration in woody organs of trees and forest communities. *J. Biol., Osaka City Univ., 16,* 15–26.

Yoneda, T. (1975a). Studies on the rate of decay of wood litter on the forest floor. I. Some physical properties of decaying wood. *Jap. J. Ecol., 25,* 40–6.

— (1975b). Studies on the rate of decay of wood litter on the forest floor. II. Dry weight loss and CO_2 evolution of decaying wood. *Jap. J. Ecol., 25,* 132–40.

Yoneda, T., Yoda, K., & Kira, T. (1974). Accumulation and decomposition of wood litter in Pasoh Forest. Malaysian IBP Synthesis Meeting, Kuala Lumpur, August 1974 (unpublished).

25

Crown characteristics of tropical trees

P. S. ASHTON

Botany Department and Institute of Southeast Asian Biology,
University of Aberdeen, Scotland

Canopy structure, through its influence on the leaf area index, has a major influence on dynamic processes within a stand. Horn (1971) described the change in canopy structure that occurs during seral succession in broad-leaved temperate deciduous forest and speculated on its relation with changes in rates of both net and gross primary production. Though foresters in the tropics have long recognized that pioneer trees, in common with those of seral succession in temperate forests, generally grow more rapidly and "demand more light" (i.e., have compensation points at higher light intensities) than those of primary lowland rain forest, the canopy characteristics of tropical secondary communities, in which the large palmately lobed leaves of trees such as *Cecropia*, *Musanga*, and *Macaranga* are so conspicuously present, would appear to differ fundamentally from those of birch regeneration described by Horn (1971). Secondary forests are rapidly extending as the old primary forests are hewn and abandoned, and in this era of total utilisation some silviculturists have even assumed that secondary regeneration following clear felling can be managed as a productive, sustainable wood crop. Nevertheless though Schimper (1903) and Richards (1939, 1952), among others, have described principal features of tropical trees and succession, very little detailed work exists on woody seral succession in the tropics; no quantitative analyses of the changes associated with it have been carried out. Some pioneer hardwood species are being grown in plantations as sources of cellulose; in the Far East these include *Anthocephalus chinensis* (Lam.) Rich. ex Walp., *Eucalyptus deglupta* Bl., and *Albizia falcataria* (L.) Fosb. The last is reported to put on as much as 20 m in height over 2 years from planting as a sapling (A. V. Revilla, pers. comm.). However, Coombe (1960) and Coombe & Hadfield (1962) analysed the growth of individual pioneer species and showed that the unit leaf rate is surprisingly low in comparison with those of temperate herbs, and Stephens (1968) found that the net photosynthetic rate of A. *chinensis* is also comparably lower.

An account is presented here of a series of simple observational studies undertaken from July to October 1975, following several years' gestation of some ideas that finally bore fruit thanks to two happy circumstances: a stimulating discussion with Francis Hallé and Tom Givnish, followed by my first opportunity to teach a course in forest botany to honours students in the University of Malaya. This class undertook the drawing and analysis of the forest profiles as part of its practical work. Studies of individual trees were made in Malaysia by Aberdeen honours student, Alastair Wallace. We aimed (1) to describe accurately the distribution of leaf surfaces, modes of growth, and branching characteristics of selected secondary (gap phase) and primary (building and mature phase; Whitmore, 1975) forest tree species at comparable heights, and (2) to describe the distribution of leaf surfaces throughout the canopy of a stand of secondary forest and of trees in the building phase beneath a canopy gap of moderate size in primary forest. These observations were compared with estimates of leaf area index (LAI, the area of leaf surface above a unit area of ground) and leaf area density (LAD, the area of leaf surface per unit volume of space) of the whole mature phase forest profile at the same site measured by K. Yoda (in press). Comparative studies, made by the author, of girth and wood volume increments as well as structure in the same forest formation, but on a variety of soils, in Sarawak, East Malaysia, also existed. The totality of information allowed the development of a hypothesis to explain the adaptive significance of the different crown and canopy structures and leaf area distributions that are found within Malaysian lowland forest trees.

Our data are still too fragmentary to do more than pose problems, but they permit identification of lines for future research. Above all, I hope they demonstrate the unequalled opportunities for simple, yet exciting, work that can be pursued by university students in the tropics.

Methods
Single trees

Up to three "typical" free-standing trees of nine cultivated or secondary forest and five primary forest mature phase species were chosen for analysis; the latter were growing in shade. Habit was classified by means of the architectural models of Hallé & Oldeman (1970). Fifty leaves, representing the full range of sizes, were taken from each species and mean leaf area estimated. Leaf shapes and inclinations were noted. LADs from the ground to the outside of the canopy were measured along transects 1 m wide from the edge of the crown up to the trunk or through to the far side of the crown, by carefully removing branches outside the transect, erecting a trellis of poles, and counting the leaves within 1-m^3 cubes; from these the LAIs were calculated metre by metre along the transects. The method of measurement constrained us to select individuals below 5 m tall, though a few trees up to 9 m tall were measured by careful felling of their crowns. The leaf scars and axillary branches on the main stems were counted and expressed as a ratio, termed the "degree of ramification." The diameter and length of the tenth internode were estimated for 5–10 shoots per tree. Moisture contents for groups of 10 leaves and their corresponding stems were calculated for each species and expressed as a percentage of oven-dry weight.

Stands

Two sites were selected at Pasoh Forest Reserve, Negeri Sembilan, West Malaysia. The first was in a 20-m by 100-m rectangle of secondary forest (see Fig. 25.6). This had established and grown following destructive sampling, early in 1971, of the original primary forest (which still surrounds it) by members of the Japanese-Malaysian project led by T. Kira as part of the International Biological Program (see Chap. 24). Here a 17-m-long transect was established using essentially the same procedure already described; every plant was identified, and the mean leaf area was calculated for each species. The first 10 m of the transect consisted of a mixed-species stand, 4–5 m tall, fully exposed, and without an overhanging emergent canopy; the remaining 7 m, of similar height, was overhung at a height of about 35 m by the diffuse hemispheric crown of an emergent *Koompassia malaccensis* Maing. ex Benth. and led up to the margin of the unfelled forest. In order to elucidate the probable nature of the early establishment phase at this site, topsoil and litter were gathered from replicate 1-m^2 quadrants within the site, at its margin, and at a distance of 100 m and 500 m into the adjacent primary forest. These samples were spread on sterile builder's sand in drums, covered with cheesecloth, placed in the sun, watered, and observed at intervals over a 2-month period. Seedlings were counted and, where possible, identified and ascribed to obligate mature or gap phase species.

The second site was in primary forest and consisted of a stand of mature phase species, of unknown age but again mostly less than 5 m tall, growing in a roughly elliptic natural canopy gap with dimensions about 10 m by 15 m (see Fig. 25.7). Here a 10-m transect was measured along the longer diameter; it was overhung by emergent crowns throughout its length, though direct sunlight was experienced for 2.5 hours around midday. LADs and LAIs were here calculated also only for the first 5 m height of foliage.

Individual leading and lateral twigs of the commoner primary and secondary forest tree species, including palms, were tagged in July and observed periodically until mid-November to ascertain whether growth was continuous or intermittent during the period.

Results

A full discussion of the results is to be published elsewhere; they are here presented in the form of two tables and five figures. Those from individual tree species are summarised in Table 25.1; specific examples are illustrated in Figs. 25.1 and 25.2. The two transects are illustrated and variation in their leaf distributions summarised in Figs. 25.3 and 25.4, and a comparison of the ranges of variability in the "effective diameter" (width) of leaves – a variable defined by Horn (1971, p. 48) and below under "Discussion" – is made in Fig. 25.5. The results of the forest soil experiment are summarised in Table 25.2. Photographs illustrating the plots and some of the species referred to are included as Figs. 25.6–25.8.

Discussion

Macaranga gigantea (Rchb. f. & Zoll.) M.A. might appear to be the example par excellence of a pioneer tropical tree (Fig. 25.6). Its pithy, continuously growing, orthotropic stems and huge, broad-lobed, peltate leaves, held almost perpendicular to incident light on long petioles, recall the South American *Cecropia* and the African *Musanga cecropioides* R. Br. – the latter the subject of an analysis of growth by Coombe & Hadfield (1962). One notes that the large leaf size is associated with a low degree of ramification. In all these characteristics the contrast with the temperate pioneer birch, discussed by Horn, is extreme, and nowhere more than in the arrangement of its leaves in the crown. Horn described how the small-leaved birch can stack a high LAI into its deep, narrow crown as a "multilayered" species. Indices for 5-m-tall birch are unavailable, but J. C. Forbes (1973, pers. comm.) found that 1 acre of 50-year-old senile Scottish *Betula pendula* Roth. had an index of 2.8, with 820 current-year shoots per square metre of ground, and a 15-year-old stand about 5 m tall had 1250 shoots per square metre. As mean shoot length was greater in the latter example, the ratio between these two would underestimate the LAI of the young stand, which was not determined but which would probably be in excess of 4.5. Satoo (1970) had an index range of 4.0–5.5 for 47-year-old *Betula maximowiczii* Rupr., almost three times our mean for *M. gigantea* (Table 25.1).

Table 25.1. *Some characteristics of selected rain forest trees*

Species	Branching model	Leaf area index Mean	Leaf area index Range	Mean leaf area (cm²)	Mode of growth[a]	Degree of ramification	Average diameter tenth internode (mm)	Average length tenth internode (mm)	Moisture content (% fresh wt.) Twig	Moisture content (% fresh wt.) Lamina
Macaranga gigantea	Rauh	1.9	1.7–2.1	2200	C	1:36	27	49	87	70
Macaranga hypoleuca	Rauh	4.1	3.4–5.3	190	C	1:7	8	30	78	58
Glochidion sericeum	Cook	2.0	1.6–2.7	23	C	1:1.6	4	55	72	74
Breynia coronata	Cook	2.6	2.3–2.9	8	C	1:1.8	4	19	67	59
Muntingia calabura	Troll	2.77	2.3–3.1	13	C	1:8.7	2	14	72	74
Dillenia suffruticosa	Attims	4.5	3.9–5.3	240	C	1:2.1	8	45	82	80
Carica papaya	Corner	2.3	2.0–2.8	2800	C	None	47	15	75	64
Alstonia angustiloba	Prévost	2.3	1.8–2.8	35	I		17	119[b]	75	73
Terminalia catappa	Aubréville	2.9	2.3–3.6	180	I		12	2[c]	63	67
Semecarpus curtisii	Chamberlain	0.9	0.8–1.0	500	I?					
Dipterocarpus cornutus	Massart	1.3	1.2–1.4	560	I	1:8.9	10	72		64
Shorea maxwelliana	Roux	0.9	0.8–1.0	36	I	1:1.5	4	39	50	53
Hopea pubescens	Roux	1.85	1.7–2.0	10	I	1:1.6	1	17	53	52
Durio griffithii	Roux	1.6	1.3–1.9	44	I	1:1.6	3	28	60	63

[a]C, continuous; I, intermittent.
[b]Between verticils.
[c]Towards twig apex.

However, not all tropical pioneer trees fit this description. One measured individual of *Macaranga hypoleuca* (Rchb. f. & Zoll.) M.A. had an LAI as high as 5.3, and the average for this species is quite comparable to that for birch. It is noteworthy that it shares the same branching model as *M. gigantea;* its smaller leaves, associated with a higher degree of ramification, would appear to explain the higher LAI (Fig. 25.6). The crown of *M. gigantea* is broad and shallow; that of *M. hypoleuca* is deep, with moderate though declining LAD maintained to the base. Horn pointed out that the distance the umbra is

Fig. 25.1. Distribution of leaf area in some Malayan trees. Leaf area densities calculated in square metre per cubic metre in cubes of 1-m depth, but varying volumes.

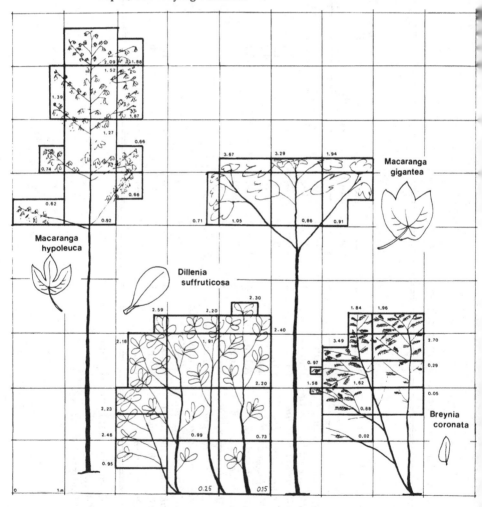

cast beneath a leaf is proportional to the diameter of the largest circle
that can be fully inscribed within its margin; this is known as the
"effective diameter." R. A. Sunderland (unpubl.) found that leaves
of seedlings at the Pasoh Forest floor failed to assimilate in full shade,
and this will almost certainly be so for light-demanding pioneers.
The mean effective diameter of *M. gigantea* leaves was 36 cm; that of
the narrower lobes of leaves of *M. hypoleuca*, only 7 cm. Nevertheless, *Glochidion sericeum* Hk. f. and *Breynia coronata* Hk. f. (Fig. 25.6),
both in the same family as *Macaranga* but belonging to Cook's
model, have small leaves with a mean effective diameter of at most
3.5 cm, a high degree of ramification, but a surprisingly low LAI.
Clearly other factors are responsible; the concentration of LAD in the

Fig. 25.2. Distribution of leaf area in some Malayan trees. Leaf area
densities calculated in square metre per cubic metre in cubes of 1-m
depth, but varying volumes.

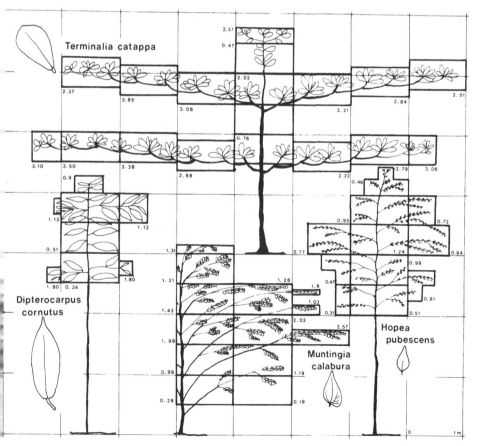

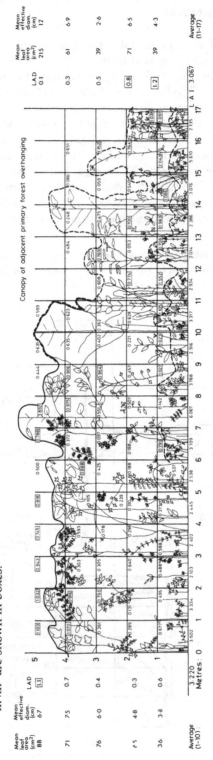

Fig. 25.3. Pasoh Forest, Japanese IBP destructive sampling plot within 2-ha plot 1. Figure represents 4-year-old secondary forest on 17-m by 1-m profile diagram, indicating leaf area density (LAD) and leaf area index (LAI) of all plants below 5 m. Quantities over 0.7 m²/m³ are shown in boxes.

Table 25.2. *Germination of seeds from forest topsoil planted July 26, 1975*

Habitat	No. of individual seedlings on				Mature phase species	Gap phase species
	Aug. 26 1975	Sept. 11 1975	Oct. 3 1975	Total		
Four-year second-ary forest						
site 1	26	33	11	70	2	68
site 2	18	1	6	25	0	20
Primary forest edge						
site 1	26	20	9	55	0	38
site 2	18	35	15	68	3	50
100 m within pri-mary forest						
site 1	8	11	6	25	2	16
site 2	36	40	8	84	4	53
½ km within pri-mary forest	14	29	12	55	4	20
Total number of individuals	146	169	67	382	15	265

Fig. 25.4. Pasoh Forest primary forest beside main path at plot 3 access. Figure represents 10-m by 1-m profile diagram, indicating leaf area density (LAD) and leaf area index (LAI) of forest regeneration to height of 5 m below gap of unknown age. Canopy of adjacent mature phase forest overhangs throughout. Quantities over 0.7 m²/m³ are shown in boxes.

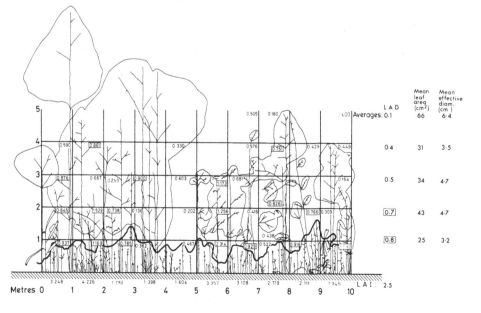

top of the crown implies high compensation points, but may also partially be explained by the densely distichous leaf arrangement on nearly horizontal twigs, with the blades on average hanging at 5° below the horizontal in *Glochidion* and 7° in *Breynia,* as compared with 32° in *M. gigantea.* Leaf blades of *M. hypoleuca* were observed hanging at 67° in the sun at midday, but were versatile and noticeably more horizontal in the shade, morning, and evening. *Muntingia calabura* L., with leaves of a similar effective diameter as *Glochidion,* though a lower degree of ramification on the similarly horizontal twigs, has a somewhat higher LAI, which may be ascribable to the arrangement of its branches into distinct, more or less horizontal layers, each placed where a decline will have occurred in the area occupied by umbrae cast by the leaves above; the twigs and leaves in direct sun at the sides of the crown are descending, and their upper surfaces are, therefore, also nearly perpendicular to incident radiation. *Dillenia suffruticosa* (Griff.) Martelli has an exceptionally high LAI in view of its large leaf size (Fig. 25.8). In this tree the internodes are long and the leaves are short-stalked and inclined

Fig. 25.5. Change in proportion of leaf sizes through profile of secondary forest and some primary forest regeneration in Malaya.

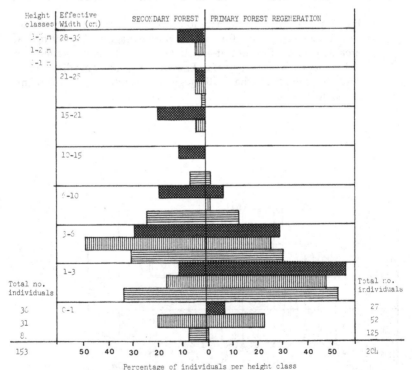

Percentage of individuals per height class

obliquely upwards to incident radiation, thus reducing their umbrae. Rather high leaf densities are sustained for some depth into the
crown. In contrast, though the large leaves of *Carica papaya* L. are
deeply dissected, they are borne almost perpendicular to radiation
on long petioles arising between very short internodes, and LAI is
expectedly low.

Fig. 25.6. Pioneer species of artificial gap studied at Pasoh Forest,
including *Macaranga hypoleuca* (pale stem, right centre), *Macaranga
gigantea* (centre, large leaves), and *Breynia coronata* (distichous
leaves above figure).

Fig. 25.7. Path passing through corner of regeneration of mature phase species in natural gap studied.

Plagiotropic branching systems allow a tree to acquire stature rapidly. Pagoda trees such as *Terminalia* and *Alstonia* and many others are a distinctive feature, especially of later seral succession, but they ultimately develop broad, hemispheric, emergent crowns. Such wide-spaced plagiotropic branching patterns allow the canopy of leaves, borne densely on the horizontal branches, to be extended above competitors by the economy of a single, bold, vertical shoot

Fig. 25.8. Pioneer species of xeric conditions on waste land: *Dillenia suffruticosa* (foreground); *Fragaea fragrans,* a monopodial species with high leaf area index (above right), and pagoda tree *Alstonia angustiloba* (above left) beginning to overtop canopy.

extension, and the leaves are thereby exposed to high light intensity from all directions. As T. J. Givnish (pers. comm.) predicted, they are generally obovate, tightly whorled, and held steeply above or below the horizontal. Nevertheless, these dense horizontal swards would be expected to sustain a high heat load around midday, and it may be significant that many such species are deciduous in a predominantly evergreen tree flora.

As all the mature phase individuals studied were growing at least partially in the forest shade, it is not surprising that they have lower LAIs than the pioneers. Mean leaf area is clearly variable among mature phase trees in the understorey too, though broad and lobed leaves are conspicuously absent. What first distinguishes these understorey crowns is the absence of a dense outer layer of leaves and their rather even distribution through the deep, narrow crowns to achieve, in fact, surprisingly high LAIs. This must in some measure, of course, be ascribed to their generally low compensation points, but the analogy with birch is obvious.

Once again, leaf area of mature phase species is correlated with the degree of ramification; for all trees studied, both gap and mature phase, this correlation was calculated at $r = 0.968$ (99.9% confidence limits). Twig diameter was similarly correlated with degree of ramification ($r = 0.927$) and leaf area ($r = 0.978$). Internode length was not found to be correlated with any of these characters, but was very variable within individuals so that this result might be due to sampling error.

The profile of secondary forest suggests how crowns are modified in competition to contribute to the overall canopy characters of a forest. In the first 10 m there is a striking concentration of high LAD in the top of the canopy, where it is as high as in herbaceous vegetation, but there is a rapid decline beneath. There is a slight increase near the ground, which is contributed by herbs and a few reinvading mature phase seedlings. The mean LAI would have been as low as 2.9 were it not for one *M. gigantea*, at 8 m, with strong apical dominance, which has stolen a march on its rivals and built up an LAI of 3.9 above 4 m (this individual figure represents total leaf area above 4 m in a space exceeding 1 m³). This compares with LAIs of 8 for the whole forest that originally occupied the site, of about 4 for the main canopy above 35 m (R. Kato, Y. Tadaki & H. Ogawa, pers. comm.), and of about 3 for an individual emergent leptophyll crown of *Shorea curtisii* Dyer ex King (J. B. Kenworthy, pers. comm.). Between 10 and 17 m, where the secondary forest abuts primary forest and an emergent overhangs, the mean LAI is only slightly lower, but the LAD is markedly differently distributed. The canopy is uneven and

LAD lowest at the top, increasing to the highest figures in the lowest metre, where there is a high density of primary forest species establishing from seed dispersing in from the margin. These reinvading species will be more shade-tolerant, and LAD will be augmented by stagnant mature phase seedlings persisting by means of their stored seed reserves. The upper levels of the canopy here are still composed overwhelmingly of pioneers, and it is, therefore, surprising how similar the distribution of foliage is here to that in the profile through the primary forest building phase, where no strictly pioneer species were represented. This latter also resembles the estimates of Kato et al. concerning variation in LAD at these heights in the original mature phase stand, though densities are higher at all levels in our examples, as should be expected.

Mean leaf areas and effective diameters are lower at all heights in the profile of primary forest regeneration than in that of the secondary forest, and if the high uppermost figure caused by domination of the small sample by a single *Lithocarpus* is ignored, there is a slight suggestion of a peak immediately above the ground layer. Comparison of the range in effective leaf diameter between the first 10 m of the secondary forest profile and that of the primary forest building phase reveals that it is greater in the former at all heights, although it is there skewed towards the larger diameters. The peaks in the distributions are at a higher diameter in the canopy of the secondary forest, but unexpectedly, they are the same in the ground layer in both situations.

Conclusions and some hypotheses
Architectural models and their adaptive significance

Though the primary forest species chosen conformed to architectural models different from those of the secondary forest, the models of Hallé and Oldeman were not, in general, confined to a particular stage in succession, though some of the simpler models, such as Tomlinson's and Leeuwenberg's, would appear to be rare in shady habitats. Thus the shade-tolerant *Semecarpus curtisii* Dyer ex King belongs to the same model as the secondary forest *Oroxylum indicum* Vent.; many understorey Euphorbiaceae belong, like *Macaranga*, to Rauh's model; Aubréville's model is represented by *Ploiarium alternifolium* (Vahl) Melch. in secondary, and the genus *Palaquium* in primary, forest; and Troll's model is found in every conceivable tropical habitat within the ubiquitous family Leguminosae.

However, Hallé and Oldeman's classification of architectural models is based primarily on three variables: whether growth is continuous or intermittent; whether all axes are equivalent and orthotropic or whether they are differentiated (i.e., some plagiotropic, some orthotropic); and whether inflorescences terminate the modules or are axillary and thus do not restrict their continued elongation. In our secondary forest profile sample, the leading shoots of 41% of the 39 tree species included grew continuously during the period they were under observation. Probably all mature phase emergents grow intermittently, and only 13% of the 55 species in our building phase profile grew continuously during the period of observation. Nevertheless, the rhythmic branching pattern of *Macaranga*, for instance, conforming to Rauh's model, was produced by an apex that was continuous in its growth. Rhythmic growth was judged to exist by the distribution of axillary buds that had formed lateral shoots. Conversely, though the plagiotropic branches of *Shorea maxwelliana* King and *Hopea pubescens* Ridl. were not regularised and conformed to Roux's model, in which branching is acclaimedly continuous, the actual process of shoot extension was markedly intermittent and, in the latter at least, synchronised throughout the sapling populations. Undifferentiated orthotropic branching can be seen as a means of rapidly, yet economically, extending the height of a broad crown, advantageous in the competitive early stages of species-rich tropical forest succession. Givnish & Vermeij (1976) pointed out that vertical and inclined branches bear less lateral stress for a given load than horizontal. Differentiated branching systems, however, can be seen to confer advantages, though different, in both the successional canopy and in the mature phase understorey. In the former, they allow the successive extension of a sequence of dense photosynthesizing planes above neighbouring competitors; in the latter, as Givnish & Vermeij (1976) have predicted, they provide an economic means of arranging small horizontally oriented leaves more or less randomly in space. The positions of inflorescences would not appear to bestow differential advantages in different seral stages. The direct adaptive significance of the diversity of architectural models in tropical forests, therefore, remains for the most part complex and obscure, but it may be explained more readily when reiteration is considered (see Chap. 23). It would appear that Hallé & Oldeman (1970) may be right in inferring, as has Corner (1949), that angiospermous trees evolved from unbranched forms, perhaps monocarpic and with terminal inflorescences, and that the variation that exists among modern tropical trees in the positioning of inflorescences reflects constraints imposed by their ancestry. The fact that a given leaf arrangement can be achieved by a variety of branching patterns may be explained, at least partially, in terms of the variety of evolutionary stages reached in the positioning of inflorescences.

Adaptive arrangements of leaf surfaces

Horn (1971) observed that the understorey of temperate deciduous forest on some sites tends to be dominated by species with single dense layers of leaves (monolayers). Givnish (pers. comm.) suggested that large thin leaves have the advantage in the understorey of increasing the likelihood of at least some chloroplasts interrupting sunflecks and that large leaves are more economic energetically than many small leaves wherever there is an advantage in developing temporary structural members. Whereas Horn discussed variation in leaf size and arrangement during seral succession in relation to their effectiveness in utilising light, Parkhurst & Loucks (1972) assumed that leaf sizes are adjusted to minimise the amount of water transpired per unit of CO_2 taken up; they also concluded that shade leaves should be larger than sun leaves. Givnish & Vermeij (1976) have created a mathematic model that, by weighing transpirational costs against photosynthetic gains, predicts the variation that might be expected in leaf size, shape, and inclination among lianes from sunny to shady and xeric to mesic habitats. In mesic and intermediate conditions, they predict that the largest leaves should occur in moderate shade in the middle layers of the forest; the smallest should be in the understorey, as larger leaves will impede transpiration in the deep shade, whereas the effects on photosynthesis of leaf temperature and resistance are likely to be small. Further, they predict that leaf sizes will be most variable in the understorey, as the absolute difference between photosynthetic gains and transpirational costs is likely to depend only weakly on leaf size under the prevailing conditions. In the canopy there should be a decline in mean leaf size from mesic to xeric conditions; in the understorey, a slight increase.

The building phase profile consists of species that would normally become established beneath the mature phase canopy before a gap was formed; it does not, in fact, fit Givnish and Vermeij's prediction for a mesic or intermediate site, as effective leaf diameters are neither more variable nor smaller near the ground than higher up in that profile. Systematic collections of leaves from emergent and main canopy trees in the course of enumeration nearby in the same forest demonstrate that the mean leaf area and effective width are lower, as is generally stated, at least than at middle elevations in our profile. Kato et al. (pers. comm.) obtained a lower LAI estimate (about 1) than we did for the understorey below 5 m tall in the mature phase of the same forest. We see, then, that the conditions of low light and high humidity that prevailed before canopy opening have favoured establishment of a structurally heterogeneous understorey of deep-

crowned, rather small-leaved trees, which, under the changed conditions created by a break in the canopy, have been able to build up an
unexpectedly high LAI.

But how can the secondary forest, which was growing in an essentially similar habitat throughout the length of our profile (the felled
area being but 20 m wide), be explained by Givnish and Vermeij's
model? Clearly, the difference between the understorey in the two
profiles must be due to differences in the two canopies above; the
higher mean leaf size in the canopy of the more exposed secondary
succession suggests that the photosynthetic gain is great relative to
the transpirational cost here, whereas the higher variance of both effective leaf diameter and shape implies that here too the balance between photosynthetic gain and transpirational costs is only weakly
dependent on leaf size; the balance, by inference, can be achieved by
a range of alternative combinations of characteristics. We might even
deduce that the secondary forest canopy experiences less water stress
than that of the primary building phase. This would appear unlikely;
is there an alternative explanation?

The mature leaves of an emergent *Shorea curtisii* are small, sclerophyllous, shiny above and pale waxy beneath, and held at a steep inclination to incident radiation. Kenworthy (1971) found that they apparently lacked stomatal control, and he calculated indirectly (pers.
comm.) that an individual crown would have an LAI of approximately 3, equivalent to the whole of our secondary forest. Leaves of
its shade-tolerant sapling are large, thin, mesophyllous, slightly pen-
dent, and with sensitive stomatal control. We have found stomatal
control to be highly sensitive in *Macaranga tanarius* (L) M. A. Kenworthy (pers. comm.) found that stomata of several other pioneer
species studied in Malaya closed at night, opened in the morning,
and closed again gradually through the afternoon. Most secondary
forest species have stout, pithy, and energetically cheap stems, confirmed by our determinations of percent water content, and often
seem to be furnished with larger vessels than those of mature phase
canopy trees. They are generally short lived. The great hemispheric
crown of the emergent tree allows the sun to intercept only part of it
at a time and thus spreads the heat and transpirational load through
the day. Kenworthy has also shown that the columnar trunk accumulates water during the night sufficient to act as a reservoir
through the day following. The pioneer tree stem can hardly provide
such a store, particularly in view of its leaf and crown characters. Yet
the leading shoots of many pioneer species grow continuously.

Givnish & Vermeij (1976) would argue that where transpiration
costs are high – as in any canopy – it may become advantageous for
a plant to trade off root costs for drier leaves; if a plant does this,

however, it must be at the price of reduced photosynthetic potential. I suggest that the mature phase canopy trees are adapted principally to long-term survival at maturity, during which the principal limiting factor will be water. This they can only do at the cost of losing the ability to respond in the short term, which will reduce their unit leaf rate (net assimilation rate). Individuals of our pioneer species are adapted to maximise unit leaf rate, which is high in comparison with other trees even if not with herbs, by having large leaves, flat dense crowns, stomatal sensitivity, large vessels, and continuous growth; but prolonged drought would prove disastrous, though I know of no observations to support this. The inference that increase in mean diameter of vessels increases rate of water flow is contentious, as this will also depend on the number of active vessels and has never been confirmed; yet it will certainly increase the risk of embolism in tall trees. I have often wondered how both temperate and tropical gymnosperms, though lacking vessels and often evergreen in seasonally arid climates, have no peers once they have overtopped the angiosperm canopy. Givnish and Vermeij also alluded to the increase in respiratory demands with increasing heights of crowns. It would be particularly interesting to undertake a detailed study of the changes that occur in species such as *Alstonia* or *Bombax* that have the characteristics of pioneers when young, yet finally develop those of the mature phase emergent.

Coombe & Hadfield (1962) showed that unit leaf rate in *Musanga cecropioides* is greater when LAI is low, and in mixed stands competition combines with continuous growth to ensure that this is generally so. But this does not imply that net primary production is high in such secondary forest; quite the converse, for the LAI may more than compensate for high unit leaf rate. So far as I am aware, no data have been collected on this vital subject. It can be seen in Table 25.2 that the number of small size classes is greater in the primary forest building phase than in the secondary forest. Foresters know that more wood can be grown in plantations of many small than a few big trees. How can this unexpectedly low productivity in young secondary forest be explained in terms of natural selection? If we assume that interspecific selection dominates secondary succession, then the adaptations described would be expected, for instead of sustaining a high rate of net primary production, such selection might favour those species that can most rapidly establish a dense light-excluding crown above their competitors; this accomplished, high net primary production becomes irrelevant. Such a hypothesis presupposes a level of determinism at seedling establishment. In the mature phase understorey this can be provided through the maintenance of a diversity of seedlings on the ground surviving on stored seed reserves, but pioneers produce many small seeds lacking such a store. Our investigation of forest soil showed that even deep within the primary forest there are plenty of seeds of a variety of pioneer species dormant in the soil (though it is not known how long this dormancy lasts), and these respond as soon as light intensity increases and/or humidity decreases. Such seed dormancy could even provide a means of survival during occasional severe drought.

Givnish & Vermeij (1976) considered the importance of interspecific competition in determining the leaf characteristics of lianes, and the competitive as well as photosynthetic advantages of large broad-based leaves arranged perpendicularly to incident radiation. However, lianes do not conform with the system here described and are at first sight an enigma in that they often bear exceptionally large leaves high in the canopy. Many, though of great stature, are noted for the large size of their vessels. Are these properties associated in them too with a short life span? Givnish and Vermeij's estimates of the optimal size, shape, and orientation of liane leaves are calculated on the assumption that their positions are determined by light conditions created by the trees on which they grow, and not by each other, in marked contrast to those of free-standing trees. They contend that in lianes, which are not self-supporting, a premium is set more than in other plants on producing photosynthetic tissues that can simultaneously maximise growth and shade out competitors, but this does not explain how a spendthrift economy can be maintained at great heights above the ground (unless the vines are provided with aerial roots). Nevertheless, it would appear that the leaf characteristics of pioneer tropical trees are largely determined by the same factors and are free from the same constraints, whereas "within-tree leaf competition" may have selective value in long-lived mature phase trees, at least when young. Thus the broad peltate leaf, inclined perpendicular to incident radiation on a long petiole that is analogous to a branch of limited growth, and weak, hardly branching stems are found in short-lived *Macaranga gigantea* as much as in herbaceous climbing *Merremia*. It is significant that a huge, almost round peltate leaf on a long petiole, the ideal shape for rapid photosynthesis in the absence of transpirational constraints, is attained by *Macaranga tanarius*, confined to the most fertile soils and mesic sites. By contrast, the branching habit and leaf arrangements of mature phase saplings may be explained by the need to build structural members that continue to contribute to the tree as it slowly grows into the canopy. Here again the pagoda tree shape provides a versatile branching system that fits it alone for both pioneer and climax.

Why, then, does our thrifty Scottish birch not emulate baroque

Macaranga, for evaporation will be low in the cool boreal summer? The leaf surface boundary layer is diminished by air movement to such an extent that it is independent of leaf size at wind speeds in excess of 5 m/second, above which speed size has mechanical disadvantages. The absence of persistent wind in the humid tropics of West Malaysia would appear to be a crucial factor that may also explain the rarity of macrophyll pioneer species in continuously or seasonally windy tropical humid climates such as the windward eastern slopes of Luzon's Sierra Madre or parts of Sri Lanka.

An analogy with the canopy characteristics of the secondary and primary regeneration of Pasoh is provided by comparing the structure and growth of primary rain forests on a variety of soils in Sarawak; a full description of this study must await publication elsewhere. It was found that net production of wood by volume on a shallow and very freely draining leached soil and a deep fertile loam was almost the same, even though the standing volume on the latter was almost double that on the former. There was no significant difference in mean girth increment in any size class between the two forests, though sample error among the relatively few larger trees may have obscured the lower girth increments that might be predicted on xeric sites. These results can be explained by the fact that the mesic site carried a dense stand of large emergent trees that accounted for most of the wood production, the understorey being diffuse; the xeric infertile site carried such a dense stand of small trees that their wood production offset the absence of forest giants. Mortality, of course, was also very high among this dense understorey. Students are presently in the field attempting to estimate the LAI and LAD in these contrasting forests whose structural differences appear closely analogous to those observed by Horn in beech forest on mesic and oak-hickory forest on xeric sites in New Jersey. It is noteworthy that secondary forest on shallow, infertile, xeric tropical soils includes few macrophylls and shares with the mature canopy the dominance of small, sclerophyllous, short-stalked leaves steeply set above the horizontal. One recalls also that tropical pines and eucalypts, which are among the most favored fast-growing timber trees and noted for their high LAI, are xerophytic pioneers, and that of the trees studied by us, *Dillenia suffruticosa*, *Macaranga hypoleuca*, and *Muntingia calabura* have the greater edaphic tolerance.

Horn observed the dominance of species with dense monolayered crowns, such as *Viburnum acerifolium* Bong. and *Podophyllum peltatum* L., in the understorey of New Jersey oak-hickory forest. One might anticipate that natural selection would favour evolution of similar species in the tropics. Whereas they may be deciduous in the temperate zone and thus potentially allow a period during which canopy tree seedlings can compete, this would not be likely in a rain forest, and such a dense understorey could lead to the eventual extinction of the canopy species. Burmese teak forests frequently have a dense understorey of bamboos, which prevent regeneration; but the bamboos flower gregariously and are monocarpic, so that the whole understorey periodically dies and tree regeneration occurs at such times. In the mountains of tropical Asia the genus *Strobilanthes* behaves similarly. The only example that comes to mind in lowland evergreen forest is *Eugeissona tristis* Griff., the stemless bertam palm that forms dense thickets of large, narrowly pinnate leaves on upper slopes and ridges in Malaya. It has long been known to inhibit regeneration of many canopy trees, but what prevents such a palm from eventually breaking the forest cycle? One suspects that the palm itself cannot persist in direct sunlight. An autecologic study of this species is overdue.

General discussion

Whitmore: *Macaranga hypoleuca, Macaranga pruinosa,* and *Macaranga hosei* in Malaya, which can be found as small secondary forest treelets, reach a height of 40 m (i.e., they occur in the two different niches). How can this be squared with your ecophysiologic approach?

Ashton: They do not maintain the dense flat crown characteristics when they become large.

Givnish: Did you find that the small-leaved *Macaranga* is restricted to the more xeric habitats?

Ashton: It is not restricted, but it occurs also on the more xeric habitats, as far as I could see, and more so than *M. gigantea.* I was surprised how wide the range of *M. gigantea* was; this is difficult to understand.

Givnish: Is the leaf thickness of the small-leaved species greater than that of the broad-leaved species?

Ashton: No.

Hallé: First, do you agree that within one particular architecture the main source of ecologic variability is the possibility of reiteration? Second, if the pagoda tree is so efficient, why is it nearly excluded from temperate forests?

Ashton: To the first, I cannot give an answer because I have been mainly studying in this example the young vegetation where the level of reiteration is low. The second point is important. It seems to me that we underestimate the mean wind velocity in temperate forests. Even a wind of 5 m/second is sufficient to eliminate the advantage of a large over a small leaf. Another aspect of wind is that broad, flat, large-leaved crowns are unstable in a windy climate. I have seen a *Cecropia* plantation in Mindanao flattened by quite a light typhoon. It would be the same with a *Terminalia*-like tree with a flat, large-leaved crown.

Baker: And yet *Terminalia catappa* is found on windy seashores.

Whitmore: And the tallest tree of the Solomon Islands is *Terminalia calamansani*, which is also characteristic of regrowth forest after cyclones.

Ashton: The point about *T. catappa* is an interesting one because of Dr. Fisher's photographs in presumably a rather windier climate than that of coastal Malaya. The shape is very much modified, and the mature crown configuration develops much earlier, so I believe wind modifies crown shape. I commented earlier that the distribution of leaves might have an effect on their aerodynamics. In regard to Dr. Whitmore's comment, I don't know the full circumstances. Can you tell me, is this a tree which does, in fact, suffer a lot from wind damage?

Whitmore: In a cyclone that hit Kolombangara Island in the western Solomons we investigated which tree species were most damaged. More trees of *Campnosperma brevipetiolatum* were blown over than any other species. The rest were all about the same (Whitmore, 1974). So the conspicuously damaged species was not *Terminalia*.

Ng: Have we, in fact, established that reiteration is less common in pioneer species than in the primary forest species?

Oldeman: In Ecuador, for instance, where tropical growth is shown below 1800 m, pioneer species are little reiterated; this seems to be a general tendency. *Rhus typhina*, a pioneer species, provides an example in New England. It always quite closely conforms to Leeuwenberg's model.

Tomlinson: This species produces root suckers; is that reiteration?

Oldeman: That is a form of reiteration that one finds in pioneer trees. Root suckers in *Macaranga*, sometimes in *Cecropia*, and basal suckers in *Casearia*.

Givnish: Perhaps with the "paucilayers" or pagoda trees one might have to look for an explanation other than Horn's multilayer model in which mutual shading is avoided. The advantage of a paucilayered shelf-type tree over something that has its leaves evenly spread out is that such a tree can take great advantage of side lighting.

Ashton: Yes, particularly if its leaves are arranged in tiers.

Givnish: It may be by the nature of the disturbances that typically occur in tropical forests (i.e., gaps forming because of tree falls as compared with the larger area disturbances that tend to occur, say, in New England) that tropical species may have evolved more specific adaptations to the margins of small gaps.

Kira: I was much impressed by your idea of using the leaf area density as a way of describing architecture of trees. How did you calculate the leaf area index for a tree?

Ashton: We took free-standing trees and cut a transect, as it were, through to the middle of the trunk and sometimes right through the tree to the other side. Obviously, to do this we had to remove the branches on one side of the tree – and if in the process we removed some of the leaves from within the area we were going to calculate, of course, we had to put those to one side – but basically we were counting the leaves attached to the tree as far as we possibly could. We calculated the mean leaf area from a sample of 50 leaves. We did this for several trees of each species. We calculated the leaf area densities on a per unit volume basis. The leaf area index, of course, is not calculated for the whole crown projection. We wanted to get a uniform figure for leaf area index for comparison, so in these small trees, where the trunks were never greater than 4–5 cm in diameter, we calculated the square meter about the trunk. We were, in fact, calculating the leaf area index for the flattest part of the top of the crown only, for several trees, and getting an average.

Ng: Dr. Longman and I have brought up the point that the amount of leaves on a tree, even if it is evergreen, can be variable, depending on whether it has just flushed and whether it is holding two or more flush generations at the time it is measured.

Ashton: We have considered this in our analyses. I would like to get back to architectural models and ask Francis Hallé a couple of questions, related to my interest in the distribution of different models in different parts of the forest cycle. First, you discuss continuous and intermittent branching, and I had understood this to mean that the growth of the tree was continuous or intermittent, but this appears not to be the case. *Macaranga*, for instance, conforms to Rauh's model, with the branches coming off at certain levels, although the actual growth of the apices is continuous, as far as I could see. The axillary buds that develop are not randomly distributed along the continuously growing shoot. Second, young dipterocarps usually conform to Massart's model, in which the growth is said to be continuous, but in which the plagiotropic lateral branches are more or less randomly distributed up the central axis. In fact, the trunk grows, not continuously, but in flushes that are synchronized throughout the sapling and young tree populations. Finally, how do you, in fact, identify what architectural model a tree belongs to in the mature phase of the forest when it is in a young stage but is a species that changes from plagiotropic to orthotropic branching at maturity?

Hallé: As to the first point, do you have any curve of the length of the leader against time?

Ashton: No, I do not, but the length of the internodes does not seem to vary.

Hallé: I agree, but in my opinion if you make a curve of length against time you will find rhythmic growth.

Oldeman: If a big canopy tree of Massart's or Roux's model changes to the formation of orthotropic branches with age, this represents reiteration of the model. Consequently, the initial model cannot be recognized any more. You have to examine the development of a tree to determine its model.

Ashton: The point is that in the young stages it will not flower. And, therefore, if the definition of the model depends on the position of the inflorescence at maturity, the tree may conform to a different model.

Hallé: You cannot identify the model before flowering. In the case of the dipterocarp, if it is Massart's model when the tree is young, it will also be Massart's model when it becomes adult and flowering. In this particular model the topographic position of the flowering is irrelevant.

Whitmore: Is it possible to recognize these models when they are sterile?

Tomlinson: It depends on the model. You obviously cannot distinguish between Corner's and Holttum's models until they flower.

Ashton: In the case of dipterocarps, the young tree conforms to Massart's model. But when the tree is fully grown, instead of the leaves being alternate along the branches and the branches plagiotropic, the branches become orthotropic and the leaves spiral. It is a common feature, not true of all dipterocarps, but of many of them. Now I can visualize another tree of a different family that has a model when it's young that can only be defined in flower; but, like a dipterocarp, it changes its model at maturity. So what do you do when you describe those young stages? Does the concept of the architectural model have different levels of accuracy?

Tomlinson: No, it would be called reiteration.

Hallé: I would prefer to give the answer after my next visit to Malaya.

References

Coombe, D. E. (1960). An analysis of the growth of *Trema guineensis. J. Ecol.,* 48, 219–31.

Coombe, D. E., & Hadfield, W. (1962). An analysis of the growth of *Musanga cecropioides. J. Ecol., 50,* 221–34.

Corner, E. J. H. (1949). The durian theory or the origin of the modern tree. *Ann. Bot. Lond., N.S. 52,* 367–414.

Forbes, J. C. (1973). Production and nutrient cycling in two birchwood ecosystems. Ph.D. thesis, University of Aberdeen.

Givnish, T. J., & Vermeij, G. J. (1976). Size and shapes of liane leaves. *Amer. Nat., 110,* 743–78.

Hallé, F. & Oldeman, R. A. A. (1970). *Essai sur l'architecture et la dynamique de croissance des arbres tropicaux.* Paris: Masson.

Horn, H. S. (1971). *The Adaptive Geometry of Trees.* Princeton: Princeton University Press.

Kenworthy, J. B. (1971). Water and nutrient cycling in a tropical forest. In *Transactions, First Aberdeen-Hull Symposium on Malesian Ecology,* pp. 49–59.

Parkhurst, D. F., and Loucks, O. L. (1972). Optimal leaf size in relation to environment. *J. Ecol. 60,* 505–37.

Richards, P. W. (1939). Ecological studies on the rain forest of Southern Nigeria. I. The structure and floristic composition of the primary forest. *J. Ecol., 27,* 1–61.

– (1952). *The Tropical Rain Forest.* Cambridge: Cambridge University Press.

Satoo, T. (1970). A synthesis of studies by the harvest method: Primary production relations in the temperate deciduous forests of Japan. In *Analysis of Temperate Forest Ecosystems,* vol. 2, *Ecological Studies,* ed. D. E. Reichle, pp. 55–72. New York: Springer-Verlag.

Schimper, A. F. W. (1903). *Plant Geography on a Physiological Basis.* Tr. W. R. Fisher, P. Groom, & I. B. Balfour. London: Oxford University Press.

Stephens, G. R. (1968). Assimilation by Kadam (*Anthocephalus cadamba*) in laboratory and field. *Turrialba, 18,* 60–23.

Whitmore, T. C. (1974). *Change with Time and the Role of Cyclones in Tropical Rain Forest on Kolombangara, Solomon Islands.* Commonwealth Forestry Institute, Paper 46. Oxford: Holywell Press.

– (1975). *Tropical Rain Forests of the Far East.* Oxford: Clarendon Press.

26
Tree falls and tropical forest dynamics

GARY S. HARTSHORN
Department of Zoology, University of Washington, Seattle, Washington

This chapter is offered in response to the oft-asked question: Where are all the large trees, if this is a virgin tropical forest? Here I will document the size-class distribution of trees in a tropical wet forest and examine several possible reasons for low densities of large trees. On the basis of factual and theoretic considerations of tropical forest dynamics, several hypotheses relevant to the successful colonization of disturbed areas are offered.

Tropical forest data used in this chapter are from Finca La Selva, the Organization for Tropical Studies' biological station in northeastern Costa Rica. Descriptions of the La Selva forests and environments can be found elsewhere (Holdridge et al., 1971; Hartshorn, 1972; Bourgeois, Cole, Riekerk, & Gessel, 1972).

Size-class distributions

Diameter measurements taken during a 1970–71 inventory of three 4-ha permanent study plots were used to calculate the average number of stems in 1-in. (2.5-cm) size classes. The size-class distribution of tree diameters at breast height (DBH) in La Selva (Fig. 26.1) shows the reverse J-shape or negative exponential distribution characteristic of most forests (Richards, 1966). There is a clear preponderance of stems in the small-diameter classes. The definition of what is a large tree is certainly subjective, but if we arbitrarily define "large" as 1 m DBH or more, there are relatively few large trees in the La Selva forest ($\bar{x} = 2.6 \pm 1.7$/ha, N = 12, range 0–6). Many relatively undisturbed temperate forests on favorable sites have higher densities of large trees (Fig. 26.1), despite the fact that the total density of a temperate forest may be as low as half the density of a tropical forest.

Present address: Organization for Tropical Studies, Universidad de Costa Rica, San José, Costa Rica

617

Fig. 26.1. Size-class distribution of stem diameters (≥ 0.25 dm) in virgin La Selva forest and in virgin temperate forest in Warren's Woods, Michigan. (Warren's Woods data from Cain, 1935)

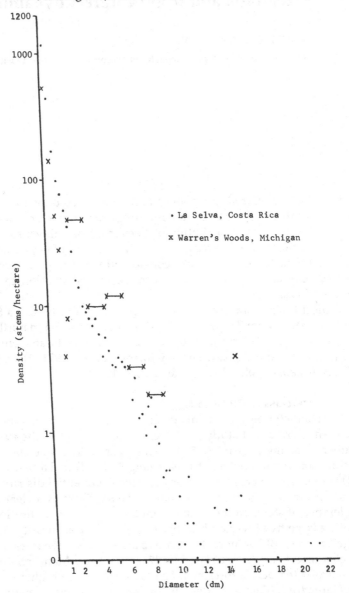

The paucity of large trees in tropical forests is even more striking if relative density is considered (Fig. 26.2). In the La Selva forest, fewer than 1% of the stems ≥10 cm DBH are 1 m or more in diameter. My observations of other rain forests in Costa Rica, Colombia, and Peru suggest that the size-class distribution of tree diameters in the La Selva forest is representative for species-rich neotropical forests. Exceptions do occur, but usually where restrictive site factors exist, often resulting in species-poor or single-species stands: e.g., *Prioria copaifera* Griseb. (Caesalpiniaceae) swamp forest, tropical montane *Quercus* spp. (Fagaceae), or *Podocarpus* spp. (Taxaceae) forests. The La Selva size-class distribution is comparable to that of other paleotropical rain forests (Baur, 1968).

Tropical trees are usually slender (i.e., for a given diameter the temperate tree usually is not nearly as tall as its tropical counterpart). The physiognomic aspect of tall trees, slim boles, smooth bark, and

Fig. 26.2. Relative densities of stem diameters (≥10 cm) in virgin La Selva forest and in virgin temperate forest in Warren's Woods, Michigan. (Relative densities in Warren's Woods calculated from data in Cain, 1935)

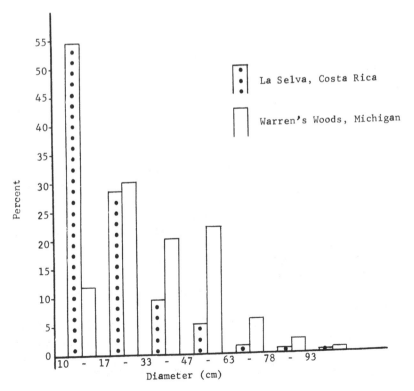

small crowns accentuates the visual impression of small diameters. When the temperate zone visitor is told or sees that the density of large trees is quite low in the tropical rain forest, he almost invariably concludes that he is in a disturbed or secondary forest. Obviously, the visitor may be correct, for it is not easy to distinguish between a late secondary and a mature forest. Although the two stages usually have a similar physiognomy, they are best differentiated floristically (Budowski, 1965; Holdridge, 1967), but anyone who has attempted it will attest that identification of tropical trees is not as simple as it sounds.

The virgin forest of La Selva shows no visual signs of disturbance in this century. Demographic and growth studies of the populations of *Pentaclethra macroloba* (Willd.) Ktze. (Mimosaceae) and other species (Hartshorn, 1972, 1975) offer strong evidence for its undisturbed status. It is incorrect to attribute the low density of large trees in this forest to a late successional stage or to past human disturbance. A study of forest dynamics currently in progress makes it increasingly obvious that the La Selva forest is much more dynamic than previously thought and that the high frequency of natural tree falls probably prevents most trees from attaining large diameters.

Turnover rates

The permanent intensive study plots at La Selva are particularly good for quantifying the frequency of natural tree falls, the size of canopy openings (or gaps), and turnover rates. All gaps occurring from February 1970 through March 1976 in area I and from March 1971 through March 1976 in areas II and III were drawn on 20-cm by 20-cm stem maps of each 20-m by 20-m subplot in the 4-ha intensive study plots. (Base maps were prepared by the College of Forest Resources, University of Washington, Seattle.) Turnover rates were calculated by dividing the total area covered by gaps during the time period into the total area of the plot, then multiplying the quotient by the time period to estimate the total number of years necessary to cover the entire plot with gaps. It should be noted that a new tree fall occasionally covered part of an older gap. Such occurrences were noted, but only the area of new gap that did not overlap the older gap was added to the total area in gaps for the time period.

As it is possible for a gap to be hit by successive tree falls during the time period, the number of gaps is not necessarily equivalent to the number of tree falls.

The calculated turnover rates (Table 26.1) are at best crude estimates; yet they are reasonably close, with a mean turnover rate of 118 ± 27 years. The faster turnover rate in the swampy half of area II is reasonable in view of the very poor structure of the poorly drained, occasionally flooded soils. The very fast turnover rate of 80 years in area I is not as easily explained. It partly due to the occurrence of several gaps in an extremely rainy December 1970. Although the monitoring of gaps in areas II and III did not commence until 1971, the same heavy rains did not appear to cause a comparable number of gaps in areas II and III. Area I does contain a narrow treeless slough of about 1800 m², but the slopes on either side are not nearly as steep as the slopes in area III. The incidence of gaps in area I appears to be representative of the relatively large, level area of old alluvium in the northwestern part of La Selva. The area is rather intensively used by researchers, who are well aware of the frequency of tree falls in their study areas. (I do not believe the researcher traffic is a causative factor.) It is possible that some characteristics of the physical structure of the old alluvial soils in area I are simply not as supportive of trees as are the soils of areas II and III.

If the particular forest in area I is in equilibrium (Hartshorn, 1972, 1975) with a turnover rate of 80 years, one might expect the size of gaps to be significantly smaller in area I or the area I forest to regenerate faster (i.e., have trees with faster growth rates). The mean gap size in area I is intermediate compared with the mean values in areas II and III (Table 26.1). Comparative growth rate data are not yet available to test the prediction of faster growth rates of area I trees compared with trees in areas II and III.

In the literature on tropical forest vegetation a paradox exists concerning growth rates of tropical trees. On the one hand, there is the amazingly rapid recuperative power of tropical vegetation, which can clothe an abandoned clearing, or milpa, with fast-growing spe-

Table 26.1. *Forest turnover rates on intensive study plots at La Selva (see text for explanation of calculations)*

	Intensive study plot			
	I	IIa	IIb	III
Soil type	Old alluvium	Alluvium	Old alluvium	Residual
Topography	Plateau	Swamp	Rolling	Steep
Plot size (m²)	40,000	20,000	20,000	40,000
Area in gaps (m²)	3,017	837	723	1,483
Time period (yr)	6	5	5	5
Gap area/year	503	167	145	297
Number of gaps	29	7	13	17
Mean gap size (m²)	104 ± 95	120 ± 125	54 ± 43	87 ± 90
Turnover rate (yr)	80	119	138	135

cies such as *Ochroma lagopus* Swartz (Bombacaceae), *Cecropia* spp. (Moraceae), or *Trema micrantha* (L.) Blume (Ulmaceae). In an abandoned pasture at La Selva, for example, *O. lagopus* attained 35 cm DBH in only 5 years. In contrast to the extremely fast growth rates of pioneer species, there is a considerable body of data documenting very slow growth rates of trees in old-growth forests (Wadsworth, 1958; Richards, 1966; Hartshorn, 1972). The rapid growth of early successional species and the slow growth of mature forest species are generally explained by postulating a very long developmental period to mature forest (Richards, 1966).

It is of interest to consider the status of the cuipo tree, *Cavanillesia platanifolia* HBK. (Bombacaceae), a conspicuously abundant canopy tree in eastern Panama. Very little cuipo regeneration is observable, suggesting that it is a successional species and that the forests with abundant large cuipo trees are successional. Bennett (1968) suggests that the present abundance of cuipo trees had its origin in the large-scale abandonment of cultivated land by Indians following the arrival of Spanish conquistadors. This would mean that the extensive forests of eastern Panama dominated by cuipo have yet to reach maturity after more than 400 years. However, I should like to point out that the cuipo tree is extremely fast growing; two trees > 30 m tall and 1 m in DBH in Turrialba, Costa Rica, were planted only 25–30 years ago (L. R. Holdridge, pers. comm.). *Cavanillesia* and the related genera *Ceiba, Chorisia, Bombax,* and *Pseudobombax* have a similar growth form, often with a pot-bellied or barrel-shaped bole. If the shade-intolerant, fast-growing members of this group can attain the canopy in 25–50 years, it is demographically unreasonable that more than a small proportion of the population (or cohort) survives for 400 years, even though they may be some of the giant trees in the forest. It is much more likely that the apparent lack of regeneration is due to the very fast growth rate of juvenile cuipo trees and the persistence of mature individuals for possibly a century or more.

The absence of regeneration of some species in mature tropical forests is well documented (Aubréville, 1938; Schulz, 1960; Richards, 1966; Baur, 1968). Aubréville's mosaic theory of tropical forest regeneration was developed to explain the absence of regeneration of species present as large trees in the virgin forest. He found the climax tropical forests of West Africa to be spatially dynamic, so that it was difficult to predict the dominant species of a particular climax community. For example, species X may dominate a particular community today, but the lack of regeneration indicates it will be replaced by other species and it may be regenerating in another area dominated by other species that it will eventually displace.

In a study of forest succession on Barro Colorado Island (BCI), Panama, Knight (1975) recognized two main population structure patterns: (1) species demonstrating the typical negative exponential distribution of size classes and (2) species with several large individuals but none in the smaller size classes. Pattern 1 species occur in similar proportions (37% and 39%) and densities (59% and 65%) in the two main forest types: one about 60 years old and the other much older – possibly 150 years. Pattern 2 species occur in the same proportion (15%) in the two types of forests, but the relative density of pattern 2 species is twice as high (9%) in the younger forest. Knight's population structure analysis shows that the younger forest has a higher density of individuals of pattern 2 species. Because of the abundance of pattern 2 species in both types of forests, Knight suggests that the older BCI forest is still not a mature or climax forest. But I find it more remarkable that two adjacent forests, approximately 60 and 150 years old, have such similar abundance patterns of reproducing and nonreproducing species. It is not unreasonable to expect that the "nonreproducing" species are in fact regenerating in gaps – those nasty, vine-tangled, often impenetrable patches that one tries to avoid when walking through the forest.

Knight (1975) mentions that the density of *Cecropia* spp. is higher in the old BCI forest than in the young forest and that gaps appear to be more common in the older forest. I think he is correct in suggesting a relation between *Cecropia* abundance and the frequency of tree falls. A middle-aged secondary forest probably has very low mortality of canopy trees and very few tree falls, so there may be a lengthy successional period to late secondary forest with a low incidence of new gaps.

A similar period of few tree falls may also occur in high-graded forests. High grading, or the selective cutting of valuable timber species, clearly removes large trees. The sparse scars of high grading are quickly covered by regeneration, so that for some subsequent period (perhaps 10–50 years), the frequency of tree falls is lower than normal. There is little doubt that a large proportion of tropical forests remaining today have been high-graded. The reduced frequency of tree falls in high-graded forest or in mid to late secondary forest has often caused erroneous interpretations of the successional status of some tropical forests. For example, the high-graded forest with its few gaps is interpreted as virgin, whereas the true virgin forest with a normal frequency of tree falls is considered secondary forest because it has more canopy openings than the old, high-graded forest!

Theoretic considerations

If the La Selva turnover rates are representative of humid neotropical forests, and I believe they are, we can clearly generalize that tropical forests are much more dynamic than temperate forests. Because our meager understanding of tropical forests has had a strong temperate bias, a theoretical exploration of tropical forest dynamics may enhance our efforts to understand tropical forest ecosystems.

Turnover rates of tropical forests should be an important selective force on the adaptive strategies of tropical tree species. The faster the turnover rate, the greater the selective pressures should be for colonizing species. Evolutionary processes have produced some excellent pioneer species: e.g., *Cecropia* spp., *Musanga cecropioides* R. Br. (Moraceae), *Ochroma lagopus*, *Trema micrantha*. Such species as these might be considered the prototype gap tree, characterized by tremendous invasiveness, very fast growth rate, precocity, and prolific seed production. One might expect a single species to fill the role of primary woody colonizer, and this general pattern holds fairly well for large disturbed areas in a given life zone or habitat. *Cecropia obtusifolia* Bertol. in tropical wet forest, *Trema micrantha* in tropical premontane rain forest, *Alnus jorullensis* HBK. (Betulaceae) in tropical montane rain forest and lower montane wet forest, and *Ochroma lagopus* on newly exposed riverine deposits often form single-species-dominated stands. Yet pioneer species occur in very low densities in mature forest (see Table 26.2), even though new gaps are available each year. In the competitive milieu associated with successful regeneration in gaps, the weedy species seldom compete successfully.

In contrast to the usual dominance of a single pioneer species in large disturbed areas, the species composition of forest gaps is remarkably varied (Hartshorn, unpubl.). Approximately 75% of the 105 canopy tree species at La Selva are dependent on light gaps for successful regeneration. Why do so many species utilize an apparently uniform resource – a gap? The unpredictability of gap occurrence appears to be a key factor in multispecies occupancy of gaps. It is useful to explore first what would happen if trees of predictable size fell at predictable times in predictable places. Over evolutionary time, one species would have been selected for colonizing the predictably occurring gaps in a given habitat. Thus, instead of 80 potentially colonizing species, only one or possibly a few species would predominate in gaps, and the evolutionary end product would be a monospecific stand.

Seed dormancy

Several factors should be important determinants of which species successfully colonize a gap. Dormancy mechanisms could be operative, allowing viable seeds to persist on the forest floor until the canopy is opened by a tree fall. From Malaysian forest litter samples removed to areas receiving full sunlight, seeds of several second-growth tree species germinated (Symington, 1933), but it is not known how long the seeds had been in the litter or soil before the samples were collected. Several weedy herbaceous species – e.g., *Erechtites hieracifolia* (L.) Raf. (Compositae), *Phytolacca rivinoides* Kunth & Bouche (Phytolaccaceae) – and weedy tree species – e.g., *Ochroma lagopus* (Vasquez-Yanes, 1974), *Cecropia obtusifolia* (P. Harcombe, pers. comm.) – have good dormancy mechanisms. Several tropical dry forest species that mature fruit at the end of the rainy season have dormancy mechanisms that inhibit germination during the dry season (Croat, 1969; G. Frankie, pers. comm.). Much less is known about the seed ecology of tree species in the wetter tropical forests, but it appears that seed dormancy mechanisms are less common in primary forest species (Guevara & Gómez-Pompa, 1972). One of the very few primary forest species at La Selva that has seed dormancy is *Sacoglottis trichogyna* Cuatr. (Humiriaceae), which takes 18–24 months to germinate. Dormancy of *S. trichogyna* seeds is most likely an integral part of seed dispersal by flotation.

Alternatively, shade-tolerant seedlings could persist in a suppressed state until the canopy is opened by a tree fall. If the suppressed seedling or juvenile escapes destruction by the falling tree (e.g., in the area along the bole between the root mass and where the crown lands), it receives considerably more light and usually responds with a burst of vegetative growth. At La Selva, this strategy is used by some of the rare canopy tree species (e.g., *Licania macrophylla* Benth., Chrysobalanaceae) that produce seed once every 2–5 years. Growth-rate studies (Hartshorn, unpubl.) indicate that, in general, the potential growth rate of shade-tolerant species is not as great as for the shade-intolerant species. It is not known if the length of suppression affects the growth potential of the seedling or juvenile.

Time of gap occurrence

If seed dormancy and seedling suppression are not common features of gap species in humid tropical forests, then one might predict that seed production should be essentially continuous. *Cecropia obtusifolia* is an excellent example of such a species, an individual tree producing mature seed for 10 months of a year (Hartshorn, unpubl.). Yet *C. obtusifolia* seldom occurs in gaps in the primary forest. Seed production by most primary forest species is more intermittent than in *C. obtusifolia* (Frankie, Baker, & Opler, 1974). The majority of species has at least one fruit crop per year, but several species may skip 1–5 years between fruit crops. Although the nonfruiting interval is not as long or as synchronized as in Southeast Asian Dipterocarpaceae (Janzen, 1974), it no doubt effectively excludes those shade-intolerant species from the potential colonizing species pool for some portion of the nonfruiting interval.

Thus, the time of gap occurrence is a very important determinant with respect to the available seed pool. Even for annual reproducers, time may be a determining factor, because most species have fairly short fruiting periods of 1–3 months (Frankie et al., 1974). The many species with little or no endosperm that germinate quickly are also probably unavailable to the colonizing pool for some portion of each year.

There should be a direct relation between seed size and the length of time a gap is colonizable by that seed (i.e., a larger seed can push the seedling to a greater initial height before the seedling must produce its own photosynthate). For example, if species A has a small seed that enables the seedling to be 10 cm tall before it is photosynthetically independent and species B has a large seed that puts the seedling up to 50 cm tall before it is independent, and seeds of both species arrive in a gap that already has a dense layer of seedling leaves between 20 and 30 cm above the ground, then the gap is still colonizable by species B but not by species A. In other words, the larger the seed is, the longer the gap is colonizable by that species.

Some species are notoriously poorly synchronized in their seed production (Frankie et al., 1974): Some individuals may produce fruit for 1–2 months, and other individuals of the same species may produce fruit 5–9 months later. Such sporadic seed production by individuals or segments of a population would clearly be advantageous for a gap species because its seeds would be potential colonizers for a greater portion of the year. The sporadic reproductive effort is probably quite complex if cross-pollination is necessary.

Seed dispersal

Proximity of seed source to a gap and diaspore dispersal mechanisms should be important factors in determining which species get their seeds into a gap. It is generally accepted that the mature individuals of a given species are often far apart or widely scattered in the tropical rain forest (Richards, 1966), yet no test of the "hyper-dispersed theory" has appeared in the literature. Even if it is eventually shown that most species have clumped or more regular spatial distributions, the densities of most species are very low (Table 26.2) so that most seed sources may be considerable distances from a colonizable gap or from each other. Proximity of seed source is hindered by the much greater prevalence of dioecism in tropical forests than in temperate forests (Bawa & Opler, 1975). With dioecious species, only the female trees bear fruit; thus, the density of seed trees is approximately one-half the density of all individuals. Skewed sex ratios of dioecious species may further reduce the density of female trees (Opler & Bawa, 1977).

Not all seeds of colonizing species simply come from intact trees around a gap. Biotic dispersal of diaspores is extremely important in tropical forests. At La Selva, about 49% of the tree species have diaspores of the type dispersed by birds, 13% have bat-dispersed diaspores, 3% have diaspores that are probably dispersed equally well by bats and birds, and 9% have wind-dispersed diaspores. Most gap species have diaspores of the bird or bat type. The volant animals

Table 26.2. *Density (per hectare) and importance value (percent in parentheses) of most important, exemplary gap, and pioneer tree species in intensive study plots at La Selva* [a]

	Intensive study plot			
	I	IIa	IIb	III
Most important species				
Pentaclethra macroloba	49(23)	61(19)	61(18)	71(21)
Welfia georgii	54(10)	10(2.6)	26(4.4)	34(5.3)
Socratea durissima	40(7.6)	+	10(1.7)	49(6.4)
Protium pittieri	14(3.9)	+	6(1.2)	24(4.0)
Warscewiczia coccinea	14(3.1)	+	−	−
Iriartea gigantea	−	18(3.8)	47(7.0)	34(5.0)
Carapa guianensis	−	31(11)	5(2.1)	−
Pterocarpus hayesii	−	13(8.0)	+	−
Astrocaryum alatum	−	32(5.4)	+	−
Exemplary gap species				
Apeiba membranacea	4(1.8)	7(2.4)	6(3.0)	1(0.6)
Casearia arborea	9(2.3)	−	2(0.4)	3(0.6)
Dipteryx panamensis	1.8(2.0)	−	−	−
Goethalsia meiantha	8(2.6)	6(1.8)	19(5.5)	3(0.7)
Hampea appendiculata	0.8(0.2)	6(1.3)	2(0.4)	0.3(0.1)
Hyeronima oblonga	0.3(0.1)	2(0.4)	0.5(0.4)	0.3(0.8)
Laetia procera	5(2.3)	−	2(0.4)	2.5(1.0)
Pourouma aspera	3.5(1.0)	3(0.8)	9(2.6)	12(2.7)
Rollinia microsepala	−	−	1(0.2)	−
Simarouba amara	0.5(0.1)	−	0.5(0.2)	2.3(0.6)
Xylopia sericophylla	0.2(0.1)	−	−	0.2(0.2)
Pioneer species				
Cecropia insignis	−	0.5(0.2)	1.5(0.4)	−
Cecropia obtusifolia	0.5(0.2)	5(1.2)	1.5(0.4)	0.2(0.1)
Ochroma lagopus	−	0.5(0.3)	−	−
Trema micrantha	−	−	−	−
Total	391(100)	353(100)	422(100)	536(100)

[a] +, less important species present; −, less important species absent.

should effectively disperse seeds over a much larger area than could either arboreal or terrestrial animals or wind.

Substrate conditions

Substrate conditions within a gap may have an important effect on which species colonize a gap. When a canopy tree falls, a small area (< 20 m²) of mineral soil is exposed by the upturned root mass. The area of crown impact is usually covered to a considerable depth by branches and leaf litter. The fallen bole and larger branches provide a different substrate for germination. As it is unlikely that a species germinates or survives equally well on mineral soil and on 10 cm of decaying leaf litter, there may be site preferences within a gap. In swampy areas, the upturned root masses and fallen logs may be the only safe sites available for germination for many species. The large seeds of such swamp species as *Mora* spp. (Caesalpiniaceae), *Carapa guianensis* Aubl. (Meliaceae), and *Pachira aquatica* Aubl. (Bombacaceae) may enable the seedlings more quickly to establish a good root system in the poorly structured, occasionally flooded swamp soils (McHargue & Hartshorn, unpubl.).

Size of gap

The size of a gap may influence which species do or do not colonize a gap. Weedy species (e.g., *Cecropia, Trema*) are generally found only in large gaps (> 500 m²), such as in abandoned clearings or when one large tree takes down another large tree in domino fashion. It is not clear if the invasion of weedy species follows a simple probabilistic model or if microclimatic or substrate conditions are mitigating. Considering the reproductive capacity of weedy species, it seems reasonable to assume that their seeds do get into most gaps, rather than just in the large gaps. A more plausible explanation may be that small gaps have less favorable microclimatic conditions than do large gaps for the germination of weedy species. The average gap size in the La Selva plots is only 89 ± 88 m² (N = 59, range 8–376), and as the colonizable surface, whether mineral soil or crown debris, is at the base of a crude irregularly shaped cylinder with "sides" 35–45 m tall, only a small amount of direct sunlight reaches the colonizable surface. In small gaps the intensity or amount of light and/or heat may be insufficient to stimulate germination of weedy species. If this is true, the very infrequently occurring large gaps, such as hurricane blowdowns or river meander scars, may be the primeval habitat of weedy tree species.

Gap size may also cause important changes in microclimate and root competition, which in turn may determine which gap species can successfully colonize a gap. The death of trees in a gap temporarily lessens the root competition toward the center of the gap. The reduction of root competition in large gaps is probably quite substantial. A similar pattern may also occur with vesicular-arbuscular mycorrhizal infection. Death of many roots in a large gap may substantially reduce the probability of infection for colonizing tree species. The survival probability of a seedling of a gap species may be greatly reduced if it does not become infected with vesicular-arbuscular mycorrhiza (Janos, 1975).

Plant-herbivore relations

Plant-herbivore relations should play an important role in determining which species successfully colonize gaps. The general patterns of seed predation and its probable consequences for diversity, spatial patterning, and reproductive biology of tropical tree species have been elucidated by Janzen (1970, 1971a). The few seeds that arrive in a gap are usually a considerable ecologic distance from the parent tree(s), so it is unlikely that density-dependent or distance-responsive seed predation is an important mortality factor of seeds in a gap.

Grazing by herbivores on seedlings and saplings in gaps is likely to be space-related. There is considerable evidence from temperate regions that plant species are not equally palatable to generalist herbivores (Cates & Orians, 1975). Leaves of approximately 100 plant species at La Selva have been experimentally fed to caged katydids (*Orophus conspersus*, Tettigoniidae), a native generalist herbivore, to test the hypothesis that tropical plant species are not equally palatable to a generalist herbivore. Preliminary results show that weedy species are most palatable and climax species least palatable, with gap species intermediate (Waltz & Hartshorn, unpubl.). There is also an indication that the common gap species are less palatable than the rare or less common ones. Differential grazing pressures by herbivores should favor those species suffering less leaf or meristem damage.

A variety of plant features act to reduce herbivore damage. Physical characteristics such as pubescence, thorns, spines, urticating hairs, trichomes, and toughness are often effective in reducing grazing pressures. Mutualistic associations between plants and animals afford considerable protection of plant tissues. Some plant species provide domiciles and/or proteinaceous food bodies for predatory ants that protect the host plant's tissues and organs from herbivores and competing plant species (Janzen, 1966, 1967). Many other plant species produce extrafloral nectar that is actively gathered by predatory ants (e.g., Ponerinae). The ants are not resident on the plants, but their presence on the meristem and leaf surface, feeding on extrafloral nectar, probably acts as a deterrent to herbivorous insects (Hartshorn, 1972).

It is well known that plants possess an imposing array of chemicals known or suspected to have antiherbivore properties (Whittaker, 1970; Rhoades & Cates, 1976). These range from acutely toxic compounds such as alkaloids and cyanogenic glycosides (James, 1950; Levin, 1971) to digestibility-reducing substances such as tannins (Feeny, 1968, 1970; Hathaway, 1960). Not all plant species rely on physical, chemical, and/or mutualistic means to reduce herbivore pressure. Some plants, particularly ephemerals, may escape temporally or spatially because they are unpredictable in space and available for only a short time (Janzen, 1971b,c). These species should expend less energy on physical, chemical, or mutualistic defenses because their rarity and/or ephemerality reduce grazing pressures even if they are undefended.

Several predictions (summarized in Table 26.3) are possible concerning the patterns of defense against herbivores. In general, the level of defense (i.e., the amount of energy invested in defensive compounds or structures) should be positively correlated with the successional status of the plant species. Fast-growing, invasive species should suffer less herbivore damage because of their unpredictable occurrence in space and/or time, even though they put less energy into defense. The diversion of energetic resources for defense probably has a greater negative effect on total fitness of a pioneer species than a comparable diversion in a later successional species (Cates & Orians, 1975).

Defenses should be greater in evergreen than in deciduous species of the same successional status. Evergreen leaves are available for a longer time, have potentially longer income times, and are usually constructed to withstand periods of drought stress (Orians & Solbrig, 1977). Selection favors better defense of the potentially more productive leaves, regardless of the level of grazing.

Defenses should be greater in leaves of species that continually produce new leaves than in species that intermittently produce flushes of new leaves. Young leaves usually have different defenses (e.g., chemical compounds, greater pubescence, nectar production) than older leaves (D. Rhoades, pers. comm.). If young leaves are present intermittently, the herbivores should have more difficulty locating and completing their development on them.

The level of defense should be positively correlated with abundance. The shorter the distance between individuals, the easier, in general, it should be for specialist herbivores to find them (Janzen, 1970; Tahvanainen & Root, 1972). Therefore, a plant should receive more herbivore pressure where it is common than where it is rare.

Table 26.3. *Predictions concerning tropical forest dynamics and successful establishment of regeneration in disturbed areas*

1 Tropical forests are more dynamic than extratropical forests

2 Humid tropical forests are more dynamic than dry tropical forests

3 There should be a positive relation between turnover rate of tropical forests and proportion of shade-intolerant species dependent on gaps for successful regeneration

4 Seed dormancy occurs in very few primary forest species in nonseasonal tropical forests

5 Successful colonization of gaps should be dependent on seed availability, seed size, proximity of seed source, diaspore dispersal mechanisms, substrate conditions, size of gap and plant-herbivore relations

6 In absence of seed dormancy, continuous seed production, and seedling shade tolerance, available seed pool of potential colonizing species is restricted to those species that fruited just before or during time period a gap is available for colonization

7 The larger the seed size, the longer the time a gap is colonizable

8 Probability of a seed's arriving in a gap decreases with increasing distance from seed source

9 Species dependent on gaps for successful regeneration should have more effective diaspore dispersal mechanisms than do climax species (i.e., most gap species should have bat-, bird-, or wind-dispersed diaspores)

10 Some gap species should be restricted to particular substrate conditions within a gap (e.g., mineral soil exposed by upturned root mass)

11 Probability of colonization by weedy pioneer species should be positively correlated with size of gap

12 Tropical plant species are differentially palatable to generalized herbivores

13 Level of defense against herbivores should be positively correlated with successional status of plant species

14 Gap species should be intermediate in palatability between very palatable pioneer species and inedible climax species

15 Rare gap species should be more dependent on escape in time or space than common gap species for successful establishment in a gap

16 Rare gap species should have faster growth rates than common gap species

17 Defenses should be greater in evergreen than in deciduous species of same successional stage

18 Defenses should be greater in leaves of those species that continually produce new leaves than in species that intermittently produce flushes of new leaves

19 Level of defense should be positively correlated with abundance

These predictions are particularly pertinent to the success of species establishment in gaps. If a sizable cohort of a species' seedlings colonize a gap, that cohort should receive much more intensive herbivore pressure than if only one or a few seedlings were present. Preliminary results show that gap species are intermediate in their level of defense between the "defenseless" pioneer species and the highly inedible climax species. Furthermore, there is some indication that rare gap species are more palatable to a generalist herbivore than are the common gap species. This suggests that the common gap species have a higher level of defense (i.e., a greater proportion of the total energy available is used for defensive purposes). Other things being equal (e.g, leaf area, size, incident radiation), the rare species should have a greater proportion of energy available for nondefensive purposes, such as competitive interactions or reproductive effort. Thus, in the absence of grazing, the competitive ability of the rare gap species should be greater than that of the common gap species. This, of course, holds only as long as the rare species remains rare. If the density of the rare species increases, it should be subject to greater herbivore pressures than a better-defended common species at similar densities.

Density-dependent plant-herbivore interactions are probably important factors in the maintenance of many rare species in tropical forests. Rare species may not colonize many gaps, but when they do successfully colonize a gap (at low densities), the probability of reaching maturity is greater than for a common species in the same gap. Furthermore, because the probability of success increases as the species becomes rarer, the competitive ability of a rare species is inversely proportional to its abundance. This may help account for the persistence of so many rare species in tropical forests.

Conclusions

The size-class distribution of tree diameters in the La Selva virgin tropical rain forest shows the reverse J-shape characteristic of most forests. Both the absolute and relative densities of trees 1 m DBH or more are quite low: 2.6/ha and 0.6%, respectively. The low density of large trees is due primarily to the dynamic nature of the virgin forest: The high frequency of natural tree falls prevents most trees from attaining large size. Data from a monitoring of natural gaps on three permanent plots were used to calculate a mean turnover rate of 118 years for the La Selva forest. This fast turnover rate appears to be representative of other species-rich neotropical forests. In general, tropical forests appear to be much more dynamic than extratropical forests.

A lack of understanding of tropical forest dynamics has led to misinterpretation of the successional status of several tropical forests, especially in distinguishing between late secondary and mature forests. A major problem is the frequent occurrence of populations of large individuals of several species with an apparent absence of regeneration. Such a population structure is usually taken to indicate that the forest has not reached maturity. It is likely that many of the "nonregenerating" species are dependent on gaps in the forest for successful regeneration; thus, they should be considered normal components of the mature forest rather than successional species that must die out as the forest matures.

The fast turnover rates of tropical forests should be an important selective force on the adaptive strategies of tropical tree species. As many as 75% of the canopy tree species may be dependent on gaps for successful regeneration. For a particular gap, the sizable pool of potential colonizing species may be reduced drastically as a consequence of several biotic and abiotic factors. Theoretic considerations of tropical forest dynamics suggest that the following factors are important to the successful regeneration of tree species in gaps:

1. Time of gap occurrence is important because of the intermittent production of seeds by virtually all tropical forest species. Most of the tree species dependent on gaps do not appear to have seed dormancy mechanisms or shade-tolerant seedlings that would enable the progeny to persist in or near the forest floor until a gap occurs. There should be a direct relation between seed size of a species and length of time a gap is colonizable by that species.

2. Proximity of seed source to a gap and mechanisms of diaspore dispersal should be important determinants of which species get their seeds into a gap. Low densities of mature individuals for most tree species and the presence of many dioecious species may result in substantial distances between seed sources and a gap. The majority of gap species have diaspores dispersed by birds and bats.

3. Different substrate conditions within a gap (e.g., mineral soil, fallen bole, or crown debris) may have a selective effect on germination or seedling establishment.

4. The size of gap influences which species do or do not colonize a gap. The larger a gap, the more weedy or pioneer species there are in a gap. Gap size probably also has an important effect on changes in microclimate, root competition, and probability of endomycorrhizal infection.

5. Plant-herbivore relations are probably important in determining which species successfully colonize a gap. Experimental evidence indicates a broad range in leaf palatability of some 100 species at La Selva. The level of defense should be positively correlated with the

successional status of the plant species. Defenses should be greater in evergreen than in deciduous leaves and greater in leaves of species that continually produce new leaves than in species that intermittently produce flushes of new leaves. The level of defense should be positively correlated with abundance: A plant should receive more herbivore pressure where it is common than where it is rare. There is some indication that rare gap species are more palatable than common gap species. If common gap species have higher levels of defense than do rare species, the latter may have a greater proportion of energy available for competitive interactions. The greater competitive ability of rare species in the absence of grazing may help explain the persistence of so many rare species in tropical forests.

Acknowledgments

The research discussed in this chapter and the author were supported by NSF Grant GB 40747 administered by the Organization for Tropical Studies and the University of Washington, G. H. Orians, principal investigator. Several OTS courses served as fertile testing grounds for many of the ideas presented here. The helpful and constructive comments of Drs. D. Janos, D. Janzen, G. Orians, and J. Vandermeer on an earlier version of this chapter are greatly appreciated.

General discussion

Jeník: Tropical forests are definitely more dynamic than temperate forests, as is evident to anyone who has camped in tropical rain forests. The sound of falling trees and falling branches is often heard. The relation of root competition is rarely discussed, although it is probably very important in the development of vegetation in a gap.

Whitmore: I once measured several individuals of a very vigorous near-pioneer species, *Campnosperma brevipetiolatum*, growing in a gap of about 400 m² and was very surprised, on taking height measurements every 6 months for 6.5 years, to discover that they grew only 0.5 m. Two of the same species in a plantation (equivalent to a big gap) added 2.14 m (Whitmore, 1974).

Hartshorn: I agree with both of you that many environmental factors other than light are very important in determining which species enter and survive in gaps. Unfortunately, we really do not have much information on how these factors affect successful colonization.

Oldeman: The term "chablis" does not imply any single reason for the tree fall and includes all resulting environmental factors. It indicates that very complex phenomena are involved in destruction of a part of the forest.

Sarukhán: There are a number of regional names, both in Spanish and several native tongues, that refer rather precisely to this situation in a very descriptive form. I see no need to import exotic terminology.

Givnish: I am interested in your proposed explanation for the phenomenon of leaf flushing, as it predicts a difference between early and late successional species that accords with the observed pattern. For example, Dr. Ashton found that species growing in early succession grow continuously, whereas those growing in late succession grow in flushes. We have now had two explanations for the phenomenon of leaf flushing. Dr. Borchert's physiologic model would predict that, under the greater rates of growth and transpiration favored by conditions in early succession, there should be more leaf flushing than in late successional trees. In contrast, the "escape in time" hypothesis predicts that there should be more leaf flushing in the later phases.

Hartshorn: I agree with your principle, but we have definitional problems here; we now have had several definitions, although not necessarily stated, concerning "pioneer" species versus "gap" and "mature forest" species. Certainly pioneer species do characteristically grow continuously (i.e., continually put out new leaves). We find gap species doing either one or the other – flushing or continually growing – and my impression is that most of them flush intermittently.

Borchert: Quite a number of species grow continuously only as long as they are young; as they get older they show a progressive tendency to flush (i.e., growth becomes more and more intermittent). This could be happening in the two contrasted natural situations. In a new opening trees are young and they therefore grow continuously; the larger they get (and I ascribe the change to size), the more intermittently or periodically they grow.

Tomlinson: My question is really about Dr. Hartshorn's chapter, but it is probably best directed to Dr. Sarukhán. I believe Dr. Hartshorn indicated that species are different in their ability to resist predator pressure. And I was curious to know if Dr. Sarukhán could tell me if this would be reflected in the survivorship curve? In other words, could you recognize from the survivorship curve the phenomenon Dr. Hartshorn is talking about?

Sarukhán: It is not easily evident just from a survivorship curve whether a species is capable or not of escaping predation. One needs to know more details about the dynamics of numerical change in the life history of a species to be able to assess more confidently predator effects upon populations.

Baker: Would it not be easier to make a chemical analysis?

Tomlinson: Yes, but I wondered if there might be some sort of correlation. Dr. Hartshorn, did I understand you correctly when you suggested that there could be selection for rareness: i.e., the rare occurrence of a species may be an adaptive function in that species?

Hartshorn: There is not selection for rarity per se; there is selection for particular characteristics or syndromes that may enable a species to persist or to exist as a rare species in certain situations. I have encountered a very interesting phenomenon: In one particular area or type of forest certain species may be exceedingly rare, represented by very few individuals; 50 km distant, in what appears to be a comparable forest with comparable habitat, the same species may be very abundant.

Whitmore: May I give a very dramatic example of a rare species suddenly becoming very common? *Glochidion tetrapteron,* named because it has a square, winged stem, was described from a single collection. It was relocated after days of searching at Fraser's Hill by the industrious Columbo Plan forest botanist in Malaya – so we then had two collections. A new road was built through the mountains of Malaya some 30 miles further south. A stand of this *Glochidion* about 1 mile long developed with perhaps 12,000 plants. I believe this happened because there was a mature plant seeding nearby at the right moment to get in as a pioneer, and this filled the length of the new bare roadside.

Baker: Is there any evidence that it will suffer from now being abundant? *Whitmore:* Not during the first 4–5 years. Maybe somebody who is in Malaya can have another look at it.

Longman: Another possible explanation could be the presence of dormant seeds in the soil. As regards light effects on seed dormancy, an extremely small quantity is needed to stimulate light-sensitive seeds to germinate. In normal forest situations there must be either a covering structure or possibly a lot of litter on top to suppress germination of light-sensitive seeds. My own guess is that the most likely mechanism to look for is diurnal fluctuation of temperature. In addition, the tree *Hildegardia barteri* provides an example of secondary dormancy. This species grows on rocky granite outcrops in West Africa, and the freshly collected seeds (fruits) germinate in about 24 hours (Enti, 1968). Seed kept at 50%–70% relative humidity develops a dormancy that can be broken by restoring it to 90% relative humidity for a few weeks. If seed is kept all the time at 90% relative humidity, it germinates freely any time.

References

Aubréville, A. (1938). Les forêts de l'Afrique occidentale française. *Ann. Acad. Sci. Col., 9,* 1–245.

Baur, G. N. (1968). *The Ecological Basis of Rainforest Management.* Sydney: Forestry Commission of New South Wales, Australia. 499 pp.

Bawa, K. S., & Opler, P. A. (1975). Dioecism in tropical forest trees. *Evolution, 29,* 167–79.

Bennett, C. F. (1968). Human influences on the zoogeography of Panama. *Ibero-americana, 51,* 1–112.

Bourgeois, W. M., Cole, D. W., Riekerk, H., & Gessel, S. P. (1972). Geology and soils of comparative ecosystem study areas, Costa Rica. *Trop. For. Ser., 11,* 1–37.

Budowski, G. (1965). Distribution of tropical American rain forest species in the light of successional processes. *Turrialba, 15,* 40–2.

Cain, S. A. (1935). Studies on virgin hardwood forest. III. Warren's Woods, a beech-maple climax forest in Berrien County, Michigan. *Ecology, 16,* 500–13.

Cates, R. G., & Orians, G. H. (1975). Successional status and palatability of plants to generalized herbivores. *Ecology, 56,* 410–18.

Croat, T. B. (1969). Seasonal flowering behavior in central Panama. *Ann. Missouri Bot. Gard., 56,* 295–307.

Enti, A. A. (1968). Distribution and ecology of *Hildegardia barteri* (Mast.) Kosterm. *Bull. IFAN (Ser. A), 30,* 881–95.

Feeny, P. B. (1968). Effect of oakleaf tannins on larval growth of the winter moth *Operophthera brunnata. J. Insect Physiol., 14,* 1805–17.

– (1970). Seasonal changes in oakleaf tannins and nutrients as a cause of spring feeding by winter moth caterpillars. *Ecology, 51,* 656–81.

Frankie, G. W., Baker, H. G., & Opler, P. A. (1974). Comparative phenological studies of trees in tropical wet and dry forests in the lowlands of Costa Rica. *J. Ecol., 62,* 881–919.

Guevara, S., & Gómez-Pompa, A. (1972). Seeds from surface soils in a tropical region of Veracruz, Mexico. *J. Arnold Arbor., 53,* 312–35.

Hartshorn, G. S. (1972). The ecological life history and population dynamics of *Pentaclethra macroloba,* a tropical wet forest dominant, and *Stryph-*

nodendron excelsum, an occasional associate. Ph.D. dissertation, University of Washington.
- (1975). A matrix model of tree population dynamics. In *Tropical Ecological Systems: Trends in Terrestrial and Aquatic Research,* ed. F. B. Golley & E. Medina, pp. 41–51. New York: Springer-Verlag.
Hathaway, D. E. (1960). Plant phenols and tannins. In *Chromatographic and Electrophoretic Techniques,* ed. I. Smith, pp. 308–54. New York: Wiley.
Holdridge, L. R. (1967). *Life Zone Ecology,* 2nd ed. San José, Costa Rica: Tropical Science Center. 206 pp.
Holdridge, L. R., Grenke, W. C., Hatheway, W. H., Liang, T., & Tosi, J. A., Jr. (1971). *Forest Environments in Tropical Life Zones: A Pilot Study,* Oxford: Pergamon Press. 747 pp.
James, W. O. (1950). Alkaloids in the plant. In *The Alkaloids,* ed. R. Manske, vol. 1, pp. 15–90. New York: Academic Press.
Janos, D. P. (1975). Vesicular-arbuscular mycorrhizal fungi and plant growth in a Costa Rican lowland rainforest. Ph.D. dissertation, University of Michigan.
Janzen, D. H. (1966). Coevolution of mutualism between ants and acacias in Central America. *Evolution, 20,* 249–75.
- (1967). Interaction of the bull's-horn acacia (*Acacia cornigera* L.) with an ant inhabitant (*Pseudomyrmex ferruginea* F. Smith) in eastern Mexico. *Univ. Kansas Sci. Bull., 47,* 315–558.
- (1970). Herbivores and the number of tree species in tropical forests. *Amer. Nat., 104,* 501–28.
- (1971a). Seed predation by animals. *Annu. Rev. Ecol. Syst., 2,* 465–92.
- (1971b). Escape of juvenile *Dioclea megacarpa* (Leguminosae) vines from predators in a deciduous tropical forest. *Amer. Nat., 105,* 97–112.
- (1971c). Escape of *Cassia grandis* L. beans from predators in time and space. *Ecology, 52,* 964–79.
- (1974). Tropical blackwater rivers, animals, and mast fruiting by the Dipterocarpaceae. *Biotropica, 6,* 69–103.
Knight, D. H. (1975). An analysis of late secondary succession in species-rich tropical forest. In *Tropical Ecological Systems: Trends in Terrestrial and Aquatic Research,* ed. F. B. Golley & E. Medina, pp. 53–9. New York: Springer-Verlag.
Levin, D. A. (1971). Plant phenolics: An ecological perspective *Amer. Nat., 105,* 1805–17.
Opler, P. A., & Bawa, K. S. (1977). Sex ratios of tropical dioecious trees: Selective pressures and ecological fitness. *Evolution* (in press).
Orians, G. H., & Solbrig, O. T. (1977). A cost-income model of leaves and roots with special reference to arid and semi-arid areas. *Amer. Nat., 111* (in press).
Rhoades, D. F., & Cates, R. G. (1976). Toward a general theory of plant antiherbivore chemistry. *Rec. Adv. Phytochem., 10,* 168–213.
Richards, P. W. (1966). *The Tropical Rain Forest: An Ecological Study,* Cambridge: Cambridge University Press. 450 pp.
Schulz, J. P. (1960). Ecological studies on forests in northern Suriname. *Verh. K. Nederl. Akad. Wet. A. Natkd., 53,* 1–267.
Symington, C. F. (1933). The study of secondary growth on rain forest sites in Malaya. *Malayan Forester, 2,* 107–17.
Tahvanainen, J. O., & Root, R. B. (1972). The influence of vegetational diversity on the population ecology of a specialized herbivore *Phylotreta cruciferae. Oecologia, 10,* 321–46.

Vasquez-Yanes, C. (1974). Studies on the germination of seeds of *Ochroma lagopus* Swartz. *Turrialba*, *24*, 176–9.

Wadsworth, F. H. (1958). Status of forestry and forest research in Puerto Rico and the Virgin Islands. *Caribbean Forester*, *19*, 1–24.

Whitmore, T. C. (1974). *Change with Time and the Role of Cyclones in Tropical Rain Forest on Kolombangara, Solomon Islands*. Commonwealth Forestry Institute, Paper No. 46. Oxford: Holywell Press.

Whittaker, R. H. (1970). The biochemical ecology of higher plants. In *Chemical Ecology*, ed. E. Sondheimer & J. B. Simeone, pp. 43–70. New York: Academic Press.

27

Gaps in the forest canopy

T. C. WHITMORE

Commonwealth Forestry Institute, University of Oxford, Oxford, England

This volume's title, "Tropical Trees as Living Systems," reflects an attempt to escape from an earlier static, typologic approach. In this chapter I hope to try to relate the numerous presentations we have heard on the morphology or physiology of individual trees to the concept of the forest canopy as a dynamic, continually changing entity. This also reflects an attempt to study processes, already drawn to our attention in Chapters 23 and 26.

What follows is largely a resumé of matters pertaining to one of the central unifying themes of my new synthesis on rain forest ecology (Whitmore, 1975): the idea of the forest growth cycle; an idea that crystallises concepts which have been developing for some time amongst rain forest ecologists and scientific foresters and that was formulated in a slightly different context by Watt (1947).

The forest growth cycle

The canopy of a forest changes continually as trees grow up and die and others replace them. This state of dynamic equilibrium may conveniently be subdivided into a forest growth cycle of three phases: the gap phase, the building phase, and the mature phase (Whitmore, 1975). The phases are abstractions, not separate entities. The gap phase, comprising juvenile (seedling and sapling) trees, passes by growth into the building phase, which is a pole forest and itself matures by continual growth of its constituent trees.

Figure 27.1 shows the mosaic of the three phases on 2.02 ha of tropical lowland evergreen mixed dipterocarp rain forest at Sungei Menyala, Malaya. Long narrow gaps created by the windfall of single trees are clearly seen. The extensive area of building phase at the north end is forest regrown after clearance 54 years earlier. Figures 27.2 and 27.3 show forests in profile with patches of building phase

set within mature phase canopy. On the classic concept of stratification within the tropical rain forest canopy, Fig. 27.2 has A stratum
emergents and a nearly continuous B stratum, which merges into C
except at two points near the left end and completely into D. (The E
stratum of forest floor herbs and seedlings is omitted from the diagram.) Note how application of the stratum concept loses sight of the
significance of a gap in the A (emergent) stratum, which represents a
patch of building phase. Stratification is a static, typologic concept of
forest structure that gives no recognition to the dynamic nature of the
canopy. Aspects of tree architecture and physiology change as trees
grow up. It follows that these are more likely to be susceptible to analytic interpretation if considered with reference to the forest growth
cycle than to canopy stratification. This applies, for example, to stratification of leaf division, size, and shape (Chap. 15). It also applies to
tree crown shape. Chapters 8, 10, and 23 discussed change of crown
form as a tree matures. Chapters 7, 9, and 23 discussed how it is determined both by behaviour of dormant buds and by changes in differentiation levels of existing active meristems, triggered by changes
in the environment. The profiles in Figs. 27.2 and 27.3 clearly show
how change of crown form is related to the forest growth cycle. The
deep conical crowns of young trees in the building phase are distinguished from trees with crowns broader than deep in mature forest.

Fig. 27.1. Canopy phases on 2.02 ha of tropical lowland evergreen
mixed dipterocarp rain forest at Sungei Menyala, Malaya, in 1971.
(From Whitmore, 1975)

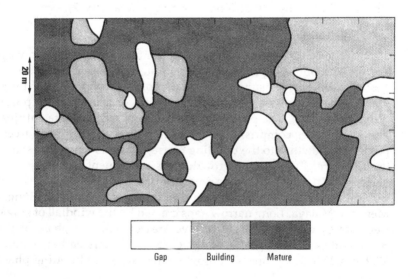

Gap   Building   Mature

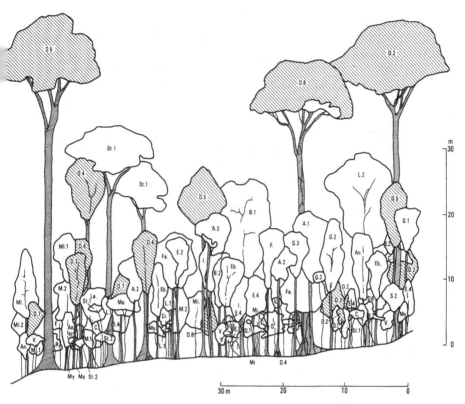

Fig. 27.2. Mature and building phases on ridge at Belalong, Brunei, in tropical lowland evergreen mixed dipterocarp rain forest. Plot area 60 m by 7.5 m, all trees over 4.5 m tall shown. Dipterocarps hatched. Note that in building phase these retain tall, narrow (monopodial) crowns of youth (see text). Species and stem numbers. Anacardiaceae: *A.1, Buchanania insignis* (1); *A.2, Gluta laxiflora* (2); *A.3, Swintonia glauca* (2). Annonaceae: *An, Popowia pisocarpa* (3). Burseraceae: *B.1, Dacryodes rostrata* (2); *B.2, Dacryodes laxa* (2). Celastraceae: *C, Lophopetalum javanicum* (3). Crypteroniaceae: *Cr, Crypteronia griffithii* (1). Dipterocarpaceae. *D.1, Dipterocarpus candatus* (1); *D.2, Hopea bracteata* (4); *D.3, Shorea dolichocarpa* (1); *D.4, Shorea glaucescens* (4); *D.5, Shorea laevis* (3); *D.6, Shorea parvifolia* (3); *D.7, Vatica odorata* (2). Ebenaceae: *Eb, Diospyros sumatrana* var. *decipiens* (3). Fagaceae: *Fa, Quercus argentata* (2). Flacourtiaceae: *F, Hydnocarpus pentagyna* (2). Euphorbiaceae: *E.1, Aporusa nitida* (1); *E.2, Aporusa prainiana* (1); *E.3, Mallotus* sp. (2); *E.4, Pimelodendron griffithianum* (4). Guttiferae *G.1, Calophyllum depressinervosum* (1); *G.2, Calophyllum rubiginosum* (2); *G.3, Calophyllum* sp. (1). Hypericaceae: *H, Cratoxylum sumatranum* (1). Icacinaceae: *I, Stemonurus umbellatus* (2). Lauraceae: *La, Litsea* sp. (2). Leguminosae: *L.1, Fordia*

n most species this maturation represents a loss of a vigorous leader, with imposed sympodiality. As trees grow up, their crowns eventually extend above the relatively humid, windless, and shaded lower canopy levels to become exposed, at the level of the top of mature phase canopy, to lower humidity, more wind, and greater insolation. A few species retain the youthful monopodial construction throughout life even when they reach the height of the mature phase canopy (*Mezzettia* and *Monocarpia* amongst Malayan Annonaceae: Kochummen, 1972) or even when they become emergent (*Araucaria cunninghamii* and *Araucaria hunsteinii* in New Guinea: Whitmore, 1975, Figs. 14.6, 14.8; Chap. 10). This group includes all forest palms (Chaps. 11, 23).

Importance of gap size

Any forest is a mosaic of patches at different stages of maturity. The forest growth cycle starts with a gap, so the size of gaps determines the size of the patches in the forest (i.e., it determines the coarseness of the mosaic of structural phases). The larger a gap, the more the microclimate within it differs from that under a closed forest canopy. The various facets of microclimate are interlinked. Principal differences between the environment in gaps and that below forest canopy are an increase in light and change in its quality, an increase in temperature, and a saturation deficit. There are also an increase in nutrients as dead plants decay, a temporary decrease in root competition, and sometimes also changes in microrelief and soil profile. The reality of root competition in a small gap was shown by healthy-appearing small trees of *Campnosperma brevipetiolatum*, which during 6.6 years of observation added only 0.5 m height per year in a small gap of 0.04-ha extent, whereas similar trees grew 2.14 m per year in an open plantation (Whitmore, 1974).

The amount of new plant matter produced per unit area per unit time, which is the new primary productivity of the forest, differs in the various phases of the growth cycle. It is low in the gap phase,

Fig. 27.2 (*cont.*)
filipes (3); L.2, *Koompassia malaccensis* (1). Meliaceae: *Ml.1, Dysoxylum motleyanum* (1); *Ml.2. Sandoricum maingayi* (1). Moraceae: *Mor, Prainea frutescens* (2). Myrsinaceae: *My, Ardisia* sp. (3). Myristicaceae: *Mi, Knema latericia* (3). Myrtaceae: *M.1, Eugenia cuneiforme* (3); *M.2, Eugenia rosulenta* (2). Olacaceae: *O, Ochanostachys amentacea* (1). Sapotaceae: *S.1, Ganua palembanica* (2); *S.2, Palaquium rostratum* (1). Sterculiaceae: *St.1, Scaphium macropodum* (4); *St.2, Heritiera sumatrana* (2). Tiliaceae: *T, Pentace macrophylla* (1). Verbenaceae: *V, Teijsmanniodendron coriaceum* (3). Rubiaceae: *R, Timonius borneensis* (1). (From Whitmore, 1975; after Ashton, 1964)

increases to a maximum in the building phase as the sapling trees enter their grand period of growth, and declines during the mature phase (Whitmore, 1975, Fig. 8.3). It clearly follows that different estimates of the productivity of a forest will be obtained, depending on which phase is sampled or which predominates. This point must be remembered when assessing the significance of published figures (Whitmore, 1975, Chap. 8) and those for Malayan lowland evergreen dipterocarp rain forest presented by Kira (Chap. 24).

Gap size
Different species are successful in gaps of different size. Gap size, therefore, has an important influence on species composition and spatial arrangement in the forest.

The tiniest gaps are formed by the slow death of a single big tree whose crown and bole fall to the ground, often in pieces. The crowns of many big emergent trees are 15–18 m in diameter. Many trees are blown over by wind before they die of old age. The falling bole cuts a swathe through the forest and the crown smashes down. Trees are

Fig. 27.3. Ridgetop grove of *Agathis macrophylla* in lowland rain forest, Vanikolo, Santa Cruz Islands. Plot area 60 m by 7.5 m, all trees over 6 m tall shown. *Agathis* stippled, *Macaranga polyadenia* hatched horizontally, *Campnosperma brevipetiolatum* hatched vertically. (From Whitmore, 1966)

especially prone to wind-throw when the ground is sodden after heavy rain and has lost part of its cohesion, and this is especially likely in soils with impeded drainage. Wind-throw gaps of 80 ha created by a single squall have been recorded in the Sarawak peat swamp forest, and rare, exceptionally strong squalls seem responsible. Many wind gaps are much smaller. Existing wind gaps are sometimes widened in area by further wind-throw at their edges. Lightning also creates gaps, and these are also especially common in peat swamp forest (Malaya and Sarawak), where they attain 0.6 ha, and in mangrove forest (Malaya and Papua New Guinea: White, 1975). In hilly country, landslides create gaps. Some soils are less stable than others; for example, landslides are noticeably more frequent over granite than over sedimentary rocks in Malaya. Earth tremors may be an important cause; White (1975) considers gaps caused by tremors to be widespread in Papua New Guinea.

The most extensive gaps in tropical rain forest are those created by the cyclones that occur between latitudes 10° and 20° north and south of the equator. Many square kilometers can be destroyed in a single storm. For example, three cyclones (in 1961, 1966, 1968) on Savaii Island, Samoa, destroyed over half the potentially exploitable forest, reducing the stocking by 90% to 15 stems/hectare. It is worth emphasising that not all the world's tropical rain forests lie within the two cyclone belts. Part of the structural and floristic complexity of the Sunda shelf rain forest of Sumatra, Malaya, and Borneo results from the absence of cyclones, or indeed of any factors that cause widescale devastation. Hence, such forests show an absence of the relative structural and floristic uniformity over large areas that results as huge gaps are simultaneously recolonized. (The absence of man's influence is also important in these Sunda shelf forests, especially in Malaya.)

The size and position of gaps are largely due to chance (Webb, Tracey, & Williams, 1972), but certain places are more likely to have them than others. Gaps are caused by what are essentially rare catastrophic events. As is so often the case in ecology, extremes are more important than means.

Establishment in gaps

There are two broad alternative starts to the building phase of the forest growth cycle. Either existing seedlings and saplings commence upward growth, or new trees establish from seeds germinating in the gap.

As a first generalization, it is found that established seedlings and saplings most often grow up to maturity in small gaps. The external stimulus of the change in microclimate caused by gap formation stimulates apical growth (Chap. 20). By contrast, the drastic changes in microclimate that follow the formation of a big gap result in death of all or many of the young plants previously growing there under forest shade. Big gaps are colonized by a group of species absent from the undergrowth of high forest and equipped successfully to exploit open sites. In an interesting early experiment on Gunung Gede, in Java, which has apparently never been repeated or extended, artificial small gaps of 0.1 ha in primary forest were soon colonized by surviving young individuals of primary forest tree species. By contrast, in larger gaps of 0.2–0.3 ha these persistent individuals were suppressed by a lush growth of invading tree species (Kramer, 1926, 1933).

Species of these two contrasting ecologic groups are, respectively, shade-tolerant and light-demanding in their early life. Foresters have long known of the existence of these two classes. Indeed, an important basis of silviculture is to manipulate the forest canopy to create gaps of a size that favours growth of chosen species. Other characteristics are linked, and each class possesses a whole syndrome of properties.

Extremely light-demanding species are often called "pioneers" because they can establish only in the open and are unable to do so under shade, including their own shade (Fig. 27.4a). Seeds are usually small, are continually and prolifically produced without regard to any season, and are efficiently dispersed by wind or animals. Tiny seeds produce tiny seedlings, whose tiny roots can penetrate compacted ground (Chap. 5). Height growth in youth is very

Fig. 27.4. Stand tables for (a) light-demanding species, *Endospermum medullosum*, and (b) shade-tolerant species, *Parinari salomonensis*. Tropical lowland evergreen rain forest, 13.35 ha, Kolombangara, Solomon Islands. (From Whitmore, 1974)

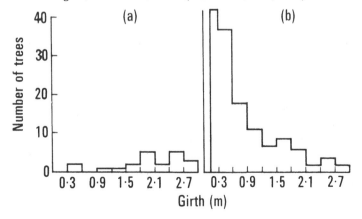

rapid; individuals falling behind become shaded and die (hence a one-layer canopy forms). Wood is of low density and pale in color. The classic example is balsa, *Ochroma lagopus*.

In contrast, shade-tolerant species are apparently characterised as a class by big seeds with substantial food reserves and are able to establish in deep shade; characteristically, seeds are produced periodically, in response to climatic stimuli and have either no or brief dormancy. Seedlings can often persist, growing slowly or not at all, in dense undergrowth shade, marking time till a gap occurs above them. The timber is typically dense and dark. These species can replace themselves in situ (Fig. 27.4b).

By analogy with the contemporary jargon of evolutionary zoologists, shade-tolerant species have a K-oriented character syndrome; pioneer species, an r-oriented syndrome.

An important question on which we at present have very little information is: Is there a fundamental difference between shade-tolerant and light-demanding species in their efficiency in utilizing light? A second related question is: Do they differ in the ratio of photosynthesis to respiration? A review is given by Whitmore (1975, Chap. 8.4). Briefly, what is known so far is that woody plants as a class have lower unit leaf (net assimilation) rates and lower relative growth rates than herbs. Net photosynthetic rates are also low compared with herbs, and no trees are known with the Kranz syndrome (photosynthesis via the C_4 carboxylic acid pathway). The very small number of tropical trees studied are similar to the better known temperate ones in this respect. Two tropical pioneer species tested (*Anthocephalus chinensis, Cecropia peltata*) have been shown to have a higher ratio of daytime fixation of CO_2 to nighttime respiration than shade leaves of climax forest tree species, and this is due to higher night respiration in the latter (Odum et al., 1970). It has been speculated that this implies that pioneer species maximise production of chemical defences against predators. But there remains frustratingly little evidence, and far more observations are required. Nevertheless, it is highly probable that tropical tree species will be found to differ in these basic physiologic respects both between sun and shade leaves and between species with different light requirements. It is clearly necessary, therefore, to consider carefully what significance can be attached to measures of productivity or its facets, respiration and photosynthesis, based on small sample plots in entire, species-rich rain forest communities (Chap. 24). Although the methodology for the quantitative analysis of plant growth is now well established (Evans, 1972), many pitfalls have been discovered, and Evans (1976), reviewing a lifetime of investigations on a shade-tolerant temperate herb species, suggested by an initial study in the Nigerian rain forest, outlines how formidable they are. It is tempting to speculate, with Oldeman and Ashton (Chaps 23, 25), that crown shape and disposition of leaves within the crown have a functional significance, and Oldeman (Chap. 23) has pointed out that pioneer species which spend their life in one microclimate do not change crown structure. But actual experiments will be fiendishly difficult to design (see Evans, 1976) and may yield surprises. Notably, the work of Evans and his associates has shown complex compensations between different facets of the growth of a species when it is exposed to different environments. This is a virgin field for enquiry in forest ecology and is fundamental to production studies.

The light-demanding and shade-tolerant syndromes are not, in fact, two exclusive, sharply defined categories. A study in the Solomon Islands, in forests that are species-poor by tropical standards, showed that amongst the 12 big tree species that constitute most of the top of the canopy there are two other classes between these extremes (Whitmore, 1974, 1975, Chap. 6.3). In details all species differed. In richer forests even more kinds of responses to canopy gaps would be expected. The Solomon Islands study consisted of monitoring the total population of the 12 species on 13.35 ha for 6.6 years, and it continues. Sarukhán (Chap. 6) outlined the first results from an even more intensive study in Mexico aimed also at giving a model of populations whose change in response to key regulating factors (principally cyclones in the case of the Solomons) can be predicted. For many species one of the major regulating factors determining population structure is the chance occurrence of a gap above seedlings to allow them to grow up. Without such a gap seedlings of all plants, except undergrowth species that spend their whole lives in dense shade, will eventually die. The progressive death of individual families of seedlings is shown in Fig. 27.5. Perhaps the palm *As-*

Fig. 27.5. Fluctuation of seedling populations of dipterocarp, *Parashorea tomentella*, in tropical lowland evergreen rain forest at Sepilok, Sabah. (*a*) Fate of individual families of seedlings. (*b*) Total population. Numbers on eight plots of 4×10^{-4} ha. (From Whitmore, 1975; after Fox, 1972)

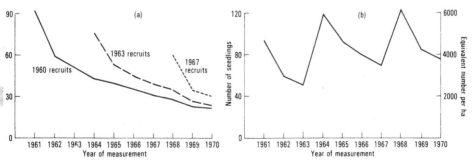

trocaryum mexicanum, analysed in detail by Sarukhán (Chap. 6), is an example of the extreme shade-tolerant class. We may predict that the importance of gaps will be revealed later as other species are studied by him and his associates.

Amongst about 200 Malayan tree species (about 8% of the tree flora) regarded as representative, differences have been observed in rate of germination, initial seedling morphology, and initial seedling size. Different combinations of characters to some extent reflect different predation-escape mechanisms (Chap. 5). Janzen (Chap. 4) described different frequencies and abundances of seed production and interpreted them as modes to ensure that some seeds escape predation. Sarukhán (Chapt. 6) showed that predation is a major source of loss and reduction in seed and seedling population size. Escape from predation contains an element of chance that comes in the life cycle mainly before the separate chance of gap formation, essential to the continued growth of many species. We get a glimpse of some of the reasons for floristic complexity and heterogeneity.

Pioneer species colonise big gaps by introducing seeds that germinate after the gap has formed. The seed may be lying dormant in the soil and germinate in response to high temperatures and/or bright light in the gap. Alternatively, it may be carried or blown in after the gap forms. In the eastern tropics adjacent logging areas commonly develop pure stands of different pioneer species, suggesting that here the second alternative holds and invasion is on a first-come, first-served basis. The classic experiment in Malaya by Symington (1933), which showed germination of dormant pioneer seeds after clearance, did not preclude recent arrival of other seeds; the area studied lay only a few metres from a stand of pioneer trees. There are in fact important lacunae in our knowledge of the longevity of tropical tree seed, especially whether shade-tolerant species differ as a class from pioneers. Current work by Dr. Ng will help to improve this situation (Ng, 1974 and Chap. 5). Unequivocal but simple experiments could be conducted to discover the size and composition of any seed bank in the soil under virgin rain forest.

Many, perhaps most, big woody climbers are light-demanders that establish in gaps and are carried up with the forest canopy as it grows. This was clearly demonstrated in the Solomon Islands forest survey already mentioned (Whitmore, 1974). Many, but not all, climbing palms belong to this class, notably *Korthalsia, Myrialepis*, and *Plectocomiopsis* (Chap. 11). Very few palms except climbers are light-demanding pioneers, and palms are conspicuously absent from secondary rain forest (Whitmore, 1973).

The different classes of species, adapted to different niches in the forest growth cycle, can accommodate comments at this symposium on the evolution of tropical trees and forests. Doyle (Chap. 1) believes the first of the angiosperms that survive today were understorey species of the Aptian and Lower Albian (Lower Cretaceous, 120–125 × 10⁶ years ago) – small shade-tolerant trees. He considers that pioneers evolved by the middle or upper Albian (Lower Cretaceous, 110–120 × 10⁶ years ago), but that replacement of conifers in climax forest by big, shade-tolerant angiospermous trees only occurred well into the Upper Cretaceous (65–110 × 10⁶ years ago).

Conclusions

I have tried to show that the forest growth cycle is a feature central to many aspects of the biology of forest plants. Despite much recent and current work, our understanding of tropical rain forest dynamics is incomplete, but only simple experiment or observation is required to clarify several important facets. Much of what has been said could equally well be illustrated by examples from the temperate regions, though the simplicity of forest structure and the poorer floras make it less easy to appreciate the full range of possibilities. This is especially true of northwestern Europe, most markedly the British Isles.

The mosaic of structural phases of the forest growth cycle and the floristic difference between patches of different sizes has an important bearing on the concept of the plant community. What is a "climax forest" in this kaleidoscopic mixture I have described? Is it only a hypothetic entity – the ultimate community of self-perpetuating, shade-tolerant species that regenerate under their own shade – and in practice is never attained over big areas because of repeated catastrophes at lesser intervals than the life span of pioneer species? There is clear evidence that temperate deciduous forest in North America has had a continuing history of disasters dating back before European penetration (Raup, 1964). But this is subject for another symposium, where temperate and tropical forest ecologists explore the extent to which forests throughout the world have an essential unity, resulting from the operation of the same factors.

Acknowledgments

I should like to thank Dr. Peter Stevens for drawing to my attention the teleologic implications of the term "strategy," as applied to aspects of the biology of plant species. In connection with the material of this chapter I have previously referred to "shade-bearing" and "pioneer strategies." Here they are replaced by the neutral term "character syndromes," which seems preferable. I likewise readjusted the terminology of other writers.

General discussion

Givnish: Do you agree with Whittaker's notion of a climax as being not so much a stage in the successional process, but the landscape mosaic that results from the interplay of disturbances and succession?

Whitmore: Yes, this is so. As long as gap forming remains on average the same, then, of course, the forest will remain on average the same for that factor. For example, if hurricanes continue to occur about once a century here in New England, the forests will remain much as they are today. Superimposed upon this we have the phenomenon, in the species-rich eastern tropics, of insufficient room in each individual gap for the numerous species capable of occupying it. Of all the species that could potentially grow in it, there will be only one or a few available at the right time. If the gap-forming processes change, the whole forest mosaic will change. The tropical cyclone belt is currently moving closer to the equator. It has reached the Solomon Islands during the last 25 years. This is probably having a very dramatic effect on forest composition, because old, climax, shade-tolerant forest is being increasingly knocked over and replaced with seral forest, including species of pioneers and near-pioneers (Whitmore, 1974).

Ashton: You said that seed dormancy has not been proved. A very large proportion of what we call gap phase or pioneer species flower and fruit more or less continuously. It is not necessarily important whether the seed was dormant for a long time or only for a few days. The important question is whether it is present but does not germinate under the conditions that prevail in the mature phase.

Whitmore: Yes, this is what I am saying. One should be able to distinguish between a continuous seed rain of pioneer species and old seeds, triggered to germinate by the gap conditions. One might experimentally distinguish between these alternatives with a screen that kept out the seed rain.

Ashton: This does alter the degree of chance involved in whether or not plants establish themselves. You said that the size of gap is very important in determining the sequence of seral succession. What evidence do you have from actual measurements in the field? In my experience this is a very difficult thing to confirm. In the north of Malaya there is an area over 100 mi² of what is called "storm forest," a dipterocarp forest that was blown down sometime late in the nineteenth century. It is now a magnificent even-age stand of essentially building and mature phase dipterocarp species, but not pioneers. This implies that the trees must have been on the ground as saplings after the storm in order to grow because there no longer would have been a source for seed. In other words, here we have a huge gap that was, I would guess, not colonized by biologic nomads.

Whitmore: We really do not know enough about the Kelantan storm forest to allow a very detailed discussion about what happened. In the northeast corner of Malaya there was a forest entirely cut about 1969, which had regrown after an exceptional cyclone in 1880, followed by a fire. This probably eliminated dipterocarp seedlings. The trees had a kink in them about 8 ft up, which strongly suggests that they had grown up through a climber tangle (Whitmore, 1975, Plate 15.5). Shorea spp. (*S. leproslula, S. parviflora*), which were commonest there, are both near the extreme pioneer end of the ecologic spectrum of *Shorea*. The *Dipterocarpus crinitus* that grew there is also known to be a species that apparently does not maintain itself in high forest (Whitmore, 1975, p. 185), but is eventually eliminated.

Concerning the evidence for my statement that different species colonize gaps of different sizes, I mentioned the Dutch experiment done in Java by Kramer (1926, 1933), where gaps of different sizes were made. Dr. Oldeman suggests that a *series* of species appears in succession in a gap, but I prefer to say that composition of the colonizing stand depends on gap size. I do not myself think a neat floristic succession is distinguishable.

Sarukhán: The r- and K-selection concepts are very useful in certain ways, but they are utterly meaningless if used out of context. As our information on life histories of tropical trees builds up, we may reach a stage where the use of the terms would be more appropriate than it is at present. In relation to the topic in Dr. Oldeman's, Hartshorn's and Whitmore's chapters (Chaps. 23, 26, 27), I would like to ask whether they have information on the rate of recovery of these gaps in the floristic sense (i.e., the composition in terms of the arboreal components of the mature forest). We have carried out studies on floristic recovery in secondary succession after agricultural activity (Sarukhán, 1964, 1968). It was found that before the fifth year of succession about 90% of the arboreal components of the forest are already present in the secondary succession, albeit many of them in the form of small seedlings.

Whitmore: We must be very careful not to make generalizations from single observations. At the end of the Japanese occupation a forest was felled in Malaya (Sungei Kroh), but not cultivated, and the succession was studied for over 25 years. The development of pure *Macaranga gigantea* took about 10 years. The final composition of the forest did not appear in a couple of years. I suspect that this differs from place to place. At Sungei Kroh the situation was confused by the early death of survivors (mother trees). The availability of seed from mother trees within dispersal range of the site is what matters.

Ashton: I have seen comments in the literature, of the kind Dr. Sarukhán has just made. But it does not make sense from my own experience, and this may be because my experience is in a different kind of forest and probably with a different history from the one he is talking about. Southeast Asia may possibly, together with the Amazon area, be the only place in the tropics where there are extensive tracts of forest that have never been very seriously damaged by man. If there is an area of ostensibly tropical rain forest, but which perhaps 500 or 1000 years ago was part of an ancient civilization or of agricultural land, it may be very difficult to assess whether the floristic composition of the apparent mature phase reflects that of the *original* primary forest or whether it is itself in arrested secondary succession in the absence, perhaps, of a good source of mature phase species.

Givnish: Dr. Whitmore, you have essentially distinguished between pioneer and shade-tolerant species. Are you assuming that small trees (at maturity) regenerate in gaps and that big trees (at maturity) regenerate in the shade? Is it not possible for emergents to regenerate in gaps?

Whitmore: It is very convenient to distinguish between short-lived pioneers and long-lived pioneers (i.e., between small tree pioneers and big tree pioneers). For example, many *Macaranga* and *Trema* species are small tree pioneers. *Gmelina arborea* and *Triplochiton* are examples of big tree pioneers; they are important because they can be grown by foresters in plantations. In my chapter I have considered only big trees. These illustrate the difference between pioneers and shade-tolerant species. One can guess that among the 3000 or so species of trees in the Malay Peninsula there probably are light-demanding and shade-tolerant small tree species as well.

Hartshorn: In La Selva, Costa Rica, the largest trees are gap reproducers. Also, the proportion of species that depends on gaps for regeneration decreases with decreasing stature. In other words, as you move down from canopy and emergents to subcanopy and understory, the proportion of gap species decreases. As a group, the understory species are least dependent upon gaps for successful regeneration.

Givnish: I would like to raise an issue you discussed in your book, Dr. Whitmore, as to why these short-lived, early successional species do not persist. The reason you suggested was that these trees start to develop sprawling crowns and become more susceptible to shading. Could you achieve an understanding of the difference between gigantic trees, like *Ceiba,* that are pioneers but also persist into the emergent story, and shorter pioneers like *Cecropia,* by viewing them in terms of their modular construction? A short-lived species that invades and moves rapidly from gap to gap must seize available space quickly, and thus should build a spreading crown that covers a large area rapidly. If a species like *Cecropia* or *Rhus* does this by proliferating totipotent orthotropic axes, it almost immediately precludes any further upward growth. Presumably, it would be a very difficult process to get rid of all these spreading units unless something rather fundamental happened to the modular growth of the tree; a plant must cast off all but a few of these totipotent units if it is to resume terminal growth. Trees like *Ceiba* that do reproduce in gaps would seem to be required to maintain only a limited number of active shoot meristems in the initial stages of their upward growth, lest they spread out, divert their energy to a multiplicity of axes, and forfeit their potential for upward growth.

Whitmore: There is a distinction between short-lived and long-lived and between small and big, pioneer species, but part of the syndrome of the pioneer species is this ability to preempt a volume of space in order to exclude competitors. In my experience, if a big pioneer tree produces a big crown before it grows in height, it eventually continues growth by means of a leader. I think that is the answer. It is not true in my experience that short-lived pioneers differ from long-lived pioneers in their different crown volumes. *Endospermum medullosum,* for example, which is a long-lived, big pioneer species of New Guinea and the Solomons, has a crown just as big as any *Macaranga,* but it has the added potential to develop a tall trunk. This difference is interesting when one considers that the genera are closely related.

Ashton: But *Endospermum* is Aubréville's model and *Macaranga* is Rauh's.

Givnish: The idea of seizing space is appropriate, but you can seize it at a lower or a higher level.

Whitmore: But pioneer trees all grow up simultaneously in a gap. If an individual gets left behind, it becomes shaded out and disappears. This is a characteristic of light-demanders and is the reason why a forest of them always has a uniform canopy (Dawkins, 1965).

Givnish: The best way for a plant to fill out a given volume at any level is to divide its growth among several shoots that continue to grow and divide. A plant cannot fill out a volume on a given level as rapidly if it retains apical dominance and a strong leader.

Whitmore: The details need to be settled, but as a big pioneer grows up to fill a space, it still presumably has a leader.

Raup: I had some experience in Honduras a good many years ago visiting the United Fruit Company's forest tree plantations. There both native and exotic trees were planted in a great variety of sites. Some were in previously cultivated land that was still open. Others were planted in guamil, the young growth that comes up in old fields. Workmen would cut lanes through this secondary growth and plant the trees about 20–30 ft apart. They discovered that they had to clean the young trees of vines about once a month. It required a tremendous number of man-hours. Elsewhere, in new farm patches cut out of the forest, it was possible to see the forest in section if the clearing had just been done. I saw typical tall straight trees reaching up into the canopy, but these might be 100 ft apart or more. The rest of the trees were all crooked. I conceived the notion that the ones that got to the canopy were the ones that just got through the vines and managed to carry some of the vines with them as lianes. They had to be very vigorous trees and must have had a very good start in order to do this. The vines must be a major factor in determining what succeeds in a gap.

Whitmore: The situation you have described seems to be exactly the structure of the forest at Khao Chong in southern peninsula Thailand, where the first Japanese tropical forest production studies were made. They had some big trees with a solid canopy below them which included many vines.

Raup: The big trees were more successful in carrying vines. The United Fruit Company's ambition was to clear away those vines so that more trees got up. I do not think they were very successful at this because it required too much labor.

Oldeman: Rollet (1974) made a statistical inventory of about 5000 km² of rain forest in Venezuela; his figures substantiate the phenomenon of big trees carrying climbers with them as they grow up.

Raup: This must be a very real deterrent to the emergence of a tree.

Dransfield: Dr. Whitmore said there were no pioneer palms, but I said earlier that *Pigafetta* is a pioneer species. It has an incredibly fast growth rate, something in excess of 1 m/year in cultivation. It is also the only native Indonesian palm that I have observed with germination staggered between 10 days and a whole year. Seeds are small for a palm (about 5 by 4 by 3 mm) and produced in large quantities. These facts fit in with the other characteristics of pioneer plants.

Ng: It must be emphasized that there are many different types of pioneering communities. Even in a place like Malaya, which is rather small geographically, travel from one end of the country to another will demonstrate floristically different pioneer communities. For example, there are *Anthocephalus* forests, *Macaranga* forests, *Trema* forests, *Adinandra* forests. *Gleichenia* may be a pioneer suppressing later growth. It may be dangerous to generalize from one to all. The biology of pioneer species can offer surprises. For example, in *Endospermum malaccensis* and *Vernonia arborea* the proportion of viable seeds may be as low as 0.1%. How can a pioneer species succeed with this low reproductive rate?

Whitmore: What Dr. Ng says is important. The fact that in Malaya there are stands of pioneers in different gaps in high forest that are quite different is an indication to me that there is a first-come, first-served principle operating in colonizing big gaps. If there was a seed rain and the seeds had a long dormancy, there would be an accumulation in a particular place of a mixture of seeds, given the fact that the pioneers are all equally dispersed. Then, when a gap is formed a similar mixture of species would grow up everywhere. In fact, this does not occur, as can be demonstrated by looking at a logging area in which separate sections were created at different times. What one sees are fairly pure stands of different species of *Macaranga* and other pioneers differing from section to section. This suggests to me that the seeds rain in the gap and the ones that rain in first grow up and fill up the space. We perhaps should keep some different kinds of pioneer communities, such as the *Glei-*

chenia and *Adinandra* ones Dr. Ng mentioned, out of the discussion of what goes on in the forest mosaic and the forest growth cycle as I have attempted to define it. For example besides these two there is a group of pioneer tree species that come up as pure or nearly pure stands on alluvial river banks (i.e., *Eucalyptus deglupta* and *Octomeles sumatrana*). This is essentially a prisere and is a different situation from recolonization of gaps in the forest. *Gleichenia* and *Adinandra* (Holttum, 1954) characterize degraded sites. We must, therefore, distinguish different kinds of pioneer habitats.

Oldeman: In spite of the diversity of pioneer communities, pioneer trees all have certain points in common architecturally. For instance, they tend to conform quite closely to their models, they have very often rather massive branches, they all grow fast.

Tomlinson: Dr. Whitmore, how does a standing tree die?

Whitmore: I would suggest old age, assisted by fungi. These moribund trees are seen as stag-headed individuals that die back from the branch tips to produce a bare outer limb and a leafy inner limb. Then one branch drops off at a time, and the trunk falls to the ground in pieces.

Oldeman: I quite agree. A tree is not immortal though it lives a long time. If cambium production cannot be maintained, a regressive cycle of diminished cambium production is established, the cambium receives less nutrients, it produces less transport tissue . . . and so on. In French Guiana I made counts and found that about 1 tree in 10 died upright.

Whitmore: Big trees in the forest often are not growing at all, which fits in absolutely with what you are saying. In a plot on Vanikolo Island measurements were made for 15 years. Small trees of *Agathis macrophylla* were growing, but big trees were not (Whitmore, 1966).

References

Ashton, P. S. (1964). Ecological studies in the mixed dipterocarp forests of Brunei State. *Oxford Fo. Mem.*, *25*, 1–75.

Dawkins, H. C. (1965). Time as a factor in tropical rain forest. *J. Ecol.*, *53*, 836–8.

Evans, G. C. (1972). *The Quantitative Analysis of Plant Growth*. Oxford: Blackwell.

– (1976). A sack of uncut diamonds: The study of ecosystems and the future resources of mankind. *J. Ecol.*, *64*, 1–40.

Fox, J. D. (1972). The natural vegetation of Sabah and natural regeneration of the dipterocarp forest. Ph.D. thesis, University of Wales.

Holttum, R. E. (1954). *Adinandra* belukar. *J. Trop. Geogr.*, *3*, 27–32.

Kochummen, K. M. (1972). Annonaceae. In *Tree Flora of Malaya*, ed. T. C. Whitmore, vol. 1. London: Longmans.

Kramer, K. (1926). Onderzoek naar de natuurlijke verjonging in den uitkap in Preanger gebergte-bosch. *Med. Proefst. Boschw. Bogor*, *14*.

– (1933). Die natuurlijke verjonging in het Goenoeng Gedeh complex. *Tectona*, *26*, 156–85.

Ng, F. S. P. (1974). Seeds for reforestation: A strategy for sustained supply of indigenous species. *Malayan Forester*, *37*, 271–7.

Odum, H. D., Lugo, A., Cintron, G., & Jordan, C. F. (1970). Metabolism and evapotranspiration in some rain forest plants and soils. In *A Tropical Rain Forest: A Study of Irradiation and Ecology at El Verde, Puerto Rico*, pp.

H3–H52. ed. H. T. Odum & R. F. Pigeon. Oak Ridge, Tenn.: U.S. Atomic Energy Commission.

Raup, H. M. (1964). Some problems in ecological theory and their relation to conservation. *J. Ecol., 52* (Suppl.), 19–28.

Rollet, B. (1974). *L'architecture des forêts denses humides sempervirentes de plaine.* Nogent-sur-Marne: Centre Technique Forestier Tropical. 298 pp.

Sarukhán, J. (1964). Estudio sucesional de una área talada en Tuxtepec, Oax. In *Contribuciones al estudio ecológico de las zonas cálido-húmedas de México.* Inst. Nac. Inv. For. México, Pub. Esp. no. 3, pp. 107–72.

– (1968). Análisis sinecologico de las selvas de *Terminalia amazonia* en la plaincie costera del Golfo de México. Thesis, Colegio de Postgraduados, Escuela Nacional de Agricultura, Chapingo, México.

Symington, C. F. (1933). The study of secondary growth on rain forest sites in Malaya. *Malayan Forester, 2,* 107–17.

Watt, A. S. (1947). Pattern and process in the plant community. *J. Ecol., 35,* 1–22.

Webb, L. J., Tracey, J. G., & Williams, W. T. (1972). Regeneration and pattern in the subtropical rain forest. *J. Ecol., 60,* 675–95.

White, K. J. (1975). The effect of natural phenomena on the forest environment. *Papua New Guinea Sci. Soc.*

Whitmore, T. C. (1966). The social status of *Agathis* in a rain forest in Melanesia. *J. Ecol., 54,* 285–301.

– (1973). *Palms of Malaya.* London: Oxford University Press.

– (1974). *Change with Time and the Role of Cyclones in Tropical Rain Forest on Kolombangara, Solomon Islands.* Commonwealth Forestry Institute, Paper 46. Oxford: Holywell Press.

– (1975). *Tropical Rain Forests of the Far East.* Oxford: Clarendon Press.

Index to subjects
and plant genera